Studies in Computational Intelligence

Volume 798

Series editor

Janusz Kacprzyk, Polish Academy of Sciences, Warsaw, Poland
e-mail: kacprzyk@ibspan.waw.pl

The series "Studies in Computational Intelligence" (SCI) publishes new developments and advances in the various areas of computational intelligence-quickly and with a high quality. The intent is to cover the theory, applications, and design methods of computational intelligence, as embedded in the fields of engineering, computer science, physics and life sciences, as well as the methodologies behind them. The series contains monographs, lecture notes and edited volumes in computational intelligence spanning the areas of neural networks, connectionist systems, genetic algorithms, evolutionary computation, artificial intelligence, cellular automata, self-organizing systems, soft computing, fuzzy systems, and hybrid intelligent systems. Of particular value to both the contributors and the readership are the short publication timeframe and the worldwide distribution, which enable both wide and rapid dissemination of research output.

More information about this series at http://www.springer.com/series/7092

Javier Del Ser · Eneko Osaba
Miren Nekane Bilbao · Javier J. Sanchez-Medina
Massimo Vecchio · Xin-She Yang
Editors

Intelligent Distributed Computing XII

Editors
Javier Del Ser
Basque Center for Applied Mathematics (BCAM)
TECNALIA, University of the Basque Country(UPV/EHU)
Derio, Bizkaia, Spain

Eneko Osaba
TECNALIA
Derio, Bizkaia, Spain

Miren Nekane Bilbao
University of the Basque Country (UPV/EHU)
Bilbao, Bizkaia, Spain

Javier J. Sanchez-Medina
Computer Science Department
University of Las Palmas de Gran Canaria
Las Palmas, Spain

Massimo Vecchio
OpenIoT Reesarch Unit
FBK CREATE-NET
Trento, Italy

Xin-She Yang
Department of Design Engineering and Mathematics
Middlesex University
London, UK

ISSN 1860-949X ISSN 1860-9503 (electronic)
Studies in Computational Intelligence
ISBN 978-3-030-07616-0 ISBN 978-3-319-99626-4 (eBook)
https://doi.org/10.1007/978-3-319-99626-4

This Springer imprint is published by the registered company Springer Nature Switzerland AG
The registered company address is: Gewerbestrasse 11, 6330 Cham, Switzerland

Preface

Intelligent distributed computing emerged in the 70s as a result of the exploitation of synergies between different ideas coming from the fields of Intelligent and Distributed Computing. It is a stream directly derived from artificial intelligence, yielding a new generation of intelligent systems built upon the combination of approaches from this classical field with computational intelligence, distributed and multi-agent systems.

This book collects a number of research contributions gravitating on recent advances intelligent and distributed computing, comprising both architectural and algorithmic findings related to these fields. A major focus is placed on new techniques and applications for evolutionary computation, swarm intelligence, multi-agent systems, multi-criteria optimization and Deep/Shallow machine learning models, all conceived as technological drivers to enable autonomous reasoning and decision-making in complex distributed environments. Part of the book is also devoted to new scheduling and resource allocation methods for distributed computing systems. The book comprises the peer-reviewed proceedings of the 12th International Symposium on Intelligent Distributed Computing (IDC 2018), which was held in Bilbao, Spain, from 15 to 17 October 2018. IDC 2018 continues the legacy of the prior IDC Symposium Series as an initiative of two research groups: (i) Systems Research Institute, Polish Academy of Sciences, Warsaw, Poland, and (ii) Software Engineering Department, University of Craiova, Craiova, Romania.

After a rigorous peer-reviewed process, the International Program Committee selected 38 papers, which are published in these conference proceedings. The selection of these contributions was rigorous with the intention of maintaining the high quality of the conference. We would like to thank the Program Committee for their hard work during the review process. This peer-reviewed phase is very important for achieving the high-quality standards of the IDC conference. We would also like to thank Janusz Kacprzyk, editor of Studies in Computational Intelligence series and member of the Steering Committee, for his support and encouragement for the development of the IDC Symposium Series. The organization also feels indebted to Thomas Ditzinger, Viktoria Meyer and Ramamoorthy

Rajangam from Springer, for their kind support during the production of the conference proceedings.

The 38 contributions published in this volume tackle a wide variety of topics related to theory and applications of intelligent distributed computing. We also want to highlight the organization of four different special sessions: Intelligent And Distributed Optimization For Logistics, Transportation And Vehicle Routing (INDILOG), Soft Computing And Computational Sustainability (COMPSUS), Multi-Agent Systems Technology and Learning (MASTL 2018) and Machine Learning methods applied to practical Predictive Maintenance problems (ML-PdM). We want to dedicate a sincere thanks to the organizers of the three tutorials held during the conference: Multi-Objective Big Data Optimization with jMetalSP (JMETALSP), The K-means algorithm on Big Data domains (KMBD) and Applying Machine Learning to detect Android Malware: AndroPyTool and OmniDroid.

IDC 2018 enjoyed outstanding keynote speeches by distinguished guest speakers: Prof. Jose A. Lozano—Full Professor in the University of the Basque Country (UPV-EHU), Spain, and Research Professor in the Basque Center for Applied Mathematics (BCAM), Spain; Dr. David Camacho—Associate Professor in Universidad Autónoma de Madrid, Spain; Francisco Herrera—Full Professor in the University of Granada, Spain; Dr. Eleni I. Vlahogianni—Associate Professor in the National Technical University of Athens, Greece; and Prof. Albert Bifet—Full Professor in Telecom ParisTech, France.

Finally, we appreciate the efforts made by of all the organizers from the Joint Research Lab (JRL)—University of the Basque Country, Basque Center of Applied Mathematics and TECNALIA Research and Innovation, for collaborating in the organization of this event. Special thanks also to the people from the Bizkaia Aretoa, for hosting this event in such a beautiful venue.

June 2018

Javier Del Ser
Eneko Osaba
Miren Nekane Bilbao
Javier J. Sanchez-Medina
Massimo Vecchio
Xin-She Yang

Organization

General Chair

Javier Del Ser	University of the Basque Country (UPV-EHU), TECNALIA and BCAM, Spain

Steering Committee

Janusz Kacprzyk	Polish Academy of Sciences, Poland
Costin Badica	University of Craiova, Romania
David Camacho	Universidad Autonoma de Madrid, Spain
Paulo Novais	University of Minho, Portugal
Filip Zavoral	Charles University Prague, Czech Republic
Frances Brazier	Delft University of Technology, The Netherlands
George A. Papadopoulos	University of Cyprus, Cyprus
Giancarlo Fortino	University of Calabria, Italy
Kees Nieuwenhuis	Thales Research and Technology, The Netherlands
Marcin Paprzycki	Polish Academy of Sciences, Poland
Michele Malgeri	University of Catania, Italy
Mohammad Essaaidi	Abdelmalek Essaadi University in Tetuan
Mirjana Ivanovic	University of Novi Sad, Serbia
Amal El Fallah Seghrouchni	LIP6/University Pierre and Marie Curie, France

Technical Program Chairs

Xin-She Yang	Middlesex University, UK
Massimo Vecchio	FBK CREATE-NET, Italy
Javier J. Sanchez-Medina	Universidad de Las Palmas de Gran Canaria

Program Committee

Antonio J. Nebro	University of Malaga
Tomáš Pitner	Masaryk University
Martin Sarnovsky	Technical University of Košice
Iztok Jr. Fister	University of Maribor
Antonio Masegosa	Ikerbasque
Agustín Orfila	Universidad Carlos III de Madrid
Alessandro Longheu	DIEEI—University of Catania
Alfredo Cuzzocrea	ICAR-CNR and University of Calabria
Amal El Fallah Seghrouchni	LIP6—University of Pierre and Marie Curie
Idoia de la Iglesia	University of Deusto
Amparo Alonso Betanzos	Universidad Da Coruña
Andrea Omicini	Universita di Bologna
Antonio Liotta	Eindhoven University of Technology
Aquilino A. Juan	Universidad de Oviedo
Barna Laszlo Iantovics	Petru Maior University of Tg. Mures
Bertha Guijarro	Universidad Da Coruña
Camino Fernández Llamas	Universidad de León
Carlos Palau	Universidad Politécnica de Valencia (UPV)
Corrado Santoro	University of Catania
Dana Petcu	West University of Timisoara
Dariusz Krol	Wroclaw University of Technology
David Bednarek	Charles University Prague
Juan Alejo Vazquez	University of Deusto
David Fernandez Barrero	University of Alcalá
Domenico Rosaci	University Mediterranea of Reggio Calabria
Dorian Cojocaru	University of Craiova
Dumitru Dan Burdescu	University of Craiova
Emilio S. Corchado	University of Salamanca
Fernando Esteban Barril Otero	University of Kent
Filip Zavoral	Charles University Prague
Florin Leon	Technical University "Gheorghe Asachi" of Iasi
Florin Pop	University Politehnica of Bucharest
Francisco Jurado	Universidad Autónoma de Madrid

George Eleftherakis	The University of Sheffield International Faculty
Giacomo Cabri	Universita di Modena e Reggio Emilia
Giancarlo Fortino	University of Calabria
Giandomenico Spezzano	CNR-ICAR and University of Calabria
Giuseppe Di Fatta	University of Reading
Giuseppe Mangioni	University of Catania
Grzegorz J. Nalepa	AGH University of Science and Technology
Heitor Silverio Lopes	UTFPR
Igor Kotenko	SPIIRAS
Ioan Salomie	Technical University of Cluj-Napoca
Jakub Yaghob	Charles University Prague
Jason J. Jung	Chung-Ang University
Jen-Yao Chung	IBM T. J. Watson Research Center
Jorge Gomez-Sanz	Universidad Complutense de Madrid
Jose Machado	Universidade Minho
Pedro Lopez-Garcia	University of Deusto
José M. Valls	Universidad Carlos III de Madrid
Juan Manuel Corchado	University of Salamanca
Juan Pavon	Universidad Complutense Madrid
Juan Tapiador	Universidad Carlos III de Madrid
Lars Braubach	University of Hamburg
Lucian Vintan	"Lucian Blag" University of Sibiu
Marcin Paprzycki	IBS PAN and WSM
María Dolores Rodríguez Moreno	University of Alcalá
Maria Ganzha	University of Gdansk, Poland
Marie-Pierre Gleizes	IRIT
Marjan Gusev	University Sts Cyril and Methodius
Martijn Warnier	Delft University of Technology
Mert Bal	Assistant Professor at the Miami University
Mihaela Oprea	University Petroleum-Gas of Ploiesti
Mirjana Ivanovic	University of Novi Sad
Mostafa Ezziyyani	Faculty of Sciences and Technologies of Tanger
Nick Bassiliades	Aristotle University of Thessaloniki
Nik Bessis	University of Derby
Paul Davidsson	Malmo University
Paulo Novais	University of Minho
Phan Cong-Vinh	NTT University
Radu-Emil Precup	Politehnica University of Timisoara
Rainer Unland	University of Duisburg-Essen, ICB
Razvan Andonie	Central Washington University
Ricardo Aler	Universidad Carlos III de Madrid
Safeeullah Soomro	Indus University Pakistan
Salvador Abreu	Universidade de Evora and CENTRIA
Sancho Salcedo Sanz	University of Alcalá

Salvatore Venticinque	Seconda Universita di Napoli
Shahram Rahimi	Southern Illinois University
Stanimir Stoyanov	University of Plovdiv
Stefan-Gheorghe Pentiuc	University Stefan cel Mare Suceava
Stefano Galzarano	University of Calabria (UNICAL), Italy
Vadim Ermolayev	Zaporozhye National University
Vicente Matellán	Universidad de León
Vincenza Carchiolo	Universita di Catania
Viviana Mascardi	Universit degli Studi di Genova
Eneko Osaba	TECNALIA, Spain
Ibai Laña	TECNALIA, Spain
Jesus L. Lobo	TECNALIA, Spain
Izaskun Oregi	TECNALIA, Spain

Publicity Chair

Andrés Iglesias — Universidad de Cantabria, Spain

Publications Chair

Eneko Osaba — TECNALIA, Spain

Web Chair

Ibai Laña — TECNALIA, Spain

Local Organizing Committee

Javier Del Ser	University of the Basque Country (UPV-EHU), TECNALIA and BCAM
Manuel Vélez	University of the Basque Country (UPV-EHU)
Iker Sobrón	University of the Basque Country (UPV-EHU)
Miren Nekane Bilbao	University of the Basque Country (UPV-EHU)
Cristina Perfecto	University of the Basque Country (UPV-EHU)
Armando Ferro	University of the Basque Country (UPV-EHU)
Juan José Uncilla	University of the Basque Country (UPV-EHU)
Aritz Pérez	Basque Center for Applied Mathematics (BCAM)

Contents

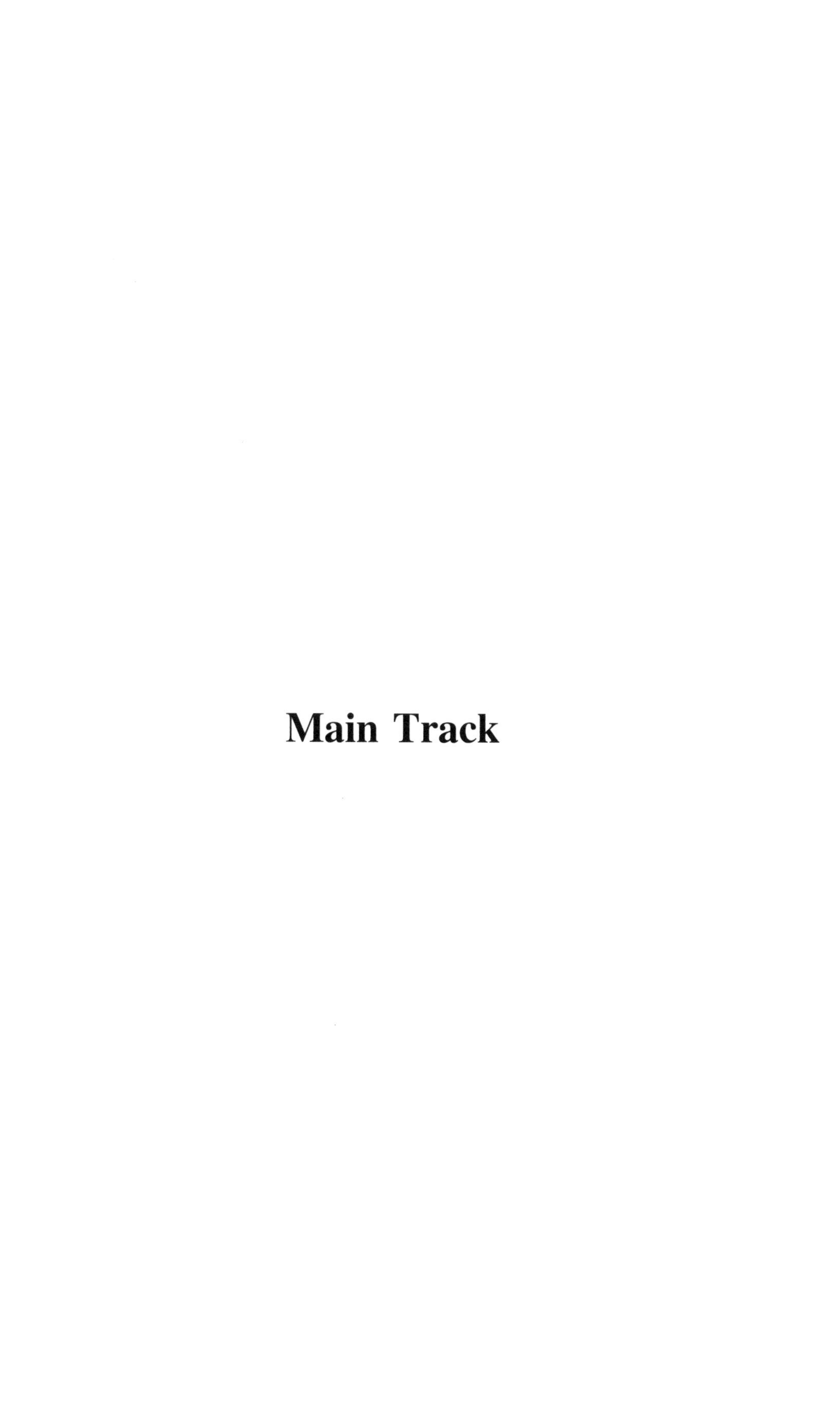

Main Track

Long Distance In-Links for Ranking Enhancement

V. Carchiolo, M. Grassia, A. Longheu, M. Malgeri, and G. Mangioni(✉)

Dip. Ingegneria Elettrica, Elettronica e Informatica,
Università degli Studi di Catania, Catania, Italy
giuseppe.mangioni@dieei.unict.it

Abstract. Ranking is a widely used technique to classify nodes in networks according to their relevance. Increasing one's rank is a desiderable feature in almost any context; several approaches have been proposed to achieve this goal by exploiting in-links and/or out-links with other existing nodes. In this paper, we focus on the impact of in-links in rank improvement (with PageRank metric) and their distance from starting link. Results for different networks both in type and size show that the best improvement comes from long distance nodes rather than neighbours, somehow subverting the commonly adopted social-based approach.

1 Introduction

Networks are nowadays extensively exploited to model many phenomena, expecially if we consider well-know prefixes, as *complex*, *social*, or *online* networks. In all these scenarios, The process of joining a network has been investigated for a while [1–3] to understand how networks arise and evolve. In particular, to model the linking mechanism between a new node (*newcomer*) and the rest of existing network, two complementary point of view exist depending on whether the connection is built using the set of out-links (where the newcomer links to existing nodes) or via in-links (nodes in the network point to the newcomer). The use of in- or out-links depends on what the network actually models and on the new node's intentions semantics being considered; both however represent the involvement of the newcomer (and consequently, of any node) to the network it is joining to.

One of the most adopted metric to quantify and weight this role is via a node's *ranking*, generically defined as the amount that allows a *weak order* of all nodes (where different nodes can have the same rank, as opposed to *total order*).

The idea of ranking is not new and its adoption spans different scenarios, as e-commerce transactions [4,5], recommendation networks [6,7] webpages relevance in Search Engines Optimization (SEO) [8–11], data envelopment analysis [12], scientific journals prestige [13] and many others.

Although different semantics is ascribed to each context in terms of what a node and its in-links/out-links represent, the common principle is that a higher

J. Del Ser et al. (Eds.): IDC 2018, SCI 798, pp. 3–10, 2018.
https://doi.org/10.1007/978-3-319-99626-4_1

rank means better reputation for that node, therefore the target for any ranking mechanism is the increase of rank. In this work we focus on the impact of in-links in rank improvement and their distance from starting link. in particular, a new node t joins a network by creating an in-link with at least one node j; intuitively we could think that in order to increase its rank, t aims at getting new in-links from j's neighbourhood, similarly to what occurs in social networks, where after knowing someone, a person usually gets introduced to his/her acquaintances. In this paper though, we show that a better rank improvement can be achieved by acquiring long distance in-links (with respect to the first node j the newcomer 'target' t connects to), specifically given t and j, we establish further in-links between t and other nodes in the network, ordering these nodes according to their distance from j, measuring the corresponding change in t's ranking via the well-known PageRank metric [14]; PageRank is chosen since it is a consolidated, extensively used ranking mechanism, even if the question considered in this paper does not depend on the metric adopted for ranking. Results for different networks both in type and size show that the best improvement comes from long distance nodes rather than from j's neighbours, somehow subverting the commonly adopted social-based approach.

The main contribution of this work hence lies in the fact that if long-distance links are preferred, the joining process of a newcomer will lead to achieve a good position in the network.

This work carries on our research [15,16] on the same topic, usually known as *best attachment* or *link building* problem. It is also related to our previous investigations on similar scenarios [17–21], where we analyzed the attachment process in trust networks with a reputation-base approach whose metric was EigenTrust [22], that exploits PageRank for trust assessment; although this paper is not about trust, the question of ranking improvement still holds.

Other works concerning the best attachment problem exist in literature, e.g. [23] shows the use of asymptotic analysis to assess how a page's pagerank can be controlled by creating new links; [24] discusses a similar approach but generalized to websites with multiple pages. In [25] the link building problem is modelled with constrained Markov decision processes, where in [26] author investigate on the influence that node outlinks changes has on the resulting PageRank; finally, in [27] authors demonstrate that the best attachment problem is NP-hard and also show that there exist both upper and lower bounds for certain classes of heuristics.

The paper is organized as follows. In Sect. 2 we briefly recall the PageRank metric, whereas in Sect. 3 we illustrate our simulations and discuss the results in ranking improvement depending on which in-links are chosen (according to their distance as discussed above), showing that long distance in-links provide better results. Our concluding remarks and future works are finally shown in Sect. 4.

2 PageRank

In this section we outline the basics of PageRank algorithm.

First, the simple and widely accepted model of a network described in literature [28–30] is a graph where the nodes can be agents, persons, devices, resources etc and the set of directed and weighted arcs represent nodes relationships (e.g. trustworthiness in a social network, or hyperlink in the web).

The other model extensively used (and which PageRank is based on) is the *random surfer*. In a few words, the idea is having a surfer that visit nodes in the network starting from a random node (generally selected via a uniform probability distribution), then moving to another by selecting one among the links outgoing from the initial random node (the selection is still guided by uniform probability of $1/k$ where k is the number of out-links, i.e. the so-called *out-degree*). At each step the random walker also has a probability of jumping to a random node (that is, it does not follows out-links); this behaviour allows to cope with cases when no outgoing links are present, or when the surfer falls into a set of nodes pointing each other but isolated from the rest of network (this is called *rank sink* in [14]). In PageRank the surfer moves into the network of web pages, selecting hyperlinks at random from page to page, and occasionally choosing a URL not related to the page he is currently visiting.

To formalize such ideas, say N is the number of nodes in the network and A the $N \times N$ *network adjacency matrix* or *link matrix*, where each a_{ij} is the weight of the arc going from node i to node j. S is the $N \times 1$ *sink vector*, defined as:

$$s_i = \begin{cases} 1 & \text{if } out_i = 0 \\ 0 & \text{otherwise} \end{cases} \qquad \forall i \leq N$$

where out_i is the number of outlinks of node i. V is the *personalization vector* of size $1 \times N$, equal to the transposed initial distribution probability vector in the Markov chain model P_0^T. While this vector can be arbitrarily chosen as long as it's stochastic, a common choice is to make each term equal to $1/N$. $T = \mathbf{1}_{N\times 1}$ is the *teleportation vector*, where the notation $\mathbf{1}_{N\times M}$ stands for a $N \times M$ matrix where each element is 1.

In the general case the Markov chain built upon the network graph is not always ergodic, so it is not used directly for the calculation of the steady state random walker probabilities. As described in [14], the *transition matrix* M, used in the associated random walker problem, is derived from the link matrix, the sinks vector, the teleportation vector and the personalization vector defined above:

$$M = d(A + SV) + (1 - d)\,TV \tag{1}$$

where $d \in [0, 1]$ is called *damping factor* and in Brin and Page's original implementation it is set to 0.85. As we know from the Markov chain theory, the random walk probability vector at step n can be calculated as:

$$P_n = M^\mathsf{T} P_{n-1} \tag{2}$$

the related random walker problem can be calculated as:

$$P = \left(\lim_{n \to \infty} M^n \right)^{\mathsf{T}} P_0 = \lim_{n \to \infty} (M^{\mathsf{T}})^n P_0 = M_{\infty}^{\mathsf{T}} P_0 \qquad (3)$$

Each element of the probability vector is the PageRank value of a page, whose interpretation is *'the probability for the random surfer of being at a page at any given point in time during the walk'*.

The semantics is that the PageRank algorithm will assign high PageRank values to nodes (pages) that would appear more often in a random surfer walk, i.e. nodes with high PageRank are *relevant* pages that will be more visited by surfers and this ranking came from the link structure of the graph.

3 Results

3.1 Simulation Settings

In this work we want to study the effect on the ranking value (evaluated via PageRank) of a target node t when it is pointed by an in-link starting from a generic node i of the network. Specifically, we want to correlate the rank increment of t with the distance between i and t. The question is:*is there some sort of correlation or not? And if so, can we quantify it?*

To answer these questions, we carried out a set of simulations on two types of directed networks, Erdös-Rényi (ER) and Scale-Free (SF), in order to discover how this phenomenon depends on the links pattern. Both networks are built with 10k nodes with an average degree per node of 15.

In all the simulations, we start by randomly choosing a node i of the network and by connecting it and 15 of its neighbours to the target node t. This is done in order to have an initial set of nodes connected to the t thus allowing to compute the initial PageRank for t. All simulations were repeated 10 times by changing the initial randomly selected node i. After the construction of the initial attachment set of nodes, for each node j of the network we compute its distance from target node t and the increment of the rank (due to the increment of its PageRank value) as a consequence of linking node j to the node t. All simulations are then repeated on 5 different instances of ER and SF networks, aiming to eliminate a possible bias related to a particular network. In the following we will show results obtained by averaging over single simulations.

3.2 Simulation Results

Figure 1 reports the probability that a generic node of the network j is far from the target node t of a given network distance, computed after the construction of the initial set of attachment nodes. In other words, it shows the fraction of nodes of the networks with a given distance from the target node t.

In the case of ER networks, for almost 70% of nodes t is at 4 hops, while about 27% of nodes reach t in 3 hops. Finally, nodes with an average distance

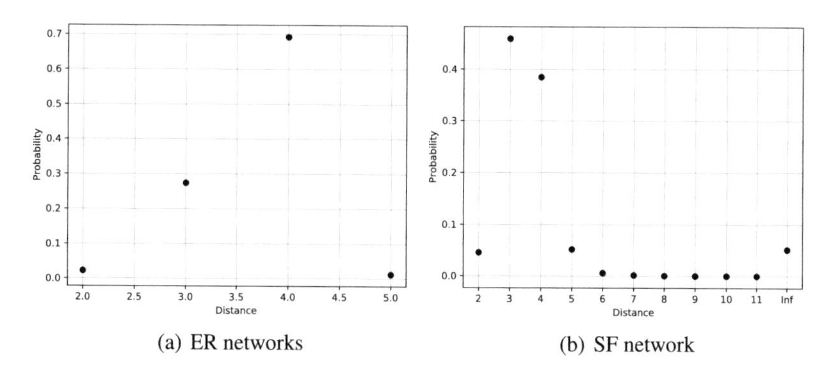

(a) ER networks (b) SF network

Fig. 1. Fraction of nodes with a given distance from node t.

to t of 2 and 5 hops are very rare. For SF networks, the most common nodes are those for which the node t is at 3 and 4 hops. There are about 6% of nodes that are far to t of 2 and 5 hops. Finally, there are about 6% of nodes for which there are no directed path from them to node t or, in other words, whose distance to t is infinity.

Figure 2 shows the rank gain (as a consequence of an increment of the PageRank value) of the target t when we connect a node j (at a given distance from t) to t. Let us note that the rank value 1 means the top position (i.e. the highest value of PageRank), so increments of the rank positions are negative values. In the figure, the min, average and max values are reported for each point.

By analyzing the figure we can note that long distance links, i.e. links from farthermost nodes, can provide a better rank gain for t. In general, the farther away is the node the greater is the rank gain. This is particularly true in SF networks in the case of inf distance, where the rank gain is maximum.

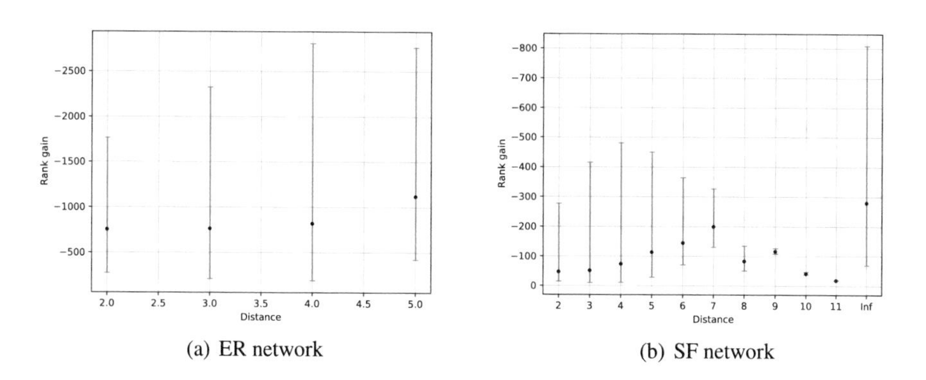

(a) ER network (b) SF network

Fig. 2. Rank gain of the node t when it is connected by an in-link from a node j that reaches it in a given number of hops

Our analysis highlights that long-distance links are better if a node wants to gain as much as possible rank. As a consequence, a very simple in-attachment strategy a node t can follow in order to increase its ranking is to select as in-neighbours those nodes for which t is very far away. This is obviously possible if and only if we know a-priori the whole network topology or, at least, the distance from each node to the target node. In a real world setting, this is not probably always true and a common strategy is to select the in-neighbours by using a random strategy.

To analyse the effect of a random neighbour selection in combination with the rank gain it leads, we make a graph of the product $probability * rank\_gain$ as a function of the distance. This metric is used to mediate the rank gain with the probability that a node with a given distance to t is randomly selected. In other words, while farthermost nodes are those with highest rank gain, on the other hand they are very rare. Therefore, such a metric allows to combine two antagonist effects in only one measure.

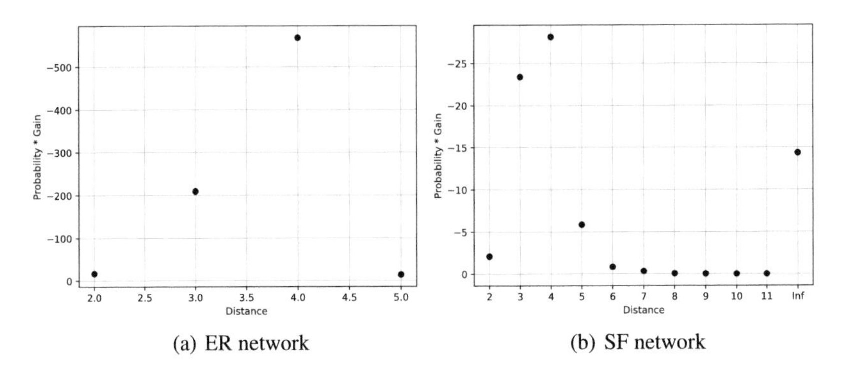

(a) ER network (b) SF network

Fig. 3. Probability * (rank_gain of the node t) when it is connected by an in-link from a node j that reaches it in a given number of hops

Considering Fig. 3, in the case of ER networks it follows the same trend of the Fig. 1, in the case of SF networks it gives a slightly different view. For example, while nodes at distance 7 give a very good rank gain (see Fig. 2), they are very rare (see Fig. 1), and this fact is well captured by the proposed metric, as depicted in Fig. 3, where it measures almost 0.

4 Conclusions

In this work we study how a node's ranking can be effectively improved by adding in-links, i.e. being linked by others existing nodes. In particular we adopted the widely used PageRank algorithm in a scenario with both Erdös-Rényi (ER) and Scale-Free (SF) 10k nodes networks. Simulations average results highlighted that the highest rank improvement can be achieved acquiring long distance nodes in-links, rather than being pointed by neighbours of first acquaintances.

These preliminary and somehow counterintuitive results suggest further investigation, with the final aim of addressing the in-link building problem (a.k.a. *best attachment* problem) with a proper heuristic.

References

1. Albert, R., Barabasi, A.L.: Statistical mechanics of complex networks. Rev. Modern Phys. **74**, 47 (2002)
2. Newman, M.: The structure and function of complex networks. SIAM Rev. **45** (2003)
3. Kunegis, J., Blattner, M., Moser, C.: Preferential attachment in online networks: measurement and explanations. In: Proceedings of the 5th Annual ACM Web Science Conference. WebSci 2013, pp. 205–214. ACM, New York (2013)
4. Fung, R., Lee, M.: Ec-trust (trust in electronic commerce): Exploring the antecedent factors. In: Proceedings of the 5th Americas Conference on Information Systems, pp. 517–519 (1999)
5. Sameerkhan, P., Shahrukh, K., Mohammad, A., Amir, A., Bali, A.: Comment based grading and rating system in e-commerce. Int. J. Eng. Res. Gen. Sci. **3**(1), 1319–1322 (2015)
6. Weng, J., Miao, C., Goh, A., Shen, Z., Gay, R.: Trust-based agent community for collaborative recommendation. In: Proceedings of the Fifth International Joint Conference on Autonomous Agents and Multiagent Systems, AAMAS 2006, pp. 1260–1262. ACM, New York (2006)
7. Liu, X.: Towards context-aware social recommendation via trust networks. In: Lin, X., Manolopoulos, Y., Srivastava, D., Huang, G. (eds.) Web Information Systems Engineering - WISE 2013. Lecture Notes in Computer Science, vol. 8180, pp. 121–134. Springer, Heidelberg (2013)
8. Pan, B., Hembrooke, H., Joachims, T., Lorigo, L., Gay, G., Granka, L.: In google we trust: Users decisions on rank, position, and relevance. J. Comput.-Med. Commun. **12**(3), 801–823 (2007)
9. Chauhan, V., Jaiswal, A., Khan, J.: Web page ranking using machine learning approach. In: 2015 Fifth International Conference on Advanced Computing Communication Technologies (ACCT), pp. 575–580, February 2015
10. Su, A.J., Hu, Y.C., Kuzmanovic, A., Koh, C.K.: How to improve your search engine ranking: Myths and reality. ACM Trans. Web **8**(2), 8:1–8:25 (2014)
11. Jiang, J.Y., Liu, J., Lin, C.Y., Cheng, P.J.: Improving ranking consistency for web search by leveraging a knowledge base and search logs. In: Proceedings of the 24th ACM International on Conference on Information and Knowledge Management, CIKM 2015, pp. 1441–1450. ACM, New York (2015)
12. de Blas, C.S., Martin, J.S., Gonzalez, D.G.: Combined social networks and data envelopment analysis for ranking. Eur. J. Oper. Res. **266**(3), 990–999 (2018)
13. Guerrero-Bote, V.P., Moya-Anegn, F.: A further step forward in measuring journals scientific prestige: the SJR2 indicator. J. Inf. **6**(4), 674–688 (2012)
14. Page, L., Brin, S., Motwani, R., Winograd, T.: The pagerank citation ranking: bringing order to the web (1998)
15. Carchiolo, V., Longheu, A., Malgeri, M., Mangioni, G.: Gain the best reputation in trust networks. In Brazier, F., Nieuwenhuis, K., Pavlin, G., Warnier, M., Badica, C. (eds.) Intelligent Distributed Computing V. Volume 382 of Studies in Computational Intelligence, pp. 213–218. Springer, Heidelberg (2012)

16. Buzzanca, M., Carchiolo, V., Longheu, A., Malgeri, M., Mangioni, G.: Dealing with the best attachment problem via heuristics. In: Badica, C., El Fallah Seghrouchni, A., Beynier, A., Camacho, D., Herpson, C., Hindriks, K., Novais, P. (eds.) Intelligent Distributed Computing X, pp. 205–214. Springer International Publishing, Cham (2017)
17. Carchiolo, V., Longheu, A., Malgeri, M., Mangioni, G.: Users' attachment in trust networks: reputation vs. effort. Int. J. Bio-Inspired Comput. **5**(4), 199–209 (2013)
18. Carchiolo, V., Longheu, A., Malgeri, M., Mangioni, G.: The cost of trust in the dynamics of best attachment. Comput. Inf. **34**, 167–184 (2015)
19. Buzzanca, M., Carchiolo, V., Longheu, A., Malgeri, M., Mangioni, G.: Direct trust assignment using social reputation and aging. J. Ambient Intell. Hum. Comput. **8**(2), 167–175 (2017)
20. Carchiolo, V., Longheu, A., Malgeri, M., Mangioni, G.: Network size and topology impact on trust-based ranking. IJBIC **10**(2), 119–126 (2017)
21. Carchiolo, V., Longheu, A., Malgeri, M., Mangioni, G.: Trust assessment: a personalized, distributed, and secure approach. Concurrency Comput. Pract. Experience **24**(6), 605–617 (2012)
22. Kamvar, S.D., Schlosser, M.T., Garcia-Molina, H.: The Eigentrust algorithm for reputation management in P2P networks. In: Proceedings of the Twelfth International World Wide Web Conference (2003)
23. Avrachenkov, K., Litvak, N.: The effect of new links on google pagerank. Stochast. Models **22**(2), 319–331 (2006)
24. de Kerchove, C., Ninove, L., Dooren, P.V.: Maximizing pagerank via outlinks. CoRR **abs/0711.2867** (2007)
25. Fercoq, O., Akian, M., Bouhtou, M., Gaubert, S.: Ergodic control and polyhedral approaches to pagerank optimization. IEEE Trans. Automat. Contr. **58**(1), 134–148 (2013)
26. Sydow, M.: Can one out-link change your pagerank? In: Szczepaniak, P.S., Kacprzyk, J., Niewiadomski, A. (eds.) AWIC. Volume 3528 of Lecture Notes in Computer Science, pp. 408–414. Springer (2005)
27. Olsen, M., Viglas, A., Zvedeniouk, I.: An approximation algorithm for the link building problem. CoRR **abs/1204.1369** (2012)
28. Berkhin, P.: A survey on pagerank computing. Internet Math. **2**, 73–120 (2005)
29. Marsh, S.: Formalising trust as a computational concept. Technical report, University of Stirling, Ph.D. thesis (1994)
30. Walter, F.E., Battiston, S., Schweitzer, F.: A model of a trust-based recommendation system on a social network. J. Autonom. Agents Multi-agent Syst. **16**, 57 (2008)

Concept Tracking and Adaptation for Drifting Data Streams under Extreme Verification Latency

Maria Arostegi[1(✉)], Ana I. Torre-Bastida[1], Jesus L. Lobo[1], Miren Nekane Bilbao[2], and Javier Del Ser[3]

[1] TECNALIA, 48160 Derio, Spain
{maria.arostegi,isabel.torre,jesus.lopez}@tecnalia.com
[2] University of the Basque Country (UPV/EHU), 48013 Bilbao, Spain
nekane.bilbao@ehu.eus
[3] TECNALIA, University of the Basque Country (UPV/EHU) and Basque Center for Applied Mathematics (BCAM), Leioa, Bizkaia, Spain
javier.delser@tecnalia.com

Abstract. When analyzing large-scale streaming data towards resolving classification problems, it is often assumed that true labels of the incoming data are available right after being predicted. This assumption allows online learning models to efficiently detect and accommodate non-stationarities in the distribution of the arriving data (concept drift). However, this assumption does not hold in many practical scenarios where a delay exists between predicted and class labels, to the point of lacking this supervision for an infinite period of time (extreme verification latency). In this case, the development of learning algorithms capable of adapting to drifting environments without any external supervision remains a challenging research area to date. In this context, this work proposes a simple yet effective learning technique to classify non-stationary data streams under extreme verification latency. The intuition motivating the design of our technique is to predict the trajectory of concepts in the feature space. The estimation of the region where concepts may reside in the future can be then exploited for producing more updated predictions for newly arriving examples, ultimately enhancing its accuracy during this unsupervised drifting period. Our approach is compared to a benchmark of incremental and static learning methods over a set of public non-stationary synthetic datasets. Results obtained by our passive learning method are promising and encourage further research aimed at generalizing its applicability to other types of drifts.

Keywords: Classification · Extreme verification latency
Concept drift · Non-stationary environments

1 Introduction

Nowadays, daily decision making processes depend on increasing volumes of largely heterogeneous data, generated at an ever-growing speed, thereby

J. Del Ser et al. (Eds.): IDC 2018, SCI 798, pp. 11–25, 2018.
https://doi.org/10.1007/978-3-319-99626-4_2

motivating the derivation of new learning models for data analysis. Indeed, the Big Data paradigm refers to data that surpass traditional scales of digital information and therefore, require a processing capacity far beyond that of conventional databases and computation mainframes. In order to gain value from data, efficient analytical models are needed, which is in turn the core of the so-called Big Data analytics [15]. Big Data analytics can be defined as the capability of extracting useful information from large datasets or streams of data, that due to its volume, variability, and velocity, cannot be performed by means of traditional methods.

In this work we focus on data produced in a streaming fashion, which must be ingested and processed in real time by fulfilling stringent constraints in terms of memory and processing power [8]. Particularly, data stream mining is used to extract knowledge in the form of patterns over non-stopping data streams [11]. The research in data stream mining has gained a high attraction due to the importance of its applications and the increasing generation of streaming information. Hence, applications of data stream analysis and algorithms, systems and frameworks that address streaming challenges have been highly developed over the last years. Scenarios where streaming data is produced abound in practice, including sensor networks, web logs, computer network traffic or industrial machinery, among many others. In response to the strict computational requirements posed by streaming data analysis, diverse methods for preprocessing and learning have been hitherto proposed in the literature, including sampling, data condensation, density-based or grid-based approaches, divide and conquer strategies and distributed computing [27]. Among them, incremental learning allows models for data analysis to be built on chunks of data, thereby updating the learned knowledge and lowering the computational demand of the model training to its minimum [23].

Nonetheless, computation constraints are not the only problem springing from data stream mining. It is often assumed that data streams are generated by stationary processes, following a fixed and unknown probability distribution to be captured by the analytical model deployed for e.g. classification. However, this assumption is far from reality, since in practice the phenomena yielding data flows are subject to seasonality, unexpected events and other factors that imprint unpredictable changes in data. Concept drift, general term used to refer to data streams whose characteristics evolve over time, has grasped the attention of the community towards the derivation of passive or active strategies to accommodate such changes in the construction of data-based models, stimulating a number of thorough reviews on this topic in the last couple of years [4,12].

Among the different types of drift that may occur over data streams in practice – infinite length, concept-drift, concept-evolution, feature-evolution and limited labeled data [16] – we concentrate on slowly evolving drifts: concepts corresponding to every class evolve over time by following a progressive trajectory in the feature space. This particular type of drifts is representative of sensing equipment whose calibration degrades progressively as a result of e.g. transducer material deterioration, the long-term impact of environmental pollution (such a dust or chemicals) and other reasons alike. When a slowly evolving drift occurs, a supervised learning algorithm designed to tackle a classification problem over the drifting data stream must adapt to the evolving concept either passively (for instance, resorting to diversity inducing techniques [22]) or actively (by detecting the drift and reconfiguring the learned model accordingly).

Interestingly, most contributions reported to date around data stream classification over drifting data assume that an immediate class supervision is available. Since the true labels for the data stream are known after they are predicted, the classifier at hand can be updated immediately. So do drift detection techniques, which can promptly identify symptoms that a concept change occurs by comparing a supervised record of past and present samples. While given as a premise in manifold studies [1,2,13], this instantaneous verification of data streams does not always hold in practice: many practical scenarios modeled as data stream classification problems are characterized by a lagged label supervision with respect to prediction time, which makes it even more involved to deal with non-stationary data streams [21]. These assumptions posed on timing and availability of information are open challenges in data stream mining research [18], for which techniques such as semi-supervised learning have been proposed in the related literature.

This verification latency can be extremely large (even infinite or *extreme*) in certain setups where information about the true labels associated to data can be costly to obtain, or even unfeasible [24]. These circumstances hinder not only the incremental knowledge update of the predictive model, but also the detection and/or adaptation to concept drifts. This work addresses the problem of learning in non-stationary data streams under extreme verification latency by proposing a classification model capable of tracking and passively adapting to evolving concept drifts occurring over non-supervised streaming data. Intuitively, the designed approach, coined as TRACE, relies on a low-complexity trajectory prediction algorithm (e.g. a Kalman filter) to track and anticipate the region of the feature space in which concepts (represented by their centroids) will be located in the future. This estimated position is used to correct the prediction made by the classification model, as well as to update future estimated positions with information labeled by the predicted samples, which is representative of the prevailing status of the concepts in the stream. The performance of the designed approach is assessed over a set of public synthetic datasets featuring evolving drifts of very diverse nature. Results of our proposed scheme are compared to the ones obtained (1) when no verification latency is held; and (2) when the model is not modified anyhow under verification latency. The obtained insights suggest further work aimed at extending the principles of our designed method to real datasets, potentially encompassing more complicated drift characteristics.

The rest of the paper is structured as follows: first a brief literature review is given in Sect. 2 so as to put our work in context, whereas Sect. 3 describes the proposed technique. Experimental results are presented and discussed in Sect. 4 and finally, Sect. 6 concludes this work and outlines future research lines stemming from our findings.

2 Background and Related Work

Concept drift can be referred to as *sudden* (abrupt, instantaneous) or *gradual* when it comes to the speed at which one concept evolves into another. Further classification criteria of gradual drifts can be found in [25], which considers moderate and slow drifts, and [30], which uses the term *virtual concept drift* to describe the need for a change of the currently utilized model due to a change of data distribution. The scheme contributed in this study is suited to deal with gradual concept drifts, particularly softly evolving concept changes.

In essence, learning in non-stationary environments under concept drift can be approached from two perspectives: active versus passive [6]. The difference resides on the adaptation mechanism: on one hand, active approaches hinge on the use of an explicit detector of the drift in the stream data, upon which the application of an adaptation mechanism is triggered. On the other hand, passive approaches preemptively modify the model over time without explicitly detecting the drift, so that the model gets prepared for any concept change to come (e.g. by inducing diversity in the learning algorithm of the model at hand [19]). Our scheme relies on a passive approach, in which the drift is not detected, but rather *tracked* so as to correct the prediction output by the model.

As explained in the introduction, the gap between the time at which a prediction is made on an incoming data sample and the availability of its corresponding true label (supervision) is denoted as verification latency [21]. There are different methods to handle verification latency, including semi-supervised learning, in which the labeled data available for the model training is used to label the remaining unlabeled data, so that the model trained on labeled and unlabeled data together performs better than the model trained on just labeled data. Another approach is active learning, by which the learning algorithm itself selects those instances that when labeled, would be most informative for the classification problem. When deployed over non-stationary data streams, in all these strategies it is necessary to design a adaptation mechanism to resolve the change induced by concept drifts in the distribution of data. This becomes a challenging issue under extreme verification latency, due to the lack of supervision that could contribute to a better – yet delayed – drift detection and adaptation.

If we leave aside the problem of extreme verification latency, the literature is rich in regards to classification models for concept drift environments [3,7,14,20,26,28,31]. However, in more demanding setups with extreme latency verification the number of related studies decreases considerably. Three works dealing with data stream classification under extreme verification latency in non-stationary environments are worth to be mentioned: Stream Classification Algorithm Guided by Clustering (SCARGC, [24]), Arbitrary Sub-Populations Tracker (APT, [17]) and Compacted Object Sample Extraction (COMPOSE [5]). SCARGC consists of clustering followed by classification applied repeatedly in a closed loop fashion. The algorithm uses the current and past cluster positions obtained by clustering the unlabeled data to track the drifting classes over time. The APT algorithm works in a two stage learning strategy: expectation-maximization to determine the optimal one-to-one assignment between the unlabeled and labeled data, and next the classifier is updated to reflect the population parameters of newly received data, as well as the drift characteristics. COMPOSE is a geometry-based framework to learn from non-stationary streaming data that follows three steps: (1) α-shapes that represent the current class conditional distribution are constructed; (2) α-shapes are compacted (shrinked) to represent the geometric center of each class distribution; and (3) from the compacted α-shapes new instances are extracted to serve as labeled data for future time steps.

Table 1. Recent literature related to concept drift and verification latency

Ref.	Learning model	Strategy	Type of drift	Verification latency
[26]	Ensembler	Active	Gradual	No
[31]	KNN	Passive	Gradual	No
[3]	Lazy learning	Active	Gradual and virtual	No
[28]	Ensembler	Passive	Virtual	No
[20]	SAM+ KNN	Passive	Heterogeneous	No
[14]	Statistical method	Active	Virtual	No (focus on imbalanced data)
[7]	Neural Network	Active	Abrupt	No
[24]	SCARGC	Passive	Gradual	Extreme
[17]	APT	Passive	Gradual	Extreme
[5]	COMPOSE	Passive	Gradual	Extreme
[29]	FAST COMPOSE	Passive	Gradual	Extreme
[9]	MASS	Passive	Heterogeneous	Extreme

The main weakness of the above approaches is that all are computationally expensive on high dimensional data. This drawback motivates the upsurge of improved versions such as FAST COMPOSE [29], a modification of COMPOSE that allows the algorithm to work without core support extraction so as to reduce its complexity; and Modular Adaptive Sensor System (MASS, [9]), another modified version of COMPOSE proposed as a solution to extreme verification latency in stream data, yet still computationally expensive for resource constrained applications (e.g. sensor networks). Table 1 summarizes the main features of the studies in this literature survey.

3 Proposed TRACE Approach

Following the schematic diagram depicted in Fig. 1, we consider a classification problem in a streaming scenario, where data samples $\mathbf{x}_t$ – each comprising N features or predictors x_t^n – are assumed to be fed to a classification model $M_{\boldsymbol{\theta}_t}(\cdot)$, which produces a predicted label $\widehat{y}_t = M_{\boldsymbol{\theta}_t}(\mathbf{x}_t)$ drawn from a set $\mathcal{Y}$ of possible class labels. By $\boldsymbol{\theta}_t$ we refer to the parameters defining the pattern learned by the model from the supervised knowledge about the stream at time t. The true label $y_t \in \mathcal{Y}$ associated to $\mathbf{x}_t$ is assumed to be known at time $t + \Delta$, where Δ denotes the verification latency. This being said, at time t the classifier receives both sample $\mathbf{x}_t$ and $y_{t-\Delta}$. Extreme verification latency occurs when $\Delta \rightarrow \infty$, i.e. there is no labeled information for the parameters $\boldsymbol{\theta}_t$ of the model to be updated in a supervised manner. Without loss of generality we assume that the lack of supervision starts at a given time instant t^*, so that model $M_{\boldsymbol{\theta}_{t^*}}(\cdot)$ is capable of discriminating among classes with good accuracy before the verification latency starts.

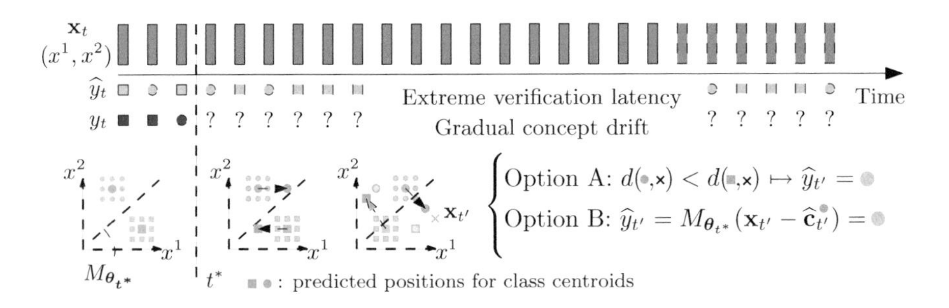

Fig. 1. Schematic diagram illustrating the main design principle of TRACE. By predicting the trajectory of concepts ■ and ● over the feature space, a straightforward distance-based prediction can be made (Option A) or samples can be shifted back to the last known concept distribution from which the predictive model was updated.

In this setup we consider an gradual drift, which can be regarded as a continuous change in the joint distribution $P_{\mathbf{X},Y}(\mathbf{x}_t, y)$ of explanatory variables and the labels over time. By gradual drift it is meant that $P(\mathbf{x}_t, y)$ and $P(\mathbf{x}_{t+1}, y)$ are not equal, but *similar* and statistically dependent on each other, i.e. $P(\mathbf{x}_t, \mathbf{x}_{t+1}|y) \neq P(\mathbf{x}_t|y) \cdot P(\mathbf{x}_{t+1}|y)$ and $D(P(\mathbf{x}_t|y), P(\mathbf{x}_{t+1}|y)) \leq \delta$, with $D(\cdot,\cdot)$ denoting a measure of distance between probability distributions, and $\delta \geq 0$ is a small constant. When such changes occur during the period when no label verification is available (i.e. $\forall t > t^*$), our designed approach must adapt the learned knowledge in $M_{\boldsymbol{\theta}_{t^*}}(\cdot)$ so as to reflect the evolved data distribution and maintain a good prediction accuracy in these circumstances.

This being said, at t^* TRACE starts tracking the evolution of every class $y \in \mathcal{Y}$ over the N-dimensional space spanned by the features of the instances along the stream. To this end, a centroid $\mathbf{c}_t^y$ is estimated and updated every time a new sample $\mathbf{x}_t$ arrives as

$$\mathbf{c}_t^y = \begin{cases} \rho \mathbf{c}_{t-1}^y + \rho \mathbf{x}_t & \text{if } M_{\boldsymbol{\theta}_t}(\mathbf{x}_t) = y, \\ \mathbf{c}_{t-1}^y & \text{otherwise}, \end{cases} \tag{1}$$

namely, as a ρ-weighted average of the accumulated centroid of past samples predicted to belong to the same class. Parameter ρ permits to gauge the inertia of the average with respect to past samples along the stream. It should be noted at this point that this näive technique to infer centroids is adopted for simplicity in the preliminary approach described in this paper. Its performance is closely linked to the suitability of the concepts in the stream to be described by a single centroid, as well as by the separability of classes. Nevertheless, as we will argue in the concluding section the proposed design accommodates more elaborated schemes for this purpose, such as incremental clustering techniques or avant-garde schemes for optimal data representation, such as Growing Neural Gas [10].

Once centroids for every class have been updated at time t, a trajectory prediction algorithm $\Psi_{\boldsymbol{\vartheta}}^y(\mathbf{c}_t^y) = \widehat{\mathbf{c}}_{t+1}^y$ is used to predict the location of the centroid of every class in the feature space at time $t+1$. Since we deal with gradual drifts this predicted trajectory can be assumed to be representative enough of the approximate position where samples belonging to the class should be at time $t+1$. Intuitively this prediction by itself could be used to replace $M_{\boldsymbol{\theta}_{t^*}}(\cdot)$ while label verification latency occurs, by just predicting newly arriving samples to belong to the label associated to the closest centroid. Thereafter, the predicted position $\widehat{\mathbf{c}}_{t+1}^y$ could be thought of as an additive bias, in the feature space, of samples received at time $t+1$. Thereby, once the aforementioned centroid-based prediction has been made, $M_{\boldsymbol{\theta}_{t^*}}(\cdot)$ could be fed with the corrected version of the sample $\mathbf{x}_{t+1} - \widehat{\mathbf{c}}_{t+1}^y$ to yield a second predicted label. As we will later discuss in light of the obtained simulations, the convenience of one prediction method or the other depends strongly on the separability of classes at time t^*. Once predictions have been produced, centroids are again updated as per (1) and the trajectory predictor is trained incrementally with this updated centroid position.

The overall working flow of TRACE is described in Algorithm 1. It is important to note that the proposed method does not update model $M_{\boldsymbol{\theta}_{t^*}}(\cdot)$ whatsoever. Instead, it relies exclusively on the predicted trajectory for the set of class centroids. This, obviously, comes along with a reduced applicability of the proposed scheme, e.g. drifts where new concepts emerge or disappear from the stream would require further developments. Nevertheless, this preliminary work poses the rationale and the main design guidelines for subsequent advances inspired by this approach. Furthermore, the algorithm above displayed can be extended to the case when data arriving in batches by applying Lines 2, 4, 5, 6, and 8 on every sample within the batch.

Algorithm 1. Proposed TRACE approach

Input : $M_{\boldsymbol{\theta}_{t^*}}(\cdot)$, $\mathbf{x}_{t+1}$ for $t > t^*$, centroids $\mathbf{c}_t^y$ and trajectory prediction models $\Psi_{\boldsymbol{\vartheta}_t}^y(\cdot)$ $\forall y \in \mathcal{Y}$.

Output: Label $\widehat{y}_{t+1}$ corresponding to $\mathbf{x}_{t+1}$.

1 **foreach** $y \in \mathcal{Y}$ **do**
2   Query the trajectory prediction algorithm $\widehat{\mathbf{c}}_{t+1}^y = \Psi_{\boldsymbol{\vartheta}_t}^y(\mathbf{c}_t^y)$ to yield an estimated position of the centroid of class y
3 **end**
4 Let $y^{\diamond} = \arg_y \min d(\widehat{\mathbf{c}}_{t+1}^y, \mathbf{x}_{t+1})$, where $d(\cdot,\cdot)$ is a distance measure
5 **Option A:** Let the prediction be $\widehat{y}_{t+1} = y^{\diamond}$
6 **Option B:** Let the prediction be $\widehat{y}_{t+1} = M_{\boldsymbol{\theta}_{t^*}}(\mathbf{x}_{t+1} - \widehat{\mathbf{c}}_{t+1}^{y^{\diamond}})$
7 **foreach** $y \in \mathcal{Y}$ **do**
8   Compute centroids $\mathbf{c}_{t+1}^y$ with $\mathbf{x}_{t+1}^y$ and its predicted label using (1)
9   Update trajectory predictor $\Psi_{\boldsymbol{\vartheta}_t}^y(\cdot)$ with $\mathbf{c}_{t+1}^y$, yielding $\Psi_{\boldsymbol{\vartheta}_{t+1}}^y(\cdot)$
10 **end**

4 Experimental Setup

A set of computer experiments has been designed to shed light on the performance of the proposed TRACE scheme. Specifically the benchmark detailed in what follows aims at providing an empirically informed answer to the following questions:

Q1: Does TRACE render any performance gain with respect to incremental learning methods when deployed on data streams subjecto to gradual concept drift and extreme verification latency?
Q2: Is there any connection between the characteristics of the classification problem at hand (e.g. drift evolution) and the performance of TRACE?
Q3: Under which circumstances should we expect a better prediction accuracy and adaptability to drift shown by TRACE?

To this end we use a public repository of non-stationary data streams commonly adopted by the stream mining community [24]. Specifically, the repository contains 15 synthetic datasets featuring gradual drift changes over time; Table 2 summarizes their main characteristics. The column *Drift interval* denotes the interval in number of examples between consecutive drifts, whereas N and $|\mathcal{Y}|$ refer to the number of features and classes, respectively.

For all simulations presented in this study a naïve k-Nearest Neighbor (kNN) approach will be used as the classification model $M_{\boldsymbol{\theta}}(\cdot)$, due to its simplicity and the straightforward forgetting mechanism that this lazy learner allows. Data instances are assumed to arrive in batches of size 200 samples, and ρ is set to 0.5 in all cases. Different schemes will be considered:

1. An incrementally updated kNN with a neighborhood size equal to $k = 3$ samples, and a forgetting mechanism consisting of a sliding window of 5

batches over which similarities with the sample to be tested is computed. In this case it is assumed that labels are verified all over the stream (`3NN-forget-nolatency`).
2. Another incremental kNN model with $k = 3$ neighbors without forgetting mechanism, again assuming that no verification latency holds during the reception of the stream (`3NN-nolatency`).
3. A similar approach to `3NN-nolatency`, but increasing the neighborhood size to $k = 10$ neighbors (`10NN-nolatency`).
4. Two kNN models operating under extreme latency verification. In these cases no modification nor update are done to the database of samples to accommodate concept drift and verification latency (`3NN-latency` and `10NN-latency`).
5. Our TRACE approach using Option A as per Algorithm 1, using a Kalman filter as trajectory detector (`KALMAN-TRACE-A`).
6. Our TRACE approach using Option B with $k = 3$ and $k = 10$ as per Algorithm 1 (`3NN-TRACE-B` and `10NN-TRACE-B`).

While other techniques could be used for this purpose, `TRACE` approaches will use a Kalman filter with $\sigma = 0.7$ as their trajectory predictor $\Psi_{\vartheta}^{y}(\cdot)$.

Table 2. Non-stationary datasets from [24] utilized in this work.

Dataset	$\|\mathcal{Y}\|$	N	Total number of instances	Drift interval
1CDT	2	2	16000	400
1CHT	2	2	16000	400
1CSurr	2	2	55283	600
2CDT	2	2	16000	400
2CHT	2	2	16000	400
FG_2C_2D	2	2	200000	2000
GEARS_2C_2D	2	2	200000	2000
MG_2C_2D	2	2	200000	2000
UG_2C_2D	2	2	100000	1000
UG_2C_3D	2	3	200000	2000
UG_2C_5D	2	5	200000	2000
4CE1CF	5	2	173250	750
4CR	4	2	144400	400
4CRE-V1	4	2	125000	1000
4CRE-V2	4	2	183000	1000
5CVT	5	2	40000	1000

5 Results and Discussion

Table 3 shows the obtained accuracy scores for each (dataset, technique) combination averaged over the entire stream. The best results among the schemes that assume extreme latency verification are underlined.

Table 3. Average accuracy over the entire stream for benchmark datasets

	No verification latency			Extreme latency verification				
	3NN-forget-nolatency	3NN-nolatency	10NN-nolatency	3NN-latency	10NN-latency	3NN-TRACE-B	10NN-TRACE-B	KALMAN-TRACE-A
1CDT	0.997	0.994	0.994	0.984	0.976	0.961	0.959	**0.997**
1CHT	0.982	0.974	0.977	0.936	0.935	0.909	0.915	**0.986**
1CSurr	0.982	0.826	0.825	0.702	0.703	**0.906**	0.902	0.873
2CDT	0.821	0.695	0.658	0.581	0.578	**0.881**	0.877	0.879
2CHT	0.790	0.680	0.649	0.575	0.573	**0.814**	0.818	0.813
FG_2C_2D	0.959	0.943	0.948	0.917	**0.921**	0.807	0.789	0.698
GEARS_2C_2D	0.998	0.957	0.959	0.942	0.946	0.943	0.947	**0.958**
MG_2C_2D	0.939	0.870	0.872	0.807	**0.809**	0.807	0.807	0.804
UG_2C_2D	0.964	0.846	0.847	0.588	0.585	0.964	0.965	**0.967**
UG_2C_3D	0.960	0.932	0.937	0.729	0.724	**0.967**	**0.967**	**0.967**
UG_2C_5D	0.953	0.917	0.922	0.755	0.751	0.955	0.955	**0.955**
4CE1CF	0.974	0.991	0.985	0.961	**0.967**	0.817	0.818	0.943
4CR	0.999	0.652	0.633	0.387	0.389	0.999	0.999	**0.999**
4CRE-V1	0.969	0.476	0.491	0.289	0.293	0.969	0.969	**0.970**
4CRE-V2	0.913	0.432	0.436	0.306	0.307	0.920	0.920	**0.925**
5CVT	0.796	0.680	0.702	0.463	0.680	0.798	**0.810**	0.795
1CHT_MOD*	0.993	0.988	0.989	0.849	0.847	**0.995**	**0.995**	**0.995**

*: 1CHT_MOD is a modified version of 1CHT where latency verification is assumed to start in the middle of the dataset.

In response to question Q1, the above scores reveals that in general, schemes relying on the proposed TRACE approach perform best for most of the considered datasets. This is particularly noticeable in certain cases such as 1CSurr, UG_2C_2D, UG_2C_3D, UG_2C_5D, 4CRE-V1 and 4CRE-V2. To delve into the reasons for this outperforming behavior we have selected six representative datasets that will help us in this regard. Furthermore, a closer look will be taken at the evolution of the accuracy provided by the techniques over the stream data corresponding to each dataset, aiming at answering the questions postulated at the beginning of this section. These selected datasets are 1CHT, 1CHT_MOD, 2CHT, 5CVT, UG_2C_3D and 4CRE-V2, which we next describe for the sake of a proper understanding of the remainder of the section:

- 1CHT: one of the two classes remains located over the same feature region, whereas the other class moves over time. At first the two classes are quite close to each other, and the *moving class* moves away from the static class up to the middle of the time period spanned by the dataset. Finally, it returns gradually to the initial location.
- 1CHT_MOD: it is the same dataset than 1CHT, but shifted in time so as to depart from a distant class arrangement over the feature space. At first the moving class gets progressively closer to the static one. Once they fall in the same feature region, the moving class begins to separate from the static one until it arrives to the initial stage.

- `2CHT`: the two classes move together over the feature space, describing a linear trajectory along the horizontal axis given by the first feature.
- `5CVT`: it comprises five classes and two features. The five classes move along the vertical axis defined by the second feature maintaining the same relative distance between their centroids.
- `UG_2C_3D`: there are two classes defined on a three-dimensional feature space ($N = 3$). First the two classes are widely separated and begin to approach each other, until they finally merge together. Thereafter, the two classes keep together while rotating twice (once per features 1 and 2) and moving linearly over one of the axis.
- `4CRE-V2`: with four classes and two features, two types of drift occur in this example: the first one consists of the rotation around the origin of the feature space, whereas the second one are linear axial movements to make all classes repeatedly farther and closer to each other.

The plots in Figs. 2a to f depict the evolution of the accuracy over the data stream for the above datasets. For a better visual inspection scores have been averaged over a sliding window of size equal to 10 batches, with vertical whiskers standing for ± standard deviation of the accuracy measured over the samples between every pair of whiskers.

Regarding Q2 and Q3, when classes are not separable in the beginning of the stream, that is, when the model $M_{\theta_{t^*}}(\cdot)$ was built (namely, `1CHT` in the selected subset of problems) no gains are yielded by our proposed `TRACE` method. Indeed, the separability of classes before label verification disappears is crucial for achieving good accuracy scores. Since `TRACE-B` displaces new data batches to the original status prior to verification latency, if classes were not clearly discriminated at this point, newly arriving data batches will not be discriminated correctly. This conclusion is buttressed by results of `1CHT` (Fig. 2a) and `1CHT_MOD` (Fig. 2b). The main difference among these datasets is the separation between classes in the initially supervised batches. Consequently, in `1CHT` the accuracy level remains almost uniform until the moving class begins to get away from the other, slightly decreasing thereafter. Contrarily, in `1CHT_MOD` accuracy is high while the moving class approaches the static one; when the two classes are close to each other, the performance degrades slightly, and recovers quickly as both classes become more easily separable.

Furthermore, the specific drift trajectory followed by the concepts along the stream is also decisive for the performance of `TRACE`. In `2CHT` and `5CVT` all classes move together, synchronously and in the same direction over the feature space, leading to uniform accuracy scores. By contrast, in `4CRE-V2` example, the drift consists of the 4 classes moving clockwise at a constant radial velocity around a fixed point, with an axial linear movement that makes them get close or far away from each other in a repeated fashion. As shown in Fig. 2f, the accuracy degrades when the combined effect of rotation and linear class displacement jeopardize the separability of classes by overlapping drifted concepts within the same feature region.

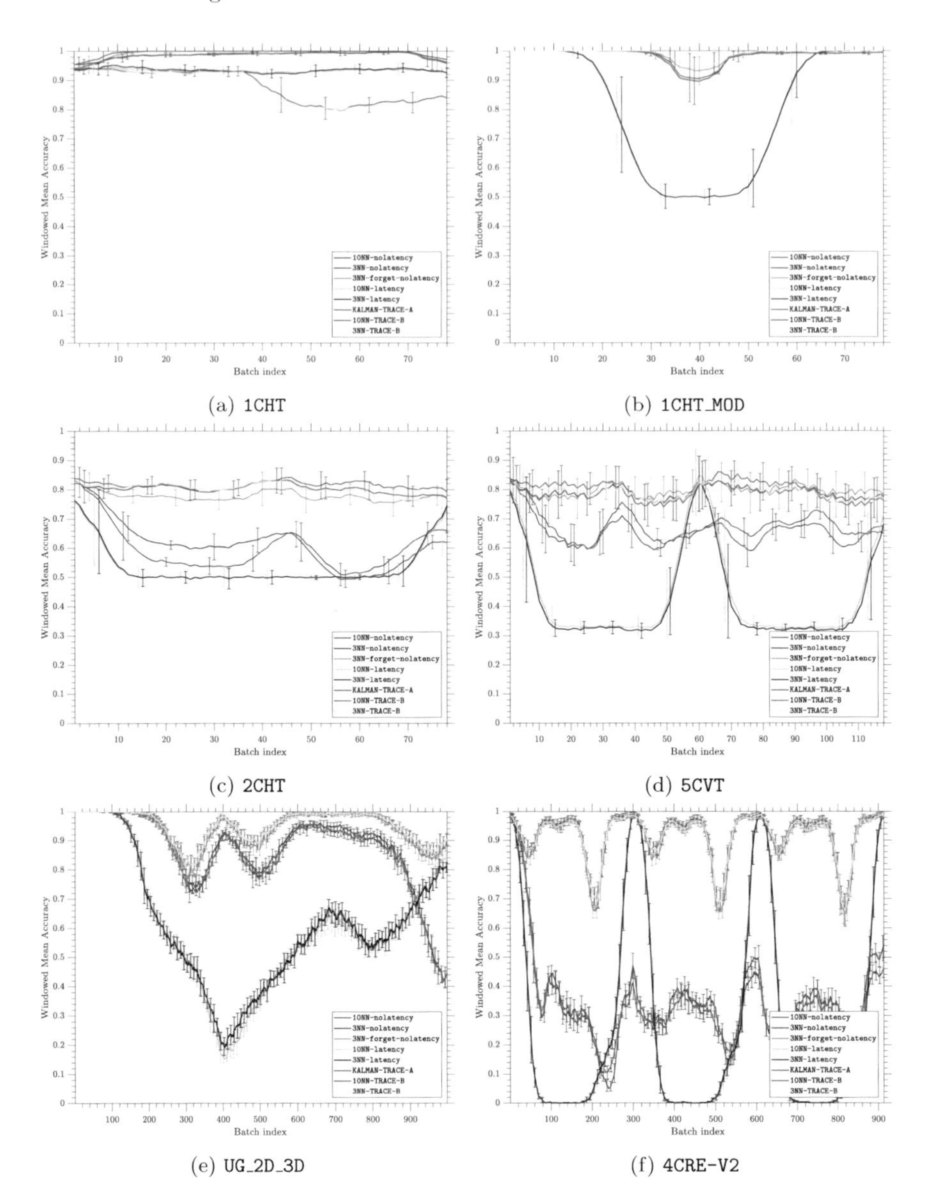

Fig. 2. Evolution of the accuracy over time for different selected datasets, averaged over a sliding window of 10 batches.

Finally, a note on computational complexity should be stated. It is clear that when scaling up to datasets with higher number of classes, the fact that a trajectory predictor is trained for every class could make the overall TRACE approach unfeasible for certain practical setups with severely constrained computational resources. Nevertheless, future workarounds are planned to alleviate this expected computational complexity.

6 Conclusions and Future Work

Among the computationally demanding scenarios addressed by Big Data Analytics, real-time streaming data classification has lately grasped good attention of the research community. This interest is particularly stimulated by non-stationary phenomena, by which classification models for streaming data become obsolete at a point due to the drift experienced by concepts to be classified over time. This computational paradigm becomes even more involved when supervised knowledge of the predicted samples is unavailable in practice, eventually making it difficult to adapt the model or detect drifts during the unsupervised stream period. In this context, this paper has elaborated on `TRACE`, a novel passive classification method for data streams with smooth drifts under extreme verification latency. In essence the proposed design traces and predicts the feature region where the different classes will reside in future time steps of the unsupervised stream period. This knowledge can be exploited to make straightforward predictions of newly arriving data samples by comparing them to the predicted centroids for every class. Alternatively, this sample-to-centroid association can be further used to correct the drift experienced by the sample and use the originally trained model under label verification. `TRACE` has been evaluated over fifteen synthetic datasets in comparison to naïve schemes for stream data classification.

The obtained results are promising, hence `TRACE` can be regarded as an experimental baseline model to be improved in the future. To this end future efforts will be conducted to extend the applicability of `TRACE` to consider more complex drifts, such as the generation and removal of classes and features. To this end, the adoption of incremental algorithms to optimally represent multidimensional datasets seems a promising path of work for the near future. Diversity-based ensemble classifiers will be also hybridized with `TRACE` so as to guide the induced diversity along the predicted drift direction.

Acknowledgements. This work was supported in part by the Basque Government under the EMAITEK funding program. Jesus L. Lobo also thanks the funding support from the EU project Pacific Atlantic Network for Technical Higher Education and Research - PANTHER (grant number 2013-5659/004-001 EMA2).

References

1. Bose, R.J.C., van der Aalst, WM., Žliobaite, I., Pechenizkiy, M.: Handling concept drift in process mining. In: International Conference on Advanced Information Systems Engineering, pp. 391–405. Springer (2011)
2. Dehghan, M., Beigy, H., ZareMoodi, P.: A novel concept drift detection method in data streams using ensemble classifiers. Intell. Data Anal. **20**(6), 1329–1350 (2016)
3. Delany, S.J., Cunningham, P., Tsymbal, A., Coyle, L.: A case-based technique for tracking concept drift in spam filtering. Knowl.-Based Syst. **18**(4–5), 187–195 (2005)
4. Ditzler, G., Roveri, M., Alippi, C., Polikar, R.: Learning in nonstationary environments: a survey. IEEE Comput. Intell. Mag. **10**(4), 12–25 (2015)

5. Dyer, K.B., Capo, R., Polikar, R.: Compose: a semisupervised learning framework for initially labeled nonstationary streaming data. IEEE Trans. Neural Netw. Learn. Syst. **25**(1), 12–26 (2014)
6. Elwell, R., Polikar, R.: Incremental learning of concept drift in nonstationary environments. IEEE Trans. Neural Netw. **22**(10), 1517–1531 (2011)
7. Escovedo, T., Koshiyama, A., da Cruz, A.A., Vellasco, M.: DetectA: abrupt concept drift detection in non-stationary environments. Appl. Soft Comput. **62**, 119–133 (2018)
8. Fan, W., Bifet, A.: Mining big data: current status, and forecast to the future. ACM SIGKDD Explor. Newsl. **14**(2), 1–5 (2013)
9. Frederickson, C., Gracie, T., Portley, S., Moore, M., Cahall, D., Polikar, R.: Adding adaptive intelligence to sensor systems with mass. In: IEEE Sensors Applications Symposium (SAS), pp. 1–6 (2017)
10. Fritzke, B.: A growing neural gas network learns topologies. In: Advances in Neural Information Processing Systems, pp. 625–632 (1995)
11. Gaber, M.M., Zaslavsky, A., Krishnaswamy, S.: Mining data streams: a review. ACM Sigmod Rec. **34**(2), 18–26 (2005)
12. Gama, J., Žliobaite, I., Bifet, A., Pechenizkiy, M., Bouchachia, A.: A survey on concept drift adaptation. ACM Comput. Surv. **46**(4), 44 (2014)
13. Gonçalves Jr., P.M., De Barros, R.S.M.: RCD: a recurring concept drift framework. Pattern Recogn. Lett. **34**(9), 1018–1025 (2013)
14. Hofer, V., Krempl, G.: Drift mining in data: a framework for addressing drift in classification. Comput. Stat. Data Anal. **57**(1), 377–391 (2013)
15. Kambatla, K., Kollias, G., Kumar, V., Grama, A.: Trends in big data analytics. J. Parallel Distrib. Comput. **74**(7), 2561–2573 (2014)
16. Khan, L., Fan, W.: Tutorial: data stream mining and its applications. In: International Conference on Database Systems for Advanced Applications, pp. 328–329 (2012)
17. Krempl, G.: The algorithm apt to classify in concurrence of latency and drift. In: International Symposium on Intelligent Data Analysis, pp. 222–233 (2011)
18. Krempl, G., Žliobaite, I., Brzeziński, D., Hüllermeier, E., Last, M., Lemaire, V., Noack, T., Shaker, A., Sievi, S., Spiliopoulou, M.: Open challenges for data stream mining research. ACM SIGKDD Explor. News. **16**(1), 1–10 (2014)
19. Lobo, J.L., Del Ser, J., Bilbao, M.N., Perfecto, C., Salcedo-Sanz, S.: DRED: an evolutionary diversity generation method for concept drift adaptation in online learning environments. Appl. Soft Comput. **68**, 693–709 (2018)
20. Losing, V., Hammer, B., Wersing, H.: Tackling heterogeneous concept drift with the self-adjusting memory (sam). Knowl. Inf. Syst. 1–31 (2018)
21. Marrs, G.R., Hickey, R.J., Black, M.M.: The impact of latency on online classification learning with concept drift. International Conference on Knowledge Science, Engineering and Management, pp. 459–469. Springer (2010)
22. Minku, L.L., White, A.P., Yao, X.: The impact of diversity on online ensemble learning in the presence of concept drift. IEEE Trans. Knowl. Data Eng. **22**(5), 730–742 (2010)
23. Pang, S., Ozawa, S., Kasabov, N.: Incremental linear discriminant analysis for classification of data streams. IEEE Trans. Syst. Man Cybern. Part B (Cybern.) **35**(5), 905–914 (2005)
24. Souza, V.M., Silva, D.F., Gama, J., Batista, G.E.: Data stream classification guided by clustering on nonstationary environments and extreme verification latency. In: SIAM International Conference on Data Mining, pp. 873–881 (2015)

25. Stanley, K.O.: Learning concept drift with a committee of decision trees. Report UT-AI-TR-03-302, Department of Computer Sciences, University of Texas at Austin, USA (2003)
26. Street, W.N., Kim, Y.: A streaming ensemble algorithm (SEA) for large-scale classification. In: ACM SIGKDD International Conference on Knowledge Discovery and Data Mining, pp. 377–382 (2001)
27. Tsai, C.W., Lai, C.F., Chao, H.C., Vasilakos, A.V.: Big data analytics: a survey. J. Big Data **2**(1), 21 (2015)
28. Tsymbal, A., Pechenizkiy, M., Cunningham, P., Puuronen, S.: Dynamic integration of classifiers for handling concept drift. Inf. Fusion **9**(1), 56–68 (2008)
29. Umer, M., Frederickson, C., Polikar, R.: Learning under extreme verification latency quickly: fast compose. In: IEEE Symposium Series on Computational Intelligence (SSCI), pp. 1–8. IEEE (2016)
30. Widmer, G., Kubat, M.: Learning in the presence of concept drift and hidden contexts. Mach. Learn. **23**(1), 69–101 (1996)
31. Xioufis, E.S., Spiliopoulou, M., Tsoumakas, G., Vlahavas, IP.: Dealing with concept drift and class imbalance in multi-label stream classification. In: International Joint Conferences on Artificial Intelligence, pp. 1583–1588 (2011)

Adversarial Sample Crafting for Time Series Classification with Elastic Similarity Measures

Izaskun Oregi[1(✉)], Javier Del Ser[2,3], Aritz Perez[3], and Jose A. Lozano[3,4]

[1] TECNALIA, 48160 Derio, Bizkaia, Spain
izaskun.oregui@tecnalia.com
[2] TECNALIA, University of the Basque Country (UPV/EHU), Leioa, Bizkaia, Spain
javier.delser@tecnalia.com
[3] Basque Center for Applied Mathematics (BCAM), 48009 Bilbao, Bizkaia, Spain
aperez@bcamath.org
[4] University of the Basque Country (UPV/EHU), Leioa, Spain
ja.lozano@ehu.eus

Abstract. Adversarial Machine Learning (AML) refers to the study of the robustness of classification models when processing data samples that have been intelligently manipulated to confuse them. Procedures aimed at furnishing such confusing samples exploit concrete vulnerabilities of the learning algorithm of the model at hand, by which perturbations can make a given data instance to be misclassified. In this context, the literature has so far gravitated on different AML strategies to modify data instances for diverse learning algorithms, in most cases for image classification. This work builds upon this background literature to address AML for distance based time series classifiers (e.g., nearest neighbors), in which attacks (i.e. modifications of the samples to be classified by the model) must be intelligently devised by taking into account the measure of similarity used to compare time series. In particular, we propose different attack strategies relying on guided perturbations of the input time series based on gradient information provided by a smoothed version of the distance based model to be attacked. Furthermore, we formulate the AML sample crafting process as an optimization problem driven by the Pareto trade-off between (1) a measure of distortion of the input sample with respect to its original version; and (2) the probability of the crafted sample to confuse the model. In this case, this formulated problem is efficiently tackled by using multi-objective heuristic solvers. Several experiments are discussed so as to assess whether the crafted adversarial time series succeed when confusing the distance based model under target.

Keywords: Adversarial Machine Learning · Time series classification
Elastic similarity measures

J. Del Ser et al. (Eds.): IDC 2018, SCI 798, pp. 26–39, 2018.
https://doi.org/10.1007/978-3-319-99626-4_3

1 Introduction

Nowadays there is no doubt on the renaissance of Machine Learning and Data Science ignited by the digitalization of many data-intensive business sectors (e.g. Industry, Energy) and the upsurge of new models and technologies for data ingestion, analysis and visualization. Within this technological outbreak, predictive modeling prevails as one of the most practical areas of Data Science to date. Indeed, classification and regression models lay at the core of many applications in diverse disciplines and fields, from failure prognosis in industrial machinery to road traffic or energy consumption forecasting, among others [12,15].

In essence predictive modeling usually assumes that data remain statistically stable over time, which allows the learning algorithm at hand to capture the pattern between the observed features and the target variable to be predicted. A primary design goal in this regard is to properly model this pattern so that it can be used to accurately predict unseen data instances (i.e. the model is capable of *generalizing*). This design principle, however, entails a serious threat in terms of the robustness of the model against intentionally confusing data instances. Indeed, it has been lately reported that adversarial examples can be miscategorized by complex predictive models by adding small perturbations to the features of the input example, so that the perturbed sample is effective to mislead the model while closely resembling its original version [9]. The study of this vulnerability of predictive models and the derivation of strategies and countermeasures to make them robust against such attacks falls within what has been coined as Adversarial Machine Learning (AML).

In this research area, a diverse collection of adversarial sample crafting techniques has been proposed so far by the community. The research activity has been specially notable in the Computer Vision field, with Deep Neural Networks (DNN) as the predictive models under study. DNN have undoubtedly shown themselves to achieve impressive results when undertaking predictive tasks with image data. However, a recent study [20] has revealed that even these advanced neuronal networks can be broken by adversarial attacks based on small, visually imperceptible perturbations to images. Not only the attacked classifier can be fooled and change its prediction when processing the adversarial sample, but also the attack can be exploited to confuse other machine learning classifiers (the so-called *transferability* of the attack [3,16,18]). Concerns with the practical consequences of these findings in terms of security has stimulated a vibrant research activity related to this paradigm.

In the last few years, a diverse collection of adversarial sample crafting techniques has been proposed [1,10,17], which rely on DNN models mostly for image classification. Such techniques resort to the gradient information provided by the DNN learning algorithm, which steers the perturbation of the adversarial sample along the most misclassifying direction of the feature space.

Despite the prevalence of time series data in biometry, forensics and other security-related scenarios, to the best of our knowledge no attention has been paid to AML for time series classification based on elastic similarity measures (ESM). When dealing with sequences, the use of such ESM allows for non-linear warps over time, so that the computation of the similarity of two time series becomes independent of local time shifts. This property allows classifiers to perform better in applications where time warps and lags between time series do not imply that their class labels differ from each other, as in energy consumption characterization [21], activity recognition [8] or speech processing [19], among others. Indeed, several studies have shown that ESM based nearest neighbors classifiers are specially competitive for time series classification tasks [6,13]. Unfortunately, the elasticity of these measures makes it not straightforward to extrapolate prior adversarial sample crafting strategies to ESM based models.

This work covers the lack of research noted above by proposing novel strategies to produce adversarial time series capable of confusing k Nearest Neighbors (kNN) classifiers under Dynamic Time Warping (DTW) [2], one of the most frequently used ESM for time series classification. For this specific similarity measure we derive three approaches to construct such adversarial samples: the first two hinge on the adaptation of the gradient based attack strategies reported previously for DNN models and image data, for which we develop an expression for the gradient over the distance space spanned by the DTW metric. The last strategy aims at circumventing the susceptibility of gradient based strategies to its parametric configuration (i.e. gradient step size) by formulating the sample crafting process as a bi-objective optimization problem. This problem is subsequently tackled by using multi-objective heuristics. Computer experiments are performed with datasets from the UCR repository [4], which are discussed in detail to validate the effectiveness of our proposed strategies when producing adversarial time series, as well as the capability of the produced samples to confuse models constructed on datasets differing from the one utilized to build the attack.

The remainder of this paper is structured as follows: Sect. 2 provides some background material on adversarial sample crafting and DTW that is deemed necessary for a proper understanding of the contribution of this work. Next, Sect. 3 and subsections therein describe the proposed strategies to construct adversarial time series for DTW based classifiers. Section 4 presents and discusses the obtained results from the performed computer experiments and, finally, Sect. 5 ends the paper by drawing concluding remarks and tracing future research directions related to this work.

2 Background and Related Work

In order to ease the understanding of our proposed strategies to attack DTW based classifiers, this section provides an overview of recently proposed samples crafting techniques, as well as the fundamentals of the DTW similarity measure.

2.1 Adversarial Sample Crafting Techniques

We consider a training set $\mathcal{D} = \{(\mathbf{x}_n, y_n)\}_{n=1}^{N}$, where $\mathbf{x}_n \in \mathcal{X}$ represents the n-th training example (feature vector) and $y_n \in \{1, \ldots, C\}$ its corresponding class label. Using $\mathcal{D}$, a classification problem consists of learning a function $f : \mathcal{X} \rightarrow \{1, \ldots, C\}$ that predicts the class label $y \in \{1, ..., C\}$ of any unseen input example $\mathbf{x} \in \mathcal{X}$. In this context, the goal of an adversarial attack is to generate a perturbed version of a legitimate example $\mathbf{x}$ that is misclassified by classifier f. The perturbed sample can be produced as $\widetilde{\mathbf{x}} = \mathbf{x} + \boldsymbol{\Lambda}_{\mathbf{x}}$, where the perturbation vector $\boldsymbol{\Lambda}_{\mathbf{x}}$ yields from:

$$\boldsymbol{\Lambda}_{\mathbf{x}} = \arg\min_{\mathbf{z}} d_p(\mathbf{x}, \mathbf{x} + \mathbf{z}), \text{ subject to } f(\mathbf{x}) \neq f(\mathbf{x} + \mathbf{z}), \tag{1}$$

where $d_p(\cdot, \cdot)$ refers to the L_p-norm distance, and $\mathbf{z}$ is a vector with the same dimension than $\mathbf{x}$. That is, the optimal distortion $\boldsymbol{\Lambda}_{\mathbf{x}}$ corresponds to the minimum perturbation that, added to an original sample, is capable of misleading the classifier at hand.

Solving Eq. (1) is non trivial. Therefore, Szegedy et al. [20] proposed to reformulate the above problem for $p = 2$ and solve it by means of a limited-memory Broyden-Fletcher-Goldfarb-Shanno (L-BFGS) solver, i.e.:

$$\boldsymbol{\Lambda}_{\mathbf{x}} = \arg\min_{\mathbf{z}} \varepsilon \cdot d_2(\mathbf{x}, \mathbf{x} + \mathbf{z}) + J_f(\mathbf{x} + \mathbf{z}, y), \tag{2}$$

where $y = f(\mathbf{x})$ (i.e. the true label of $\mathbf{x}$), $J_f(\mathbf{x}, y)$ is the loss function used to learn f from $\mathcal{D}$ (e.g. Mean Square Error in a linear regressor), and $\varepsilon > 0$ is a constant set to ensure that $d_2(\mathbf{x}, \widetilde{\mathbf{x}}) = ||\widetilde{\mathbf{x}} - \mathbf{x}||_2 \leq \varepsilon$. Line search strategy was proposed to infer the optimal value of ε.

Instead of solving (2) optimally, Goodfellow et al. [7] proposed an alternative technique to craft adversarial samples, which was coined as *Fast Gradient Sign Method* (FGSM). Given a value of ε fixed beforehand, the distortion is given by:

$$\boldsymbol{\Lambda}_{\mathbf{x}} = \varepsilon \cdot \text{sign}\left[\nabla_{\mathbf{x}} J_f(\mathbf{x}, y_{\circledast})\right], \tag{3}$$

where sign$(\cdot)$ denotes the sign function, $y_{\circledast} \neq f(\mathbf{x})$ is the target class to which the adversarial sample $\widetilde{x}$ is to be misclassified, and $\nabla_{\mathbf{x}}$ is the gradient with respect to $\mathbf{x}$. In this approach the distortion is upper bounded under the L_∞-norm, that is, $||\widetilde{\mathbf{x}} - \mathbf{x}||_\infty \leq \varepsilon$. It is important to observe that the application of FGSM only ensures that a solution to the problem in (1) is provided if a proper selection of ε is made. In this same line of work, in [14] a fast gradient strategy was proposed for $p = 2$ (i.e., to ensure the distortion meets $||\widetilde{\mathbf{x}} - \mathbf{x}||_2 \leq \varepsilon$). In this case, the distortion is:

$$\boldsymbol{\Lambda}_{\mathbf{x}} = \varepsilon \cdot \frac{\nabla_{\mathbf{x}} J_f(\mathbf{x}, y_{\circledast})}{||\nabla_{\mathbf{x}} J_f(\mathbf{x}, y_{\circledast})||_2}. \tag{4}$$

More recently, Kurakin et al. proposed an iterative version of the FGSM technique [11]. In this scheme the adversarial sample is computed by the recurrence:

$$\widetilde{\mathbf{x}}^{(t+1)} = \widetilde{\mathbf{x}}^{(t)} + \boldsymbol{\Lambda}_{\widetilde{\mathbf{x}}^{(t)}}, \tag{5}$$

where $t \in \{1, \ldots, T_{\max}\}$, $\widetilde{\mathbf{x}}^{(1)} = \mathbf{x}$ and

$$\boldsymbol{\Lambda}_{\widetilde{\mathbf{x}}^{(t)}} = \alpha \cdot \text{sign}\left[\nabla_{\mathbf{x}} J_f(\widetilde{\mathbf{x}}^{(t)}, y_{\circledast})\right]. \tag{6}$$

In order to ensure that adversarial samples satisfy that $||\widetilde{\mathbf{x}}-\mathbf{x}||_\infty$ is bounded by ε, α is set to $\varepsilon/T_{\max}$. Similarly, for the distortion to meet $||\widetilde{\mathbf{x}}-\mathbf{x}||_2 \leq \varepsilon$, it must be computed as:

$$\Lambda_{\widetilde{\mathbf{x}}^{(t)}} = \alpha \frac{\nabla_{\mathbf{x}} J_f(\widetilde{\mathbf{x}}^{(t)}, y_\otimes)}{||\nabla_{\mathbf{x}} J_f(\widetilde{\mathbf{x}}^{(t)}, y_\otimes)||_2}, \tag{7}$$

which also depends on ε and $T_{\max}$ to achieve a feasible solution to the problem in (1), similarly to what was previously noted for FGSM.

2.2 Dynamic Time Warping

As mentioned in the introduction, ESMs (including DTW) are characterized by their ability to stretch or compress the time axis in order to mitigate the influence of local time shifts in their similarity computation. This is accomplished by formulating the calculation of the similarity as an optimization problem which seeks to find, among a constrained set of possible time series alignments, the one with minimum *weight*. In the case of DTW, the sought alignment corresponds to the one with the lowest Euclidean Distance (ED). Let us denote two time series $U = (u_1, \ldots, u_i, \ldots, u_r)$ and $V = (v_1, \ldots, v_j, \ldots, v_s)$, with r and s their total number of observations. A path p is a finite sequence of $(i,j) \in [1,r] \times [1,s]$ pairs representing the alignment between u_i and v_j observations. When dealing with the DTW measure, an allowed alignment is a path $p = \{(i_1, j_1), \ldots, (i_l, j_l), \ldots, (i_L, j_L)\}$ of length Q that satisfies:

(i) $(i_1, j_1) = (1,1)$,
(ii) $(i_L, j_L) = (r,s)$; and
(iii) $(i_l - i_{l-1}, j_l - j_{l-1}) \in \{(1,0),(1,1),(0,1)\}$ for $l = 2, \ldots, L$.

The weight associated to path p is given by:

$$w(p) = \sqrt{\sum_{(i,j)\in p} (u_i - v_j)^2}, \tag{8}$$

from which the DTW measure between U and V is:

$$\mathrm{DTW}(U,V) = \min_{p \in \mathcal{P}} w(p), \tag{9}$$

where $\mathcal{P}$ denotes the set of allowed alignments fulfilling the above set of constraints. We will hereafter use p^* to refer to the path that represents the alignment with minimum weight. Although DTW can be computed between two time series with different number of observations, in what follows we will assume, without loss of generality, that all time series are of length r.

3 Attack Strategies for DTW Based kNN Classifiers

Our primary design goal is to generate adversarial samples capable of misleading DTW based kNN classifiers. To this end we adapt the adversarial sample crafting techniques reviewed in Sect. 2 to the DTW measure. At this point it is important to highlight that these methods were originally aimed at confusing DNNs designed for image classification. In such models, $J_f(\mathbf{x}, y)$ is selected to be, in most cases, a cross-entropy loss function. Given its differentiability, the application of gradient based techniques is straightforward. However, kNN is a *lazy* classifier (it requires no training phase), and has neither an associated model f nor a loss function $J_f(\mathbf{x}, y)$. Indeed, when a kNN model is queried for the class of an unlabeled sample, the classifier simply computes its distance (or similarity) to each sample in the training set, so that the predicted label is the majority class among the labels of the k closest (correspondingly, most similar) training examples.

Following the workaround used by Papernot et al. in [16], we resort to a smoothed, differentiable approximation of a kNN model. In this manner we can proceed forward with the reformulation of the adversarial sample crafting problem in Eq. (1) with the DTW similarity measure. That is, let us assume that the class of a time series U must be predicted under the DTW similarity measure. For this purpose, the kNN model is fed with a training set of the form $\mathcal{U} = \{(U_n, y_n)\}_{n=1}^{N}$, where $U_n = (u_1^n, ..., u_r^n)$ represents a training (reference) time series, and $y_n \in \{1, ..., C\}$ its associated class label. When $k = 1$ (one nearest neighbor, 1NN), instead of computing the minimum distance among the query and every training time series, we can define a *conditional probability* for a given class $c \in \{1, ..., C\}$ as:

$$\mathrm{P}(c|U) = \frac{1}{N} \sum_{n=1}^{N} \delta_{y_n,c} \cdot \sigma(U, U_n), \tag{10}$$

where $\delta_{y_n,c}$ is the Kronecker delta, and $\sigma(U, U_n)$ the soft-max function:

$$\sigma(U, U_n) = \frac{\exp\left[-\mathrm{DTW}(U, U_n)^2\right]}{\sum_{m=1}^{N} \exp\left[-\mathrm{DTW}(U, U_m)^2\right]}. \tag{11}$$

Given the conditional probability definition, the classifier predicts a label for U as $\arg\max_c \mathrm{P}(c|U)$, i.e., as the class maximizing the output conditional probability of the smoothed kNN as per (10).

Thereby, the DTW based smoothed kNN algorithm (DTW-SNN) produces an adversarial sample $\widetilde{U}$ for a time series U whose true label is c as

$$\widetilde{U} = U + \Omega_U, \tag{12}$$

where in this case, Ω_U is a time series that represents the perturbation imprinted to U that can be computed by solving an optimization problem similar to (1):

$$\Omega_U = \arg\min_{Z} \mathrm{DTW}(U, U + Z), \tag{13}$$

subject to $\mathrm{P}(c|U+Z) < 1/C$, i.e., there is at least one class in $\{1, \ldots, c-1, c+1, \ldots, C\}$ that is predicted for $U+Z$ with higher probability than c itself.

Once we have defined a smoothed version of a 1NN classifier, we now derive three different adversarial sample crafting techniques to attack this particular kind of classifiers under the DTW measure. Such techniques are inspired partly by those reviewed in Sect. 2, and can be described as follows:

1. The first technique, coined as DTW based Bi-objective Optimization (DTW-BIO), aims at jointly considering the probability of the perturbed time series to successfully confuse the target model and the amount of distortion imprinted to the original time series. Given a time series dataset $\mathcal{U}$ and the DTW-SNN model defined as per Eq. (10), let Ω_U denote the noise added to a clean sample $U \in \mathcal{U}$ whose true label is c. Expression (13) postulates that the optimal Ω_U, when applied to the original sample, is capable of misleading the classifier while minimally corrupting the clean sample. When solving such a problem, it is important to note that there is no control on the value of $\mathrm{P}(c|\widetilde{U})$. That is, we ensure that $\mathrm{P}(c|\widetilde{U}) < 1/C$, but the problem formulated in (1) does not allow the attack to be designed for arbitrarily low values of $\mathrm{P}(c|\widetilde{U})$. Therefore, we reformulate this problem as an bi-objective optimization problem that seeks a set of Pareto-optimal Ω_U differently balancing the similarity between U and $\widetilde{U}$ and the class probability $\mathrm{P}(c|\widetilde{U})$:

$$\Omega_U = \arg\min_{Z} \left\{ \mathrm{DTW}(U, U+Z), \mathrm{P}(c|U+Z) \right\}, \tag{14}$$

where, since $\mathrm{DTW}(U, U+Z)$ denotes a measure of dissimilarity, the lower it is the more similar U and $U+Z$ will be. In order to efficiently tackle this bi-objective problem, we resort to the NSGA-II heuristic solver [5].
2. The second technique (DTW-FG) is a DTW based extension of the FSGM of Goodfellow et al. (see Sect. 2). We limit the dissimilarity between samples as $\mathrm{DTW}(U, \widetilde{U}) \leq \varepsilon$, where ε is a positive constant. We derive a DTW based gradient descent technique to compute the perturbation Ω_U by moving the clean time series U along the direction of the negative gradient of $\mathrm{P}(c|U)$:

$$\Omega_U = -\varepsilon \frac{\nabla \mathrm{P}(c|U)}{||\nabla \mathrm{P}(c|U)||_2}, \tag{15}$$

where $\nabla \mathrm{P}(c|U)$ is the gradient of the class probability, and $||\cdot||_2$ is the L_2-norm. Appendix A provides further details on this gradient computation.
3. The third technique (DTW-IFG) follows the iterative version of the fast gradient method of Kurakin et al. [11] to the DTW setting. Consider an initial condition of the form $\widetilde{U}^{(0)} = U$. In this case, the adversarial sample would be computed by following the recurrence:

$$\widetilde{U}^{(t)} = \widetilde{U}^{(t-1)} + \Omega_{\widetilde{U}^{(t-1)}} \text{ for } t \in \{1, \ldots, T_{\max}\}, \tag{16}$$

where

$$\Omega_{\widetilde{U}^{(t-1)}} = -\alpha \frac{\nabla \mathrm{P}(c|\widetilde{U}^{(t-1)})}{||\nabla \mathrm{P}(c|\widetilde{U}^{(t-1)})||_2}. \tag{17}$$

and $\alpha = \varepsilon/T_{\max}$ to ensure $\text{DTW}(U, \widetilde{U}^{(t)}) \leq \varepsilon$. In this case, gradients are given also as per Appendix A.

4 Experiments

In this section we present and discuss the results obtained from a set of experiments aimed at assessing the vulnerability of DTW based 1NN classifiers against adversarial samples generated by using the techniques in Sect. 3. In this set of experiments we have used the `Synthetic Control` time series dataset from the well-known UCR archive [4]. This dataset is divided in two sets – namely, the training and the test set – both with 300 labelled time series (50 per class label). Given the training set, the challenge associated to this dataset consist of classifying every time series in the test set in one of the 6 classes. Two are the reasons for selecting this dataset: (1) the length of the time series (60 observations), which allows DTW computations within reasonable times; and (2) the performance of DTW based 1NN classifiers, which for this case reports accuracy scores of 0.993.

We will focus our discussion on the so-called *intra-technique* transferability [16]: we will evaluate the capacity of an adversarial sample crafted by our methods to mislead classification models belonging to the same family (1NN), yet trained with different datasets. Consider a set of adversarial samples $\widetilde{\mathcal{U}}$ built on a classifier f; in other words, if adversarial samples in $\widetilde{\mathcal{U}}$ have been designed properly, f should misclassify all of them. The intra-technique transferability is defined as:

$$R_{\text{intra}}(\widetilde{\mathcal{U}}, f, f') = \frac{\left|\{\widetilde{U} \in \widetilde{\mathcal{U}} : \, f'(\widetilde{U}) \neq f'(U)\}\right|}{|\widetilde{\mathcal{U}}|}, \tag{18}$$

where $|\cdot|$ denotes set cardinality, f is the classifier for which adversarial samples $\widetilde{\mathcal{U}}$ were designed, and f' represents a classifier of the same family of f but trained with a different dataset. In our case, f and f' are both DTW based 1NN classifiers with different training sets.

To evaluate the intra-technique transferability across more than one model under attack, we have split the subset of 300 test time series in 5 disjoint, stratified splits. Four of these sets are used as training sets for the DTW based 1NN models $\{\mathcal{U}_i\}_{i=1}^4$, while the remaining one $\mathcal{U}_5$ is used to craft adversarial samples. Consequently, the set of time series that are modified to confuse models trained on $\mathcal{U}_i$ do not participate at all in the training process of such models. This being said, $R_{\text{intra}}(\widetilde{\mathcal{U}}, f, f')$ will be quantified for all 16 combinations of the considered training sets, e.g. given a training dataset $\mathcal{U}_i$ for the DTW based 1NN model, the analysis will verify whether samples crafted to confuse this trained model (*source* model) are misclassified by another DTW based 1NN model (*target* model) trained with $\mathcal{U}_{i'}$, where $i, i' \in \{1, \ldots, 4\}$.

4.1 Results and Discussion

We begin by discussing the results in Figs. 1a to c, where the intra-technique transferability as per (18) is shown for all training set combinations between the source and target classifiers and for the proposed DTW-BIO, DTW-FG and DTW-IFG techniques. For the DTW-BIO approach, the NSGA-II solver has been configured with a population size of 100 individuals, tournament selection, SBX crossover and polynomial mutation with respective distribution parameters equal to 15 and 20, and a maximum of 10^4 number of fitness evaluations. Given the set of Pareto-optimal solutions for each $U \in \mathcal{U}_5$, the adversarial sample was produced by adding the perturbation whose $P(c|\widetilde{U})$ was closer to – yet smaller than – $1/C = 1/6$. To generate gradient based adversarial samples, ε has been set to 4.0, which has been tuned off-line based on off-line trials. For the DTW-IFG case, the maximum number of iterations has been set to $T_{\max} = 25$.

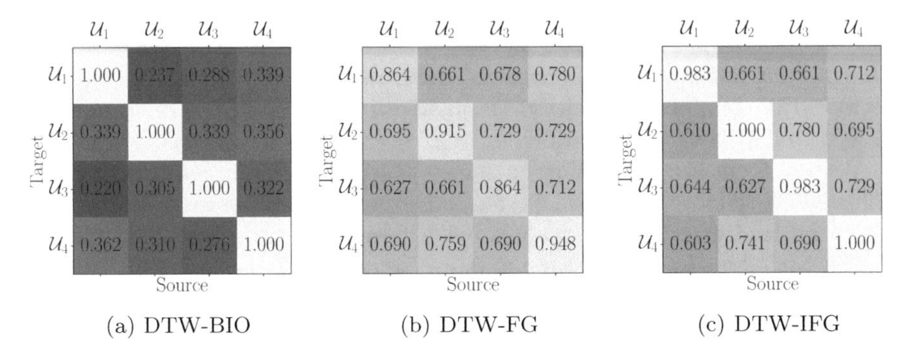

(a) DTW-BIO (b) DTW-FG (c) DTW-IFG

Fig. 1. Intra-technique transferability for adversarial samples drawn by (a) DTW-BIO; (b) DTW-FG; and (c) DTW-IFG. Results must be interpreted as e.g. *as shown in entry* $(1, 3)$ *of DTW-BIO, 28.8% of the adversarial samples crafted from source model trained with dataset* $\mathcal{U}_3$ *are misclassified by target model trained with dataset* $\mathcal{U}_1$. The average accuracy computed over 4 DTW based 1NN classifiers fed with training sets $\mathcal{U}_1$, $\mathcal{U}_2$, $\mathcal{U}_3$ or $\mathcal{U}_4$ is 0.979 measured over $\mathcal{U}_5$ as their test set.

A first inspection of these plots reveals that as expected, the crafted adversarial samples perform best when both source and target models share the same training set (diagonal entries in the matrices). This noted fact is in accordance with similar conclusions extracted by Papernot et al. in [16] where kNN models (under Euclidean distance) were found to be to a certain extend robust to intra-technique transferability. However, it is clearly shown that gradient based strategies produce more transferable adversarial samples than the DTW-BIO technique, due to the fact that this latter approach overfits the crafted perturbation to the training set of the source model. This calls for further research aimed at determining whether overfitting is due to the small sizes of the training datasets (60 time series in total, 10 per class), as well as to the disjoint split by which they were built.

We now turn the focus on Fig. 2, where we show the trade-off – achieved by the three proposed schemes – between the distortion of a corrupted sequence with respect its clean version, and the class probability $P(c|\widetilde{U})$ of the produced adversarial time series. To trace the evolution of DTW-FG as a function of the distortion, the technique has been run for $\varepsilon \in \{1.0, 2.0, 3.0, 4.0\}$. For the DTW-IFG, the maximum distortion has been set to be $\varepsilon = 4.0$ and the maximum number of iterations to $T_{\max} = 25$. The curve depicts the similarity and class probability values for the partial adversarial time series yielded by the recurrence in Expression (17). We can observe in this second plot that DTW-BIO attains a more diverse portfolio of adversarial samples than the other techniques, while requiring less parameter tuning. The downside is a higher computational complexity, which depending on the application scenario can make this approach less preferable than its gradient based counterparts.

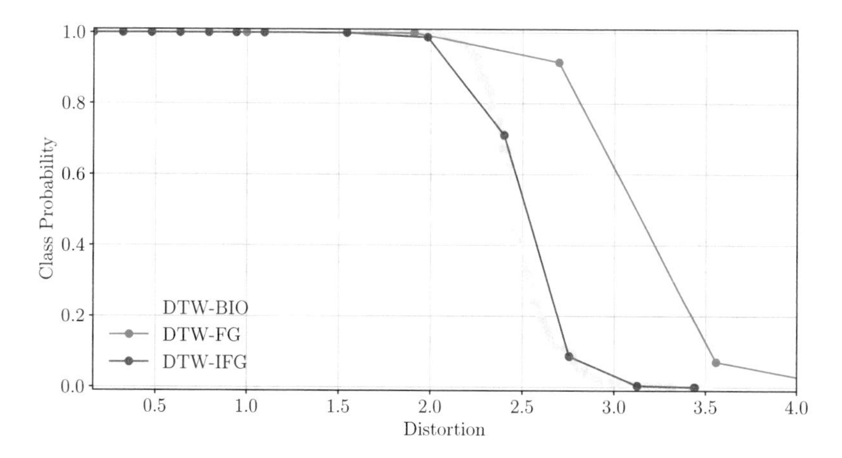

Fig. 2. Class probability of the proposed DTW based adversarial sample crafting schemes as a function of the distortion for an arbitrarily chosen time series whose true label is $c = 2$: DTW-BIO (yellow), DTW-FG (blue) and DTW-IFG (red).

Finally, Fig. 3 compares visually the original sequence and its corresponding adversarial time series. From top to bottom, the first figure depicts the original sequence (black) over the time series in $\mathcal{U}$ sharing its class label $c = 1$ (gray). The second figure illustrates the adversarial sample crafted by DTW-BIO (red solid line), as well as the reference time series of the predicted class (in this case the gray time series belong to class 3). Likewise, the third and fourth figures illustrate the adversarial samples generated by DTW-FG and DTW-IFG respectively. As opposed to the DTW-BIO example, it is interesting to observe that gradient based methods produce an adversarial sample that is misclassified by the target model as label 6. In all cases, the degradation of the adversarial example is almost negligible with respect to its clean version, which underscores the conclusion that robust models against subtle adversarial attacks are also needed for time series classifiers relying on elastic similarity measures.

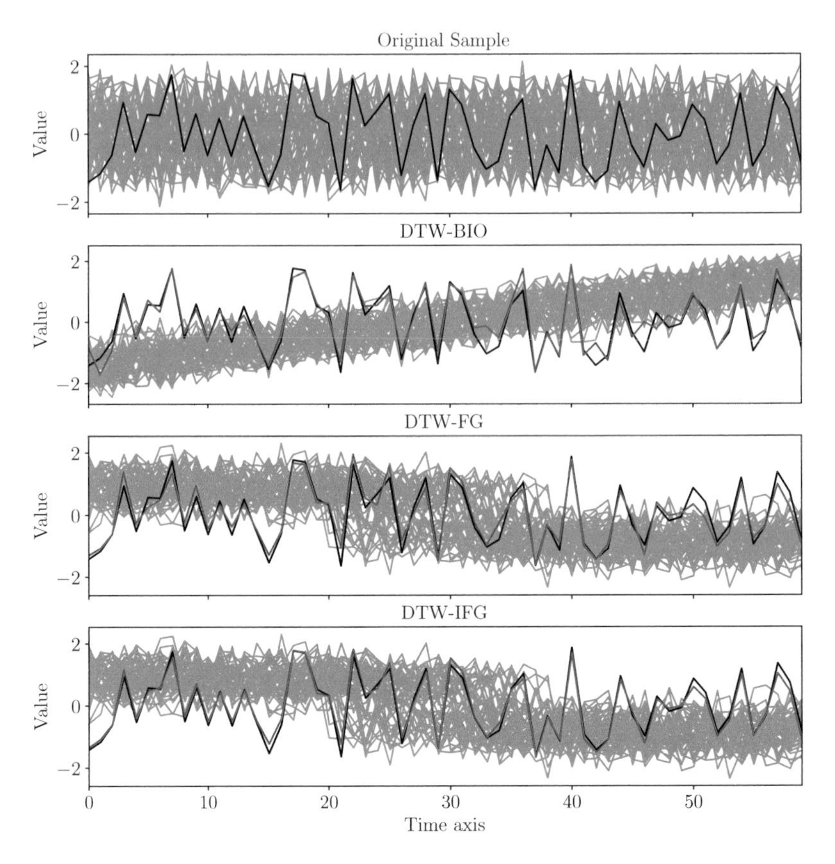

Fig. 3. Comparison between an arbitrary original sequence (black) and its adversarial samples (red) crafted by every proposed technique. The class label predicted for the original and adversarial sequences is represented by the gray set of time series depicted in the background of each subfigure: from top to bottom, class label $c = 1$ (which is the class of the original time series), $c = 3$ (DTW-BIO) and $c = 6$ (DTW-FG and DTW-IFG).

5 Conclusions and Future Work

This work has presented three adversarial time series crafting techniques aimed at misleading DTW based 1NN classifiers. Specifically, the proposed portfolio of techniques hinge on an optimization problem that aims at discovering the smallest perturbation that, when added to a legitimate time series, forces DTW based classifiers to issue erroneous predictions for the perturbed time series. To solve this problem, two gradient based techniques (DTW-FG and DTW-IFG) and a bi-objective optimization problem (DTW-BIO) have been used.

The performance of these techniques when attacking DTW based 1NN classifiers has been empirically assessed by inspecting the intra-technique transferability across 4 classifiers fed with different training data. We have utilized the

`Synthetic Control` public time series dataset from the UCR archive. Conclusions drawn from the results obtained with these experiments are listed below:

1. Regardless the adversarial sample crafting technique, the observed intra-technique transferability evinces that crafted adversarial time series are capable of misleading DTW based 1NN classifiers trained with different training sets than the one used for the crafting process.
2. Adversarial time series produced by DTW-FG and DTW-IFG are, in general, more transferable than those furnished by the BIO-DTW technique. However, BIO-DTW is more effective when attacking target classifiers whose training set is the same than that of the source model used for generating the adversarial samples.
3. The degradation of the generated samples with respect to their original version is almost negligible for the majority of time series considered in the experiments, disregarding the crafting technique.

Several future research lines are rooted on the findings reported in this study. To begin with, each adversarial sample technique should be evaluated in a benchmark encompassing a higher number of time series classification problems. Since DTW permits to compute similarity among time series of different length, we plan to extend our study to the case where the time series in source and target training sets have different number of observations. We will also analyze how the transferability of adversarial samples produced by DTW-BIO evolves with the degree of overlapping between the training sets of source and target classification models. Finally, we plan to analyze the so-called *cross-technique* transferability, which aims at verifying the capability of adversarial samples to confuse different target classification models when their learning algorithms differ from that used in the source classification model (using distance based models such as Support Vector Machines or Gaussian Mixtures).

Acknowledgments. This work has been supported by the Basque Government through the EMAITEK, BERC 2014–2017 and the ELKARTEK programs, and by the Spanish Ministry of Economy and Competitiveness MINECO: BCAM Severo Ochoa excellence accreditation SVP-2014-068574 and SEV-2013-0323, and through the project TIN2017-82626-R funded by (AEI/FEDER, UE).

A Appendix: Computation of $\mathrm{P}(c|U^v)$ gradient

In this appendix we formally derive, for the DTW setting:

$$\nabla \mathrm{P}(c|U^v) = \left(\frac{\partial \mathrm{P}(c|U)}{\partial u_d}\right)_{d \in \{1,\ldots,v\}}. \tag{19}$$

Consider the DTW-SNN model introduced in Sect. 3. The partial derivative with respect the input variable u_d is given by:

$$\frac{\partial \mathrm{P}(c|U)}{\partial u_d} = \frac{1}{N}\left[\sum_{n=1}^{N} \delta_{c,y_n} \frac{\partial \sigma(U, U_n)}{\partial u_d}\right]. \tag{20}$$

For the sake of a simpler notation, let us write the soft-max function as, $\sigma(U, U_n) = H_1/H_2$, where:

$$H_1 = \exp\left[-\mathrm{DTW}(U, U_n)^2\right] \text{ and } H_2 = \sum_{m=1}^{N} \exp\left[-\mathrm{DTW}(U, U_m)^2\right]. \tag{21}$$

Equation (20) is, therefore, rewritten as follows:

$$\frac{\partial \mathrm{P}(c|U)}{\partial u_d} = \frac{1}{N}\left[\sum_{n=1}^{N} \delta_{c,y_n} \frac{\frac{\partial H_1}{\partial u_d} H_2 - H_1 \frac{\partial H_2}{\partial u_d}}{H_2^2}\right]. \tag{22}$$

To compute the partial derivatives of H_1 and H_2, note that we need to differentiate $\mathrm{DTW}(U, U_n)$. To this end, let p_n^* be the optimal alignment between U_n and U. In other words, let p_n^* be the alignment satisfying:

$$\mathrm{DTW}(U, U_n) = \sqrt{\sum_{(i,j)\in p_n^*} (u_i - u_j^n)^2}, \tag{23}$$

where u_i and u_j^n are the i-th and j-th observations of U and U_n time series respectively (see Eqs. (8) and (9)).

Considering the equation above, the derivatives of H_1 and H_2 are given by:

$$\frac{\partial H_1}{\partial u_d} = -2 \sum_{(i,j)\in p_n^*} \delta_{i,d}(u_d - u_j^n) \exp\left[-\mathrm{DTW}(U^v, U_n)^2\right], \tag{24}$$

and

$$\frac{\partial H_2}{\partial u_d} = -2 \sum_{m=1}^{N} \left\{ \sum_{(i,j)\in p_m^*} \delta_{i,d}(u_d - u_j^m) \exp\left[-\mathrm{DTW}(U, U_m)^2\right] \right\}, \tag{25}$$

where $\delta_{i,d}$ is the Kronecker delta.

References

1. Akhtar, N., Mian, A.: Threat of adversarial attacks on deep learning in computer vision: a survey. arXiv preprint arXiv:180100553 (2018)
2. Berndt, D.J., Clifford, J.: Using dynamic time warping to find patterns in time series. In: Workshop on Knowledge Discovery in Databases, Seattle, WA, pp. 359–370 (1994)
3. Biggio, B., Corona, I., Maiorca, D., Nelson, B., Šrndić, N., Laskov, P., Giacinto, G., Roli, F.: Evasion attacks against machine learning at test time. In: Joint European Conference on Machine Learning and Knowledge Discovery in Databases, pp. 387–402. Springer (2013)
4. Chen, Y., Keogh, E., Hu, B., Begum, N., Bagnall, A., Mueen, A., Batista, G.: The UCR Time Series Classification Archive (2015). www.cs.ucr.edu/~eamonn/time_series_data/

5. Deb, K., Pratap, A., Agarwal, S., Meyarivan, T.: A fast and elitist multiobjective genetic algorithm: NSGA-II. IEEE Trans. Evol. Comput. **6**(2), 182–197 (2002)
6. Ding, H., Trajcevski, G., Scheuermann, P., Wang, X., Keogh, E.: Querying and mining of time series data: experimental comparison of representations and distance measures. Proc. VLDB Endow. **1**(2), 1542–1552 (2008)
7. Goodfellow, I.J., Shlens, J., Szegedy, C.: Explaining and harnessing adversarial examples. arXiv preprint arXiv:14126572 (2014)
8. ten Holt, G.A., Reinders, M.J., Hendriks, E.: Multi-dimensional dynamic time warping for gesture recognition. In: Conference of the Advanced School for Computing and Imaging, vol. 300, p. 1 (2007)
9. Huang, L., Joseph, AD., Nelson, B., Rubinstein, BI., Tygar, J.: Adversarial machine learning. In: Proceedings of the 4th ACM Workshop on Security and Artificial Intelligence, pp. 43–58. ACM (2011a)
10. Huang, L., Joseph, A.D., Nelson, B., Rubinstein, B.I., Tygar, J.: Adversarial machine learning. In: Proceedings of the 4th ACM Workshop on Security and Artificial Intelligence, pp. 43–58. ACM (2011b)
11. Kurakin, A., Goodfellow, I., Bengio, S.: Adversarial examples in the physical world. arXiv preprint arXiv:160702533 (2016)
12. Lana, I., Del Ser, J., Velez, M., Vlahogianni, E.I.: Road traffic forecasting: recent advances and new challenges. Proc. VLDB Endow. **10**(2), 93–109 (2018)
13. Lines, J., Bagnall, A.: Time series classification with ensembles of elastic distance measures. Data Min. Knowl. Discov. **29**(3), 565–592 (2015)
14. Miyato, T., Maeda, S., Koyama, M., Ishii, S.: Virtual adversarial training: a regularization method for supervised and semi-supervised learning. arXiv preprint arXiv:170403976 (2017)
15. Molina-Solana, M., Ros, M., Ruiz, M.D., Gómez-Romero, J., Martín-Bautista, M.J.: Data science for building energy management: a review. Renew. Sustain. Energy Rev. **70**, 598–609 (2017)
16. Papernot, N., McDaniel, P., Goodfellow, I.: Transferability in machine learning: from phenomena to black-box attacks using adversarial samples. arXiv preprint arXiv:160507277 (2016a)
17. Papernot, N., McDaniel, P., Wu, X., Jha, S., Swami, A.: Distillation as a defense to adversarial perturbations against deep neural networks. In: 2016 IEEE Symposium on Security and Privacy (SP), pp. 582–597. IEEE (2016b)
18. Papernot, N., McDaniel, P., Goodfellow, I., Jha, S., Celik, Z.B., Swami, A.: Practical black-box attacks against machine learning. In: Proceedings of the 2017 ACM on Asia Conference on Computer and Communications Security, pp. 506–519. ACM (2017)
19. Sakoe, H., Chiba, S.: Dynamic programming algorithm optimization for spoken word recognition. IEEE Trans. Acoust. Speech Signal Process. **26**(1), 43–49 (1978)
20. Szegedy, C., Zaremba, W., Sutskever, I., Bruna, J., Erhan, D., Goodfellow, I., Fergus, R.: Intriguing properties of neural networks. arXiv preprint arXiv:13126199 (2013)
21. Villar-Rodriguez, E., Del Ser, J., Oregi, I., Bilbao, M.N., Gil-Lopez, S.: Detection of non-technical losses in smart meter data based on load curve profiling and time series analysis. Energy **137**, 118–128 (2017)

Slot Co-allocation Optimization in Distributed Computing with Heterogeneous Resources

Victor Toporkov[1], Anna Toporkova[2], and Dmitry Yemelyanov[1](✉)

[1] National Research University "MPEI", ul. Krasnokazarmennaya, 14, Moscow 111250, Russia
{ToporkovVV,YemelyanovDM}@mpei.ru

[2] Higher School of Economics, National Research University, ul. Myasnitskaya, 20, Moscow 101000, Russia
AToporkova@hse.ru

Abstract. In this work, we introduce slot selection and co-allocation algorithms for parallel jobs in distributed computing with non-dedicated and heterogeneous resources. A single slot is a time span that can be assigned to a task, which is a part of a parallel job. The job launch requires a co-allocation of a specified number of slots starting and finishing synchronously. Some existing resource co-allocation algorithms assign a job to the first set of slots matching the resource request without any optimization (the first fit type), while other algorithms are based on an exhaustive search. In this paper, algorithms for efficient and dependable slot selection are studied and compared with known approaches. The novelty of the proposed approach is in a general algorithm efficiently selecting a set of slots according to the specified criterion.

Keywords: Distributed computing · Grid · Economic scheduling
Resource management · Slot · Job · Allocation · Optimization

1 Introduction and Related Works

Modern high-performance distributed computing systems (HPCS), including Grid, cloud and hybrid infrastructures provide access to large amounts of resources [1]. These resources are typically required to execute parallel jobs submitted by HPCS users and include computing nodes, data storages, network channels, software, etc. The actual requirements for resources amount and types needed to execute a job are defined in resource requests and specifications provided by users. HPCS organization and support bring certain economical expenses: purchase and installation of machinery equipment, power supplies, user support, etc. As a rule, HPCS users and service providers interact in economic terms and the resources are provided for a certain payment. Thus, as total user job execution budget is usually limited, we elaborate an actual task to

J. Del Ser et al. (Eds.): IDC 2018, SCI 798, pp. 40–49, 2018.
https://doi.org/10.1007/978-3-319-99626-4_4

optimize suitable resources selection in accordance with a job specification and a restriction to a total resources cost.

Economic mechanisms are used to solve problems like resource management and scheduling of jobs in a transparent and efficient way in distributed environments such as cloud computing and utility Grid. The significant and important feature for well-known scheduling solutions for distributed environments is the fact that the scheduling strategy is formed on a basis of efficiency criteria [1–4]. The metascheduler [4–6] implements the economic policy of a VO based on local resource schedules. The schedules are defined as sets of slots coming from resource managers or schedulers in the resource domains, i.e. time intervals when individual nodes are available to perform a part of a parallel job. In order to implement such scheduling schemes and policies, first of all, one needs an algorithm for finding sets of simultaneously available slots required for each job execution. Further we shall call such set of simultaneously available slots with the same start and finish times as execution window.

The scheduling problem in Grid is NP-hard due to its combinatorial nature and many heuristic-based solutions have been proposed. NWIRE system [3] performs a slot window allocation based on the user defined efficiency criterion under the maximum total execution cost constraint. However, the optimization occurs only on the stage of the best found offer selection. First fit slot selection algorithms (backtrack [7] and NorduGrid [8] approaches) assign any job to the first set of slots matching the resource request conditions, while other algorithms use an exhaustive search [9,10] and some of them are based on a linear integer programming (IP) [9] or mixed-integer programming (MIP) model [10]. Moab scheduler [11] implements the backfilling algorithm and during a slot window search does not take into account any additive constraints such as the minimum required storage volume or the maximum allowed total allocation cost.

Modern distributed and cloud computing simulators GridSim and CloudSim [12,13] provide tools for jobs execution and co-allocation of simultaneously available computing resources. Base simulator distributions perform First Fit allocation algorithms without any specific optimization. CloudAuction extension [13] of CloudSim implements a double auction to distribute datacenters' resources between a job flow with a fair allocation policy. All these algorithms consider price constraints on individual nodes and not on a total window allocation cost. However, as we showed in [14], algorithms with a total cost constraint are able to perform the search among a wider set of resources and increase the overall scheduling efficiency. GrAS [15] is a Grid job-flow management system built over Maui scheduler [11]. The resources co-allocation algorithm retrieves a set of simultaneously available slots with the same start and finish times even in heterogeneous environments. However the algorithm stops after finding the first suitable window and, thus, does not perform any optimization except for window start time minimization. Algorithm [16] performs job's response and finish time minimization and does not take into account constraint on a total allocation budget. [17] performs window search on a list of slots sorted by their start time, implements algorithms for window shifting and finish time minimization, does not support other optimization criteria and the overall job execution cost constraint.

AEP algorithm [18] performs window search with constraint on a total resources allocation cost, implements optimization according to a number of criteria, but does not support a general case optimization. Besides AEP does not guarantee same finish time for the window slots in heterogeneous environments and, thus, has limited practical applicability.

In this paper, we propose algorithms for an efficient slot selection based on user defined criteria. The novelty of the proposed approach consists in implementing a dynamic programming scheme to allocate a set of simultaneously available slots in heterogeneous HPCS with non-dedicated resources. The paper is organized as follows. Section 2 introduces a general scheme for searching slot sets efficient by the specified criterion. Then several implementations are proposed and considered. Section 3 contains simulation results for comparison of proposed and known algorithms. Section 4 summarizes the paper and describes further research topics.

2 Resource Selection Algorithm

2.1 Problem Statement

We consider a set R of heterogeneous computing nodes with different performance p_i and price c_i characteristics. Each node has a local utilization schedule known in advance for a considered scheduling horizon time L. A node may be turned off or on by the provider, transferred to a maintenance state, reserved to perform computational jobs. Thus, it's convenient to represent all available resources as a set of slots. Each slot corresponds to one computing node on which it's allocated and may be characterized by its performance and price.

In order to execute a parallel job one needs to allocate the specified number of simultaneously idle nodes ensuring user requirements from the resource request. The resource request specifies number n of nodes required simultaneously, their minimum applicable performance p, job's computational volume V and a maximum available resources allocation budget C. The required window length is defined based on a slot with the minimum performance. For example, if a window consists of slots with performances $p \in \{p_i, p_j\}$ and $p_i < p_j$, then we need to allocate all the slots for a time $T = \frac{V}{p_i}$. In this way V really defines a computational volume for each single node task. Common start and finish times ensure the possibility of inter-node communications during the whole job execution. The total cost of a window allocation is then calculated as $C_W = \sum_{i=1}^{n} T * c_i$.

These parameters constitute a formal generalization for resource requests common among distributed computing systems and simulators. Additionally we introduce criterion f as a user preference for the particular job execution during the scheduling horizon L. f can take a form of any additive function and vary from a simple window start time or cost minimization to a general independent parameter maximization with the restriction to a total resources allocation cost C. As an example, one may want to allocate suitable resources with the maximum possible total data storage available before the specified deadline.

2.2 Window Search Procedure

For a general window search procedure for the problem statement presented in Sect. 2.1, we combined core ideas and solutions from algorithm AEP [18] and systems [15,17]. The proposed algorithm allocates a set of n simultaneously available slots with performance $p_i > p$, for a time, required to compute V instructions on each node, with a restriction C on a total allocation cost and performs optimization according to criterion f.

The algorithm scans a list of available slots ordered by their non-decreasing start time and checks all intermediate slots combinations suitable for the job execution [18]. Generally such combinations consist of $m > n$ slots, so we introduce a special procedure to select a subset of n slots (a window) optimal according to the criterion f with a restriction on the total cost C. An intermediate window providing maximum f value is returned as a problem solution. In this a way the general algorithm scheme is similar to the maximum value search in an array of intermediate f_i values.

Let us discuss in more details the procedure which allocates an optimal (according to a criterion f) window of n slots out of an extended combination S_E of m slots. In a general case we consider a subset allocation problem according to some additive criterion: $Z = \sum_{i=1}^{n} c_z(s_i)$, where $c_z(s_i) = z_i$ is a target optimization characteristic value provided by a single slot s_i of a window.

In this way we can state the following problem of an optimal n - size window subset allocation out of m - slots set S_E:

$$Z = x_1 z_1 + x_2 z_2 + \cdots + x_m z_m, \tag{1}$$

with the following restrictions:

$$x_1 c_1 + x_2 c_2 + \cdots + x_m c_m \leq C$$
$$x_1 + x_2 + \cdots + x_m = n$$
$$x_i \in \{0, 1\}, i = 1, \ldots, m,$$

where z_i is a target characteristic value provided by slot s_i, c_i is total cost required to allocate slot s_i for a time T_i, x_i - is a decision variable determining whether to allocate slot s_i $(x_i = 1)$ or not $(x_i = 0)$ for a window.

This problem relates to the class of integer linear programming problems, which imposes obvious limitations on the practical methods to solve it. However we used 0-1 knapsack problem as a base for our implementation. Indeed, the classical 0-1 knapsack problem with a total weight C and items - slots with weights c_i and values z_i have the same formal model (1) except for extra restriction on the number of items required: $x_1 + x_2 + \cdots + x_m = n$. To take this into account we implemented the following dynamic programming recurrent scheme:

$$f_i(C_j, n_k) = \max\{f_{i-1}(C_j, n_k), f_{i-1}(C_j - c_i, n_k - 1) + z_i\},$$
$$i = 1, \ldots, m, C_j = 1, \ldots, C, n_k = 1, \ldots, n, \tag{2}$$

where $f_i(C_j, n_k)$ defines the maximum Z criterion value for n_k - size window allocated out of first i slots from S_E for a budget C_j. After the forward induction procedure (2) is finished the maximum value $Z_{max} = f_m(C, n)$. x_i values are then obtained by a backward induction procedure.

For the actual implementation we initialized $f_i(C_j, 0) = 0$, meaning $Z = 0$ when we have no items in the knapsack. Then we perform forward propagation and calculate $f_1(C_j, n_k)$ values for all C_j and n_k based on the first item and the initialized values. Then $f_2(C_j, n_k)$ is calculated taking into account second item and $f_1(C_j, n_k)$ and so on. So after the forward propagation procedure (2) is finished the maximum value $Z_{max} = f_m(C, n)$. Corresponding values for variables x_i are then obtained by a backward propagation procedure.

3 Simulation Study

3.1 Simulation Environment Setup

An experiment was prepared as follows using a custom distributed environment simulator [5,6,18]. For our purpose, it implements a heterogeneous resource domain model: nodes have different usage costs and performance levels. A space-shared resources allocation policy simulates a local queuing system (like in CloudSim [12]) and, thus, each node can process only one task at any given simulation time. The execution cost of each task depends on its execution time which is proportional to the dedicated node's performance level. The execution of a single job requires parallel execution of all its tasks.

During the main experiment series we performed a window search operation for a single job requesting $n = 7$ nodes with performance level $p_i >= 1$, computational volume $V = 800$ and a maximum budget allowed is $C = 644$. At each experiment a new instance for the computing environment was automatically generated with the following properties. The resource pool includes 100 heterogeneous computational nodes. Each node performance level is given as a uniformly distributed random value in the interval [2, 10]. So the required window length may vary from 400 to 80 time units. The scheduling interval length is 1200 time quanta which is enough to run the job on nodes with the minimum performance. However we introduce the initial resource load with advanced reservations and local jobs to complicate conditions for the search operation. The additional load is distributed hypergeometrically and results in up to 30% utilization for each node.

These simulation characteristics generally will not allow to allocate the most expensive (and usually the most efficient) CPU nodes for a job execution. However they remain quite diverse to study algorithms' efficiency and complexity features.

Additionally an independent value $q_i \in [0; 10]$ is randomly generated for each computing node i to compare algorithms against a general case $Q = \sum_{i=1}^{n} q_i$ window allocation criterion. As particular q implementations one can use nodes available disc or RAM space, power consumption, channel bandwidth, etc.

3.2 General Algorithms Comparison

We implemented the following window search algorithms based on the general window search procedure introduced in Sect. 2.2.

- *FirstFit* returns first suitable and affordable window found, performs start time minimization and represents algorithms from [15,17].
- *MultipleBest* algorithm searches for multiple non-intersecting alternative windows using FirstFit algorithm. When all possible window allocations are retrieved the algorithm searches among them for alternatives with the maximum Q value. In this way *MultipleBest* is similar to [3] approach.
- *MaxQ* implements a general square window search procedure presented in Sect. 2.2 and returns a window with maximum total Q value.
- *MaxQ Lite* follows the general square window search procedure but does not implement slots subset allocation procedure (1). Instead at each step it returns the first n cheapest slots of S_E. The total Q value of these n slots is returned as a target criterion which is then maximized during the search procedure. Thus, *MaxQ Lite* has much less computational complexity compared to *MaxQ* but does not guarantee an accurate solution.

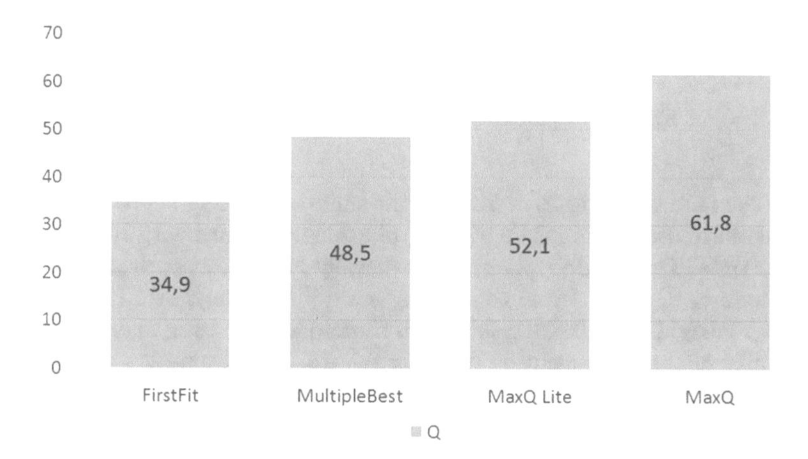

Fig. 1. Simulation results: average window Q value

Figure 1 shows average $Q = \sum_{i=1}^{n} q_i$ value obtained during the simulation. Note that as q_i was generated randomly on a [0;10] interval for a single window of 7 slots we have the following practical limit specific for our experiment: $Q \in [0; 70]$.

As can be seen from Fig. 1, *MaxQ* is indeed provided the maximum average criterion value $Q = 61.8$, which is quite close to the practical maximum, especially compared to other algorithms. The advantage over *MultipleBest* and *MaxQ Lite* is almost 20%. *MaxQ Lite* implements a simple heuristic but still is able to provide a better solution compared to the best of 31 different alternative executions retrieved by *MultipleBest*. *First Fit* provided average Q value exactly in the middle of [0;70] which is 44% less compared to $MaxQ$.

3.3 Dependable Resources Allocation

As a practical implementation for a general Q parameter maximization we propose to study a resources allocation placement problem. Figure 2 shows Gantt chart of 4 slots co-allocation (hollow rectangles) in a computing environment with resources preutilized with local and high-priority tasks (filled rectangles).

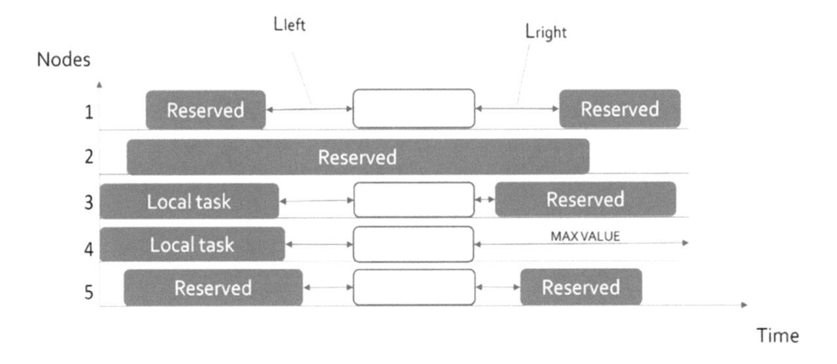

Fig. 2. Dependable window co-allocation metrics

As can be seen from Fig. 2, even using the same computing nodes (1, 3, 4, 5 on Fig. 2) there are usually multiple window placement options with respect to the slots start time. The window placement generally may affect such job execution properties as cost, finish time, computing energy efficiency, etc. Besides, slots proximity to neighboring tasks reserved on the same computing nodes may affect a probability of the job execution delay or failure. For example, a slot reserved too close to the previous task on the same node may be delayed or cancelled by an unexpected delay of the latter. Thus, dependable resources allocation may require reserving resources with some reasonable distance to the neighboring tasks.

As presented in Fig. 2, for each window slot we can estimate times to the previous task finish time: L_{left} and to the next task start time: L_{right}. Using these values we consider the following criteria for the whole window allocation optimization: $L_{\Sigma} = \frac{1}{n}\sum_{i=1}^{n}(l_{left_i} + L_{right_i})$ represents average time distance between window and the neighboring tasks reserved on the same nodes; $L_{min\Sigma} = \frac{1}{n}\sum_{i=1}^{n} min(l_{left_i}, L_{right_i})$ displays average time distance to the nearest neighboring tasks. For a dependable scheduling we are interested in maximizing L_{Σ} and $L_{min\Sigma}$ values. So, by setting $Q = L_{\Sigma}$ and $Q = L_{min\Sigma}$ we performed scheduling simulation with the same settings described in Sect. 3.1. The results of 1000 independent scheduling cycles are compiled in Table 1.

Table 1. Dependable window co-allocation simulation results

Algorithm	Average L_{Σ}	Average $L_{min\Sigma}$	Operational Time, ms
Multiple Best	719	278	156
First Fit	433	99	7
Max Q Lite	713	279	7
Max Q	842	370	2860

As a result *MaxQ* provided best values for both L_{Σ} and $L_{min\Sigma}$ criteria (highlighted in bold in Table 1). The advantage over *Multiple Best* is 15% against the total distance L_{Σ} and 25% against $L_{min\Sigma}$ distance to the nearest tasks. However due to a higher computational complexity it took *MaxQ* almost 3 seconds to find such allocation over 100 available computing nodes, which is 20 times longer compared to *Multiple Best*.

At the same time a simplified *MaxQ Lite* provided almost the same scheduling results as *Multiple Best* for even less operational time: 7 ms. *FirstFit* does not perform the target criteria optimization and, thus, provides the smallest L_{Σ} and $L_{min\Sigma}$ distances for the same operational time as *MaxQ Lite*.

4 Conclusions and Future Work

In this work, we address the problem of dependable slot selection and co-allocation for parallel jobs in distributed computing with non-dedicated resources. For this purpose a general square window allocation algorithm was proposed and considered. A special slots subset allocation procedure is implemented to support a general case optimization problem. Simulation study proved optimization efficiency of the proposed algorithms according to target criteria. A practical implementation was considered for a dependable resources allocation providing up to 25% advantage over traditional scheduling algorithms.

As a drawback, the general case algorithm has a relatively high computational complexity, especially compared to First Fit approach. In our further work, we will refine a general resource co-allocation scheme in order to decrease its computational complexity.

Acknowledgements. This work was partially supported by the Council on Grants of the President of the Russian Federation for State Support of Young Scientists (YPhD-2297.2017.9), RFBR (grants 18-07-00456 and 18-07-00534) and by the Ministry on Education and Science of the Russian Federation (project no. 2.9606.2017/8.9).

References

1. Lee, Y.C., Wang, C., Zomaya, A.Y., Zhou, B.B.: Profit-driven scheduling for cloud sevices with data access awareness. J. Parallel Distrib. Comput. **72**(4), 591–602 (2012)
2. Buyya, R., Abramson, D., Giddy, J.: Economic models for resource management and scheduling in Grid computing. J. Concurr. Comput. **14**(5), 1507–1542 (2002). https://doi.org/10.1002/cpe.690
3. Ernemann, C., Hamscher, V., Yahyapour, R.: Economic scheduling in grid computing. In: Feitelson, D.G., Rudolph, L., Schwiegelshohn, U. (eds.) JSSPP 2002. LNCS, vol. 2537, pp. 128–152. Springer, Heidelberg (2002)
4. Kurowski, K., Nabrzyski, J., Oleksiak, A., Weglarz, J.: Multicriteria aspects of grid resource management. In: Nabrzyski, J., Schopf, J.M., Weglarz, J. (eds.) Grid Resource Management. State of the Art and Future Trends, pp. 271–293. Kluwer Academic Publishers, Norwell (2003)
5. Toporkov, V., Tselishchev, A., Yemelyanov, D., Bobchenkov, A.: Composite scheduling strategies in distributed computing with non-dedicated resources. Proc. Comput. Sci. **9**, 176–185 (2012)
6. Toporkov, V., Toporkova, A., Yemelyanov, D.: Heuristic of anticipation for fair scheduling and resource allocation in grid VOs. Stud. Comput. Intell. **737**, 27–37 (2018)
7. Aida, K., Casanova, H.: Scheduling mixed-parallel applications with advance reservations. In: 17th IEEE International Symposium on HPDC, pp. 65–74. IEEE CS Press, New York (2008)
8. Elmroth, E., Tordsson, J.: A standards-based grid resource brokering service supporting advance reservations, co-allocation and cross-grid interoperability. J. Concurr. Comput. Pract. Exp. **25**(18), 2298–2335 (2009)
9. Takefusa, A., Nakada, H., Kudoh, T., Tanaka, Y.: An advance reservation-based co-allocation algorithm for distributed computers and network bandwidth on QoS-guaranteed grids. In: Frachtenberg, E., Schwiegelshohn, U. (eds.) JSSPP 2010. LNCS, vol. 6253, pp. 16–34. Springer, Heidelberg (2010)
10. Blanco, H., Guirado, F., Lrida, J.L., Albornoz, V.M.: MIP model scheduling for multiclusters. In: EuroPar 2012. LNCS, vol. 7640, pp. 196–206. Springer, Heidelberg (2013)
11. Moab Adaptive Computing Suite. http://www.adaptivecomputing.com
12. Calheiros, R.N., Ranjan, R., Beloglazov, A., De Rose, C.A.F., Buyya, R.: CloudSim: a toolkit for modeling and simulation of cloud computing environments and evaluation of resource provisioning algorithms. J. Softw. Pract. Exp **41**(1), 23–50 (2011)
13. Samimi, P., Teimouri, Y., Mukhtar, M.: A combinatorial double auction resource allocation model in cloud computing. J. Inf. Sci. **357**(C), 201–216 (2016)
14. Toporkov, V., Toporkova, A., Bobchenkov, A., Yemelyanov, D.: Resource selection algorithms for economic scheduling in distributed systems. In: Proceedings of International Conference on Computational Science, ICCS 2011, Singapore, 1–3 June 2011, vol. 4. pp. 2267–2276 (2011). Procedia Computer Science. Elsevier
15. Kovalenko, V.N., Koryagin, D.A.: The grid: analysis of basic principles and ways of application. J. Programm. Comput. Softw. **35**(1), 18–34 (2009)
16. Makhlouf, S., Yagoubi, B.: Resources co-allocation strategies in grid computing. In: CIIA, vol. 825, CEUR Workshop Proceedings (2011)

17. Netto, M.A.S., Buyya, R.: A flexible resource co-allocation model based on advance reservations with rescheduling support. In: Technical report, GRIDSTR-2007-17, Grid Computing and Distributed Systems Laboratory, The University of Melbourne, Australia (2007)
18. Toporkov, V., Toporkova, A., Tselishchev, A., Yemelyanov, D.: Slot selection algorithms in distributed computing. J. Supercomput. **69**(1), 5360 (2014)

About Designing an Observer Pattern-Based Architecture for a Multi-objective Metaheuristic Optimization Framework

Antonio Benítez-Hidalgo[1], Antonio J. Nebro[1(✉)], Juan J. Durillo[2], José García-Nieto[1], Esteban López-Camacho[1], Cristóbal Barba-González[1], and José F. Aldana-Montes[1]

[1] Dept. de Lenguajes y Ciencias de la Computación, University of Malaga, Campus de Teatinos, 29071 Malaga, Spain
antonio.b@uma.es, {antonio,jnieto,esteban,cbarba,jfam}@lcc.uma.es
[2] Leibniz Supercomputing Centre, Munich, Germany
juanjod@gmail.com

Abstract. Multi-objective optimization with metaheuristics is an active and popular research field which is supported by the availability of software frameworks providing algorithms, benchmark problems, quality indicators and other related components. Most of these tools follow a monolithic architecture that frequently leads to a lack of flexibility when a user intends to add new features to the included algorithms. In this paper, we explore a different approach by designing a component-based architecture for a multi-objective optimization framework based on the observer pattern. In this architecture, most of the algorithmic components are observable entities that naturally allows to register a number of observers. This way, a metaheuristic is composed of a set of observable and observer elements, which can be easily extended without requiring to modify the algorithm. We have developed a prototype of this architecture and implemented the NSGA-II evolutionary algorithm on top of it as a case study. Our analysis confirms the improvement of flexibility using this architecture, pointing out the requirements it imposes and how performance is affected when adopting it.

Keywords: Multi-objective optimization · Metaheuristics
Software framework · Software architecture · Observer pattern

1 Introduction

Most of real-world optimization problems can be formulated as minimizing or maximizing two or more conflicting functions simultaneously, so the result of optimizing them is not a unique solution but a set of trade-off solutions known as Pareto optimal set. In practice, obtaining this set is frequently unfeasible, so

J. Del Ser et al. (Eds.): IDC 2018, SCI 798, pp. 50–60, 2018.
https://doi.org/10.1007/978-3-319-99626-4_5

non-exact techniques providing an approximation of it are commonly used, being metaheuristics the most popular ones [3,4]. Metaheuristics comprise a family of approximate optimization algorithms including evolutionary algorithms, particle swam optimization, ant colony optimization, and many others [2].

Multi-objective optimization with metaheuristics is a very active research field since year 2000, which has been supported by the development and availability of software frameworks that not only provide implementations of state-of-the-art algorithms, but also benchmark problems, quality indicators, support for performance assessment, visualization tools, etc. Examples of these frameworks are PISA [1] (implemented in the C language), Paradiseo-MOEO [12], jMetalCpp [13]) (both implemented in C++), jMetal [6], MOEAFramework [10] (both written in Java), and Inspyred [9] and Platypus [11] (both written in Python).

In general, these frameworks have a monolithic architecture around which metaheuristics are implemented. Depending on their designs (all the aforementioned frameworks but PISA follow an object-oriented approach), they offer a certain degree of flexibility. Frequently, however, significant code changes are required to implement variants of existing algorithms or to add new components to, for example, inspect or analyze the internal behavior of an algorithm during the search.

Our motivation comes from our experience with the jMetal framework. The jMetal project started in 2006, although the framework was redesigned from scratch in 2015 [14] to improve its architecture and to make it more flexible and extensible. Remarkable included features are the provision of metaheuristic templates and the possibility of getting measurements of algorithm-specific information during its execution. Measures are based on the application of the observer pattern [8] and offers pull and push requests to get algorithm data.

Here we explore the design of a multi-objective metaheuristics framework whose components are all based on the observer pattern. In particular, there will be three entities, namely, observer, observable, and observer/observable which can be combined for implementing multi-objective optimization algorithms. We analyze the advantages and drawbacks of this approach by using jMetal as base platform.

The rest of this paper is organized as follows. Section 2 describes the observer pattern, which is the basis of the proposed architecture that is explained in Sect. 3. Section 4 contains some implementation details. We discuss the implications of using the observer-based pattern architecture in Sect. 5. Finally, we present the conclusions and lines of future work in Sect. 6.

2 The Observer Pattern

The book of Gamma *et al.* [8] popularized the concept of design patterns, which can be defined as general and reusable solutions to recurrent design situations happening in software design. A design pattern describes a problem to be solved, a solution to that problem, when to apply it, and its consequences. According to [8], design patterns can be classified in three categories: creational, structural and behavioral. The observer pattern belongs to the latter family.

The observer pattern allows modelling one-to-many relationships among objects, in such a way that when an object (observable) changes its state, the objects depending on it (observers) get notified and updated automatically. This requires to register the observers on the observables beforehand. The pattern is useful in many contexts, and it is included in many libraries and programming languages such as Java.

A simplified UML class diagram representing the pattern is included in Fig. 1. We can see that `Observer` is an interface containing an `update()` method while `Observable` is a class containing two public methods: one to register/unregister observers and a private method to notify registered observers that the state has changed. This method merely invokes the `update()` method of the observers.

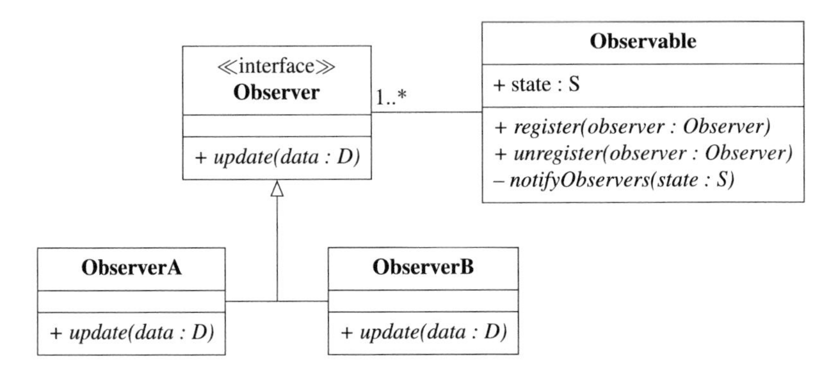

Fig. 1. Simplified UML class diagram representing the observer pattern.

3 Proposed Architecture

In this section, we propose an observer-based architecture for multi-objective metaheuristics. Without loss of generality, we focus on evolutionary algorithms, which are the most widely known and used multi-objective metaheuristics; in particular, we use the multi-objective NSGA-II [5] algorithm as a case of study.

An evolutionary algorithm follows the pseudo-code shown in Algorithm 1. The first step creates an initial population $P(0)$ which is evaluated (line 3) before starting the main loop. An iteration counter t is also initialized (line 2). The evolution process consists in iteratively selecting a mating pool of solutions M from $P(t)$ (line 5), which is used for reproduction, leading to an offspring population $Q(t)$ (line 6). This population is evaluated (line 7) and a replacement strategy is applied to get the population at the next generation $P(t+1)$ by combining $P(t)$ and $Q(t)$ (line 8). The iteration counter t is updated at the end of the loop (line 9). The algorithm stops when a finishing condition is met (line 4).

Algorithm 1. Pseudo-code of an evolutionary algorithm

1: $P(0) \leftarrow$ InitialPopulationGeneration()
2: $t \leftarrow 0$
3: Evaluation($P(0)$)
4: **while not** StoppingCriterion() **do**
5: $\quad M(t) \leftarrow$ Selection($P(t)$)
6: $\quad Q(t) \leftarrow$ Reproduction($M(t)$)
7: $\quad$ Evaluation($Q(t)$)
8: $\quad P(t+1) \leftarrow$ Replacement($P(t)$, $Q(t)$)
9: $\quad t \leftarrow t+1$
10: **end while**

3.1 Algorithm Templates in jMetal

The issue we explore in this section is how to implement the pseudo-code of Algorithm 1 to develop evolutionary algorithms. The solution adopted by jMetal is to provide a template in the form of an `AbstractEvolutionaryAlgorithm` class that closely mimics the pseudo-code:

```
@Override public void run() {
    List<S> offspringPopulation;
    List<S> matingPopulation;

    population = createInitialPopulation();
    population = evaluatePopulation(population);
    initProgress();
    while (!isStoppingConditionReached()) {
        matingPopulation = selection(population);
        offspringPopulation = reproduction(matingPopulation);
        offspringPopulation = evaluatePopulation(offspringPopulation);
        population = replacement(population, offspringPopulation);
        updateProgress();
    }
}
```

This way, an evolutionary algorithm can be implemented by extending the abstract class, i.e., by implementing all its methods, as is the case of NSGA-II.

We analyze next three different situations that can arise in practice: modify the default behavior of an algorithm, extend it to provide information during its execution (e.g., to display the current front), and adding new features such as an external archive to store all the non-dominated solutions found during search.

By using the abstract class scheme, the natural way to modify the behavior of a base algorithm is by defining subclasses of it by applying inheritance. An example is to change the stopping condition. In the default implementation of NSGA-II in jMetal, the algorithm stops when a maximum number of evaluations have been performed. Changing this condition to, for example, stop after a given time limit instead, requires to write a new subclass which simply overrides the `isStopingConditionReached()` method. This solution is simple but introduces a potential hazard: confusing users of that algorithm with many subclasses that provide slightly different behaviours. Furthermore, this new stopping condition cannot be shared by other algorithms.

By default, when an algorithm is configured and executed in jMetal, it does not show any information until it finishes. At that point it generates two files containing the solutions and the Pareto front approximation found. We might be interested, however, in getting run-time information (the iteration number, computing time, population at each iteration, etc.) for writing it to files for further analysis or plotting graphs. This can be achieved in jMetal by defining measures, which require to create a new subclass redefining the `updateProgress()` method, as measures are updated in general at the end of each algorithm iteration. As before, if different measures are needed, new subclasses need to be defined.

There may be situations where we can be interested in adding all the evaluated solutions in NSGA-II to an external archive. A reason is that NSGA-II deletes some non-dominated solutions that later could be useful, so adding them to an external unbounded archive would keep all of them. Incorporating external archives to NSGA-II can be achieved by redefining the `evaluatePopulation()` method.

Summarizing, adopting an abstract class scheme for designing evolutionary algorithms provides enough flexibility at the cost of populating the framework with a large set of variants as minor changes of the default algorithms' behaviors imply to redefine some of the methods of the template.

3.2 Observer Pattern-Based Architecture

Our alternative to the monolithic template-based approach is to decouple all the algorithm components in separate entities featuring one of three possible behaviors: observable, observer, and observer/observable. This way, a metaheuristic within this new approach is composed of components that follow the observer pattern.

The first column of Table 1 shows the components we have considered for a first prototype implementation of NSGA-II. These components can be considered as interfaces that can be implemented in many ways; the second column of the table shows the implementations currently provided. The only pure observable entity is the *CreateInitialPopulation* interface, while the rest of components of an evolutionary algorithm are both observer and observable entities. The observer category includes a *PopulationObserver* interface, representing entities that process populations somehow; in particular, we include observers for writing a population in files, storing a population in an external archive, and plotting a front.

Figure 2 shows how the aforementioned components are linked to provide an implementation of NSGA-II. All these elements, except the one that creates the initial population, must register the previous component from which it will receive data as soon as it is available. It is noteworthy that two instances of

Table 1. Entities following the observer pattern for implementing an evolutionary algorithm (NSGA-II) in jMetal.

Observable	Implementations
CreateInitialPopulation	RandomPopulationCreation
Observer/observable	Implementations
Selection	BinaryTournamentSelection
Evaluation	SequentialEvaluation
	MultithreadedEvaluation
Replacement	RankingAndCrowdingReplacement
	RankingAndHypervolumeContributionReplacement
Termination	TerminationByEvaluations
	TerminationByTime
Variation	CrossoverAndMutationVariation
Observer	Implementations
PopulationObserver	PopulationToFilesWriter
	ExternalArchiveObserver
	ParetoFrontPlot

the *Evaluation* interface are needed, as the evaluation of the initial population is observed by the *Termination* entity, while the evaluation after the variation step is observed by the replacement component. The result is a workflow where a population of solutions, which is created by the pure observable element, is processed by each component and the resulting population is notified to the registered observers.

The behavior of the components of this observer-based implementation is as follows:

1. The *CreateInitialPopulation* produces the initial population, which is notified to the *Evaluation* component. This population has a number of attributes, including an evaluation counter (initialized to 0), a flag indicating whether the computation has finished or not (initialized to False), and the start time of the computation. The population and its attributes constitute the observable data that are notified to the registered observers.
2. The first *Evaluation* component carries out the evaluation of all the individuals of the population and updates the evaluation counter attribute.
3. The *Termination* component receives the population and checks for the stopping condition by examining the corresponding attributes. If the condition is fulfilled, the attribute in the population indicating whether the computation has finished is set to True. The population and its attributes are notified to the observers.

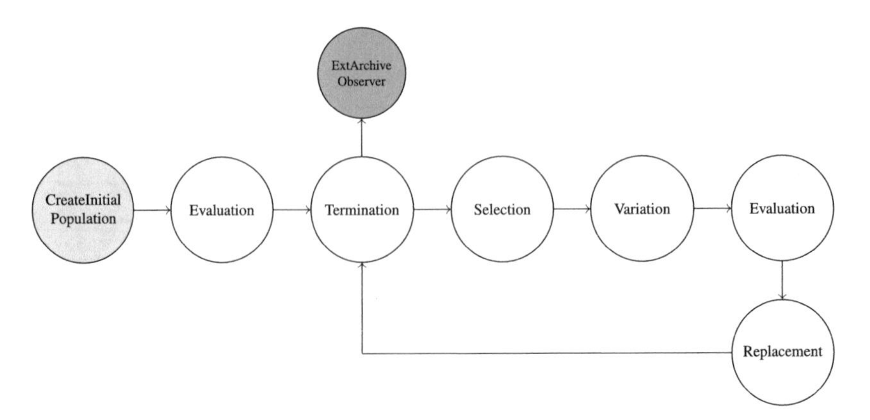

Fig. 2. Architecture of the NSGA-II evolutionary algorithm using the observer pattern. Components filled with light grey () are both observer and observable; components filled with grey () are only observable, and components filled with dark grey () are only observers.

4. The *Selection* component takes the population and generates an offspring population by applying a given selection operator (i.e., binary tournament). This offspring population is attached to the population as another attribute.
5. The *Variation* component receives the population and applies some variation operators (typically, crossover and mutation) to the offspring population. This leads to a new offspring population that replaces the original one.
6. The second *Evaluation* component checks whether the population has an offspring population attribute. If so, this last one is evaluated and the evaluation counter is updated; else, it behaves like indicated in step 2.
7. The *Replacement* component takes the population and the offspring population and creates a new population by applying some replacement strategy (e.g., ranking and crowding in the case of NSGA-II), and the offspring population attribute is eliminated.

All the observer/observable components check their termination attribute every time they are notified a population. In case of being `true`, a component notifies the population to its registered observers and finishes. In the case of observers, they also make the same check. This way, all the components finish in an ordered way.

4 Implementation Details

We have implemented a prototype of the observer-pattern architecture in jMetal. The main issue to deal with is how to implement the `Observable` class, as the *Observer* interface only requires to include the `update()` method in the classes implementing it.

A direct implementation of the observer pattern, where the observable objects directly invoke the update method of the observers, can lead to a stack overflow problem. Our approach has been to associate a thread to each component, so all of them are concurrent entities, and use a bounded buffer (with a capacity for only one element) to apply the producer/consumer concurrent scheme. This way, observers are waiting on their buffers by invoking a get operation and, when observables make a notification and invoke the `update()` method of its registered observers, they put the observable data (i.e, populations and their attributes) into the buffer, thus waking up the observers. To avoid concurrent accesses, the observed data are copied before being inserted into the buffers.

It is worth mentioning that other implementations are possible. Another possibility is to use a message passing library, what would allow to deploy the algorithm components in different nodes of a distributed systems (i.e., a cluster). This way, the most compute intensive components (typically, the evaluators) could be deployed in nodes having highest performance processors.

5 Discussion

We discuss in this section the requirements the observer-based approach imposes and how they are dealt with. We focus on the observable data, the manipulation of global data, the flexibility of the architecture, and the performance implications.

Any observable entity must offer a public interface to its potential observers defining the data that will be provided. The data always include a population and a number of attributes, which are used and updated somehow by each component. As a consequence, all the components must state clearly what they are waiting for and what they produce; otherwise, the resulting algorithm will fail.

All the components are independent entities that do not share a common memory space where global data, like the evaluation counter, can be stored. As a consequence, there is a need of using attributes which are attached to the population in each step, which can be a major issue in case of algorithms using global coarse grained shared data structures. This matter needs further development and that will determine the viability of the observer-based approach.

Our main motivation in this study is to propose a modular and flexible architecture that allows to extend an existing evolutionary algorithm without the need of modifying the current implementations nor applying inheritance to create specialized subclasses. The NSGA-II version based on this scheme (shown in Fig. 2) can be extended in many ways:

- Adding an external unbounded archive to store the non-dominated solutions found, which can be achieved by including an external archive observer and registering it to the *Evaluation* components. An external bounded archive could be also incorporated the same way.
- Changing the stopping condition (e.g., by time, by the number of evaluations, interactively by the user, etc.) only requires to change the *Termination* component.
- A graphical observer can be registered into the *Replacement* component for visualizing the current population in real-time.

- A number of population observers can be attached to the *Termination* component to write the final population in files, to generate plots with the obtained Pareto front approximations, or just to assess the quality of the front by applying a quality indicator.
- Implementing the SMS-EMOA [7] evolutionary algorithm, which is based on NSGA-II, only requires to change the *Replacement* component by one computing the ranking and hypervolume contribution of the solutions of the population.

It is worth highlighting that these changes only require to add more observers, add more observers/observables, or replace some of the components by others. The newly created components can also be used by other algorithms, thus promoting code reuse.

Table 2. Running times of NSGA-II implementation in jMetal and the one using the observer pattern based architecture (the column names represent pairs population size/number of evaluations). The target problem is ZDT1.

	100/25000	200/50000
NSGA-II (monolythic)	782 ms	1422 ms
NSGA-II (observer)	1046 ms	1768 ms
SMS-EMOA (monolythic)	32586 ms	478570 ms
SMS-EMOA (observer)	34685 ms	479237 ms

A consequence of using the proposed architecture instead of a monolithic one is a performance overhead as observed data must be copied when observers are notified by observables. To quantify that overhead, we have run the NSGA-II version included in jMetal and the one developed using the observer pattern with two configurations (population sizes of 100/200 and number of evaluations of 25000/50000) are shown in Table 2[1]. The obtained times reveal a time overhead of 120–350 ms, which is only a fraction of the overall wall time even for a problem such as ZDT1 which can be evaluated in a very short time. In case of solving a real-world problem, the differences in running time should be negligible. We also include in Table 2 the execution times of the SMS-EMOA algorithm with a population size of 100 and 25000 function evaluations. The times used for monolythic and observer versions are similar, since in this algorithm, most of the computing time is spent in the replacement step.

6 Conclusions

We have presented in this paper a study about the design of an observer pattern-based architecture for a framework for multi-objective metaheuristics. In this

[1] MacBook Pro, 2.2 GHz Intel Core i7, macOS 10.13.4, Java SE 1.8.0_101, jMetal 5.5.

architecture, all the algorithm components are classified into three categories of components: observable, observer/observable, and observer. By taking a multi-objective evolutionary algorithm, NSGA-II, as a case of study, we have shown how it can be implemented using the proposed scheme.

Our analysis indicates that the resulting implementation is very flexible, allowing to develop new variants of an algorithm by simply adding or replacing components. The computing times of two algorithms using the proposed architecture indicate a minimal time overhead regarding the original monolithic-based implementations.

The future research work is in the line of validating the observer-based architecture to cope with other multi-objective algorithms which are not extensions of NSGA-II, such as MOEA/D, and with non-evolutionary metaheuristics, such as particle swarm optimization algorithms.

Acknowledgements. This work was partially funded by Grants, TIN2017-86049-R, TIN2014-58304 (Spanish Ministry of Education and Science) and P12-TIC-1519 (Plan Andaluz de Investigación, Desarrollo e Innovación). Cristóbal Barba-González was supported by Grant BES-2015-072209 (Spanish Ministry of Economy and Competitiveness). José García-Nieto is the recipient of a Post-Doctoral fellowship of "Captación de Talento para la Investigación" Plan Propio at Universidad de Málaga.

References

1. Bleuler, S., Laumanns, M., Thiele, L., Zitzler, E.: PISA – a platform and programming language independent interface for search algorithms. In: Fonseca, C.M., Fleming, P.J., Zitzler, E., Deb, K., Thiele, L. (eds.) Evolutionary Multi-Criterion Optimization (EMO 2003). Lecture Notes in Computer Science, pp. 494–508. Springer, Heidelberg (2003)
2. Blum, C., Roli, A.: Metaheuristics in combinatorial optimization: overview and conceptual comparison. ACM Comput. Surv. **35**(3), 268–308 (2003)
3. Coello, C., Lamont, G.B., van Veldhuizen, D.A.: Multi-objective Optimization Using Evolutionary Algorithms, 2nd edn. Wiley, New York (2007)
4. Deb, K.: Multi-objective Optimization Using Evolutionary Algorithms. Wiley, New York (2001)
5. Deb, K., Pratap, A., Agarwal, S., Meyarivan, T.: A fast and elitist multiobjective genetic algorithm: NSGA-II. IEEE Trans. Evol. Comput. **6**(2), 182–197 (2002)
6. Durillo, J.J., Nebro, A.J.: jMetal: a Java framework for multi-objective optimization. Adv. Eng. Softw. **42**(10), 760–771 (2011)
7. Emmerich, M., Beume, N., Naujoks, B.: An EMO algorithm using the hypervolume measure as selection criterion. In: Coello, C.A., Hernández, A., Zitler, E. (eds.) Third International Conference on Evolutionary MultiCriterion Optimization, EMO 2005. LNCS, vol. 3410, pp. 62–76. Springer (2005)
8. Gamma, E., Helm, R., Johnson, R., Vlissides, J.M.: Design Patterns: Elements of Reusable Object-Oriented Software, 1st edn. Addison-Wesley Professional, Reading (1994)
9. Garrett, A.: Inspyred. http://aarongarrett.github.io/inspyred/. Accessed 02 May 2018
10. Hadka, D.: MOEAFramework. http://moeaframework.org/. Accessed 02 May 2018

11. Hadka, D.: Platypus. http://platypus.readthedocs.io/en/latest/. Accessed 02 May 2018
12. Liefooghe, A., Jourdan, L., Talbi, E.-G.: A software framework based on a conceptual unified model for evolutionary multiobjective optimization: ParadisEO-MOEO. Eur. J. Oper. Res. **209**(2), 104–112 (2011)
13. López-Camacho, E., García-Godoy, M.J., Nebro, A.J., Aldana-Montes, J.F.: jMetalCpp: optimizing molecular docking problems with a C++ metaheuristic framework. Bioinformatics **30**(3), 437–438 (2014)
14. Nebro, A.J., Durillo, J.J., Vergne, M.: Redesigning the jMetal multi-objective optimization framework. In: Proceedings of the Companion Publication of the 2015 Annual Conference on Genetic and Evolutionary Computation, GECCO Companion 2015, pp. 1093–1100. ACM, New York (2015)

Scalable Inference of Gene Regulatory Networks with the Spark Distributed Computing Platform

Cristóbal Barba-González, José García-Nieto(✉), Antonio Benítez-Hidalgo, Antonio J. Nebro, and José F. Aldana-Montes

Dept. de Lenguajes y Ciencias de la Computación,
Instituto de Investigación Biomédica de Málaga (IBIMA), University of Málaga,
Campus de Teatinos, 29071 Málaga, Spain
{cbarba,jnieto,antonio,jfam}@lcc.uma.es, antonio.b@uma.es

Abstract. Inference of Gene Regulatory Networks (GRNs) remains an important open challenge in computational biology. The goal of bio-model inference is to, based on time-series of gene expression data, obtain the sparse topological structure and the parameters that quantitatively understand and reproduce the dynamics of biological system. Nevertheless, the inference of a GRN is a complex optimization problem that involve processing S-System models, which include large amount of gene expression data from hundreds (even thousands) of genes in multiple time-series (essays). This complexity, along with the amount of data managed, make the inference of GRNs to be a computationally expensive task. Therefore, the generation of parallel algorithmic proposals that operate efficiently on distributed processing platforms is a must in current reconstruction of GRNs. In this paper, a parallel multi-objective approach is proposed for the optimal inference of GRNs, since minimizing the Mean Squared Error using S-System model and Topology Regularization value. A flexible and robust multi-objective cellular evolutionary algorithm is adapted to deploy parallel tasks, in form of Spark jobs. The proposed approach has been developed using the framework jMetal, so in order to perform parallel computation, we use Spark on a cluster of distributed nodes to evaluate candidate solutions modeling the interactions of genes in biological networks.

Keywords: Gene Regulatory Networks · Multi-objective Metaheuristics · Distributed Computing · jMetal · Spark

1 Introduction

In computational systems biology, the inference of Gene Regulatory Networks (GRNs) from time-series of gene expression data remains an important open challenge [2]. The main goal is to obtain the sparse topological structure and the parameters that quantitatively understand and reproduce the dynamics of

J. Del Ser et al. (Eds.): IDC 2018, SCI 798, pp. 61–70, 2018.
https://doi.org/10.1007/978-3-319-99626-4_6

biological systems. However, although it is possible to computationally predict genetic interaction networks, their precision depends on the characteristics of the model used, as well as the availability and quality of the expression data.

GRNs have been modeled with different approaches, such as Boolean [1] and Bayesian networks [6], which assign conditional probabilities to the regulation parameters of each gene. In this study, we focus on the S-System model [11], as it provides a good compromise between biological relevance and mathematical flexibility. However, S-System is based on Ordinary Differential Equations' (ODE) models, which require a large number of parameters to be tuned when inferring a target network. Therefore, its application is, to date, limited to small/medium sized networks, due to the course of dimensionality and the amount of time-series data required to properly fit the model.

For the inference of appropriate parameters in an S-System, researchers have traditionally used global optimization search techniques, such as Genetic Algorithms (GAs) [12], Differential Evolution (DE) [8] and Particle Swarm Optimization (PSO) [9]. These approaches are shown to operate efficiently when inferring small-size networks, but obtained limited performance on large scale networks involving more than 50 genes. In those cases, the running time used in the evaluation phase increases exponentially with the number of parameters to be tuned and the amount data to be managed. That is why, as noticed in recent studies [2,9], generating parallel approaches for the efficient processing of large datasets of gene expression time-series to infer GRNs remains as a major challenge.

With this motivation, a flexible and robust Multi-objective Cellular Evolutionary algorithm (MOCell) [7] is adapted in this study to deploy parallel tasks, in form of Spark jobs, which involve the evaluation of candidate solutions modeling the interactions of multiple genes in biological networks. These jobs can be now distributed on parallel processing Spark environments to simultaneously minimize two objectives: (1) the Mean Squared Error (MSE) using S-System model; and (2) a Topology Regularization (TR) value. This way, it is now possible to mitigate the computational effort derived from the inference of extensive GRNs, while taking advantage of different learning procedures induced by multi-objective optimizers.

The proposed approach has been developed using the framework jMetal [5], so in order to perform parallel computation, Spark processing jobs are deployed on a cluster of distributed nodes. For testing purposes, a series of benchmarking networks of the DREAM3 challenge [10] are used, which comprise gene expressions' time-series from 100 genes of real-world organisms: *E.coli* (*Eschericia coli*) and *Yeast* (*Saccharomyces cerevisiae*). The performance of the proposed approach is evaluated in terms of computational effort and the ability to infer GRNs, in comparison with sequential and state-of-the-art variants.

The rest of this paper is structured as follows. Section 2 describes the notation used and problem formulation. Section 3 details the proposed approach. In Sect. 4, the experimental results are reported, which include running time measures and analysis from the point of view of the GRNs obtained. Finally, conclusions and future work are presented in Sect. 5.

2 Notation and Problem Formulation

In biological networks inference, the main goal is to capture the dynamics of biological systems from the time-series of gene expression datasets obtained from a given pool of molecular species, in a given time period. Such dynamics can be mathematically modeled by the S-System [13] framework to represent a network as a set of differential equations:

$$f(t,X) = \begin{pmatrix} \alpha_1 \prod_{j=1}^{n+m} X_j^{g_{1j}} - \beta_1 \prod_{j=1}^{n+m} X_j^{h_{1j}} \\ \vdots \\ \alpha_n \prod_{j=1}^{n+m} X_j^{g_{nj}} - \beta_n \prod_{j=1}^{n+m} X_j^{h_{nj}} \end{pmatrix}, \; X(0) = X_0 \tag{1}$$

where X is an n-dimensional pool of elements and the m-dimensional independent variables are expressed as X_{n+j}, $j = 1, \cdots, m$. That is, X_i is the expression level of the ith gene. Parameters $\alpha_i, \beta_i \in \mathbb{R}_+^N$ are rate constants ($N = n+m$), and $g_{ij}, h_{ij} \in \mathbb{R}^{N \times N}$ are kinetic orders that regulate the synthesis and degradation of X_i influenced by X_j.

For the evaluation of S-System models, numerical methods such as Runge-Kutta [12] have been traditionally applied, since they are highly accurate in finding parameters that lead the model to fit time-series curves of gene expression data values. Each candidate solution generated by the parallel MOCell approach is then arranged in a vector of real variables representing the *solution encoding* of tuning parameters: kinetic orders (g_{ij}, h_{ij}) and rate constants (α_i, β_i), in the S-System.

In this regard, a widely used criterion for evaluation is the discrepancy between the gene expression levels calculated numerically and those observed from time-series of actual gene expressions of system dynamics. To measure this discrepancy, the Mean Squared Error (MSE) is used as the fitness function to evaluate each candidate solution in the S-System. MSE is formulated as a minimization function:

$$f^{MSE} = \sum_{i=1}^{N} \sum_{k=1}^{M} \sum_{t=1}^{T} \left(\frac{X_{k,i}^{cal}(t) - X_{k,i}^{exp}(t)}{X_{k,i}^{exp}(t)} \right)^2 \tag{2}$$

where $X_{k,i}^{cal}(t)$ and $X_{k,i}^{exp}(t)$ are the expression levels of gene i in the kth set of time-series at time t in the *calculated* and *experimental* data, respectively. M is the set of time-series considered in the evaluation, whereas T is the number of sampling points in the experimental data (gene expression values). The main goal is to find an optimized set of parameters θ that minimizes f^{MSE}.

MSE is an efficient measure to capture the efficient fitting of time-series curves by means of sampling values in optimization procedures. However, one of the major difficulties in the S-System based inference process lies on obtaining the proper topological structure of gene interactions that generates the dynamics observed. S-System model parameters show a high degree of freedom, which results in a great number of local minima in the search space of solutions that mimics the time courses very closely. To this end, the use of pruning or penalty terms based on the Laplacian regularization in the basic MSE fitness function, is useful for attaining a sparse network topology in the canonical optimization problem. A useful pruning term was proposed in [8] as fitness function, to capture the topological structure of the network as follows:

$$f_i = f_i^{MSE} + c \cdot f_i^{Topology} \tag{3}$$

where f_i^{MSE} is Eq. 2, but only referring to gene i; $f_i^{Topology} = \sum_{j=1}^{2N-I}(|K_{i,j}|)$, with $K_{i,j}$ is obtained by sorting kinetic orders g_{ij} and h_{ij} all together in ascending order of their absolute values $(|K_{i,1}| \leq |K_{i,2}| \leq \cdots \leq |K_{i,2N}|)$. I is the maximum allowed cardinality degree of the network and c is a penalty constant. This penalty value (second term of Eq. 3) includes the maximum allowed number of gene interactions (edges) in the network and penalizes only when the number of genes that directly affect the ith gene is higher than the maximum cardinality I. As a consequence, this penalization will cause most of the genes to disconnect when their affecting kinetic orders have low values.

It is worth noting that this aggregative formulation requires additional weight factors to be set to find a good trade-off between the different terms. Nevertheless, this goal can be formulated as a bi-objective optimization problem by considering these two terms separately, leading to a Pareto dominance scheme as follows:

- Objective 1: f^{MSE} (Eq. 2), which aims to estimate the kinetic and order parameters from a limited amount of gene expression data.
- Objective 2: $f^{Topology}$ (second term of Eq. 3), to detect the sparse topological structure that is most commonly seen in biological networks.

This way, it is now possible to take advantage of specific learning models of Pareto optimality-based techniques [4] to deal with the inference of GRNs, which will result in sets of non-dominated solutions with different choices of time-series estimations and node topologies. Furthermore, additional weight factors are avoided, thus preventing the search procedure to be biased to one of the different terms, as usually observed in single-objective approaches.

The use of Pareto-based algorithms has the advantage of providing a set of trade-off (i.e., non-dominated) solutions according to the two considered objectives. The set of optimal non-dominated solutions is known as the Pareto optimal set, whose representation in the objective space is named as the Pareto front approximation [4].

Algorithm 1 Pseudo-code of the Parallel MOCell with jMetal

1: G_{max}, maximum number of generations;
2: $P^0 \leftarrow initializePopulation()$;
3: $t \leftarrow 0$, generation counter;
4: **DO In Parallel** $evaluate(\boldsymbol{x}_i \in S^0)$; // Parallel Spark jobs
5: $A^0 \leftarrow initializeExternalArchive(S^0)$;
6: **while** $t < G_{max}$ **do**
7: $\quad \mathbf{B}^t \leftarrow select(A^t, P^t)$;
8: $\quad \mathbf{O}^t \leftarrow recombination(p_c, \mathbf{B}^t)$;
9: $\quad \mathbf{O}^t \leftarrow mutation(p_m, \mathbf{O}^t)$;
10: $\quad$ **DO In Parallel** $evaluate(\boldsymbol{x}_i \in O^t)$; // Parallel Spark jobs
11: $\quad P^{t+1} \leftarrow replacement(O^t)$;
12: $\quad A^{t+1} \leftarrow updateExternalArchive(A^t, P^t)$;
13: $\quad t \leftarrow t + 1$;
14: **end while**
15: **return** $A^{G_{max}}$;

3 Distributed Approach

As previously mentioned, the proposed distributed approach entails an adaption of MOCell [7], a multi-objective cellular Genetic Algorithm (cGA), in which the tasks of inferring GRNs are encoded by individuals and evaluated in parallel by means of Spark jobs. The main characteristic of cellular GAs is that each solution belongs to a cell (or neighborhood) and can only be recombined with a reduced number of solutions (the neighbors). This way, the algorithm is able to avoid premature convergence to local optima, which result in robust solutions in terms of kinetic order and rate constant parameters of the S-System model.

The pseudo-code of the proposed MOCell is shown in Algorithm 1. This algorithm iteratively applies the selector policy to the neighborhood to fill the mating pool population so the parents are chosen among its neighbors (line 7) by a given criterion. Crossover and mutation operators are applied to the individuals (lines 8 and 9), with probabilities p_c and p_m, respectively. Afterwards, the algorithm computes, in parallel Spark jobs, the fitness value of the new offspring (line 10), and inserts them into the equivalent place of the current individuals in a new auxiliary population following a given replacement strategy. After each generation (or loop), the auxiliary population is assumed to be the population for the next generation. This loop is repeated until a stop condition is met (line 6). A maximum number of computed fitness evaluations is considered as the stop condition.

An additional population (external archive) is incorporated to MOCell to gather the non-dominated solutions found throughout the optimization procedure. To this end, MOCell creates an empty Pareto front (line 5) and, after the replacement operation, the generated offspring is included in the external archive (line 12), if appropriate. The replacement is performed on the current individual if it is dominated by the offspring or the two are non-dominated and the current individual has the worst crowding distance (as defined in [4]) in a population composed of the neighborhood plus the offspring. This criterion is also used to decide whether the offspring solutions are added to the external archive (line 12 in Algorithm 1) or not. After each generation, the old population is replaced by the auxiliary one (line 11), and a feedback procedure is invoked to replace a number of randomly chosen individuals with a number of solutions from the archive (line 12). Finally, the "*archived*" Pareto front is returned as the algorithm's output (line 15).

The proposed MOCell is developed in jMetal [5], which is an algorithmic framework that includes a number of optimization metaheuristics of the state-of-the-art. The underlying idea is that an algorithm (metaheuristic) manipulates a number of solutions with some operators to solve an optimization problem. In order to provide MOCell with parallel processing capability, jMetal uses Apache Spark [3][1] to evaluate the population in parallel at each iteration of the optimization process.

Apache Spark [14] is based on the concept of Resilient Distributed Datasets (RDD), which are collections of elements that can be operated in parallel on the nodes of a cluster by using two types of operations: transformations (e.g., map, filter, union, etc.) and actions (e.g., reduce, collect, count, etc.). The Spark-based evaluator in jMetal creates an RDD with all the solutions to be evaluated while a map transformation is used to evaluate every solution.

In jMetal, metaheuristic algorithms are comprised of a loop where a number of tentative solutions are manipulated somehow to produce a set of new solutions that have to be evaluated. As only the evaluation step is performed in parallel, the algorithm, MOCell in this case, works in a heartbeat way: after each parallel step, a sequential step takes place. This obviously prevents to get high (quasi-linear) speedups, which is the cost for having a very simple mechanism to use jMetal's algorithms in a parallel infrastructure. Our interest in this paper is to measure which are effectively the actual time reductions that can be achieved in the context of the inference of large scale GRNs.

4 Experimental Results and Discussions

In this section, a speedup analysis is conducted on a distributed Spark environment to check the possible benefits of using the proposed parallel MOCell approach in terms of computational effort. To this end, the instance Ecoli1 with 100 genes (DREAM3 benchmark) is used for network reconstruction on different Spark cluster configurations. A further analysis is also carried out to compare the obtained networks in terms of precision (quality of solution), with regards to other baseline methods in the context of the DREAM3 Challenge [10].

The computational environment is made up of 10 virtual machines, each one having 10 cores, 10 GB RAM and 250 GB virtual storage (summing up 100 cores, 100 GBs of memory and 2.5 TB HD storage). These virtual machines are used as worker nodes with the role of TaskTracker (Spark) and DataNode (HDFS)

[1] In URL https://spark.apache.org/.

to perform fitness evaluations of algorithmic candidate solutions in parallel. The Master node, which runs the core algorithm (MOCell), is hosted in a different machine with 8 Intel Core i7 processors at 3.40 GHz, 32 GB RAM and 3 TB storage space. All these nodes are configured with a Linux CentOS 6.6 64-bit distribution.

4.1 Speedup and Efficiency

One of the most widely used indicators for measuring the performance of any parallel algorithm is the *Speedup*. The standard formula of the speedup is $S_N = \frac{T_1}{T_N}$ that calculates the ratio of T_1 over T_N, where T_1 is the running time of the analyzed algorithm in 1 processor and T_N is the running time of the parallel algorithm on N processors (cores)[2]. The *Efficiency* is then calculated with the formula $E_N = \frac{S_N}{N} \times 100$. An algorithm scales linearly (ideal) when it reaches a speedup $S_N = N$ and hence, the parallel efficiency is $E_N = 100\%$. In the execution of an algorithm with linear speedup, doubling the number of processors means doubling the speed.

As the parallel evaluations of individuals are performed synchronously throughout the entire population, the proposed approach is tested by setting the population size in accordance with the number of cores in the underlying platform, although keeping the stop condition to the same maximum number of interactions for all the experiments.

Table 1 shows the running time, speedup and efficiency used by the MOCell jMetal approach running on 1, 10, 20, 50, and 100 processors. Figure 1 plots the speedups (with regards to linear speedup) and the evolution of running times consumed by the proposal. In this figure, it is clearly observable that the highest reduction in computing time is reached when the number of processors is higher. In this sense, a complete execution of the optimization algorithm requires several days ($T_1 = 243.5$ h $\approx$ 10 days) in a platform context with less than 10 cores, although this computational time is reduced to less than one day when using more than 20 cores in our parallel model. For 100 nodes, the time spent is reduced to 8.3 h.

Table 1. Experimental results of MOCell (jMetal) executed in 1, 10, 20, 50 and 100 processors for the inference of GRNs. Time units are calculated in hours.

Running time (hours)					Speedup				Efficiency			
T_1	T_{10}	T_{20}	T_{50}	T_{100}	S_{10}	S_{20}	S_{50}	S_{100}	E_{10}	E_{20}	E_{50}	E_{100}
243.05	85.77	31.70	15.50	8.30	2.83	7.66	15.68	29.28	28.33%	38.33%	31.36%	29.28%

The speedup obtained increases along with the number of processors used, although it moves away from the linear speedup as the number of resources expand. This indeed affects to the efficiency. As can be observed in Table 1, it is

[2] We use either the term processor or core to refer the same processing unit.

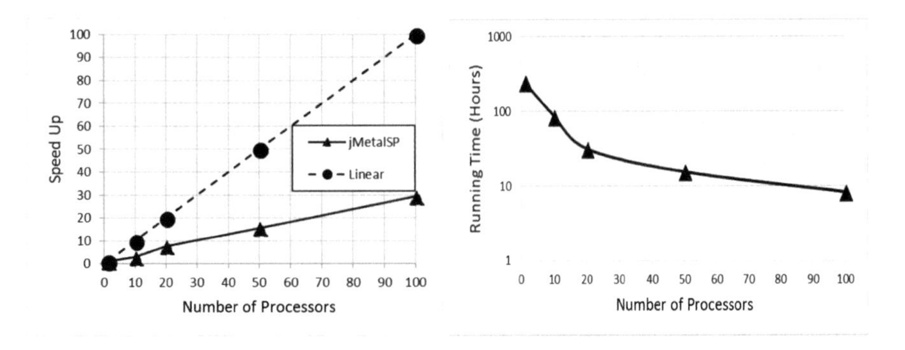

Fig. 1. Linear (ideal) versus actual speedup of the jMetal version with MOCell (left), Running time in hours (logarithmic scale, right) executed in 1, 10, 20, 50 and 100 processors.

high with 20 (38.33%) and 50 processors, but deteriorates in E_{100}. Therefore, a good trade-off between performance and resource requirement is obtained with a configuration of MOCell in jMetal with a range of 20 and 50 cores, since it finalizes all the jobs in one day approximately, although performing 50% cost efficient.

4.2 Analysis of Inferred Networks

A second set of experiments consists on evaluating the parallel MOCell in the context of the standard benchmark DREAM3 *in silico challenges*[3] [10]. This benchmark comprises two networks of *E.coli* plus three networks of *Yeast* with 100 sized genes each one. Networks of size 100 genes involve 46 time-series of 21 time points. The structure of the networks were given from the actual *E.coli* and *Yeast* organisms, which exhibit heterogeneous patters of sparsity and topology. To measure the performance of algorithms, the protocol suggested in this challenge comprises standard metrics: the Receiver Operating Characteristic curve (ROC), the area under ROC curve (AUROC) and the area under the precision-recall curve (AUPR).

The results of the parallel MOCell for the inference of network structures size 100 are given in Table 2. In these table, a series of state-of-the-art results are also incorporated, which comprise those of LASSO and the Team 239 that exclusively used the same set of time-series data in the DREAM3 challenge. The LASSO implements a baseline linear least squares regression [10].

The AUROC and AUPR metrics in Table 2 indicate that MOCell is highly competitive with regards to the approach of LASSO and Team 236. This last was the only team that used time-series datasets for size-100 challenge. Specifically, MOCell obtained outperforming AUROC values for networks Ecoli1, Yeast2 and Yeast3. It is noteworthy that, for all the algorithms, the resulting AUROC values

[3] Available at http://dreamchallenges.org.

Table 2. AUROC and AUPR for MOCell, LASSO, Team 236 (DREAM3 challenge) run on DREAM3 size-100 networks.

	Ecoli1		Ecoli2		Yeast1		Yeast2		Yeast3	
	AUROC	AUPR	AUROC	AUPR	AUROC	AUPR	AUROC	AUPR	AUROC	AUPR
LASSO	0.519	0.016	0.512	0.057	0.507	0.016	0.530	0.044	0.506	0.044
Team 236	0.527	0.019	0.546	0.042	0.532	0.035	0.508	0.046	0.508	0.065
MOCell	0.532	0.034	0.542	0.015	0.522	0.018	0.524	0.047	0.517	0.058

are kept with a high precision, whereas AUPR values declined in general with percentages lower than 7%. A reason of this deterioration in AUPR may be caused by the lower density of size-100 networks, whereas the number of non-edge elements is larger than the number of edges.

5 Conclusions

In this paper, a parallel multi-objective approach based on MOCell is proposed for the inference of Gene Regulatory Networks. This algorithm has been developed using the framework jMetal, so in order to perform parallel computation a Spark mechanism is used to deploy tasks (inference and evaluation of GRNs) on a cluster of distributed nodes. The following conclusions can be reported:

- The proposed approach is able to obtain actual reductions in computing time from several days (10) to hours (8.3) when facing complex inference of large scale GRNs. This performance have been obtained in the scope of an in-house (virtualized) computational environment with limited resources.
- A noteworthy trade-off between performance and resource requirement is obtained with a configuration of the parallel MOCell in the range of 20 and 50 cores, since it finalizes all the jobs launched in less than half a day and is 50% cost efficiently.
- Experiments on realistic benchmarking data of DREAM3 shows that MOCell is competitive and often outperforms state-of-the-art GRN inference procedures.

An important direction to take is the adaption of specific algorithmic operators to the special case of GRNs modeling and inference. The design of crossover, mutation and local search operators that include additional knowledge based on complementary data, such as microRNA expression, chromatin or protein-protein interactions, could guide the search strategy of the algorithmic proposal and enhance its potential for biologically meaningful hypothesis generation on real datasets. These specific operators would be indeed good candidates to be run in parallel, which would lead the new algorithmic models to perform even more efficiently in terms of computational effort and quality of solution.

Acknowledgements. This work was partially funded by Grants TIN2017-86049-R, TIN2014-58304 (Spanish Ministry of Education and Science) and P12-TIC-1519 (Plan Andaluz de Investigación, Desarrollo e Innovación). C. Barba-González was supported by Grant BES-2015-072209 (Spanish Ministry of Economy and Competitiveness). J. García-Nieto is the recipient of a Post-Doctoral fellowship of "Captación de Talento para la Investigación" at Universidad de Málaga.

References

1. Akutsu, T., Kuhara, S., Maruyama, O., Miyano, S.: Identification of genetic networks by strategic gene disruptions and gene overexpressions under a boolean model. Theoret. Comput. Sci. **298**(1), 235–251 (2003)
2. Angus, T.S., Yaochu, J.: Reconstructing biological gene regulatory networks: where optimization meets big data. Evol. Intell. **7**(1), 29–47 (2014)
3. Barba-Gonzaléz, C., García-Nieto, J., Nebro, A.J., Aldana-Montes, J.F.: Multi-objective big data optimization with jMetal and spark. In: International Conference on Evolutionary Multi-Criterion Optimization, pp. 16–30. Springer (2017)
4. Deb, K.: Multi-Objective Optimization Using Evolutionary Algorithms. Wiley, New York (2001)
5. Durillo, J.J., Nebro, A.J.: jMetal: a java framework for multi-objective optimization. Adv. Eng. Softw. **42**, 760–771 (2011)
6. Friedman, N., Linial, M., Nachman, I.: Using Bayesian networks to analyze expression data. J. Comput. Biol. **7**, 3–4 (2004)
7. Nebro, A.J., Durillo, J.J., Luna, F., Dorronsoro, B., Alba, E.: Design issues in a multiobjective cellular genetic algorithm, pp. 126–140. Springer, Heidelberg (2007)
8. Noman, N., Iba, H.: Inferring gene regulatory networks using differential evolution with local search heuristics. TCBB **4**(4), 634–647 (2007)
9. Palafox, L., Noman, N., Iba, H.: Reverse engineering of gene regulatory networks using dissipative particle swarm optimization. IEEE Trans. Evol. Comput. **17**(4), 577–587 (2013)
10. Prill, R.J., Marbach, D., Saez-Rodriguez, J., Sorger, P.K., Alexopoulos, L.G., Xue, X., Clarke, N.D., Altan-Bonnet, G., Stolovitzky, G.: Towards a rigorous assessment of systems biology models: the DREAM3 challenges. PLoS ONE **5**(2), 1–18 (2010)
11. Savageau, M.: Biochemical Systems Analysis: A Study of Function and Design in Molecular Biology. Addison-Wesley Educational Publishers Inc., Reading (2010)
12. Sirbu, A., Ruskin, H.J., Crane, M.: Comparison of evolutionary algorithms in gene regulatory network model inference. BMC Bioinfor. **11**(1), 59 (2010)
13. Voit, E.O.: Computational Analysis of Biochemical Systems. A Practical Guide for Biochemists and Molecular Biologists. Cambridge University Press, New York (2000)
14. Zaharia, M., Chowdhury, M., Franklin, M.J., Shenker, S., Stoica, I.: Spark: cluster computing with working sets. In: Proceedings of the 2nd USENIX Conference on Hot Topics in Cloud Computing, HotCloud 2010, Berkeley, CA, USA, p. 10. USENIX Association (2010)

Finding Best Compiler Options for Critical Software Using Parallel Algorithms

Gabriel Luque(✉) and Enrique Alba

Andalucía Tech, University of Málaga, Málaga, Spain
{gabriel,eatl}@lcc.uma.es

Abstract. The efficiency of a software piece is a key factor for many systems. Real-time programs, critical software, device drivers, kernel OS functions and many other software pieces which are executed thousands or even millions of times per day require a very efficient execution. How this software is built can significantly affect the run time for these programs, since the context is that of compile-once/run-many. In this sense, the optimization flags used during the compilation time are a crucial element for this goal and they could make a big difference in the final execution time. In this paper, we use parallel metaheuristic techniques to automatically decide which optimization flags should be activated during the compilation on a set of benchmarking programs. The using the appropriate flag configuration is a complex combinatorial problem, but our approach is able to adapt the flag tuning to the characteristics of the software, improving the final run times with respect to other spread practices.

1 Introduction

When a software package is developed, there are a lot of aspects which should be considered. The ISO/IEC 25010 standard [1] defines eight characteristics (with many sub-characteristics), such as security, usability, compatibility, efficiency... which must be taken into account when evaluating a software.

In this work, we focus on software in which the performance is one of the key features requested by the stakeholders. Many systems fall into this category. Critical and real-time systems [2] clearly need to response to the external environment changes as quick as possible. But, there are also non-critical software pieces in which performance is very relevant like program which are constantly executed and their runtime affects to the complete software ecosystem, like drives, kernel OS functions, or communication protocols.

Developers focus on the software design and on the source code to get an efficient program. These aspects are indeed crucial, but they often pay little attention how their software is built. Some world spread compiler options, like O3 or hardware platform specific ones in C/C++, are blindly applied without any additional study. The main contributions of this paper focuses on how the

J. Del Ser et al. (Eds.): IDC 2018, SCI 798, pp. 71–81, 2018.
https://doi.org/10.1007/978-3-319-99626-4_7

optimization flags used during the compilation process affect to the final software performance and how they can be optimally selected to build a more efficient program. We analyze the influence of 181 optimization flags provided by GCC compiler [3]. We model it as a complex combinatorial problem, and provide a parallel automatic tool to decide which options should be activated to get a high-performance software on a set of benchmarking problems.

The rest of this article is organized as follows. The next section introduces the problem solved, modelling it as a combinatorial problem. Section 3 explains the algorithmic approaches proposed in this paper. Section 4 shows the experimental design and Sect. 5 analyses the results from different points of view. Finally, the last section concludes and gives some open research lines.

2 Problem Description

Modern compilers have a rich set of optimization flags and a manual selection of the most adequate configuration is a hard task. Compilers support a number of basic optimization level to make easy this process. However, the question is if these basic levels are adequate for any software and any scenario. Traditional compiler optimizations define a well-studied area, and some issues are been analysed in the literature in the past. But, these works follow a different approach to the proposed in this paper: they study the compiler phase order [4], compare different compiler [5], optimize specific software [6,7], or evaluate specific flags for some hardware platforms [8].

In this paper, we have modelled the problem of selecting the optimal set of compiler flags as a combinatorial problem. Given a program P and a set of flags $F = \{f_1, f_2, \ldots, f_N\}$, this problem consists on finding a binary vector $X = (x_1, x_2, \ldots, x_N)$ (where $x_i \in \{0, 1\}$ means if the flags f_i is activated or not) which minimizes the execution time of P when it is compiled with set of flags indicated by X $\{f_i | x_i == 1 \; \forall i \in \{1, N\}\}$:

$$min_X \; Fitness(X) = executionTime(compile(P, F, X)) \quad (1)$$

where $compile(P, F, X)$ is a process which generates an executable program compiling the program P selecting the flags from F indicated by X and $executionTime(P)$ calculates the execution time of the executable program.

We focus on the optimization flags of GCC compiler, a popular C language compiler. This language is the most prominent one for the developing this kind of critical software and GCC is the most used compiler for this language. We will consider all the 181 flags activated by level O3, the most aggressive and commonly option used to get efficient executable codes.

3 Our Proposed Approach

This problem is very complex in manifold ways. On the one hand, as we showed in Eq. 1, the evaluation of a candidate solution requires the compilation and execution of a program. Even in the smallest software pieces, these activities requires between 0.5 and 1 second. On the other hand, the search space is quite large: we have two options for each flags (it is activated or not), therefore, there are $2^{181} \approx 3.1 \times 10^{54}$ candidate solutions.

The combination of these two factors makes hard the utilization of methods which require a large number of evaluations. Therefore, we have chosen a trajectory-based metaheuristic with a fast convergence to get some accurate solutions in a reasonable time. We have used Variable Neighborhood Search (VNS) [9]. This technique solves optimization problems by doing systematic changes of neighbourhood within a local search. VNS is a descendent method which does not follow a single trajectory since it explores different predefined neighbourhoods of the current solution using a local search (LS). The current solution is changed by a new one if and only if an improvement has been made. The basic idea is to change the neighbourhood structure when the local search is trapped on a local optimum. There exist several parallel models for VNS [10,11]. In this work, we have used two parallel variants:

- **Parallel moves model:** The parallel moves model is a kind of farmer/worker model allowing to speed up the exploration of the possible moves. At the beginning of each iteration of the algorithm, the farmer sends the current solution to a pool of workers. Each worker explores some neighbouring candidates, and returns back the results to the farmer.
- **Parallel multi-start cooperative model:** The model consists in launching in parallel several cooperative homogeneous VNS. Each VNS is initialized with a different solution. VNSs of the parallel multi-start model periodically interchange the current solution during execution.

Now, we show will how this generic template has been instantiated for our concrete problem. To do this, we have to describe the solution representation, the fitness function and the definition of the different neighbourhoods.

Solution Encoding. In the previous section, we define the solution of this problem as a binary vector. With this representation, a solution is a vector X where $x_i = 1$ indicates if the flags f_i is activated while $x_i = 0$ means that the flag f_i does not appear in the compilation command. But our preliminary experiments showed that this approach was not the most appropriate representation. According to the documentation, the O3 option activates all the flags considered in this work and no more, but the results of our preliminary experiments (see Table 1) showed a significant difference between using the O3 option and the activation of all the flags (without using the O3 option).

Based on these results, we decide to change the meaning of each vector component. Now, the compilation is always performed with the O3 option but our algorithm is able to deactivate some flags from O3 option. Therefore, $x_i = 1$ indicates if the flag f_i is deactivated[1] while $x_i = 0$ means that we allow to O3 to

[1] In GCC, you can activate flag with `-f`*flag* but you can also deactivate it with `-fno-`*flag* if another option (O3 in our case) has previously activate it.

Table 1. CLBG corpus of programs and execution time (in μsec) of the compiled programs with different options

Benchmark	Description	O3	All the flags
binary-trees	Allocate/traverse/deallocate binary trees	75579	84924
chameneos-redux	Symmetrical thread rendezvous requests	390695	436925
fannkuch-redux	Indexed access to tiny integer sequence	78313	128914
fasta	Generate and write random DNA sequences	18425	59158
k-nucleotide	Hashtable update and k-nucleotide strings	39271	50916
mandelbrot	Generate Mandelbrot set bitmap file	13926	23865
meteor-contest	Search for solutions to shape packing puzzle	45314	72318
n-body	Double precision N-body simulation	46661	184670
pidigits	Streaming arbitrary precision arithmetic	21742	39185
regex-redux	Match DNA 8mers and substitute magic patterns	25712	32673
rev-complement	Read DNA sequences, write their rev-complement	34159	53811
spectral-norm	Eigenvalue using the power method	61149	69154
thread-ring	Switch from thread to thread passing one token	714616	736524

use the flag f_i. Preliminary experiments show that the order in which the flags are deactivated does not change the overall performance.

Fitness Function. We use the (Eq. 1) as fitness function. The main complication found in our preliminary experiments is that although the hardware platform is dedicated to this work and the tested programs are deterministic, in some cases the time consumed by the same program (compiled with the same flags) is significant different. This is due to the operating system, that is composed by a quite large number of internal processes which are executed in a non-controllable way (by the user). This behaviour could provoke misleading conclusions. To deal with this difficulty, we execute each program five times and get the lowest value. We use this value (instead of the mean, median, or other statistical rate) since it is the most accurate representation of the execution time of the program without any interruption of any external software. However, this approach makes even harder the search process since the evaluation of a candidate solution is now more expensive.

Neighbourhood. Since we encode the solutions as a bit string, we can apply traditional variation operators. In concrete, we use a variant of bit-flip mutation. We define that X' is in the k-neighbourhood of X if the hamming distance (number of different elements between the bit string of X and the one of X') is exactly k. Then, our VNS starts changing only one bit in the solution ($k = 1$) to get a neighbour. When the convergence is detected (we explore 5 neighbours and all these candidate solutions are worse than the current one, in our experiments), it changes the neighbourhood and varies at most two bits ($k = 2$). This process continues until the $N/2$-neighbourhood is reached and in that moment it backs to the first neighbourhood. VNS also backs to the first neighbourhood when a new best solution is found.

4 Experimental Design

This section is mainly devoted to describe the benchmark problems used to test our algorithm but before describing them, some comments about the setting of our approaches and the testing platform.

Our sequential proposed algorithm has only a single parameter: the stop condition. Since the fitness function is a very time-consuming task, we use a maximum number of evaluation (4000 in our experiments) as stopping criterion. It is a quite low value but most of the runs have already converged before reaching this number of fitness evaluations. In parallel techniques, we use 8 processes which cooperate (in multi-start approach) every 25 evaluations using an unidirectional ring topology. Finally, we have performed 30 independent runs of each experiment to gather statistical information and then apply Kruskal-Wallis test to validate if the results are statistically different. We use these values to decide in the results' tables when the values are similar or not (the actual values are not shown due to the lack of space).

In this study, we propose three VNS variants: the sequential version, **VNS**$_{Seq}$, and two parallel ones; one following the farmer/worker scheme, **VNS**$_{F-W}$, and a cooperative multi-start approach, **VNS**$_{CMS}$. We compare them against four static and commonly used compiler configurations:

- **None**: No flags are activated.
- **All**: All the considered flags in this study are activated.
- **O2**: It uses the O2 option to compile the program.
- **O3**: It uses the O3 option to compile the program.

In tables reporting results (next section), we will present how the execution time is reduced (percentage) with respect to the most basic configuration (**None** one). This value is calculated using the next equation:

$$reduction(Algorithm) = \frac{t_{None} - t_{Algorithm}}{t_{None}} \times 100 \quad (2)$$

A larger value indicates a higher reduction in the execution time and it is better. The hardware platform is composed by 8 machines with Intel i-core7 processor at 2.6 GHz with 8 GB of RAM memory. The operating system is Ubuntu/Linux 14.04.5 LTS and the GCC version is 4.8.4.

In order to obtain a comparable, representative and extensive set of programs we have explored The Computer Language Benchmarks Game (CLBG) [12] which has been used in previous studies in the literature such as [13]. The CLBG has gathered solutions for 13 benchmark problems (Table 1). An interesting feature of this set of programs is most of them (with the exception of `meteor-contest`) can be easily tunable with a parameter or file in order to generate faster or slower executions. Also, this benchmark includes programs with different characteristics; in concrete, five of the programs are also using some kind of parallelism (`k-nucleotide`, `spectral-norm`, `fasta`, `rev-complement`, and `thread-ring`).

5 Experimental Analysis

In this section, we will show the results obtained in our experiments. We have performed two main experiments and analyses: in our first analysis (Sect. 5.1), we study the numerical performance of the techniques, while in our second study (Sect. 5.2), we analyse the solutions obtained by our proposed tool, and discuss about the optimization flags activated or deactivated according to the problem tested.

5.1 Quality Analysis

Our main goal in this section is to study the quality of the solutions provided by our tool. In the Table 2, we show the reduction obtained by each of the configurations calculated according to the Eq. 2 with respect to the **None** version. Several conclusion can be obtained from that table.

Table 2. Average time reduction (percentage) with respect to the **None** configuration. Boldfaced values represent the best values with statistical significance.

Benchmark	All	O2	O3	VNS_{Seq}	VNS_{F-W}	VNS_{CMS}
`binary-trees`	6.48	12.41	16.77	20.66	19.13	**23.78**
`chameneos-redux`	25.58	27.94	33.45	45.43	50.31	**53.41**
`fannkuch-redux`	25.42	50.21	54.69	**73.51**	**72.97**	**73.38**
`fasta`	3.86	35.85	70.06	74.65	**76.32**	**77.2**
`k-nucleotide`	28.69	40.37	45.00	55.99	50.12	**58.39**
`mandelbrot`	11.09	37.41	48.12	50.86	**54.18**	**54.78**
`meteor-contest`	13.76	25.61	45.96	50.58	50.84	**52.34**
`n-body`	29.95	74.20	80.41	**83.30**	**82.57**	**82.40**
`pidigits`	26.57	51.47	59.25	66.58	71.53	**77.31**
`regex-redux`	30.72	43.83	45.48	52.42	**55.31**	**56.85**
`rev-complement`	12.20	18.42	44.26	50.80	53.51	**67.37**
`spectral-norm`	3.12	10.83	14.34	42.56	**50.74**	**51.34**
`thread-ring`	3.64	4.87	6.51	**12.54**	9.34	10.75
Average	12.77	26.93	36.39	47.25	47.42	50.99

First, the utilization of the optimization flags allows to reduce the final execution time in all the benchmark problems tested in this work. In fact, this reduction is very important in most of the cases and we can observe a reduction around 50% (or even larger) for 11 out 13 cases. Only in a specific benchmark problem, `thread-ring`, the benefit of the optimization flags is minor (between 3.6% and 12.5%). This is due to this application creates many threads and switches constantly among them. Then, the influence of the code generated by the compiler is smaller since most of the time is spent changing among thread without executing real calculations. This confirms the intuitive idea that the efficacy of the compiler depends on the amount of non-calculation operations in the code (switching context or IO operations).

Second, as we stated in Sect. 3, the O3 configuration obtains different results than the configuration which activate all the flags and this difference is quite important in most of the scenarios. These figures means that the O3 adds some additional flags which are not mentioned in the documentation.

Third, our approaches outperform in all the instances to the classical O3 configuration. The difference varies from 4–5% (for `binary-trees` or `n-body`) to 41.5% (for `fannkuch-redux`). The average reduction provided by VNS with respect to the O3 option is 18%, a significant reduction. This gives us some hints that the blindly utilization of O3 is not the most adequate configuration in general and some additional studies are need to select the most beneficial flags for each program. Even in the cases in which the reduction is small, these gains can be very important for critical systems.

Analyzing our proposals, we observe that the $\mathbf{VNS}_{CMS}$ version outperforms (or it is equal to) the other two version in 12 out of 13 test problems. This is due to the parallel versions maintains a better diversity, and while the serial version is stuck in a local optimum, the parallel versions is able to get out of it and to continue the search for better solutions. However, this slows that the convergence (see time with one processor in Table 3) provoking that parallel versions consume their computational badge without improving the sequential solution or even providing a worse one when the problem requires a high number of evaluations to get the local optimum (like in `thread-ring`).

Finally, in Table 3, we show the average execution time required by our approaches. We show the execution time in a single processor (with 8 threads) aganist a real parallel configuration (with 8 machines) in order to calculate the *speedup*. Comparing the versions on a single processor, we notice that the parallel versions spend more time that serial one. As we said, parallel version have slower convergence but it allows to get better solutions. When the parallel versions are run in an actual parallel platform, the execution time is reduced in a very significant way, with a very good speedup. The $\mathbf{VNS}_{F-W}$ usually is slower than $\mathbf{VNS}_{CMS}$ since they require more synchronization (it waits to the finalization of all the workers before moving to the next iteration).

5.2 Analysis of the Selected Flags

Now, we study the flags selected by our proposals. The Fig. 1 shows how many times is selected each flags in the best found solution of each independent run (as percentage) in some representative problems. Since the number of flags is

Table 3. Average execution time (in sec) of our approaches and the speedup

Benchmark	VNS_{Seq}	VNS_{F-W}			VNS_{CMS}		
		1 proc.	8 proc.	Speedup	1 proc.	8 proc.	Speedup
`binary-trees`	2735.20	3397.12	463.45	7.33	3042.64	435.28	6.99
`chameneos-redux`	2142.90	2800.34	386.25	7.25	2494.12	346.41	7.2
`fannkuch-redux`	3175.00	4011.93	583.98	6.87	3597.28	510.25	7.05
`fasta`	1995.40	2650.69	372.29	7.12	2404.66	328.95	7.31
`k-nucleotide`	4293.60	5147.17	698.39	7.37	4643.53	639.6	7.26
`mandelbrot`	2269.30	2892	429.72	6.73	2711.36	381.34	7.11
`meteor-contest`	2974.00	3854.3	559.4	6.89	3583.97	506.21	7.08
`n-body`	3006.90	3734.57	528.23	7.07	3592.64	526.78	6.82
`pidigits`	1189.60	1605.96	224.92	7.14	1507.1	207.3	7.27
`regex-redux`	2323.80	3086.94	410.5	7.52	2872.22	402.27	7.14
`rev-complement`	2153.00	2883.3	402.7	7.16	2727.64	380.96	7.16
`spectral-norm`	2018.50	2768.57	394.38	7.02	2494.87	351.89	7.09
`thread-ring`	4550.40	5061.86	729.37	6.94	4733.78	660.22	7.17
`Average`	2533.18	3240.36	456.05	7.11	2983.49	418.69	7.13

too high to be shown in the figures, only the extremes are presented (the first five and the last five considering they are sorted by percentage).

First, we can notice that depending on the problem, a different set of flags is selected and these flags are chosen according to the characteristic of the problem. For example, we can observe that in `fannkuch-redux` (Fig. 1b), the flag `-finline` is always active. This is logical selection since the source code of this program has five inline functions and they are used constantly.

Second, we can notice that in the programs in which we obtained large runtime reductions, `fannkuch-redux`, `fasta`, and `n-body` (see Table 2), there are a clear distinction between the appropriate flags (with percentages close to 100%) and the non-beneficial flags (around 20% or even lower). This can also help to explain why in some test cases the **All** configuration offers a very low performance. This poor performance is due to it activates these non-beneficial flags (which can even harm the final runtime in specific programs).

Also, we can observe in some applications (like `thread-ring`, Fig. 1f) that the distinction of the appropriate flags is not so clear (percentages lower than 80%). This is an expected result since, as we said before, for this problem the optimization flags has only a minor effect in the final runtime.

Finally, in the last figure, we aggregate the values for all the functions. We can notice that there is not a clear set of flags which should be chosen always (the difference between the most selected one and the last one is lower than 30%) since, as we said above, it depends on the features of the problem. However, we can extract some information about our benchmark analysing this figure.

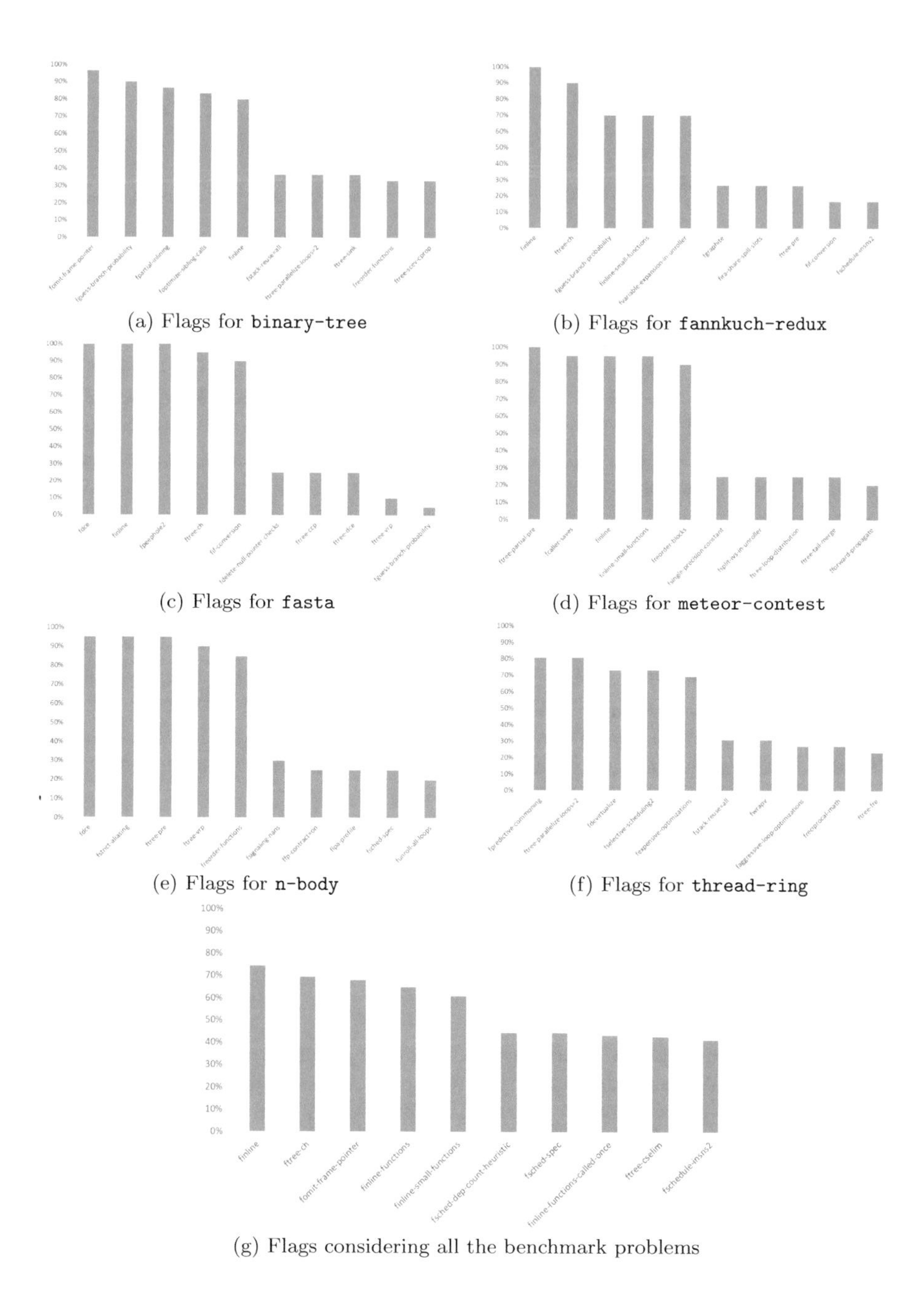

Fig. 1. Flags selected by our approach for different problems.

For example, since the `-finline` is the most selected flag, we can deduce that our programs use this kind of inline functions frequently. Examining the source codes, we can observe than 9 out 13 programs include some inline functions confirming our deduction from the figure.

6 Conclusions

In this work, we have analysed how the optimization flags can affect to the final performance of a software system. Here, we have modelled the compilation process as a combinatorial problem and we have proposed three techniques to automatically decide which optimizations flags should be activated.

Our results show that world spread levels (like O3 or O2) produce very accurate results but they are not the optimal ones, and the flags chosen can be refined to get better performance. Our proposed techniques are able to outperform the results of O3 in all the tested programs (a wide set of 13 different software pieces). The parallelism (specially, using the multi-start approach) does not only allow to reduce the execution time but it also improves the quality of the solutions. We have also observed that the selection of the flags can be performed with the programs executed in small scenarios and later, you can use that configuration to other scenarios (with larger inputs).

There are several open research lines in this domain. First, besides to flags activated by O3, modern compilers offer a wider set of flags and parametric options. Second, other compilers like LLVM or Microsoft Visual Compiler are also very popular nowadays, and we are interesting in performing a similar analysis for them. Finally, we also plan to analyse other metaheuristics in order to perform a more efficient search.

Acknowledgement. This research has been partially funded by the Spanish MINECO and FEDER projects (TIN2014-57341-R (http://moveon.lcc.uma.es), TIN2016-81766-REDT (http://cirti.es), and TIN2017-88213-R (http://6city.lcc.uma.es).

References

1. Software engineering software product quality requirements and evaluation (SQuaRE) Software product quality and system quality in use models. Standard, International Organization for Standardization, Geneva, CH (2011)
2. Hassan, M.M., Afzal, W., Lindström, B., Shah, S.M.A., Andler, S.F., Blom, M.: Testability and software performance: a systematic mapping study. In: Proceedings of the 31st Annual ACM Symposium on Applied Computing, pp. 1566–1569. ACM (2016)
3. Stallman, R.M.: GCC DeveloperCommunity: Using the GNU Compiler Collection: A GNU Manual for GCC Version 4.3. 3. CreateSpace, Paramount (2009)
4. Nobre, R., Reis, L., Cardoso, J.: Compiler phase ordering as an orthogonal approach for reducing energy consumption. In: Proceedings of the 19th Workshop on Compilers for Parallel Computing (CPC16) (2016)

5. Machado, R.S., Almeida, R.B., Jardim, A.D., Pernas, A.M., Yamin, A.C., Cavalheiro, G.G.H.: Comparing erformance of C compilers optimizations on different multicore architectures. In: Computer Architecture and High Performance Computing Workshops (SBAC-PADW), pp. 25–30. IEEE (2017)
6. Hoste, K., Eeckhout, L.: Cole: Compiler optimization level exploration. In: Proceedings of the 6th Annual IEEE/ACM International Symposium on Code Generation and Optimization, CGO 2008, pp. 165–174. ACM, New York (2008)
7. Zhong, S., Shen, Y., Hao, F.: Tuning compiler optimization options via simulated annealing. In: Second International Conference on Future Information Technology and Management Engineering, FITME 2009, pp. 305–308. IEEE (2009)
8. Kumar, T.S., Sakthivel, S., Kumar, S.: Optimizing code by selecting compiler flags using parallel GA on multicore CPUs. Int. J. Eng. Technol. **6**, 544–551 (2014)
9. Mladenovic, N., Hansen, P.: Variable neighborhood search. Comput. Oper. Res. **24**(11), 1097–1100 (1997)
10. Alba, E.: Parallel Metaheuristics: A New Class of Algorithms. Wiley, New York (2005)
11. Crainic, T.G., Toulouse, M.: Parallel meta-heuristics. In: Handbook of Metaheuristics, pp. 497–541. Springer, Heidelberg (2010)
12. Fulgham, B., Gouy, I.: The computer language benchmarks game. http://shootout.alioth.debian.org (2012)
13. Pereira, R., Couto, M., Ribeiro, F., Rua, R., Cunha, J., Fernandes, J.P., Saraiva, J.: Energy efficiency across programming languages: how do energy, time, and memory relate? In: Proceedings of the 10th ACM SIGPLAN International Conference on Software Language Engineering, SLE 2017, pp. 256–267. ACM, New York (2017)

Drift Detection over Non-stationary Data Streams Using Evolving Spiking Neural Networks

Jesus L. Lobo[1(✉)], Javier Del Ser[1,2,3], Ibai Laña[1], Miren Nekane Bilbao[2], and Nikola Kasabov[4]

[1] TECNALIA, 48160 Derio, Spain
{jesus.lopez,javier.delser,ibai.lana}@tecnalia.com
[2] University of the Basque Country UPV/EHU, 48013 Bilbao, Spain
{javier.delser,nekane.bilbao}@ehu.eus
[3] Basque Center for Applied Mathematics (BCAM), 48009 Bilbao, Spain
[4] KEDRI - Auckland University of Technology (AUT), 1010 Auckland, New Zealand
nkasabov@aut.ac.nz

Abstract. Drift detection in changing environments is a key factor for those active adaptive methods which require trigger mechanisms for drift adaptation. Most approaches are relied on a base learner that provides accuracies or error rates to be analyzed by an algorithm. In this work we propose the use of evolving spiking neural networks as a new form of drift detection, which resorts to the own architectural changes of this particular class of models to estimate the drift location without requiring any external base learner. By virtue of its inherent simplicity and lower computational cost, this embedded approach can be suitable for its adoption in online learning scenarios with severe resource constraints. Experiments with synthetic datasets show that the proposed technique is very competitive when compared to other drift detection techniques.

Keywords: Online learning · Concept drift · Spiking neural networks

1 Introduction

Data stream scenarios are characterized by huge amounts of data flowing fast and continuously, often without any bound in their duration [1,2]. These computing constraints impose a set of restrictions [3] on memory usage and runtime processing that should be considered when designing models for data stream classification. In some particular cases dealing with these scenarios, classifiers only receive one instance at a time (online learning), which yields at even more complex design challenges. Furthermore, data distribution may vary over time as a result of non-stationary phenomena (concept drift [4]), which imprints changes over the received data samples that should be also accounted at design time. Under these conditions, drift detection algorithms become relevant for *active* model design approaches, i.e. those where the learning algorithm updates the

J. Del Ser et al. (Eds.): IDC 2018, SCI 798, pp. 82–94, 2018.
https://doi.org/10.1007/978-3-319-99626-4_8

knowledge captured by the model depending on the output of the drift detection. Contrarily, in *passive* schemes the concepts learned by the model is updated continuously so as to prevent their concepts from becoming obsolete after a drift occurs [5].

Several model approaches have been presented in the literature to deal with concept drift [6,7]. Some of them base their strategy on adapting the internal structure of their classifiers [8,9], while others utilize an ensemble of models [10] to which techniques to induce diversity are applied [11,12]. When focused on drift detection approaches, they can be used as a module inside other classifiers to adapt either the classifiers internal structure or the number of classifiers within the ensemble. All of them usually use a specific classifier called *base learner*, and analyze its performance score (in general, accuracy or error rate) to indicate whether a drift has occurred. This base learner is trained on the current instance, an incremental process repeated for each incoming sample from the stream.

This work proposes a novel drift detection method coined as `eSNN-DD` (Evolving Spiking Neural Network for Drift Detection), which hinges on the so-called family of Spiking Neural Networks (SNNs, [13]). Such learning models leverage the representation of information as temporal spikes, based on which a number of spike-time association are learned to that capture temporal associations between a large number of variables in streaming data. One of the most promising flavors of SNNs is the Evolving Spiking Neural Network (eSNN, [14,15]), where the number of spiking neurons evolves incrementally over time to unveil temporal patterns from data. Our `eSNN-DD` embraces eSNNs to derive a drift detection mechanism inspired by how spiking neurons evolve along time, which is an inner architectural part of the eSNN learning method. Indeed, as opposed to the majority of drift detection mechanisms, `eSNN-DD` does not require any additional *base learner* to analyze its performance metrics towards indicating when a drift has occurred. Similarly to other drift detection methods, `eSNN-DD` is independent of the learning/adapting algorithm and can therefore be seamlessly combined with any algorithmic choice in this regard. We assess the performance of the proposed drift detection method by comparing it to that of a number of consolidated methods from the state of the art, in terms of true positives, false alarms (false positives), missed detection (false negatives) and distance to the drift point. As will be shown by the results of our experiments with synthetic datasets, `eSNN-DD` performs competitively as a drift detector yet it does not need to compare distributions or monitor performance statistics to detect the drift, thus making it easier and computationally lighter to detect drifts.

The rest of this article is organized as follows: Sect. 2 surveys related work, emphasizing on current drift detectors and eSNNs. Section 3 provides an general introduction to eSNNs, whereas Sect. 4 details the proposed approach. Section 5 presents the experimental setup designed to assess `eSNN-DD` performance. Section 6 presents and discusses the experimental results, and finally Sect. 7 concludes the paper and outlines future research paths.

2 Background

Drift detection mechanisms quantitatively characterize concept drift events by identifying change points or small-sized periods of time (windows) during which these changes may occur. In [16] detection mechanisms for online learning are categorized in three different categories. To begin with, sequential analysis methods are based on the Sequential Probability Ratio Test [17], with the Cumulative Sum (CUSUM) and Page-Hinkley Test (PHT) [18] as their most representative examples. On the other hand, statistical process control is a technique that monitors and controls the evolution of the learning process, advancing on similar ideas to the seminal Exponentially Weighted Moving Average (EWMA) approach [19]. Finally, monitoring the distribution of data on two different windows has lately emerged as one of the most popular drift detection approaches: such a comparison is carried out by using statistical tests to assess the null hypothesis that the distributions in both are equivalent in terms of a certain statistic (e.g. equal medians). If the null hypothesis is rejected, a drift is assumed to be occurring at the start of the detection window. The ADaptive sliding WINdow (ADWIN) [20] is one of the most relevant examples of this last category. As we will show in Sect. 4, our `eSNN-DD` detector does not belong to any of the previous categories because it is based on structural changes stemming from the model itself (particularly, in the output layer of the eSNN model), thereby not relying on any external *base learner* to analyze its metrics over time.

Concept drift may be categorized in terms of speed, severity and reason of the change [6,21]. In the case of speed, an *abrupt* drift may occur when a change happens suddenly between two classification contexts, whereas a *gradual* drift represents the case when dealing with a smooth transition between two concepts. Severity can be regarded as the amount of changes that a new concept causes; therefore, a measure of severity can be computed as the percentage of the input space whose target class has changed after the drift [8]. Finally, the reason of change can be divided into real or virtual drift [4]. In the experiments of this work we will test our approach under different drift speed and severity levels.

This being said, the literature has been certainly rich in what refers to comparative studies of wide range of detectors in datasets with different drift characteristics and by using diverse performance metrics. The benchmark in [6] compares some of the most utilized detectors in the literature, such as Drift Detection Method (DDM) [22], Early Drift Detection Method (EDDM) [23], PHT [18], ADWIN [20], Paired Learners (PL) [24], Concept Drift Detection (ECDD) [19], Degree of Drift (DOF) [25] and Statistical Test of Equal Proportions (STEPD) [26]. In [27] a new approach called Reactive Drift Detection Method (RDDM) is proposed and compared to DDM, EDDM and STEPD in several synthetic and real streaming scenarios. This manuscript will undertake a similar study by comparing available drift detectors to the proposed `eSNN-DD` technique.

Finally, the momentum of eSNNs has recently increased due to their renowned success in modeling the behavior and learning process of the human brain towards understanding how it develops learning tasks in practice. As a result, these models have been so far applied to several supervised learning problems in online learning environments [28,29]. Recently, in [30] a new method called SpikeTemp was proposed, which consists of a SNN with adaptive structure that can achieve better classification performance and is much faster than existing rank-order-based learning approaches. However, to the best of our knowledge this paper represents the first work dealing with the exploitation of the evolving learning procedure of eSNNs to detect drifts. This work presents the `eSNN-DD` drift detection, inspired by spiking neurons evolution, which does not require a *base learner* to analyze its performance metrics to detect the change.

3 Evolving Spiking Neural Networks

The learning procedure of eSNNs is composed by two clearly differentiated parts: (1) an encoding mechanism, which transforms real values into spikes generated over time; and (2) a neural model that connect pre-synaptic neurons to a set of output neurons through weighted neural connections. We next inspect in more detail these parts for the reader to grasp a better understanding of the basis of the drift detection mechanism. For the sake of space only the fundamentals are given; we recommend [13,15] for deeper insights on these procedures.

Architecturally an eSNN builds upon three different processing layers. The first one corresponds to the input data samples that the stream feeds to the model in a continuous fashion. The second layer encodes the input data by transforming the real values of each feature into train spikes, resulting in pre-synaptic neurons. The third layer is a repository of evolving output neurons, which represents the input data and evolve every time a new input sample arrives. In this last layer, each neuron in the repository is connected with all pre-synaptic neurons through weighted connections, whose values are learned from input data. Figure 1 depicts the eSNN architecture.

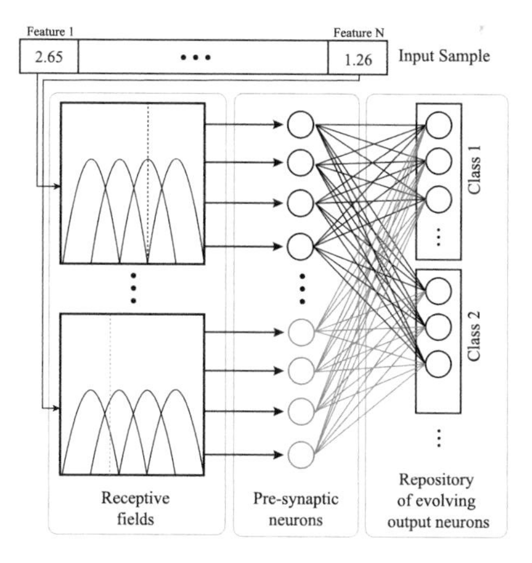

Fig. 1. Scheme of an eSNN [31]

In the encoding process each sample (namely, a vector of real-valued elements) is mapped to a sequence of spikes using, in the case of eSNNs, the so-called rank order population technique [32,33]. This neural encoding method uses Gaussian Receptive Fields (GRF) for encoding continuous values by using a collection of neurons with overlapping sensitivity profile. Each sample is encoded independently by a number N of one-dimensional GRFs, whose value impacts on the amplitude of the input neuron and must be optimized for every tackled problem. The GRF of a neuron is given by its center and width [15]. The $\beta \in \mathbb{R}[1,2]$ parameter (i.e. overlap factor) establishes the width of GRFs, and thereby their amount of overlapping and the firing time of the pre-synaptic neuron.

The neural model of the eSNNs is based on the simplified Leaky Integrate-and-Fire (LIF) model which was presented in [34]. The LIF model poses that the spike response of a neuron depends only on the arrival time of pre-synaptic spikes, that is, spikes which are fired earlier have a stronger impact on the post-synaptic potential than spikes which are fired later. In this model, each neuron fires at most once during the time coding interval T, only when its Post-Synaptic Potential (PSP) exceeds its threshold value. The model dynamics are given by:

$$PSP_i = \begin{cases} \sum_j w_{ji} \cdot mod^{order(j)}, & \text{if not fired,} \\ 0, & \text{otherwise,} \end{cases} \quad (1)$$

where $w_{j,i}$ is the weight of the connection between pre-synaptic neuron j weight and output neuron i; $mod \in \mathbb{R}[0,1]$ is the so-called *modulation factor*; and function $order(j)$ defines the rank of the spike emitted by neuron j.

When used to solve a supervised learning problem, the eSNN learning method generates and updates a repository of output neurons, in a procedure that is well described in [15] and next summarized: for each incoming sample, a new output neuron is created and added to the repository (or, instead merged with a similar output neuron). Then it is fully connected to the previous layer of pre-synaptic (input) neurons through weights $\boldsymbol{w}_j = \{w_{ji}\}_{\forall j}$ (see Fig. 1). The value of $w_{j,i}$ is computed according to the spike order through synapse j as $w_{j,i} = mod^{order(j)}$, where j is the pre-synaptic neuron of output neuron i. A threshold γ_i is considered for i as the fraction $C \in \mathbb{R}(0,1)$ of its maximum post-synaptic potential $PSP_{max,i}$, which is computed as $\gamma_i = C \cdot PSP_{max,i}$. After that, the weight vector of output neuron i is compared with all output neurons in the repository. If the Euclidean distance between the weight vector of output neuron i and that of any of the output neurons is lower than the value of parameter *SIM*, both neurons are declared to be similar to each other, and consequently their thresholds and weight vectors are merged together. Otherwise, the newly produced neuron i is added to the repository.

The testing phase is accomplished by propagating the spike trains of the test sample to all output neurons. After this process, a group of fired neurons is considered (those whose threshold γ_i has been reached): the class label for the test sample is decided according to the class label of their k-nearest neighbors.

4 Proposed Approach

The proposed eSNN-DD method is based on the number of merges that occur in a sliding window **W** with size $|\mathcal{W}| = W$. While the data distribution remains stable, it is more likely that any of the output neurons in the repository is more similar to the incoming one, and thus there will be more merging processes. In contrast, when a drift occurs the data distribution changes and the resulting neurons start to be different from those in the output neuron repository, correspondingly decreasing the number of merging processes. With this intuition in mind, the proposed detection mechanism considers that a change has taken place when there has not been any merging process W time ticks after the last fusion of neurons in repository $\mathcal{R}$.

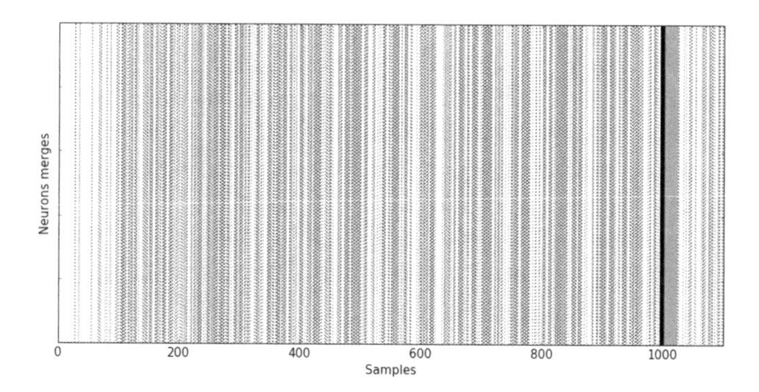

Fig. 2. Time instants at which output neurons are merged for the CIRCLE synthetic dataset with high severity and high speed. Before drift occurs at instant 1000, merging processes are held recurrently, whereas right after the drift (area shaded in gray) merges do not occur until the new concept is captured by the eSNN learning algorithm.

Figure 2 depicts the differences of the merging processes before and after the drift (sample 1000) for CIRCLE, one of the synthetic datasets compounding the landmark repository contributed in [35] for concept drift. This dataset contains a single drift starting at sample 1000. During the first 100 samples the input data distribution is frequently new for the eSNN and the merging process is carried out only occasionally as a result of the transient to the stable period. From the first 100 samples onwards, and until drift occurs, the merging process is more usual because the input data distribution remains stable. When the drift occurs, the input data distribution does not resemble the previous concept any longer, and the merging process is not triggered again until the eSNN captures the new distribution (concept). This new knowledge is acquired by storing output neurons that represent the new data distribution.

It should be noted that this detection mechanism overrides any need for further processing elements in the model pipeline, as it exploits changes in the structural change of the eSNN in response to newly arriving concepts. Algorithm 1 describes the drift detection mechanism along with the learning procedure of eSNNs. If at least one neuron has been merged in the previous W time instants or the repository size limit $|\mathcal{R}|_{max}$ has not been reached, no drift is declared. In case of a full repository and when a neuron must be merged, the oldest output neuron is discarded yielding one free slot for the newly produced output neuron. When no neuron has been merged over W time instants, then a drift is declared.

Algorithm 1. Proposed `eSNN-DD` drift detection algorithm

```
1  Set an empty repository: R = ∅
2  Set mod ∈ ℝ[0,1], C ∈ ℝ[0,1], SIM ∈ ℝ[0,1]
3  Set eSNN encoding parameters β, T, N and
4  Set sliding window size W
5  Set neuron repository maximum size |R|_max
6  Set number of merging operations as #_merges = W
7  foreach input sample x_i from the stream do
8      Update 𝒲 with x_i by discarding the oldest sample in the window
9      Calculate C_j and w_{j,i} with the samples in 𝒲 to encode x into firing
       time of multiple pre-synaptic neurons j
10     Create a new output neuron i and the connection weight vectors
11     Calculate PSP_{max,i} and γ_i = C · PSP_{max,i}
12     if min_{r∈R} EuclideanDistance(i, r) ≤ SIM then
13         Update weight vector & threshold of most similar neuron in R
14         #_merges = #_merges + 1
15     else
16         if |R| < |R|_max then
17             Add the new neuron to the repository: R = R ∪ {i}
18         else
19             Remove the oldest neuron in R
20             Insert the new neuron: R = R ∪ {i}
21         end
22         #_merges = #_merges + 0 (no merging performed)
23     end
24     if #_merges = 0 then
25         Issue a drift alarm
26         Set #_merges = W
27     end
28 end
```

Some of the eSNN parameters impact significantly on the performance of `eSNN-DD`. The size W of $\mathcal{W}$ is a key point due to the fact that the method declares a drift when there are no merges during W consecutive time instants. In the case of the repository size $|\mathcal{R}|$, the higher this parameter is, the more likely a similar neuron will be found, thus the number of merges is also affected by this parameter. Finally, *SIM* defines the threshold under which two neurons are regarded as similar and by which the merging process is triggered, hence it drives directly the detection behavior of eSNN-DD.

5 Computer Experiments

A set of experiments has been designed to analyze the behavior of eSNN-DD over several synthetic datasets. To this end, we will first find suitable values for the eSNN parameters, and then we will discuss on its detection performance in comparison with drift detectors from the literature. Before proceeding forward details on the considered datasets and drift detectors are next given.

When dealing with real data it is not possible to determine exactly when a drift occurs, its type, or if it is actually a drift. Therefore, it is not possible to carry out an analysis of the behavior of a new drift detector by focusing on real-world datasets. For this rationale we have chosen 4 synthetic datasets described in [35] and labeled as CIRCLE, LINE, SINEH and SINEV. These datasets represent different types of drift and they are very popular as concept drift benchmarks.

These datasets include one drift simulated by varying among low and high severity levels (defined as the percentage of the input space which has its label changed after the drift is completed), and among low and high speeds (correspondingly, the inverse of the time taken for a new concept to completely replace the previous one). This results in $4 \times 4 = 16$ different dataset, each containing 2000 samples and 2 normalized features $\{X_1, X_2\}$ with a binary label. Drift occurs at time instant $t = 1000$ in all of them, whereas the drifting period is 50 (high speed) and 500 (low speed).

Drift detector methods for the benchmark were selected based on their impact on the field (e.g. high number of citations), and their thorough description in the literature, which eased their implementation and minimized misunderstanding. All of them resort to a Naive Bayes classifier as their *base learner* [6,20] due to the efficiency and simplicity of this class of models. As such, the first considered method is DDM [22], which departs from the intuitive fact that, when the concept changes, the base learner will incorrectly predict the incoming samples that are shaped on a different data distribution. Therefore, when the error rate increases DDM asumes that a new concept emerges from the stream. EDDM [23] is partly based on DDM, but differs in the monitored performance metric (i.e. instead of error rate, DDM uses the so-called distance-error-rate). ADWIN [20] operates with two sub-windows of a variable-sized window, and it makes a comparison to find differences between them; in such a case it will consider that a drift has occurred. Finally, drift detection methods based on the Hoeffding's bound (HDDM, [36]) monitor the mean of the base learner's performance over time, but without assuming any distribution for the incoming data. We will refer as HDDM_A to the HDDM approach that uses a window and a moving average as its estimator. Likewise, HDDM_W does not use a window and uses EWMA as its estimator, giving a higher weight to the most recent instances.

When it comes to the evaluation of the obtained results, drift detection metrics vary considerably. Nevertheless, we have embraced the recommendations provided in [6]; therefore, comparisons are carried out in terms of false positives (FP), true positives (TP), false negatives (FN), and distance to the drift (DT). FP indicate the number of drifts identified at time instants when no drift was present in the dataset whatsoever: therefore, lower values mean that the method does not detect false drift events. Likewise, a high TP value indicates that the method has identified true drifts during a specific period of time, which in this research is given by the drifting time of every dataset. FN indicates the number of missed drift detections, which can easily reach 100% because the methods are based on one instance training (test-then-train). For this reason TP are computed during the drifting time. Finally, DT indicates the average distance to the true drift whenever TP > 1. The accuracy metric is not analyzed because in our experiments `eSNN-DD` is not hybridized with any *base learner* so as to focus exclusive on its performance as a drift detector. Only if hybridized with an adaptation method a comparison analysis on accuracy scores could be undertaken. Due to lack of space this analysis will be left out from the scope of this paper.

In order to find suitable values for the parameters related to `DD-eSNN` encoding (T, N and β) and learning (mod and C), a set of offline experiments was designed to empirically find the configuration best balancing between all detection performance metrics, which resulted to be $T = 20$, $N = 10$, $\beta = 1.5$, $mod = 0.85$ and $C = 0.95$. For those parameters with high impact on the detection (W, $|\mathcal{R}|$ and SIM), different configurations were tested for each of the 16 datasets: $W \in \{20, 25, 30\}$, $|\mathcal{R}| = \{20, 25, 30\}$ and $SIM = \{0.20, 0.25, 0.30\}$. Comparing their results in terms of TP, FP, FN, and DT, the more suitable configuration was $W = 20$, $|\mathcal{R}| = 30$ and $SIM = 0.25$. For the rest of drift detectors, their configuration was retrieved from the literature using each technique. In the case of `DDM`, the minimum number of instances before the detection of a drift is permitted ($min\_samples$) was set to 20, 25 and 30 [22]. As for `EDDM`, after occurring 30 classification errors this detector uses thresholds $\alpha = 0.95$ and $\beta = 0.9$ to detect a concept drift, and as in `DDM` $min\_samples$ equals 30 as suggested in [23]. Regarding `ADWIN`, the confidence level is $\delta = 0.1$ and the minimum frequency of instances needed for the window size to be reduced is set to $f = 16$ [6]. For `HDDM_W` the confidence interval for the drift detection is $cid = 0.001$, the confidence interval for the warning is $cid = 0.001$, and the parameter controlling the weight given to the most recent data with respect to older data is $\lambda = 0.05$. Finally, for `HDDM_A`, $cid = 0.001$, $cid = 0.001$, and the window sizes are set to 5, 20, and 50, which is included in the label as `HDDM_A`-ω (with $\omega \in \{5, 20, 50\}$).

6 Results and Discussion

A first look at the obtained statistics presented in Table 1 reveals that `eSNN-DD` is very competitive in comparison with the rest of drift detectors. Indeed, `eSNN-DD` shows good results in terms of detection statistics for both abrupt (datasets with high speed drift dynamics) and gradual drift dynamics (correspondingly, those with low values of drift speed). Specifically, our method never suffers from a non-detection (ND) and FN, a fact that evinces its robustness to missed

Table 1. Drift detection statistics for the synthetic datasets. Those statistics in which eSNN-DD shows better results are in bold, and in italics when the results are just competitive. *ND* refers to a simulation where no drift is detected within the drifting period.

CIRCLE

Methods	Low severity, high speed				Low severity, low speed				High severity, high speed				High severity, low speed			
	DT	FP	TP	FN	DT	FP	TP	FN	DT	FP	TP	FN	DT	FP	TP	FN
ADWIN	20 ± 13.17	26	11	0	273.89 ± 103.92	19	28	0	48 ± 1.41	27	5	0	220.36 ± 93.50	29	25	0
DDM-20	49 ± 0.0	0	1	0	265 ± 0.0	0	1	0	23 ± 0.0	0	1	0	227 ± 0.0	0	1	0
DDM-25	49 ± 0.0	0	1	0	265 ± 0.0	1	1	0	23 ± 0.0	0	1	0	227 ± 0.0	0	1	0
DDM-30	49 ± 0.0	0	1	0	265 ± 0.0	1	1	0	23 ± 0.0	0	1	0	227 ± 0.0	0	1	0
EDDM	29 ± 0.0	0	1	0	339 ± 116	2	2	0	11 ± 0.0	6	1	0	314.33 ± 138.88	4	3	0
HDDM_A-5	17.66 ± 15.86	38	3	0	275.9 ± 138.82	23	10	0	25 ± 24	36	2	0	272.23 ± 126.56	28	13	0
HDDM_A-20	11 ± 0.0	10	1	0	310.5 ± 127.76	4	4	0	5 ± 0.0	5	1	0	255.8 ± 142.37	5	5	0
HDDM_A-50	9 ± 0.0	2	1	0	271 ± 114	2	2	0	25 ± 22	3	2	0	209 ± 123.28	3	3	0
HDDM_W	*ND*	2	0	1	*ND*	3	0	1	23 ± 8	25	2	0	420.66 ± 55.40	18	6	0
eSNN-DD	**3 ± 0.0**	21	1	0	**191.80 ± 107.35**	19	5	0	*22.5 ± 19.5*	*16*	2	0	**163 ± 184.11**	*14*	4	0

LINE

Methods	Low severity, high speed				Low severity, low speed				High severity, high speed				High severity, low speed			
	DT	FP	TP	FN	DT	FP	TP	FN	DT	FP	TP	FN	DT	FP	TP	FN
ADWIN	20.81 ± 8.36	24	11	0	240.92 ± 152.32	26	14	0	35.66 ± 3.74	36	12	0	260.25 ± 115.33	29	24	0
DDM-20	*ND*	1	0	1	325.5 ± 33.5	0	2	0	21 ± 0.0	0	1	0	156.5 ± 13.5	0	2	0
DDM-25	*ND*	1	0	1	325.5 ± 33.6	0	2	0	21 ± 0.0	0	1	0	164 ± 21	0	2	0
DDM-30	*ND*	1	0	1	325.5 ± 33.6	0	2	0	21 ± 0.0	0	1	0	143 ± 0.0	0	1	0
EDDM	*ND*	3	0	1	206 ± 0.0	1	1	0	9 ± 0.0	0	1	0	258.66 ± 144.98	1	3	0
HDDM_A-5	19 ± 10	32	2	0	216.25 ± 136.34	23	8	0	15 ± 14	42	2	0	235.84 ± 140.90	30	13	0
HDDM_A-20	21 ± 0.0	7	1	0	232 ± 108.07	6	4	0	9 ± 0.0	8	1	0	250.14 ± 119.37	6	7	0
HDDM_A-50	27 ± 0.0	2	1	0	193 ± 0.0	2	1	0	24 ± 19	2	2	0	245.66 ± 156.06	2	3	0
HDDM_W	*ND*	0	0	0	*ND*	0	0	0	39 ± 0.0	17	1	0	439 ± 20	12	2	0
eSNN-DD	27 ± 10	30	2	0	*230.92 ± 134.58*	26	12	0	**9 ± 0.0**	7	1	0	271.14 ± 130.39	7	7	0

SINEH

Methods	Low severity, high speed				Low severity, low speed				High severity, high speed				High severity, low speed			
	DT	FP	TP	FN	DT	FP	TP	FN	DT	FP	TP	FN	DT	FP	TP	FN
ADWIN	*ND*	48	0	1	*ND*	32	0	1	*ND*	49	0	1	438.91 ± 24.18	35	12	0
DDM-20	*ND*	0	0	0	*ND*	0	0	0	*ND*	1	0	1	*ND*	1	0	1
DDM-25	*ND*	0	0	0	*ND*	0	0	0	*ND*	1	0	1	*ND*	1	0	1
DDM-30	*ND*	0	0	0	*ND*	0	0	0	*ND*	1	0	1	*ND*	1	0	1
EDDM	*ND*	0	0	0	*ND*	0	0	0	*ND*	10	0	1	*ND*	3	0	1
HDDM_A-5	33.33 ± 10.20	67	3	0	271.95 ± 133.97	56	20	0	42 ± 6	71	2	0	315.66 ± 131.99	57	15	0
HDDM_A-20	34 ± 0.0	18	1	0	247.5 ± 120.32	12	6	0	*ND*	20	0	1	291.4 ± 113.99	15	5	0
HDDM_A-50	34 ± 0.0	8	1	0	375 ± 0.0	7	1	0	*ND*	9	1	0	245.33 ± 91.22	6	3	0
HDDM_W	7 ± 0.0	32	1	0	190.4 ± 81.18	26	5	0	25.5 ± 19.5	36	2	0	272.84 ± 128.39	28	13	0
eSNN-DD	**7 ± 0.0**	33	1	0	*244.08 ± 119.38*	27	13	0	**7 ± 0.0**	**36**	1	0	**242.86 ± 130.97**	26	14	0

SINEV

Methods	Low severity, high speed				Low severity, low speed				High severity, high speed				High severity, low speed			
	DT	FP	TP	FN	DT	FP	TP	FN	DT	FP	TP	FN	DT	FP	TP	FN
ADWIN	40 ± 4.96	37	3	0	347.93 ± 97.49	23	16	0	31.78 ± 9.11	47	14	0	278.89 ± 142.38	27	29	0
DDM-20	*ND*	2	0	1	495 ± 0.0	1	1	0	19 ± 0.0	0	1	0	174.5 ± 12.5	0	2	0
DDM-25	*ND*	2	0	1	495 ± 0.0	1	1	0	19 ± 0.0	0	1	0	186.5 ± 24.5	0	2	0
DDM-30	*ND*	2	0	1	495 ± 0.0	0	1	0	19 ± 0.0	0	1	0	187 ± 25	0	2	0
EDDM	31 ± 0.0	2	1	0	349 ± 0.0	0	1	0	*ND*	5	0	1	357.33 ± 106.55	4	3	0
HDDM_A-5	19 ± 10	38	2	0	244.3 ± 139.61	26	10	0	*ND*	39	0	1	256.07 ± 130.20	28	13	0
HDDM_A-20	17 ± 0.0	9	1	0	273 ± 82	5	2	0	1 ± 0.0	10	1	0	254.66 ± 140.66	7	6	0
HDDM_A-50	*ND*	1	0	1	359 ± 0.0	0	1	0	22 ± 21	3	2	0	269.33 ± 159.70	3	3	0
HDDM_W	*ND*	0	0	0	*ND*	0	0	0	41 ± 0.0	20	1	0	382.33 ± 84.21	11	6	0
eSNN-DD	**5 ± 0.0**	*28*	1	0	*264.55 ± 158.74*	*22*	11	0	*24.50 ± 11.50*	*11*	2	0	*310.40 ± 91.79*	*5*	5	0

detections. It can be observed that eSNN-DD yields a higher number of false positives than the rest of approaches, but in return it detects the drift earlier, mostly in abrupt datasets. This noted fact suggests the existence of a trade-off between the number of FP and the distance to the drift (detection delay), a particularity that was already revealed in [37,38] for other detectors. Finally, it is worth mentioning that although eSNN-DD operates with a smaller window ($W = 30$) than HDDM_A_50 (respectively, $W = 50$), our proposed detector outperforms this latter counterpart in what regards to the distance to the drift over many datasets, such as CIRCLE, SINEH, LINE (high severity, high speed), SINEV (low severity, high speed) and SINEV (low severity, low speed).

7 Conclusions and Future Work

This work has elaborated on a novel drift detection technique coined as eSNN-DD, which does not rely on the performance of any *base learner* as opposed to other approaches from the literature utilized for drift detection. Instead, our approach exploits the structural changes (neural merges) of a eSNN predictive model to identify the drift point, hence it can be regarded as an *embedded* drift detection approach that stems as a byproduct of the particular learning algorithm of this class of neural networks. Contributions dealing with drift detection often build upon the hybridization of a classifier and a detection criteria. The embedded strategy suggested in this work can be useful to simplify the drift detection process in those applications with stringent computational constraints. Nevertheless, eSNN-DD can also work in collaboration with any other adaptive technique, roughly like most of the other detection approaches. A set of computer experiments with synthetic datasets has shed light on the competitive behavior of eSNN-DD when compared to other well-known detectors.

The extensive offline search for suitable eSNN-DD parameters suggests that in the near future research efforts should be invested towards the use of online optimization techniques to tailor the parametric configuration of the detector while in operation. A first portion of the dataset could be used to find a proper initial configuration, after which parameters could be tuned slightly during the stream mining process. Besides this research path, it would be interesting to find out whether the learning procedure of eSNN models can unveil more characteristics of the drift than its occurrence (e.g. duration, severity). Finally, we deem it necessary to assess the performance of eSNN-DD over more diverse datasets, encompassing not only more complex concepts (number of features and/or number of classes), but also different types of drifts (such as recurring or incremental).

Acknowledgements. This work was supported by the EU project Pacific Atlantic Network for Technical Higher Education and Research - PANTHER (grant number 2013-5659/004-001 EMA2), and by the Basque Government through the EMAITEK program.

References

1. Zhou, Z.H., Chawla, N.V., Jin, Y., Williams, G.J.: Big data opportunities and challenges: discussions from data analytics perspectives. IEEE Comput. Intell. Mag. **9**(4), 62–74 (2014)
2. Alippi, C.: Intelligence for Embedded Systems. Springer, Heidelberg (2014)
3. Domingos, P., Hulten, G.: A general framework for mining massive data streams. J. Comput. Graph. Stat. **12**(4), 945–949 (2003)
4. Ditzler, G., Roveri, M., Alippi, C., Polikar, R.: Learning in nonstationary environments: a survey. IEEE Comput. Intell. Mag. **10**(4), 12–25 (2015)
5. Khamassi, I., Sayed-Mouchaweh, M., Hammami, M., Ghédira, K.: Discussion and review on evolving data streams and concept drift adapting. Evolving Syst. **9**(1), 1–23 (2018)
6. Gonçalves Jr., P.M., de Carvalho Santos, S.G., Barros, R.S., Vieira, D.C.: A comparative study on concept drift detectors. Expert Syst. Appl. **41**(18), 8144–8156 (2014)
7. Demšar, J., Bosnić, Z.: Detecting concept drift in data streams using model explanation. Expert Syst. Appl. **92**, 546–559 (2018)
8. Minku, L.L., Yao, X.: DDD: a new ensemble approach for dealing with concept drift. IEEE Trans. Knowl. Data Eng. **24**(4), 619–633 (2012)
9. Gonçalves Jr., P.M., De Barros, R.S.M.: RCD: a recurring concept drift framework. Pattern Recogn. Lett. **34**(9), 1018–1025 (2013)
10. Dehghan, M., Beigy, H., ZareMoodi, P.: A novel concept drift detection method in data streams using ensemble classifiers. Intell. Data Anal. **20**(6), 1329–1350 (2016)
11. Brzezinski, D., Stefanowski, J.: Ensemble diversity in evolving data streams. In: International Conference on Discovery Science, pp. 229–244. Springer, Heidelberg (2016)
12. Lobo, J.L., Del Ser, J., Bilbao, M.N., Perfecto, C., Salcedo-Sanz, S.: DRED: an evolutionary diversity generation method for concept drift adaptation in online learning environments. Appl. Soft Comput. **68**, 693–709 (2017)
13. Gerstner, W., Kistler, W.M.: Spiking Neuron Models: Single Neurons, Populations, Plasticity. Cambridge University Press, Cambridge (2002)
14. Soltic, S., Kasabov, N.: Knowledge extraction from evolving spiking neural networks with rank order population coding. Int. J. Neural Syst. **20**(06), 437–445 (2010)
15. Schliebs, S., Kasabov, N.: Evolving spiking neural network: a survey. Evolving Syst. **4**(2), 87–98 (2013)
16. Gama, J., Zliobaitė, I., Bifet, A., Pechenizkiy, M., Bouchachia, A.: A survey on concept drift adaptation. ACM Comput. Surv. (CSUR) **46**(4), 44 (2014)
17. Wald, A.: Sequential Analysis. Courier Corporation, New York City (1973)
18. Page, E.S.: Continuous inspection schemes. Biometrika **41**(1/2), 100–115 (1954)
19. Ross, G.J., Adams, N.M., Tasoulis, D.K., Hand, D.J.: Exponentially weighted moving average charts for detecting concept drift. Pattern Recogn. Lett. **33**(2), 191–198 (2012)
20. Bifet, A., Gavalda, R.: Learning from time-changing data with adaptive windowing. In: Proceedings of the 2007 SIAM International Conference on Data Mining, SIAM, pp. 443–448 (2007)
21. Minku, L.L.: Online ensemble learning in the presence of concept drift. Ph.D. thesis, University of Birmingham (2011)

22. Gama, J., Medas, P., Castillo, G., Rodrigues, P.: Learning with drift detection. In: Brazilian symposium on artificial intelligence, pp. 286–295. Springer, Heidelberg (2004)
23. Baena-García, M., del Campo-Ávila, J., Fidalgo, R., Bifet, A., Gavaldà, R., Morales-Bueno, R.: Early drift detection method. In: Fourth International Workshop on Knowledge Discovery from Data Streams (2006)
24. Bach, S.H., Maloof, M.A.: Paired learners for concept drift. In: Eighth IEEE International Conference on Data Mining, ICDM 2008, pp. 23–32. IEEE (2008)
25. Sobhani, P., Beigy, H.: New drift detection method for data streams. In: Adaptive and intelligent systems, pp. 88–97. Springer, Heidelberg (2011)
26. Nishida, K., Yamauchi, K.: Detecting concept drift using statistical testing. In: International Conference on Discovery Science, pp. 264–269. Springer, Heidelberg (2007)
27. Barros, R.S., Cabral, D.R., Gonçalves Jr., P.M., Santos, S.G.: RDDM: reactive drift detection method. Expert Syst. Appl. **90**, 344–355 (2017)
28. Wang, J., Belatreche, A., Maguire, L., Mcginnity, T.M.: An online supervised learning method for spiking neural networks with adaptive structure. Neurocomputing **144**, 526–536 (2014)
29. Wang, J., Belatreche, A., Maguire, L., McGinnity, M.: Online versus offline learning for spiking neural networks: a review and new strategies. In: 2010 IEEE 9th International Conference on Cybernetic Intelligent Systems (CIS), pp. 1–6. IEEE (2010)
30. Wang, J., Belatreche, A., Maguire, L.P., McGinnity, T.M.: SpikeTemp: an enhanced rank-order-based learning approach for spiking neural networks with adaptive structure. IEEE Trans. Neural Netw. Learn. Syst. **28**(1), 30–43 (2017)
31. Kasabov, N.K.: Evolving Connectionist Systems: The Knowledge Engineering Approach. Springer, Heidelberg (2007)
32. Thorpe, S.J., Gautrais, J.: Rapid visual processing using spike asynchrony. In: Advances in Neural Information Processing Systems, pp. 901–907 (1997)
33. Bohte, S.M., Kok, J.N., La Poutre, H.: Error-backpropagation in temporally encoded networks of spiking neurons. Neurocomputing **48**(1–4), 17–37 (2002)
34. Thorpe, S., Gautrais, J.: Rank order coding. In: Computational Neuroscience, pp. 113–118. Springer, Heidelberg (1998)
35. Minku, L.L., White, A.P., Yao, X.: The impact of diversity on online ensemble learning in the presence of concept drift. IEEE Trans. Knowl. Data Eng. **22**(5), 730–742 (2010)
36. Frías-Blanco, I., del Campo-Ávila, J., Ramos-Jiménez, G., Morales-Bueno, R., Ortiz-Díaz, A., Caballero-Mota, Y.: Online and non-parametric drift detection methods based on hoeffdings bounds. IEEE Trans. Knowl. Data Eng. **27**(3), 810–823 (2015)
37. Gao, J., Ding, B., Fan, W., Han, J., Philip, S.Y.: Classifying data streams with skewed class distributions and concept drifts. IEEE Internet Comput. **12**(6) (2008)
38. Pears, R., Sakthithasan, S., Koh, Y.S.: Detecting concept change in dynamic data streams. Mach. Learn. **97**(3), 259–293 (2014)

Energy

A Hybrid Ensemble of Heterogeneous Regressors for Wind Speed Estimation in Wind Farms

L. Cornejo-Bueno, J. Acevedo-Rodríguez, L. Prieto, and S. Salcedo-Sanz(✉)

Department of Signal Processing and Communications, Universidad de Alcalá, 28805 Alcalá de Henares, Madrid, Spain
sancho.salcedo@uah.es

Abstract. This paper focuses on a problem of wind speed estimation in wind farms by proposing an ensemble of regressors in which the output of four different systems (Neural Networks (NNs), Suppor Vector Regressors (SVRs) and Gaussian Processes (GPRs)) will be the input of a final prediction system (An Extreme Learning Machine (ELM) in this case). Moreover, we propose to use variables from atmospheric reanalysis data as predictive inputs for the systems, which gives us the possibility of hybridizing numerical weather models with ML techniques for wind speed prediction in real systems. The experimental evaluation of the proposed system in real data from a wind farm in Spain has been carried out, with the subsequent discussion about the performance of the different ML regressors and the ensemble method tested in this wind speed prediction problem.

1 Introduction

Nowadays, wind energy is one of the most important sustainable energy source in the world, because of its annual growing and penetration in the power system and economic impact [1]. The main problem of wind energy is that it exhibits intermittent generation (depending on the weather conditions) [2], which makes difficult its integration in the system due to the impossibility to know the exact amount of energy produced by wind energy facilities in a specific period of time. Wind speed forecasting is therefore a key factor to improve this integration [3]. Moreover, it is important to work on different approaches which improve wind speed/power prediction in wind energy facilities. There are different ways to obtain reliable wind speed predictions: on one hand, physical models, that are usually based on numerical weather prediction models (NWM), and, also statistical approaches, usually based on Computational Intelligence models. NWMs simulate the physics of the atmosphere utilizing physical laws and boundary conditions, whereas statistical models are usually trained using historical data. Hybrid approaches mixing NWMs with computational intelligence algorithms have been also successfully applied to short-term forecasting of wind speed [4,5].

Many different works have proposed NWMs to obtain accurate wind speed predictions [3], including Computational Intelligence approaches such as neural

J. Del Ser et al. (Eds.): IDC 2018, SCI 798, pp. 97–106, 2018.
https://doi.org/10.1007/978-3-319-99626-4_9

computation techniques [6], neuro-fuzzy techniques [7], evolutionary neural networks [8] or kernel methods [9]. Specially important are hybrid approaches, i.e. those which have mixed statistical approaches with NWMs [4,10]. These modern forecasting models for power prediction on wind farms are based on the combination of physical and statistical models of atmospheric dynamics. The physical models can be global scale, mesoscale or even local models that take into account the specific local orography surrounding the wind farm. Regarding the statistical part of these approaches, Computational Techniques such as neural networks [11], Support Vector Machines [12] and different evolutionary-based approaches [8,13] have been applied to form these hybrid algorithms.

In this paper we propose a hybrid approach for wind speed estimation in wind farms, based on the combination of different Machine Learning regressors (Multi-layer perceptrons, Extreme Learning Machines, Support Vector Regression and Gaussian Processes) in different sub-layers. The final result is an heterogeneous bank of regressors, in which the input data are meteorological variables obtained from the ERA-Interim reanalysis, and the information processing is carried out in different layers including different regressors. The proposed methodology has similarities with the system proposed in [4], in which a bank of neural networks was proposed. In this case, however, the proposal involves different types of regressors cooperating in the different layers of the regressors bank. More in detail, we present an ensemble method for wind speed prediction in wind farms where the inputs of a final regressor will be composed by the forecasted outputs obtained by regressors at a previous layer. The final out of the system (wind speed prediction) is formed by an ELM approach. Hence, the main novelty of this study lies in the use of reanalysis data as the input of the regressors bank, and in the structure of this regression bank, which is the combination of different Machine Learning techniques. Moreover, the model proposed is used in a real wind speed prediction problem in a wind farm in Spain.

The rest of the paper is structured as follows: next section describes the data and predictive variables used in the experiments carried out in this work. Section 3 describes the Machine Learning regression techniques we use to construct the heterogeneous regressor bank. Section 4 shows the experimental part of the work, where the results are obtained by the different tested algorithms in a wind speed prediction problem located at a wind farm in Spain. Finally, Sect. 5 closes the paper by giving some concluding remarks on the research carried out.

2 Data and Predictive Variables

A *reanalysis* project is a methodology carried out by some weather forecasting centres, which consists of combining past observations with a modern meteorological forecast model, in order to produce regular gridded datasets of atmospheric and oceanic variables within a certain past period, with a temporal resolution of a few hours. There are several reanalysis projects currently in operation, but maybe the most important one is the *ERA-Interim reanalysis project*, which is the latest global atmospheric reanalysis produced by the European Centre for Medium-Range Weather Forecasts (ECMWF) [14].

ERA-Interim is a global atmospheric reanalysis with data from 1979, and continuously updated in real time. The spatial resolution of the data set is approximately 12.5 km, on 60 vertical levels from the surface up to 0.1 hPa. ERA-Interim provides 6-hourly atmospheric fields on model levels, pressure levels, potential temperature and potential vorticity, and 3-hourly surface fields.

Table 1. Predictive variables considered at each node from the ERA-Interim reanalysis

Variable name	ERA-Interim variable
skt	Surface temperature
sp	Surface pression
u_{10}	Zonal wind component (u) at 10 m
v_{10}	Meridional wind component (v) at 10 m
temp1	Temperature at 500 hPa
up1	Zonal wind component (u) at 500 hPa
vp1	Meridional wind component (v) 500 hPa
wp1	Vertical wind component (ω) at 500 hPa
temp2	Temperature at 850 hPa
up2	Zonal wind component (u) at 850 hPa
vp2	Meridional wind component (v) at 850 hPa
wp2	Vertical wind component (ω) at 850 hPa

In order to tackle the wind speed prediction problem in this paper, we consider wind and temperature-related predictive variables from ERA-Interim at some specific points in the neighborhood of the area under study (4 points surrounding the wind farm). The variables considered as predictors (Table 1) are taken at different pressure levels (surface, 850 hPa and 500 hPa), in such a way that different atmospheric processes can be taken into account. A total of 12 prediction variables per ERA-Interim node, and four nodes surrounding the area under study (wind farm) are considered at time t, i.e. in this problem we have $N = 48$ predictive variables.

3 A Hybrid Ensemble of Heterogeneous Regressors

This section describes the proposal of this work. First, the different Machine Learning regression techniques used in this paper are summarized: Support Vector Regression (SVR), Multi-Layer Perceptrons (MLP), Extreme Learning Machines (ELM) and Gaussian Processes for Regression (GPR) are the state-of-the-art algorithms selected to be compared in this problem of wind speed prediction. Then, the final hybrid ensemble is detailed.

3.1 Machine Learning Regressors Involved

Support Vector regression (SVR) [15] is one of the state-of-the-art algorithms for regression and function approximation. The SVR approach takes into account the error approximation to the data and also the generalization of the model, i.e. its capability to improve the prediction of the model when a new dataset is evaluated by it. Although there are several versions of the SVR, the classical model, ε-SVR, described in detail in [15] and used in a large number of application in Science and Engineering [16], is considered in this work.

The ε-SVR method for regression consists of, given a set of training vectors $\mathbb{T} = \{(\mathbf{x}_i, y_i), i = 1, \ldots, l\}$, training a model of the form $y(\mathbf{x}) = f(\mathbf{x}) + b = \mathbf{w}^T\phi(\mathbf{x}) + b$, to minimize a general risk function of the form

$$R[f] = \frac{1}{2}\left\|\mathbf{w}\right\|^2 + C\sum_{i=1}^{l} L\left(y_i, f(\mathbf{x}_i)\right) \tag{1}$$

where the norm of $\mathbf{w}$ controls the smoothness of the model, $\phi(\mathbf{x})$ is a function of projection of the input space to the feature space, b is a parameter of bias, $\mathbf{x}_i$ is a feature vector of the input space with dimension N, y_i is the output value to be estimated, C is a regularization parameter and $L\left(y_i, f(\mathbf{x}_i)\right)$ is the loss function selected. In this paper, we use the L1-SVRr (L1 support vector regression), characterized by an ε-insensitive loss function [15]. In order to train this model, it is necessary to solve a optimization problem that the reader can find in [15]. In this work we have used two different kernel functions. On one hand, the linear kernel is simply a dot product in the input space. On the other hand, the RBF kernel, also known as gaussian, follows the following formula: $K\left(x, x'\right) = exp^{\left(-\gamma||x-x'||^2\right)}$ being γ a hyperparameter that must be previously selected. SVR parameters γ and C must be tuned to obtain a good quality solution with this algorithm. In this case we use a grid search procedure from a grid of possible pre-defined values $C \in 2^{[-5,\ldots,12]}$ and $\gamma \in 2^{[-15,\ldots,0]}$.

A Multi-Layer Perceptron (MLP) is a particular kind of artificial neural network, successfully applied in modeling a large variety of nonlinear problems [17,18]. The MLP is a parallel information processing network consisting of an *input layer*, a number of *hidden layers*, and an *output layer*. All the layers forming an MLP are basically composed of a number of especial processing units, called *neurons*. As important as the processing units themselves, is the connectivity among them: the neurons within a given layer are connected to those of other layers by means of weighted links. The value of each weight is related to the MLP ability to learn and generalize from a sufficiently long number of examples. Such a learning process demands a proper database containing a variety of input examples or patterns and their corresponding known outputs. The adequate weight values are just those that minimize the error between the output generated by the MLP (when fed with input patterns in the database) and the corresponding expected known one in the database. The number of neurons in the hidden layer is a parameter to be optimized when using this type of neural networks [17,18]. Usually, the well-known Levenberg-Marquardt algorithm is applied to train the MLP [19]. In this paper we have therefore used the Matlab implementation of the MLP with the Levenberg-Marquardt training algorithm. In this algorithm, we must indicate a number of predefined values to train the algorithm, such as the number of epochs that is 1000 in the case under study, or the number of layers, 2 in our case.

An *Extreme Learning Machine* (ELM) [20] is a novel and fast learning method based on the structure of MLPs. The ELM approach is a novel way of training feed-forward neural networks, with perceptron structure. The most significant characteristic of the ELM training is that it is carried out just by randomly setting the network weights, and then obtaining a pseudo-inverse of the hidden-layer output matrix. The advantages of this technique are its simplicity, which makes the training algorithm extremely fast, and also its outstanding performance when compared to avantgarde learning methods, usually better than other established approaches such as classical MLPs or SVRs. Moreover, the universal approximation capability of the ELM network, as well as its classification capability, have been already proven [21]. As in the previous algorithms, we need to fix a very important parameter, the number of neurons in the hidden layer. It is estimated for obtaining good results as a part of a validation set in the learning process. Hence, scanning a range of $\tilde{N}$ values is the solution for this problem. The range of values used in this work goes to 50 until 150 because it used to the normal values for a good performance.

Gaussian Process Regression (GPR) has recently attracted a lot of attention because of their good performance in regression tasks [22]. We give here a short description of the most important characteristics of the GPR approach, the interested reader being referred to the more exhaustive reviews [23] or [22]: Given a set of N-dimensional inputs $\mathbf{x}_n$ and their corresponding scalar outputs y_n, that is, the data set $\mathscr{D}_S \equiv \{\mathbf{x}_n, y_n\}_{i=1}^{l}$, the regression task consists in obtaining the predictive distribution for the corresponding observation y_* based on $\mathscr{D}_S$ given a new input $\mathbf{x}_*$. The GPR model assumes that the observations can be modeled as some noiseless latent function of the inputs plus independent noise, $y = f(\mathbf{x}) + \varepsilon$, and then sets a zero-mean GP prior on the latent function $f(\mathbf{x}) \sim \mathscr{GP}(0, k(\mathbf{x}, \mathbf{x}'))$ and a Gaussian prior on $\varepsilon \sim \mathscr{N}(0, \sigma^2)$ on the noise, where $k(\mathbf{x}, \mathbf{x}')$ is a covariance function and σ^2 is a hyperparameter that specifies the noise power. The joint distribution of the available observations (collected in $\mathbf{y}$) and some unknown output $y(\mathbf{x}_*)$ is a multivariate Gaussian distribution, with parameters specified by the covariance function. In this case, the hyperparameters of the kernel function are obtained from these values: Signal Power that is the variance of y and Noise Power defined as (Signal Power/4); being y the target (objective function in the experiments).

3.2 Final Hybrid Heterogeneous Regressors Bank

Figure 1 shows the final structure of the proposed heterogeneous ensemble method. As can be seen, the final structure consists of a first layer of different regressors, which process the information from the input layer. The output of this layer is passed to a final regressor (an ELM in this case) to obtain the final wind speed prediction. The idea of this structure is that the final regressor is able to extract some more information out of the initial outputs of the previous layer.

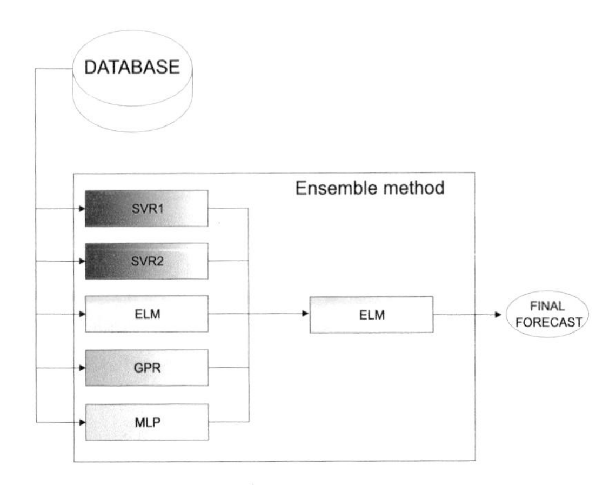

Fig. 1. Wind speed estimation by means of the ensemble method.

4 Experimental Part

This section presents the experimental evaluation of the heterogeneous regressor bank proposed in a real problem of wind speed prediction. Specifically, we consider a wind farm in Spain, which location is shown in Fig. 2. The wind farm chosen (A in the figure) is a medium-size facility, with 28 wind turbines installed. Note that the wind farm selected is located in the center of Spain, and data from the following periods have been considered: 23/11/2000 to 17/02/2013. A preprocessing step to remove missing and corrupted data was carried out. Note that we only kept data every 6 h (00 h, 06 h, 12 h and 18 h), to match the predictive variables from the ERA-Interim reanalysis to the objective variables.

The performance of the four ML regressors in the input layer (described in Sect. 3), and the final ensemble method for the wind speed prediction problem at wind farm A are shown in terms of Root Mean Square Error (RMSE) and Pearson's Correlation Coefficient (r^2). A partition of the data into *training* (60%), *validation* (20%) and *test* (20%) is carried out. This validation set is used to obtain the best SVR hyper-parameters C, ε and γ, by means of a *grid search* [15]. The validation set is also used in the training of the MLP approach, in order to avoid neural network overtraining, and in ELM to fix the number of hidden neurons. Both training and test sets have been randomly constructed from the available data after the cleaning pre-processing.

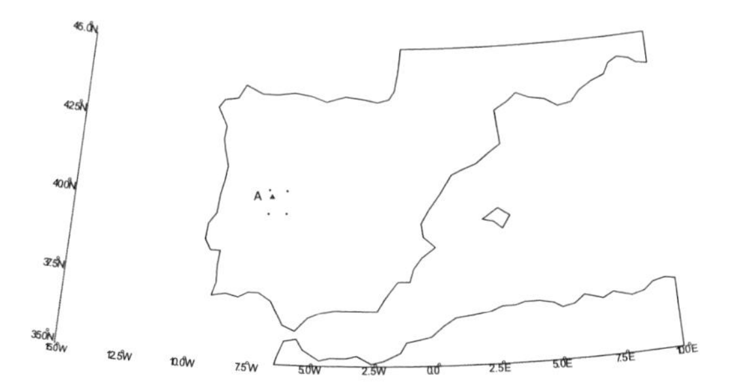

Fig. 2. Wind farm considered for the experiments. The four closest nodes from the Era-Interim reanalysis (predictive variables) are also displayed in the picture

4.1 Results

Table 2 shows the results obtained in this problem of wind speed prediction in the wind farm under study. In general, the performance of the ML regressors is good in terms of RMSE and r^2 measured. The GPR obtains the best results of all the regressors tested in this case, with an accurate reconstruction of wind speed from the ERA-Interim variables. The worst result corresponds to the SVR2 with the lineal kernel, providing a RMSE around 3.16 MW. In any case, when the final ensemble method is applied, the result obtained is better than those obtained by the rest of the algorithms used separately. We can see how the RMSE goes down until 2.5783 MW with the ensemble method if we compare it with the other regressors. The same behaviour is obtained for the case of r^2 with a best value around 71% (the worst prediction given by SVR2 provided a value of r^2 around 61%). The GPR obtains similar correlation but worse RMSE than the ensemble method, 2.61 MW.

Figure 3 shows the wind speed estimation obtained by the ensemble method and SVR with linear kernel algorithm (the best and worst approaches tested in our experiments) in this case, for the wind farm A. In order to clearly show the algorithms' performance, only the first 300 samples of the test set have been depicted in these figures. Moreover, Fig. 4 presents the scatter plot of the predictions carried out with the ensemble method and SVR with lineal kernel, respectively. As we can see, the good performance achieved with the ensemble method translates into less dispersion between real and predicted samples when compared with the SVR2 scatter plot.

Table 2. Results obtained with the different ML regressors considered and the ensemble method.

Wind farm A	RMSE [MW]	r^2 [%]
SVR1	2.7781	0.6714
SVR2	3.1556	0.6071
ELM	2.9371	0.6324
GPR	2.6135	0.7097
MLP	2.7254	0.6855
Ensemble method	2.5783	0.7089

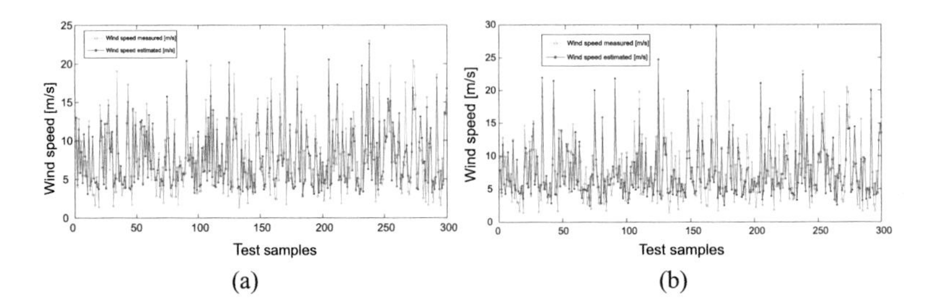

Fig. 3. Temporal wind speed estimation by means of: (a) ensemble method and (b) SVR with lineal kernel.

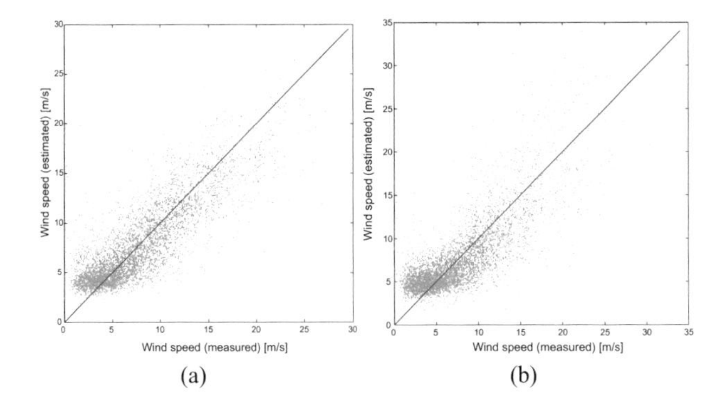

Fig. 4. Scatter of the wind speed estimation by means of: (a) ensemble method and (b) SVR with lineal kernel.

5 Conclusions

In this paper several Machine Learning regression methods have been hybridized to from a multi-regression bank for a problem of wind speed prediction. Specifically, a Multi-layer perceptron (MLP) neural network, an Extreme Learning Machine (ELM), a Gaussian Process Regression (GPR) approach and a Support Vector Regression (SVR) algorithm (with 2 different kernels) have been applied to solve this problem in a wind farm in Spain. All these techniques have been included in a bank-structure with different layers: first all the regressors provide a first wind speed estimation. A second layer formed by an ELM collect this information and tries to process it further, to get an improved prediction of the wind speed. The results obtained show that the proposed bank of ML regressors is a good option to deal with wind speed prediction in wind farms. Among the four state-of-the-art ML regression techniques tested, we have shown that the GPR approach has obtained the best results, but it is the ensemble method which gets the best accuracy in terms of RMSE and r^2 once all the experiments are made. The proposed approach (ensemble method) allows to reduce the error obtained by the different regression techniques used in this work separately.

Finally, we propose the following future work: First, we can extend the experiments to others wind farms, to generalize the proposed model. And other improvement could be a previous feature selection by some evolutionary algorithm to reduce the noise in the regressors used in the experiments.

Acknowledgements. This work has been partially supported by *Comunidad de Madrid*, under project number S2013/ICE-2933, and by project TIN2014-54583-C2-2-R of the *Spanish Ministerial Commission of Science and Technology (MICYT)*.

References

1. Kumar, Y., Ringenberg, J., Depuru, S.S., Devabhaktuni, V.K., Lee, J.W., Nikolaidis, E., Andersen, B., Afjeh, A.: Wind energy: trends and enabling technologies. Renew. Sustain. Energy Rev. **53**, 209–224 (2016)
2. Yan, J., Liu, Y., Han, S., Wang, Y., Feng, S.: Reviews on uncertainty analysis of wind power forecasting. Renew. Sustain. Energy Rev. **52**, 1322–1330 (2015)
3. Costa, A., Crespo, A., Navarro, J., Lizcano, G., Madsen, H., Feitosa, E.: A review on the young history of the wind power short-term prediction. Renew. Sustain. Energy Rev. **12**, 1725–1744 (2008)
4. Salcedo-Sanz, S., Pérez-Bellido, Á.M., Ortiz-García, E.G., Portilla-Figueras, A., Prieto, L., Paredes, D.: Hybridizing the fifth generation mesoscale model with artificial neural networks for short-term wind speed prediction. Renew. Energy **34**(6), 1451–1457 (2009)
5. Feng, C., Cui, M., Hodge, B.M., Zhang, J.: A data-driven multi-model methodology with deep feature selection for short-term wind forecasting. Appl. Energy **190**, 1245–1257 (2017)
6. Bilgili, M., Sahin, B., Yasar, A.: Application of artificial neural networks for the wind speed prediction of target station using reference stations data. Renew. Energy **32**, 2350–2360 (2007)

7. Monfared, M., Rastegar, H., Kojabadi, H.M.: A new strategy for wind speed forecasting using artificial intelligent methods. Renew. Energy **34**(3), 845–848 (2009)
8. Salcedo-Sanz, S., Ortiz-Garcá, E.G., Pérez-Bellido, A.M., Portilla-Figueras, A., Prieto, L.: Short term wind speed prediction based on evolutionary support vector regression algorithms. Expert Syst. Appl. **38**(4), 4052–4057 (2011)
9. Mohandes, M.A., Halawani, T.O., Rehman, S., Hussain, A.A.: Support vector machines for wind speed prediction. Renew. Energy **29**(6), 939–947 (2004)
10. Kusiak, A., Zheng, H., Song, Z.: Wind farm power prediction: a data-mining approach. Wind Energy **12**(3), 275–293 (2009)
11. Salcedo-Sanz, S., Perez-Bellido, A.M., Ortiz-GarcÃa, E.G., Portilla-Figueras, A., Prieto, L., Correoso, F.: Accurate short-term wind speed prediction by exploiting diversity in input data using banks of artificial neural networks. Neurocomputing **72**, 1336–1341 (2009)
12. Ortiz-García, E.G., Salcedo-Sanz, S., Pérez-Bellido, A.M., Gascón-Moreno, J., Portilla-Figueras, A., Prieto, L.: Short-term wind speed prediction in wind farms based on banks of support vector machines. Wind Energy **14**(2), 193–207 (2011)
13. Salcedo-Sanz, S., Pastor-Sánchez, A., Prieto, L., Blanco-Aguilera, A., García-Herrera, R.: Feature selection in wind speed prediction systems based on a hybrid coral reefs optimization-extreme learning machine approach. Energy Convers. Manage. **87**, 10–18 (2014)
14. Dee, D.P., Uppala, S.M., Simmons, A.J., Berrisford, P., Poli, P.: The ERA-interim reanalysis: configuration and performance of the data assimilation system. Q. J. Roy. Meteorol. Soc. **137**, 553–597 (2011)
15. Smola, A.J., Schölkopf, B.B.: A tutorial on support vector regression. Stat. Comput. **14**, 199–222 (2004)
16. Salcedo-Sanz, S., Rojo, J.L., Martínez-Ramón, M., Camps-Valls, G.: Support vector machines in engineering: an overview. WIREs Data-Mining Knowl. Discov. **4**(3), 234–267 (2014)
17. Haykin, S.: Neural Networks: A Comprenhensive Foundation. Prentice Hall, Upper Saddle River (1998)
18. Bishop, C.M.: Neural Networks for Pattern Recognition. Oxford University Press, Oxford (1995)
19. Hagan, M.T., Menhaj, M.B.: Training feed forward network with the Marquardt algorithm. IEEE Trans. Neural Netw. **5**(6), 989–993 (1994)
20. Huang, G.B., Zhu, Q.Y.: Extreme learning machine: theory and applications. Neurocomputing **70**, 489–501 (2006)
21. Huang, G.B., Zhou, H., Ding, X., Zhang, R.: Extreme learning machine for regression and multiclass classification. IEEE Trans. Syst. Man Cybern. Part B **42**(2), 513–529 (2012)
22. Rasmussen, C.E., Williams, K.H.: Gaussian Processes for Machine Learning. MIT Press, Cambridge (2006)
23. Lázaro-Gredilla, M., van Vaerenbergh, S., Lawrence, N.: Overlapping mixtures of gaussian processes for the data association problem. Pattern Recogn. **45**(4), 1386–1395 (2012)

Bio-inspired Approximation to MPPT Under Real Irradiation Conditions

Cristian Olivares-Rodríguez[1(✉)], Tony Castillo-Calzadilla[2], and Oihane Kamara-Esteban[2]

[1] Instituto de Informática, Universidad Austral de Chile, Av. General Lagos 2086, Valdivia, Chile
colivares@inf.uach.cl

[2] DeustoTech Energy and Environment, Universidad de Deusto, Av. Universidades 24, Bilbao, Spain
tony.castillo@opendeusto.es, oihane.esteban@deusto.es

Abstract. The aim of this paper is to study the possibilities of increasing the renewable power of a photovoltaic system through a barely tested bio-inspired algorithm. Photovoltaic energy has a high potential to grows but it has a strong dependence on climate conditions. Particularly, the power production of panels is reduced under partial shading conditions, which is a very common situation in several cities around the world. Therefore, the Maximum Power Point Tracker (MPPT) algorithm becomes critical to control the photovoltaic system. In this paper, we propose a novel MPPT algorithm based on the Artificial Bee Colony (ABC) bio-inspired method and we explicitly define a fitness function based on power production. The ABC algorithm is more attractive than any other bio-inspired methods due to its simplicity and ability to resolve the problem of choosing an ideal duty cycle. Specifically, it requires a reduced number of control parameters and the initial conditions have no influence over the convergence. The analysis has been carried out using real meteorological and consumption data and testing the behavior of the algorithm on a standalone photovoltaic system operating only with direct current.

1 Introduction

Energy is as important and necessary as water. Both elements are part of our daily life up to the point that humanity have a high consumption and dependence on these resources. A building without neither energy nor water may well become an uninhabitable place to live or work. The needs of energy have been supplied until now thanks to traditional non-renewable fossil sources, like coal, gas, and petroleum. Although these energy sources have been supporting all the energy demand, it is important to highlight that their exploitation has been dominated by a small percentage of countries around the world [16].

Nowadays, there are several problems regarding fossil fuels. Their high consumption leads to their depletion as well as to a great amount of pollution,

J. Del Ser et al. (Eds.): IDC 2018, SCI 798, pp. 107–118, 2018.
https://doi.org/10.1007/978-3-319-99626-4_10

consequence of the release of CO_2 to the atmosphere. Additionally, as consumption increases, so does the prices of energy [13,16]. Consequently, in order to deal with all these drawbacks, science and technology have fostered the use of renewable sources. This renewable revolution has been encouraged primarily by the most industrialized countries, maybe because they cannot allow their economies to fluctuate by the unsteady prices variation of fossil fuels on the international market.

Among renewable sources, photovoltaics (PV) is the one growing exponentially. However, there is still ground to research. It is well known that PV panels have a performance not superior to 21% [8], therefore an optimization model on solar harvesting is necessary. That is the reason behind the great amount of development of Maximum Power Point Tracker (MPPT) algorithms to keep solar panels working at their maximum performance and increase the profitability of the solar PV system. It is important to highlight that a PV installation without a MPPT algorithm would likely have losses near to 30% to 40% [3,10].

Nowadays, there are more than 40 types of algorithms developed in order to maintain operation on the MPPT as a way to achieve the profits that make the systems viable [17]. In this paper, we will just mention the most common since it is not our purpose to make a review of them. The most widely used is the Perturb and Observe (P&O) algorithm due to its simplicity and low dependency on the variables of control [5]. Also, the P&O is the most commercially used because it only requires one voltage sensor, making this feature the most attractive due to its good performance and low cost. However, it is necessary to pinpoint the existence of the following methods: Incremental Conductance (IC), Fuzzy-logic (FLZ), Artificial Neural Network (ANN) [12], Voltage-based, Current-based, etc. All MPPT control schemes follow the same goal which is to extract as much electric energy from the PV array as possible. This is only feasible using converts such as Buck, Boost, or Buck-Boost.

All the previous algorithms use a Pulse Width Modulation (PWM) in order to change the duty cycle (dc) of the converter. Based on this value, the characteristic slope of the P-V curve is altered. The problem with these algorithms is the high convergence that occurs concerning Maximum Power Point (MPP) since they cannot properly deal with real shading conditions [15]. Despite that, all methods have successfully demonstrated the ability of tracking the MPP. Unfortunately, this has been done under uniform radiation where there is only one single MPP on the P-V curve, which is dismissed to identify the Global Point (GP) on the partial shading where there are multiple MPPs on the P-V curve, which does not succeed in distinguishing the Global Point (GP) between multiple local MPPs. Many researchers have been able to address the partial shading effect by improving the traditional MPPT or by using Artificial Intelligence algorithms, such as Artificial Neural Networks [12], Particle Swarm Optimization [14], and Fuzzy logic [12]. In summary, all these methods include the development of modified MPPT techniques, array re-configurations, and use of different converter topologies [11].

Within this scope, the aim of the paper is to improve the MPPT under real conditions by implementing an evolutionary algorithm known as Artificial Bee Colony (ABC). The ABC algorithm is a bio-inspired method discussed in [1,4,15] which is simple, uses very few controlled parameters and its convergence criteria does not depend on the initial conditions of the system. It is a swarm based meta-heuristic algorithm capable of solving multidimensional and multimodal optimization problems very easily. The artificial bees are responsible for looking for the best solution through an iterative process until the algorithm research convergence. The ABC algorithm was chosen due to its ability to deal with the presence of local minimums, which traditional MPPT algorithms are not able to deal with when shading conditions appear. The main novelties of this study are detailed below:

- To analyze the behaviour of the ABC algorithm which has been barely studied in solar harvesting.
- To widen the purpose of this approach and produce another tool for the improvement of solar generation.
- To test the features of the system on a PV powered facility that operates in standalone mode by Direct Current (DC), using real parameters of irradiance, temperature, and energy consumption.

By no means does this new approach pretend to replace the methods already developed. It intends to contribute for further exploitation in this field.

2 A Photovoltaic Model Under Real Conditions

This section presents the main architecture of the microgrid modeled. The test building is located in Bilbao, in the north of Spain. The building provides housing to students and researchers alike. It is composed of 3 towers of 10 floors each one, 304 bedrooms in total, so as all common areas for entertainment as well the operational requirements necessary in this kind of buildings. It is important to point out that this installation has been modeled from a model fed only with direct current, and that it has also been necessary to lead it to a hybrid system (Solar PV and Chemical Batteries) in order to isolate it and avoid the connection to the grid, making it completely independent from the commercial network.

The case study presented in this paper addresses the following points, also shown in Fig. 1.

1. The DC microgrid has been developed to feed a building in an urban environment. The building is of moderate size with an average demand of 385 kWh/day.
2. The microgrid is inverter-less, and therefore rectifier-less, for that condition is totally for a DC building.
3. Its load profile is non-manipulated, and by consequence it is dynamic.
4. It was designed for real meteorological variables.
5. The building is in one of the poorest irradiance zones of Spain.

6. The output rated voltage is about 24 V.
7. The microgrid panels will operate under shade patterns.
8. The microgrid is based on an hybrid system (electrochemical and PV) [6].

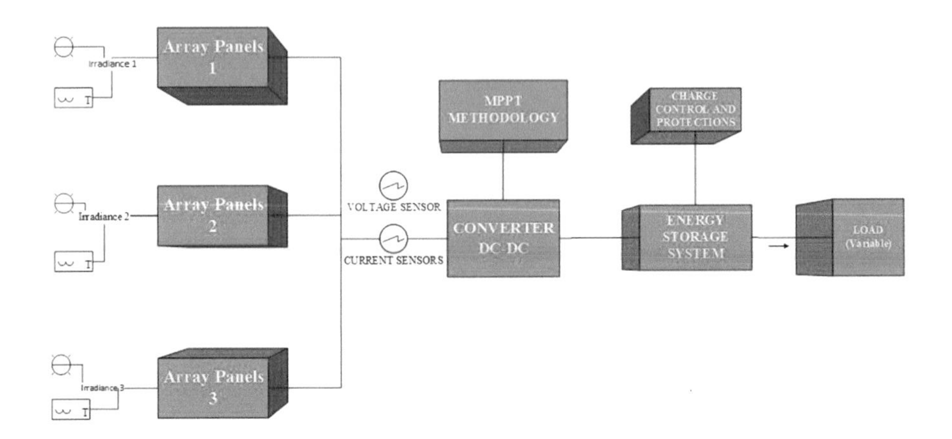

Fig. 1. Representation from the microgrid modelled.

The PV generation subsystem was configured with a commercial panel that uses monocrystalline technology, brand Top Sun, model TS-S420TA1, with a maximum efficiency of 16.38%; available in the SymPowerSystem library. The sizing algorithm determined that 1189 parallel panels are required. For sizing panels subsystem, some of the exposed criteria in the bibliography, complemented with own experience were used, taking into account the sun hours, power consumption in kWh, optimum orientation of the solar panels, and losses factor, whereas in only the consumption was considered. Figure 2 shows the behaviour of the panel under three different irradiance real patterns. The pink dots mark the MPP for each of the irradiance levels.

A converter in continuous mode was implemented. The type is known as a buck converter, because its aim is to reduce the input voltage from 48 V to a level nearby 24 V in DC both to charge the batteries and to feed the building in a direct way. Within this section of the model will focus the analysis of the paper, because it is where the tracker system of maximum power point (MPPT) is located.

The load was modeled as a variable resistance that can emulate the real active power demanded ($R = V^2/P$). The power curve is built from the quarter-hourly data supplied by the energy meter from the electric company, and the voltage is supplied by the model on every instance of the simulation. In this particular point, it was necessary to set a window control methodology to avoid maximums and minimums of voltage that could damage the appliances connected to the building due to undesired fluctuations [6].

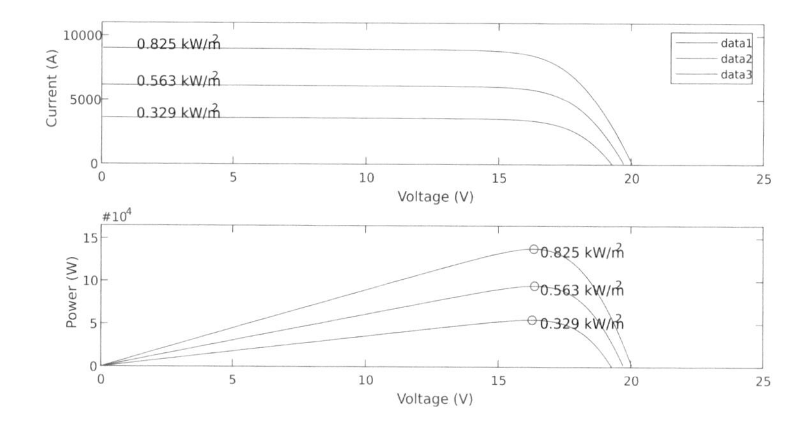

Fig. 2. Energy patterns and profiles of generation from panel modelled.

Surplus energy (if any) during the solar cycle is stored in order to supply energy to the building when consumption is superior to generation. By convention, the storage subsystem is usually designed for giving the installation an autonomy between 3 and 8 days. In this research, after analyzing the solar irradiation data of the last 30 years in the selected location, an autonomy of 3 days was established. The battery bank could cover the 100% of the energy needs during 72 h of highest consumption with the minimum generation. Such battery bank consists of 205 lithium-iron-phosphate of RAYLITE brand, model 3MIL 25S of 24 V which have four cells in series arrays of 6 V each one [7].

This subsystem allows to monitor and control the correct operation of the rest of the subsystem. It integrates the components related to the electronic instrumentation (variables measuring) and is deployed throughout the whole system. The main improvement regarding what was presented in [2] is based on the incorporation of commutation systems that protects the system from voltage peak as well as from valley values.

3 Experimental Design

The system has been modeled under three different levels of real irradiance, so as to reproduce three different maximum points at a time. In order to reach such irradiance levels from real data, these have been selected by a k-means analysis over the most common irradiance levels for the more representative meteorological conditions at Bilbao over the period since 1985 up to 2015. After the clustering analysis, the result obtained was as follows, three scenarios like winter (cluster 1), a mix of autumn-spring (cluster 2), and summer (cluster 3) with 47.2%, 31.84%, and 21.03% respectively. To know which data were used to test the algorithm see Table 1.

Afterwards, the data was loaded into every panels system that mimics the whole generation system divided into three sections. In this way, it is possible to carry out the simulation under real variation of solar irradiance, temperature, under a variable consumption in a grid-off building.

The main challenge of this proceeding is to make possible the conversion of a fossil hall of residence into a renewable one, powered only with DC electricity produced by PV installation operating on standalone mode. Additionally, this modeling has been carried out using such as data consumption and also of the temperature and irradiation from the coordinates where the hall of residence is located.

Table 1. Experimental data.

Data clusters	K1 (max values)	K2 (max values)	K3 (max values)
Irradiance	329	563	825
Temperature	17.81	17.81	17.81
Season	Winter	Spring-Fall	Summer

4 MPPT Based on the P&O Algorithm

The P&O algorithm is well known for its simplicity, its independence from the rest of the variables of the installation, and its efficiency. This algorithm [11] keeps the array working near to the MPPT of the chosen panels (48.73 V). This algorithm was chosen by the above-mentioned characteristics. Also, the P&O is the most commercially used, because it only requires one voltage sensor, making this feature more attractive due to its good performance to low cost [7]. This controller must correct the power differential (ΔP) at both high and low irradiance levels which could then produce a change in the MPPT goal of the system.

Equation 1 defines the behavior of the MPPT based on P&O Algorithm. The following bullets explain how the MPPT is working:

1. If $\Delta P = 0$, the MPPT maintain the last value of dc registered.
2. If $\Delta P < 0$ and $\Delta V < 0$, the MPPT will adjust its value for trying to keep the system operating under MPPT conditions. It is reached by adjusting the value of dc. This means a reduction of dc.
3. If $\Delta P < 0$ and $\Delta V > 0$, the MPPT will adjust its value for trying to keep the system operating under MPPT conditions. It is reached by adjusting the value of dc. This means an increase of dc.
4. If $\Delta P > 0$ and $\Delta V < 0$, the MPPT will adjust its value for trying to keep the system operating under MPPT conditions. It is reached by adjusting the value of dc. This means an increase of dc.
5. If $\Delta P > 0$ and $\Delta V > 0$, the MPPT will adjust its value for trying to keep the system operating under MPPT conditions. It is reached by adjusting the value of dc. This means a reduction of dc.

$$\frac{\Delta P}{\Delta V} = \begin{cases} > 0, V < V_{mpp} \\ = 0, V = V_{mpp} \\ < 0, V > V_{mpp} \end{cases}, \quad (1)$$

5 MPPT Based on the ABC Algorithm

The ABC algorithm was proposed by Karaboga in order to analyze the foraging behavior of honeybees colonies [9]. The ABC algorithm is based on the behavior of three types of artificial bees: employed bees, onlooker bees, and scouts bees. In such a way, a bee searching for food is called an employed bee. Also, a bee that makes decisions to choose a food source at the hive is considered as an onlooker. Finally, scout bees are bees whose food sources cannot be improved through a predefined number of trials, thus, their food sources are abandoned. The number of food sources is established as equal to the number of employed bees and also equal to the number of onlooker bees. The optimization problem is defined by the position of a food source, which represents a candidate solution; and the nectar amount of a food source, which represents the fitness of every candidate solution. All three groups work together by communication and coordination to get the optimal solution in lesser time.

We propose an MPPT algorithm based on ABC for PV generation system operating under partial shading conditions with direct control technique. To provide a direct control ABC-based MPPT, the duty cycle of the DC-DC converter corresponds to the food position and maximum power as the food source of ABC algorithm. In such a way, the optimization problem has only one parameter to be optimized duty cycle. Corresponding, the initialization phase of ABC defines that a randomly distributed population of N solutions is generated, and every solution is defined based on the Eq. 2:

$$d_i = d_{min} + rand[0, 1] * (d_{max} - d_{min}), \quad (2)$$

where d_{min} and d_{max} correspond to the minimum and maximum possible values for every candidate solution. The ABC algorithm defines the main search phase based on an iterative process of the artificial bees behavior in order to reach an optimal solution. Concretely, a set of I iteration is previously defined and by each iteration, the employee bees are responsible for producing new solutions (duty cycles) according to the Eq. 3:

$$d_{new} = d_i + \phi * (d_i - d_k), \quad (3)$$

where k are randomly selected indexes but different from i, meanwhile ϕ is a random number in the range $[0, 1]$ which defines the convergence speed. Every new candidate solution is constrained to the boundary values of duty cycle and, also, it requires a quality value that is shared by the employee bees at the hive during communication phase. In such a way, a fitness function is required, and we propose a function based on generated power according to the candidate solution (duty cycle) such as follows in Eq. 4:

$$fit(d_i) = (V_s * D) * \frac{I_s}{(1 - D)}, \tag{4}$$

After sharing phase, each onlooker bee finds a new solution based on the probability P_i such as in Eq. 5:

$$P_i = \frac{fit_i}{\sum_{n=1}^{SN} fit_n}, \tag{5}$$

where SN is the number of candidate solutions and fit_i is the fitness function of each candidate solution d_i that we have proposed in Eq. 2. The fitness of each candidate solution is compared with the previous one. If the new value is at least equal to the fitness value of the previous one, then it is replaced by such better-quality value. At the end of each search phase, the abandonment criteria are checked. It means, if a fitness value of a candidate solution does not improve then such solution is not used anymore, and a scout bee selects a new random solution in its replacement. The new solution is selected by using the initialization equation.

Finally, we have defined 50 as the maximum number of iterations and 30 as the number of candidate solutions, which is equal to the population size. Both parameters have been defined based on a trade-off between convergence speed and convergence rate. This is an exploratory study on ABC contribution on solar harvesting, therefore a deeper study is required in order to analyze the influence of parametrization on generated power.

6 Results of the Simulation

This section of the paper has been adapted to show the results of the corresponding to 60% of a whole day, it is about 14.5 h. The graphs will show a comparison in between P&O and ABC algorithms respectively. Figure 3 shows how both P&O and ABC algorithm reach a SOC of 82.67%. If the same data is widened, to highlight the performance of ABC, and it is obtained the Fig. 4, where it is possible to observe the difference in between both algorithms, taking an advance the ABC in front of P&O.

Figures 5 and 6 show the power output of the solar PV subsystem, with a difference of 2.34 kW in power generation between the P&O and ABC, having the best performance the ABC (see Fig. 5). Onto Fig. 6 it is easier to see the power difference on energy production; this ABC algorithm might have a better way to find the right dc value to adjust the power drawn of the microgrid DC modeled.

Figure 7a shows how the ABC algorithm is more steady in comparison with the P&O, which is constantly oscillating without finding a good operation point, that might has had the desired steadiness. Regarding the electrical behavior of

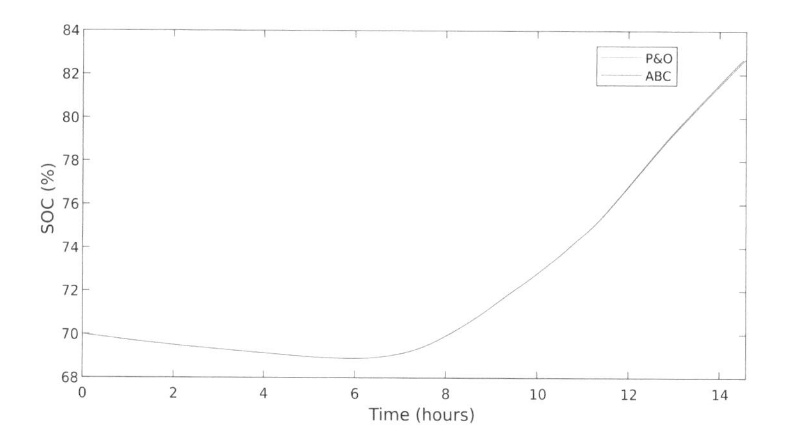

Fig. 3. SOC comparison between ABC and P&O algorithms.

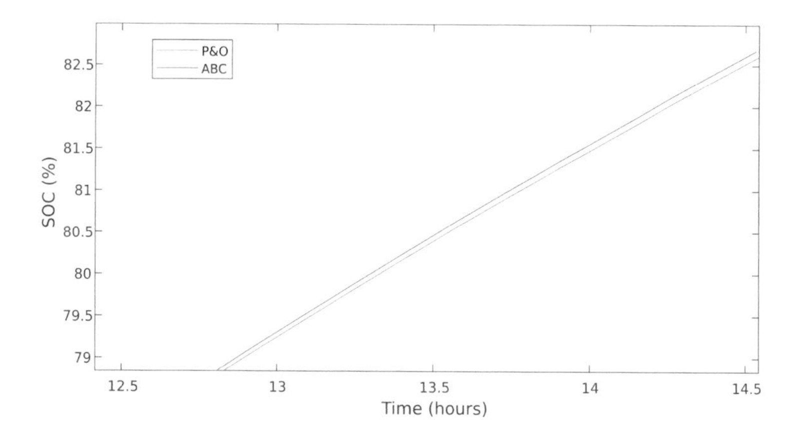

Fig. 4. SOC comparison between ABC and P&O algorithms (enlarged image between the highest generation hours).

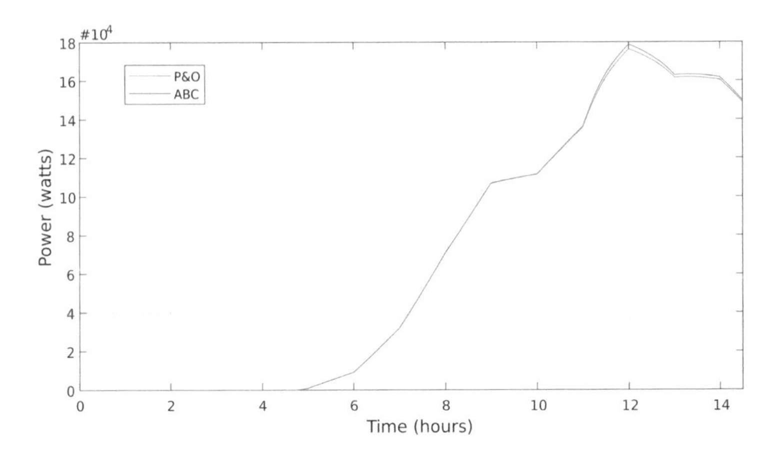

Fig. 5. Power comparison between ABC and P&O algorithms.

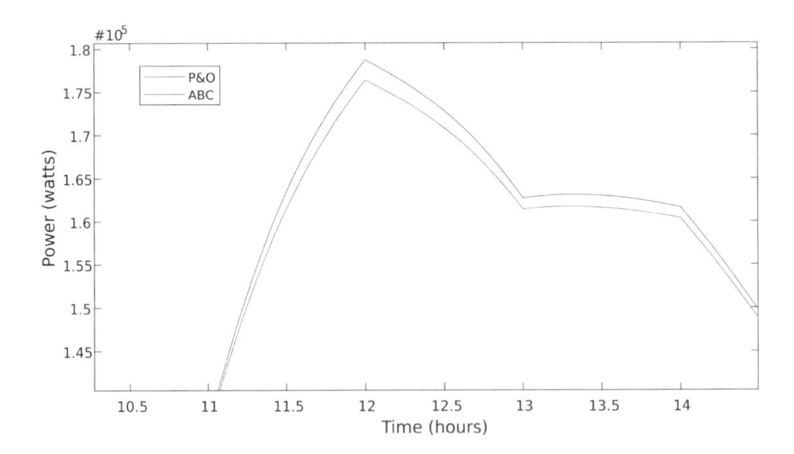

Fig. 6. Power comparison between ABC and P&O algorithms (enlarged image between the highest generation hours).

the facility (Fig. 7b), for power on the whole building, it is validated that the building could be perfectly fed on during the whole day without suffering any shortage due to lack of energy. The system is strong enough to maintain the building under operation without producing a leakage of energy.

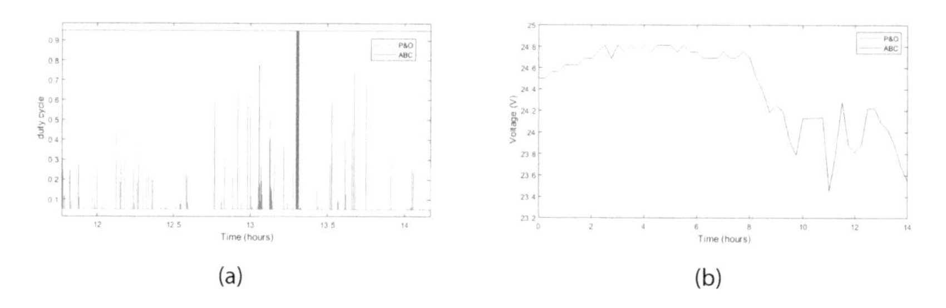

Fig. 7. dc and V comparison between ABC and P&O algorithms.

7 Conclusions

Optimization models on solar harvesting are necessary because the PV panels have performance limitation based on climate conditions. There are a high number of proposed models for maximum power point tracker (MPPT) in order to increase the profitability of the solar system. In this work, we have tested the most classical MPPT algorithm, the P&O, against a novel bio-inspired model based on Artificial Bee Colony meta-heuristic with an explicit fitness function. Both algorithms were tested on a PV model which uses batteries, taking into account real condition data (consumption and meteorological) and considering a standalone photovoltaic system operating only with DC.

Summarizing, after analyzing the results of these algorithms, it is concluded that ABC algorithm outperforms P&O algorithm in such a way that the system provides more power and stability, mainly at hours of a more solar generation where ABC generates more power than P&O. This analysis shows the potential of ABC algorithm in front of P&O, because ABC allows to increase the power harvested of the microgrid developed in this paper. Also, ABC is able to deal with a photovoltaic system working with batteries, such that is highly independent of the network. Nevertheless, it is necessary to keep on with ABC studies on a DC microgrid in order to reach a stronger methodology about MPPT bio-inspired techniques based on artificial intelligence. Concretely, we will work on both a more robust parameter control in order to precisely define an ABC algorithm setting, and a more flexible fitness function based on contextual data such as consumption and storage.

References

1. Amirjamshidi, Z., Mokhtari, Z., Moussavi, Z., Amiri, P.: MPPT controller design using fuzzy-BBBC method. Adv. Comput. Sci. Int. J. **4**(2), 122–127 (2015)
2. European Photovoltaic Industry Association, et al.: Global Market Outlook for Photovoltaics 2014–2018. Belgium, Brussels (2014)
3. Bendib, B., Belmili, H., Krim, F.: A survey of the most used MPPT methods: conventional and advanced algorithms applied for photovoltaic systems. Renew. Sustain. Energy Rev. **45**, 637–648 (2015)
4. Soufyane Benyoucef, A., Chouder, A., Kara, K., Silvestre, S.: Artificial bee colony based algorithm for maximum power point tracking (MPPT) for PV systems operating under partial shaded conditions. Appl. Soft Comput. **32**, 38–48 (2015)
5. Biabani, M.A.K.A., Ahmed, F.: Maximum power point tracking of photovoltaic panels using perturbation and observation method and fuzzy logic control based method. In: International Conference on Electrical, Electronics, and Optimization Techniques (ICEEOT), pp. 1614–1620. IEEE (2016)
6. Castillo, T., Macarulla, A.M., Kamara-Esteban, O., Borges, C.E.: Analysis and assessment of an off-grid services building through the usage of a DC photovoltaic microgrid. Sustain. Cities Soc. **38**, 405–419 (2018)
7. Castillo-Calzadilla, T., Macarulla, A.M., Borges, C.E., Alonso-Vicario, A.: Feasibility and simulation of a solar photovoltaic installation in DC for standalone services buildings (2018)
8. First, R.M.: Solar's cells break efficiency record. MIT Technology Review (2016)
9. Karaboga, D.: An idea based on honey bee swarm for numerical optimization. Technical report, Technical report-tr06. Computer Engineering Department, Engineering Faculty, Erciyes University (2005)
10. Lengoiboni, V.N.: Simulation of a maximum power point tracking system for improved performance of photovoltaic systems. Ph.D. thesis, JKUAT (2016)
11. Liu, L., Meng, X., Liu, C.: A review of maximum power point tracking methods of PV power system at uniform and partial shading. Renew. Sustain. Energy Rev. **53**, 1500–1507 (2016)
12. Mellit, A., Kalogirou, S.A.: Artificial intelligence techniques for photovoltaic applications: a review. Prog. Energy Combust. Sci. **34**(5), 574–632 (2008)
13. Mofor, L., Nuttall, P., Newell, A.: Renewable energy options for shipping. Technology Brief, p. 60 (2015)
14. Mohapatra, A., Nayak, B., Das, P., Mohanty, K.B.: A review on MPPT techniques of PV system under partial shading condition. Renew. Sustain. Energy Rev. **80**, 854–867 (2017)
15. Sundareswaran, K., Peddapati, S., Palani, S.: MPPT of PV systems under partial shaded conditions through a colony of flashing fireflies. IEEE Trans. Energy Convers. **29**(2), 463–472 (2014)
16. Teske, S., Sawyer, S., Schäfer, O., Pregger, T., Simon, S., Naegler, T., Schmid, S., Özdemir, E.D., Pagenkopf, J., Kleiner, F., et al.: Energy [r]evolution-a sustainable world energy outlook 2015 (2015)
17. Verma, D., Nema, S., Shandilya, A., Dash, S.K.: Maximum power point tracking (MPPT) techniques: recapitulation in solar photovoltaic systems. Renew. Sustain. Energy Rev. **54**, 1018–1034 (2016)

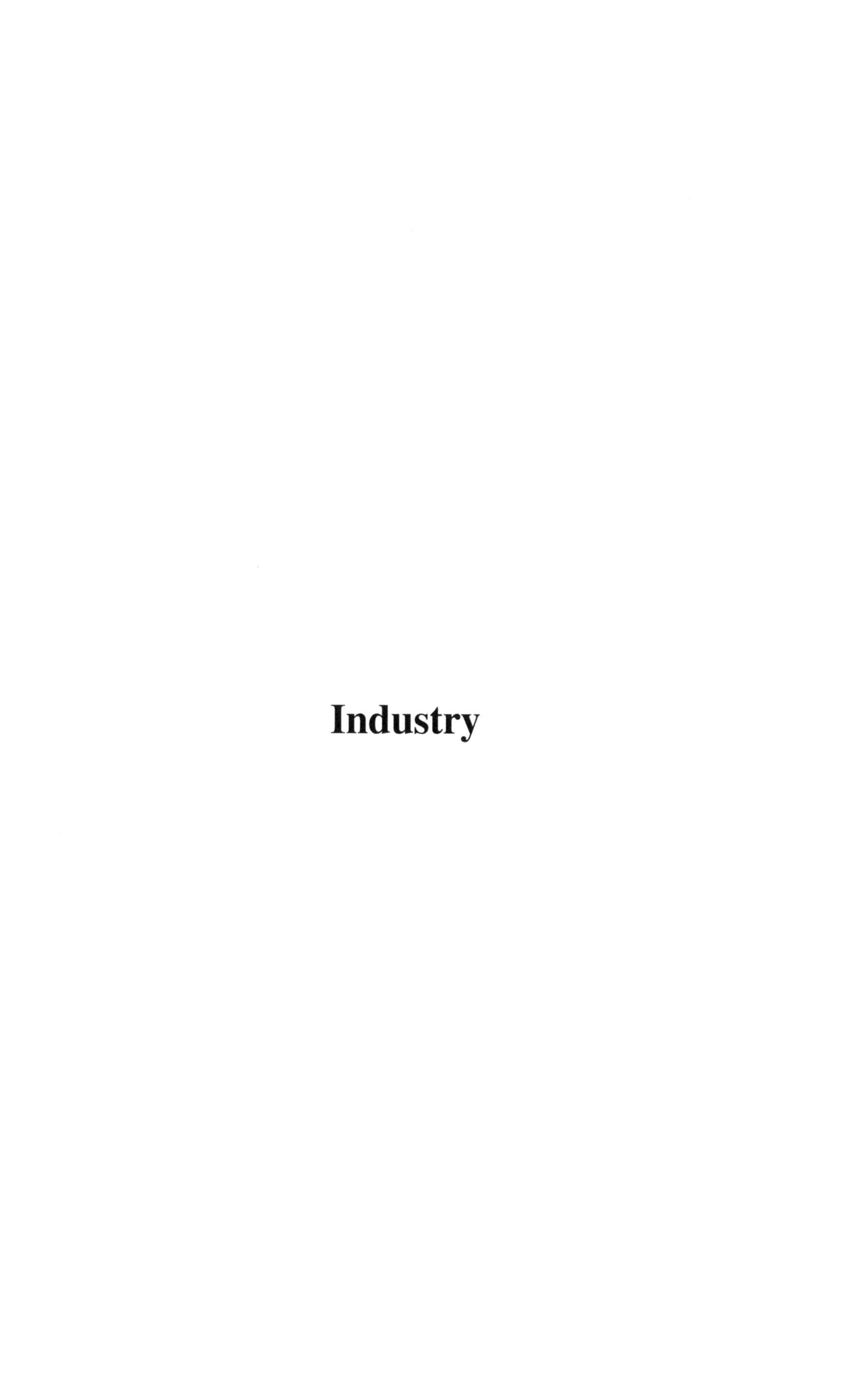

Industry

Decision Making in Industry 4.0 Scenarios Supported by Imbalanced Data Classification

Jesus Para[1](✉), Javier Del Ser[2], Aitor Aiguirre[1], and Antonio J. Nebro[3]

[1] Mondragon University, 20500 Arrasate-Mondragon, Spain
{jpara,aaguirre}@mondragon.edu
[2] TECNALIA, University of the Basque Country (UPV/EHU) and Basque Center for Applied Mathematics (BCAM), Bizkaia, Spain
javier.delser@tecnalia.com
[3] University of Málaga, 29071 Málaga, Spain
ajnebro@uma.es

Abstract. In the last years Data Science has emerged as one of the main technological enablers in many business sectors, including the manufacturing industry. Process engineers, who traditionally resorted to engineering tools for troubleshooting, have now embraced the support of data analysis to unveil complex patterns between process parameters and the quality of products and/or the performance of the production assets in plant. This work elaborates on a practical methodology to conduct data analysis within an industrial environment. The most important contribution of the proposed method is to focus on the importance of hypothesis generation dynamics among multidisciplinary experts in the process, prior to data capture itself. To exemplify the practical utility of this prescribed procedure, evidences from a real industrial case study are provided, departing from the dynamic generation of the hypothesis around the reduction of defects in the delivered products. Interestingly, this process leads to a imbalanced data classification problem, for which an extensive benchmark of supervised learning algorithm and balancing preprocessing techniques is performed to accurately predict whether parts are defective. Insights are drawn from this analysis so as to yield recommended parameter values for different stages of the production process, thereby achieving a lower defective rate and ultimately, a higher manufacturing quality of the industrial process.

Keywords: Industry 4.0 · Process monitoring
Imbalanced classification · Smart factories

1 Introduction

The new industrial revolution, coined as Industry 4.0 [24], has found one of its main technological pillars in Data Science. This discipline is based on the extraction of knowledge through the processing and analysis of data. Indeed,

J. Del Ser et al. (Eds.): IDC 2018, SCI 798, pp. 121–134, 2018.
https://doi.org/10.1007/978-3-319-99626-4_11

Data Science is present in all types of industry nowadays, helping managers and engineers in their decision making based on the knowledge gained by virtue of data analysis. As such, data-based models allow finding relationships between variables (e.g. production variables) that at first sight can not be inferred in a straightforward fashion. Disciplines such as Visual Computing [49], Decision Sciences [41] and Logistics [21], among many others, are already leveraging the benefits that Data Science provides in diverse scenarios.

In this context, the recent literature is certainly rich in what refers to data in industrial environments, mostly referred under the smart factories paradigm [37]. This flurry of contributions can be classified in two main branches: infrastructure and data analysis. On one hand, studies related to infrastructure focus on establishing the type of underlying communications, hardware and software solutions and tools needed to meet the stringent requirements for ingesting, storing and securing data generated at unprecedented scales in terms of speed, volume and variety. Examples of this first category are given in [3,20]. On the other hand, the core mission of data analysis is to obtain value from data through preprocessing, modeling and descriptive, predictive and/or prescriptive analysis. In the particular case of industrial processes, data analysis often leads to the determination of relevant parameters along the process at hand, and the prescription of optimal operational values for process improvement. For instance, in [43] a benchmark of algorithms is performed to discover relevant parameters on wood defect images. Likewise, time series analysis is undertaken in [44] for the discovery of relevant features towards their repeated use in manufacturing processes. A hybrid machine learning algorithm is proposed in [48] for recommending features in additive manufacturing (AM) at the conceptual design phase. Energy efficiency in industrial processes is addressed in [2,29] by resorting to data analysis. Finally, predictive maintenance is also one of the operational areas where data analysis has provided most quantifiable benefits [28].

Steps taken in the works belonging to this latter literature strand do not differ significantly from each other. Starting from a given problem statement, it is often necessary to preprocess the data so as to arrange it in a proper form for subsequent modeling [32]. Once data has been cleansed and shaped properly, a benchmark of mathematical models and learning algorithms is run so as to find the data analysis pipeline capable of providing the *best performance* for the posed problem. Intuitively, the measure of performance that this process aims to maximize depends roughly on the problem to be solved. A model for predictive maintenance is optimized to produce generalizable, accurate predictions on the monitored asset, whereas the performance of prescriptive models is gauged in terms of the estimated improvement of the prescribed action over the plant (as in e.g. production scheduling or stock management). The study of a tool wear prediction reported in [46] is illustrative of the generation of mathematical models and the extraction of optimal values of relevant parameters to reach the target production improvement.

This manuscript presents a real use case where data analysis is shown to provide a substantially enhanced productivity in a relevant manufacturing company from the automotive sector in Spain. However, the contribution of this paper must be understood beyond the design of the data-based models themselves: a thorough description of the whole process is given, including the generation of hypotheses. Indeed, this work capitalizes on the utmost relevance of this initial stage for the sake of a data analysis phase accounting for all variables, parameters and practicalities of the industrial process. For this to occur, brainstorming sessions must be organized to gather experts of the process and data scientists, so that the latter are fed with all information and details required to understand the whole industrial process and consequently, tightly couple their designed models to the use case under analysis. There lies the true contribution of this work, which is further put to practice in the aforementioned use case.

The rest of the paper is structured as follows: first Sect. 2 overviews the industrial use case considered in this work, whereas Sect. 3 describes the methodology used to determine what data collected from the industrial process should be taken into account for the subsequent data analysis. Next, Sect. 4 presents several classification techniques for imbalanced data used along the process. In Sect. 5 results of the data analysis are presented and discussed, as well as the optimal parameter setting of the industrial process inferred from this data analysis phase. Finally, Sect. 6 ends the paper by drawing conclusions and outlining related research lines to be pursued in the near future.

2 Description of the Use Case

The use case considered in this work takes place over one of the production lines of a automotive sector supplier in Spain[1]. The production process addressed in this work consists of low-pressure aluminum casting machines, which operate as schematically depicted in Fig. 1: cast aluminum is injected at low pressure into a mold shaped according to the part to be manufactured. Once pieces are molded and cooled, they are passed through a unitary quality control, which determines whether the quality of the parts is good (`OK` parts) or bad (`NOK` parts).

From the casting furniture several parameters can be recorded, such as chemical composition and aluminum density. From the molding machine aluminum temperature inside the mold and cooling can be registered at a rate of 1 measure per second. A 100% pass-through quality control is done, so every `OK` and `NOK` part reference is captured with perfect traceability of previous steps. The relatively low defective rate makes the record of produced parts highly imbalanced, with significantly lower number of `NOK` parts than those produced with no faults nor defects (i.e. an imbalance rate equal to 5%).

The main goal of the supplier is to achieve a parameter recommender that will help process engineers in their decision making processes. To this end, the company deemed important to find key parameters that affect the quality of the production process, so that the identification of critical variables and the prescription of their optimal levels of operation would reduce the rates of produced

[1] Further details are kept confidential as per request of the supplier.

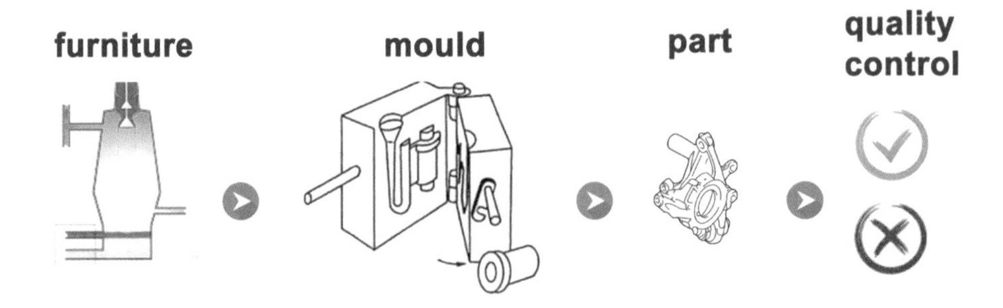

Fig. 1. Schematic diagram of the casting process flow.

NOK parts, thus reducing their impact on the profitability of the entire production plant. Besides reducing the rate of produced NOK parts, another objective pursued by the company was to verify that Data Science adds value in its decision making processes by addressing a hands-on practical scenario placed in its own facilities. As evinced throughout the rest of sections, this goal is achieved by sensing crucial phases and assets along the process and performing advanced data analysis on the captured data, the latter conforming the rationale of this research work.

For the experiments a server equipped with a Intel Xeon E5-2699V3/2.3 GHz Processor and 128 GB RAM has been used. In this server several software frameworks were deployed: NiFi for the ETL process, uxDB as the time series database, SQL Server as the SQL database and R for the development of the benchmark of machine learning models.

3 Hypothesis Generation Process

This section describes the methodology used in this work to select production variables than can be a priori relevant in order to reduce the defect rate of the plant. The work of a data scientist should not reside only in extracting the value from a dataset, but also in helping managers to understand the productive process. A key task is to ensure that the provided data source has only the variables that are truly linked (correlated) to the variable of interest. That is to say, variables that have an impact in the final result. This variable identification process, which has been approached in multiple methodologies such as 6sigma [22], is not always given as much importance as it should.

For starters, it is essential to generate multidisciplinary teams, including not only experts in data analysis, but also experts in the process. Previous knowledge, experience, and the support of the bibliography of other related studies become fundamental in the generation of hypotheses of causes that may affect the proposed problem. A brainstorming exercise with the process experts becomes fundamental, generating hypotheses about what factors may affect the process. This step, which may seem trivial, is critical and therefore very important to guide this work. Skipping it may cause experts involved in the process to generate simplistic hypotheses, that is, to postulate only hypotheses about the already used variables in this regard, forgetting others that may be fundamental.

One tool that helps to enrich this generation of hypotheses is the so-called Ishikawa diagram [18], which groups different sources of hypothesis into main topics. Expert's role in the process should be to nest ideas about each of these global concepts. Once the Ishikawa diagram is populated with ideas, priority should be given to variables on which realistic hypotheses can be formulated and verified in the short term, leaving the rest of variables initially discarded. These selected variables will be input to subsequent mathematical models, sensing them if necessary. If the initially selected variables allow to generate sufficiently explanatory mathematical models, then the exercise will be over and there would be no need for further iterating. However, if mathematical models are not conclusive enough with respect to the pursued goal (due to e.g. a lack of variables), then a new round of the Ishikawa methodology is commanded in order to select other variables that were initially discarded.

This is the correct way of sensing production processes in Industry 4.0 environments. Massively deploying sensors over the plant without performing this initial reflection among multidisciplinary members of the staff could entail higher costs, for example, due to high amounts of recorded data that do not possess any value. For the case study, the Ishikawa method concluded that the selected variables should be the following:

- Mold temperatures [°C], which are measured inside the mold cavity.
- Aluminum temperature [°C] when the material arrives at the mold.
- Pressure [bar] at which aluminum is injected into the mold.
- Physical position of the produced part inside the mold, which could make a difference in terms of the phenomena causing the defect.
- Flow $[m^3]$ of refrigerant liquid that enters the mold per second.
- Identifier of the operator working with the machine when the part at hand was produced.

All these variables are recorded every second during the part production process and stored into both SQL and time series databases. Once data are stored, the next step is to assess the influence of these variables to explain the rooting cause of the defects obtained in the produced parts. This is done by modeling this assessment as a supervised classification problem, which is approached by using different learning techniques as it is shown next.

4 Imbalanced Classification

Data contained in the databases after the capture campaign include time series corresponding to the variables recorded during the production of every part, and a label indicating whether the part was defective once it was produced. As a result, the set of supervised data examples is strongly imbalanced with respect to their class labels: `NOK` parts represent 5% of the total number of cases in the dataset. Constructing a classifier by learning from this dataset without properly accounting for the noted class imbalance may result in poor classification accuracy scores for the minority class (NOK), as the model will tend to assign all examples to the majority class (OK), which has lower practical interest. To overcome this issue three different types of approaches can be utilized: data preprocessing, cost-sensitive techniques and classification ensembles, which are described in what follows.

4.1 Preprocessing Techniques for Imbalanced Data

Data preprocessing techniques [34] hinge on modifying the initial dataset so as to a achieve a better class balance. Preprocessing can be done by oversampling the minority class, undersampling the majority class, or mixing both strategies. In the case under study different variants of the so-called Synthetic Minority Oversampling Technique (SMOTE, [9]) will be utilized. In general SMOTE oversamples the minority class by creating synthetic examples rather than by oversampling it with replacement. Results attained by this technique can be improved by combining it with other techniques: (1) Tomek links (SMOTE + TOMEK, [42]), which detects whether samples lie near the border between classes and therefore, can be removed so as to clean up the overlap between class regions; (2) Borderline (SMOTE + BORDERLINE [17]), in which the number of majority neighbors of each minority instance is employed to divide minority instances into 3 groups, safe, danger and noise, the latter being used to generate synthetic instances; and Edited Nearest Neighbors (SMOTE + ENN, [45]), which removes every example whose label differs from the class of at least two of its three nearest neighbors.

Once the initial dataset has been balanced, a classification algorithm can be applied. The following portfolio of classifiers has been considered:

- Decision Tree Classifier: in these models, the population is split in two or more homogeneous sets based on the most significant splitter. If the model is constructed with an only tree, then a decision tree is constructed. RPART [6] has been selected for this study, which creates recursive partitioning trees; C50 [31], an evolution of C4.5 trees that reduces overfitting; C4.5 algorithm and Ctree [19], which create conditional inference trees by taking into account the distributional properties of the measures; LMT [23], that generate logistic model trees; and Decision Stump [11], which implements only one split on the trees.
  Ensemble tree methods are a practical workaround to reduce bias and variance in classification. Different boosting and bagging ensembles has been considered. They are based on applying a weak learner (i.e. a data-sensitive classifier) to different *views* or versions of the training dataset, thereby producing a prediction by weighted majority voting of the produced set of classifiers. The considered boosting ensembles rely on decision trees as their weak learning algorithm: Adaboost [14], which reduce bias; and ADABOOST.M1 [13], a generalization of the previous one. Bagging operates in a similar fashion to

boosting to reduce the variance and thus, the chances of the overall model to overfit. Random Forest [5] is arguably the most utilized bagging approach in the literature when it comes to decision tree ensembles, reason for which it is included in the benchmark.

- Support Vector Machines (SVM) [40], which essentially construct an hyperplane that separates the classification regions in a highly dimensional projection of the dataset.
- K Nearest Neighbor classifiers, an instance based algorithm that classify instances based on the labels of the K closest instances in the dataset (under a given measure of closeness, distance or similarity [35]).
- Probabilistic classifiers (in particular Naïve Bayes [27]) which are based on fitting a probability distribution of the instances over the set of classes.
- Linear Discriminant Analysis [4] and its variants, in particular QDA [26], BINDA [16], DQDA [12], DLDA [12], MDEB [39] designed for small-sample based on minimum distance empirical Bayesian estimator, and MDMP [39].

4.2 Cost-Sensitive Classifiers

In conventional classification problems the aim is to minimize a measure of misclassification rate, for which every misclassification is valued equally when designing (training) the model. Cost-sensitive techniques [38] apply weights to the class to be predicted, so algorithms are modified to take into account the imbalanced data, focusing on the minority class. The benchmark will consider cost-sensitive versions of the aforementioned Random Forest [50] and SVM classifiers [8], as well as a cost-sensitive variant of the Extreme Gradient Boosting ensemble (XGBoost), an improvement of gradient boosting trees that use a more strict regularization to control overfitting [47].

4.3 Ensemble Techniques for Imbalanced Data

As anticipated in Subsect. 4.1, ensemble techniques consists of the union of several classifiers, whose overall prediction is produced by the combination of their predicted outcomes. There are two kind of approaches suited to deal with class imbalance [15]: cost-sensitive ensembles, whose training process take into account weights related to the occurrence of every class to be predicted; and ensemble imbalanced techniques [1], which embed a data preprocessing technique in a ensemble. In the experiments several methods from this second family of ensembles will be assessed: (1) RUSBoost [36], comprising Random Oversampling and Boosting; (2) SmoteBagging [25], combining SMOTE with Bagging; (3) SmoteBoost [10], correspondingly blending together SMOTE and Boosting; and (4) UnderBagging [30], which uses under-sampling embedded in a Bagging ensemble.

5 Results and Discussion

Once the benchmark was designed, a experimental assessment of the considered techniques was undertaken over a large manufacturing period over which 15234 parts were produced. After the brainstorming sessions, 17 predictors or features were engineered from the retrieved data captured in the plant. Due to confidentiality clauses such predictors cannot be made public to the community, and will be hereafter labeled as A to Q. However, it is important to note that their definition are a clear exponent of the claimed need for fusing together knowledge from different disciplines: the experience of the managers of the production chain and the capability of data scientists to model such a experience mathematically in the form of statistics and analytical expressions computed over the captured data.

This being stated, it is important to properly quantify the predictive potential of every model in the benchmark. To this end, a nested cross-validation [7] was conducted, which avoids optimistically biased estimates of performance that result from using the same cross-validation to set the values of the hyper-parameters of the model at hand. Nested cross-validation uses two loops: the inner loop conducts a grid search for model fitting (e.g. a conventional stratified K-fold cross-validation for every combination of hyper-parameters), whereas the outer loop measures the performance of the selected model that won in the inner loop on a separate external fold. The size of the outer and inner loops has been set to five folds, whereas the performance metric is the F1 score [33]:

$$F1 = 2 \cdot \frac{precision \cdot recall}{precision + recall}, \tag{1}$$

where precision is the fraction of `NOK` instances among the retrieved instances, and recall the fraction of `NOK` instances that have been retrieved over the total amount of `NOK` parts. It is important to observe that this selected score considers precision and recall over the class of interest, thereby is a good choice for imbalanced dataset, as it is the case in this work.

The reason for using the F1 score is to account for the relevance of the minority class in the class imbalance classification problem at hand.

Table 1 shows the F1 scores obtained when applying the preprocessing techniques, cost-sensitive schemes and ensembles enumerated in Sect. 4. A first look to the reported scores reveals that they fall within a narrow value range (0.60 to 0.72) given the high number of combinations considered in the benchmark. Nevertheless, it can be observed that some of the models perform notably better than others, yet differences are not high enough to be conclusive without any further statistical analysis.

The tight value interval in which the above scores are found to lie calls for a test of statistical significance. First an Anova test is performed to see if there are any differences between the different algorithms. If any, a Tukey test will be performed to see which algorithm(s) have the highest ranking. The Anova test rendered a p-value less than 2e−16, so differences exist between the models. Once this is done, differences between groups are sought to find the one that has the highest F1 score. To this end a Tukey test must be conducted, to clarify, based on confidence intervals, if there are significant pairwise differences among the models. Such differences resulted to be bounded between 0.3 and −0.3, but

Table 1. Mean± standard deviation of the F1 scores achieved by every classifier in the benchmark, measured over the outer folds

Preprocessing techniques for imbalanced data				
Algorithm	SMOTE	SMOTE + TOMEK	SMOTE + BORDERLINE	SMOTE + ENN
RandomForest	0.62 ± 0.04	0.57 ± 0.04	0.69 ± 0.02	0.68 ± 0.02
KNN	0.60 ± 0.02	0.54 ± 0.02	0.62 ± 0.02	0.61 ± 0.03
SVM	0.66 ± 0.01	0.70 ± 0.01	0.68 ± 0.04	0.68 ± 0.03
RPART	0.67 ± 0.02	0.72 ± 0.02	0.72 ± 0.02	0.69 ± 0.02
LMT	0.61 ± 0.02	0.61 ± 0.03	0.71 ± 0.03	0.67 ± 0.01
DecisionStump	0.72 ± 0.02	0.71 ± 0.02	0.68 ± 0.02	0.64 ± 0.01
NaiveBayes	0.67 ± 0.02	0.69 ± 0.02	0.69 ± 0.03	0.66 ± 0.02
QDA	0.66 ± 0.03	0.66 ± 0.01	0.67 ± 0.02	0.66 ± 0.01
LDA	0.68 ± 0.02	0.68 ± 0.01	0.66 ± 0.02	0.68 ± 0.03
BINDA	0.71 ± 0.02	0.71 ± 0.02	0.70 ± 0.03	0.51 ± 0.02
C50	0.62 ± 0.03	0.62 ± 0.03	0.69 ± 0.01	0.68 ± 0.02
CTree	0.68 ± 0.02	0.70 ± 0.03	0.71 ± 0.03	0.70 ± 0.02
DQDA	0.67 ± 0.02	0.69 ± 0.02	0.69 ± 0.03	0.66 ± 0.02
DLDA	0.70 ± 0.02	0.69 ± 0.02	0.70 ± 0.02	0.60 ± 0.02
MDEB	0.69 ± 0.02	0.69 ± 0.02	0.69 ± 0.03	0.68 ± 0.02
MDMEB	0.64 ± 0.02	0.63 ± 0.01	0.63 ± 0.03	0.64 ± 0.02
MDMP	0.68 ± 0.01	0.67 ± 0.02	0.68 ± 0.03	0.66 ± 0.02
EARTH	0.66 ± 0.02	0.71 ± 0.02	0.71 ± 0.02	0.69 ± 0.02
Adaboost	0.61 ± 0.03	0.53 ± 0.03	0.66 ± 0.01	0.67 ± 0.02
Adaboost.M1	0.60 ± 0.03	0.67 ± 0.02	0.66 ± 0.03	0.67 ± 0.04
Adabag	0.68 ± 0.02	0.72 ± 0.02	0.72 ± 0.01	0.70 ± 0.02
Cost-sensitive classifiers				
SVM	0.61 ± 0.03			
RF	0.72 ± 0.03			
XGBoost	0.68 ± 0.05			
Ensemble techniques for imbalanced data				
UnderBagging	0.64 ± 0.01			
SmoteBagging	0.68 ± 0.03			
RUSBoost	0.66 ± 0.02			
SmoteBoost	0.63 ± 0.06			

most were found to be less than 0.1. To inspect in depth these narrow differences, the model with the highest f1 score and the one with lowest f1 score were chosen based on the reported results in Table 1. An Anova test is again done among them to check differences. If there exists any, a Tukey test must be performed again to verify if the difference between them is significant. As shown in Fig. 2a, the confidence intervals of the best and worst models do not overlap with each other, hence concluding that Cost-Sensitive Random Forest is a model performing better (with statistical significance) than other models in this case study. Other options in the considered benchmark were found to perform similarly without statistically significant differences (e.g. Adabag + SMOTE + BORDERLINE or DecisionStump + SMOTE). However, the straightforward parametrization of Cost-Sensitive Random Forests and the provision of a measure of predictive importance of the engineered feature set as a byproduct of its training algorithm were decisive factors for the adoption of this model.

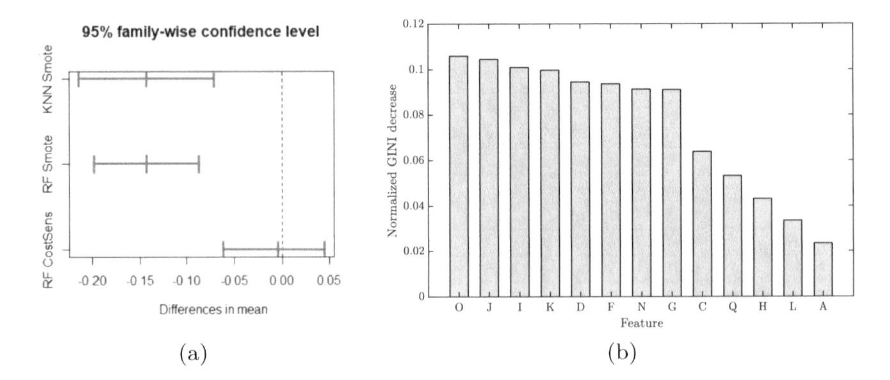

Fig. 2. (a) Confidence levels of Cost-Sensitive Random Forest ($F1 = 0.72 \pm 0.03$), RF + SMOTE($F1 = 0.62 \pm 0.04$) and KNN + SMOTE ($F1 = 0.60 \pm 0.02$); (b) Normalized GINI decrease of every feature provided by Cost-Sensitive Random Forest.

Once the best model has been selected and trained, actionable insights from the algorithm's performance are needed so that process engineers can have a system that helps them making informed decisions. In regards to the particular use case tackled in this paper, prescribed rules are recommended values for the parameters driving the machinery of the casting process, so that it runs under the most optimal conditions for the generation of as few scraps as possible.

For this purpose, first the variables evincing a highest predictive importance were identified based on the decreased Gini mean across all splits of the trees compounding the Cost-Sensitive Random Forest (Fig. 2b). Thereby a naïve decision tree is constructed using an imbalanced preprocessing technique (SMOTE + BORDERLINE) in order to construct a set of rules discriminating `OK` and `NOK` parts within the training dataset in hands. Such a tree is depicted in Fig. 3.

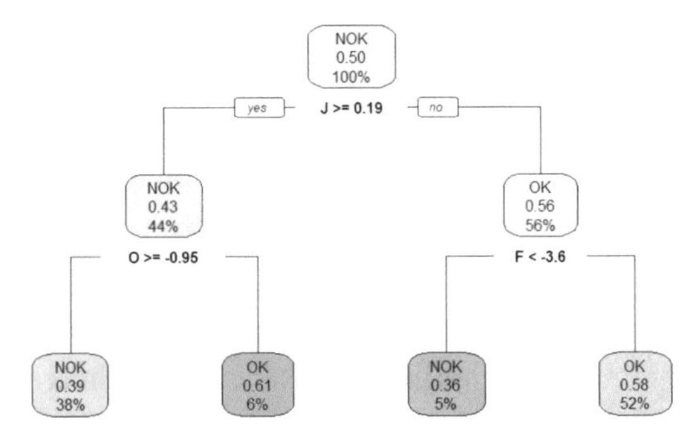

Fig. 3. Constructed decision tree from which to extract prescription rules for the manufacturing process under study.

Since features and machine parameters are closely linked to each other, the decision tree grown on the available data is by itself an actionable outcome to setup correctly the parameters of different elements of the machinery. Indeed, defective is reduced if the parameter configuration is set so as to yield values for the features within the limits imposed by the splits of the tree branches with prevalence of OK parts. From Fig. 3 two possible machine setups can be discriminated: (1) variable J must be kept below 0.19 and F must be equal to or above −3.6; and (2) variable J must be higher than or equal to 0.19, and O lower than −0.95. These prescribed rules were elevated to the managing board of the plant for its empirical testing in the plant. Results in terms of defect rate reduction were acknowledged to be highly satisfactory, lowering down the incidence of defective more than 5% with respect to the previous working regime of the casting pipeline.

6 Conclusions and Future Research

This paper has elaborated on a practical methodology to extract relevant parameters in an Industry 4.0 manufacturing process, which has been put to practice in a real aluminum injection chain process of a Spanish supplier of automotive parts. The described methodology departs from a brainstorming session aimed at the generation of hypotheses about the causes that may impact on the production of defective parts. Variables supporting the assessment of prioritized hypothesis are then identified and sensed from the monitored production chain, thereupon producing supervised time series data. Once stored in a database, the imbalanced nature of data made it necessary to resort to specialized models for supervised learning accounting for the relative low number of positive examples (OK parts) in the stored data. A benchmark of different alternatives was designed and run in order to select the most appropriate model for the detection of defect parts, which was finally chosen to be a cost-sensitive version of the Random Forest ensemble. The predictive importance of the variables produced by the learning algorithm of the selected model was further exploited to filter out the most valuable predictors. A decision tree grown on these selected feature subset served as a guideline for engineers and plant managers to settle optimal operational regimes over the machinery of the plant.

Future work will be devoted towards spanning the portfolio of models considered in this case beyond shallow learning, including time series analysis and Deep Learning. In connection with this latter area, a focus will be placed on studying how the predictive knowledge contained in the models trained over data produced in a certain plant can be extrapolated to other different albeit related processes in other plants, possibly by resorting to the latest advances in Transfer and Federated Learning.

Acknowledgements. This work has been partially developed by the Intelligent Systems for Industrial Systems research group supported by the Department of Education, Language policy and Culture of the Basque Government. Javier Del Ser also thanks the Basque Government for its funding support through the EMAITEK program.

References

1. Abolkarlou, N.A., Niknafs, A.A., Ebrahimpour, M.K.: Ensemble imbalance classification: using data preprocessing, clustering algorithm and genetic algorithm. In: 2014 4th International eConference on Computer and Knowledge Engineering (ICCKE), pp. 171–176. IEEE (2014)
2. Ang, J.H., Goh, C., Saldivar, A.A.F., Li, Y.: Energy-efficient through-life smart design, manufacturing and operation of ships in an industry 4.0 environment. Energies **10**(5), 610 (2017)
3. Babiceanu, R.F., Seker, R.: Big data and virtualization for manufacturing cyber-physical systems: a survey of the current status and future outlook. Comput. Ind. **81**, 128–137 (2016)
4. Balakrishnama, S., Ganapathiraju, A.: Linear discriminant analysis-a brief tutorial. Inst. Signal Inf. Process. **18**, 1–8 (1998)
5. Breiman, L.: Random forests. Mach. Learn. **45**(1), 5–32 (2001)
6. Breiman, L., Friedman, J., Olshen, R.A., Stone, C.J.: Classification and Regression Trees. Wadsworth, Belmont (1984). Google Scholar (1993)
7. Browne, M.W.: Cross-validation methods. J. Math. Psychol. **44**(1), 108–132 (2000)
8. Cao, P., Zhao, D., Zaiane, O.: An optimized cost-sensitive SVM for imbalanced data learning. In: Pacific-Asia Conference on Knowledge Discovery and Data Mining, pp. 280–292. Springer, Heidelberg (2013)
9. Chawla, N.V., Bowyer, K.W., Hall, L.O., Kegelmeyer, W.P.: SMOTE: synthetic minority over-sampling technique. J. Artif. Intell. Res. **16**, 321–357 (2002)
10. Chawla, N.V., Lazarevic, A., Hall, L.O., Bowyer, K.W.: SMOTEBoost: improving prediction of the minority class in boosting. In: European Conference on Principles of Data Mining and Knowledge Discovery, pp. 107–119. Springer, Heidelberg (2003)
11. Cunningham, SJ., Holmes, G.: Developing innovative applications in agriculture using data mining. In: The proceedings of the Southeast Asia Regional Computer Confederation Conference, pp. 25–29. Citeseer (1999)
12. Dudoit, S., Fridlyand, J., Speed, T.P.: Comparison of discrimination methods for the classification of tumors using gene expression data. J. Am. Stat. Assoc. **97**(457), 77–87 (2002)
13. Freund, Y., Schapire, R.E. et al.: Experiments with a new boosting algorithm. In: International Conference on Machine Learning, Bari, Italy, vol. 96, pp. 148–156 (1996)
14. Friedman, J.H.: Greedy function approximation: a gradient boosting machine. Ann. Stat. **29**(5), 1189–1232 (2001)
15. Galar, M., Fernandez, A., Barrenechea, E., Bustince, H., Herrera, F.: A review on ensembles for the class imbalance problem: bagging, boosting, and hybrid-based approaches. IEEE Trans. Syst. Man Cybern. Part C (Appl. Rev.) **42**(4), 463–484 (2012)
16. Gibb, S., Strimmer, K.: Differential protein expression and peak selection in mass spectrometry data by binary discriminant analysis. Bioinformatics **31**(19), 3156–3162 (2015)
17. Han, H., Wang, W.Y., Mao, B.H.: Borderline-SMOTE: a new over-sampling method in imbalanced data sets learning. In: International Conference on Intelligent Computing, pp. 878–887. Springer, Heidelberg (2005)
18. Hauser, S.: Analysis of requirement problems regarding their causes and effects for projects with the objective to model qualitative PRIs-empirical study (2018)

19. Hothorn, T., Hornik, K., Zeileis, A.: ctree: cponditional inference trees. The Comprehensive R Archive Network (2015)
20. Iqbal, R., Doctor, F., More, B., Mahmud, S., Yousuf, U.: Big data analytics and computational intelligence for cyber–physical systems: recent trends and state of the art applications. Future Gener. Comput. Syst. (2017, in Press)
21. Kretzschmar, J., Gebhardt, K., Theiß, C., Schau, V.: Range prediction models for e-vehicles in urban freight logistics based on machine learning. In: International Conference on Data Mining and Big Data, pp. 175–184. Springer, Heidelberg (2016)
22. Kwak, Y.H., Anbari, F.T.: Benefits, obstacles, and future of six sigma approach. Technovation **26**(5–6), 708–715 (2006)
23. Landwehr, N.: Logistic model trees. Master's thesis, Institute for Computer Science, University of Freiburg, Germany (2003)
24. Lee, J., Kao, H.A., Yang, S.: Service innovation and smart analytics for industry 4.0 and big data environment. Procedia Cirp **16**, 3–8 (2014)
25. Lertampaiporn, S., Thammarongtham, C., Nukoolkit, C., Kaewkamnerdpong, B., Ruengjitchatchawalya, M.: Heterogeneous ensemble approach with discriminative features and modified-smotebagging for pre-mirna classification. Nucleic Acids Res. **41**(1), e21 (2012)
26. Mika, S., Ratsch, G., Weston, J., Scholkopf, B., Mullers, KR.: Fisher discriminant analysis with kernels. In: Neural Networks for Signal Processing IX, pp. 41–48. IEEE (1999)
27. Murphy, KP.: Naive Bayes classifiers, p. 18. University of British Columbia (2006)
28. Nikolic, B., Ignjatic, J., Suzic, N., Stevanov, B., Rikalovic, A.: Predictive manufacturing systems in industry 4.0: trends, benefits and challenges. Annals DAAAM Proc. **28** (2017)
29. Park, C.W., Kwon, K.S., Kim, W.B., Min, B.K., Park, S.J., Sung, I.H., Yoon, Y.S., Lee, K.S., Lee, J.H., Seok, J.: Energy consumption reduction technology in manufacturing - a selective review of policies, standards, and research. Int. J. Precision Eng. Manuf. **10**(5), 151–173 (2009)
30. Qian, M., Zhang, D., Yue, X., Wang, S., Li, X., Teng, Y.: Analysis of different pigmentation patterns in mantianhong (pyrus pyrifolia nakai) and cascade (pyrus communis l.) under bagging treatment and postharvest uv-b/visible irradiation conditions. Scientia Horticulturae **151**, 75–82 (2013)
31. Quinlan, J.R.: C4. 5: Programming for Machine Learning, vol. 38, p. 48. Morgan Kauffmann, San Francisco (1993)
32. Ramírez-Gallego, S., Krawczyk, B., García, S., Woźniak, M., Herrera, F.: A survey on data preprocessing for data stream mining: current status and future directions. Neurocomputing **239**, 39–57 (2017)
33. Ricardo, B.Y., et al.: Modern Information Retrieval. Pearson Education India (1999)
34. del Río, S., Benítez, J.M., Herrera, F.: Analysis of data preprocessing increasing the oversampling ratio for extremely imbalanced big data classification. Trustcom/BigDataSE/ISPA, IEEE **2**, 180–185 (2015)
35. Ripley, B.D.: Pattern Recognition and Neural Networks. Cambridge University Press, Cambridge (2007)
36. Seiffert, C., Khoshgoftaar, T.M., Van Hulse, J., Napolitano, A.: Rusboost: a hybrid approach to alleviating class imbalance. IEEE Trans. Syst. Man Cybern. Part A Syst. Humans **40**(1), 185–197 (2010)

37. Shrouf, F., Ordieres, J., Miragliotta, G.: Smart factories in industry 4.0: a review of the concept and of energy management approached in production based on the internet of things paradigm. In: IEEE International Conference on Industrial Engineering and Engineering Management (IEEM), pp. 697–701 (2014)
38. Siers, M.J., Islam, M.Z.: Software defect prediction using a cost sensitive decision forest and voting, and a potential solution to the class imbalance problem. Inf. Syst. **51**, 62–71 (2015)
39. Srivastava, M.S., Kubokawa, T.: Comparison of discrimination methods for high dimensional data. J. Jpn. Stat. Soc. **37**(1), 123–134 (2007)
40. Suykens, J.A., Vandewalle, J.: Least squares support vector machine classifiers. Neural Process. Lett. **9**(3), 293–300 (1999)
41. Thirumalai, C., Duba, A., Reddy, R.: Decision making system using machine learning and Pearson for heart attack. In: International Conference of Electronics, Communication and Aerospace Technology (ICECA), vol. 2, pp. 206–210 (2017)
42. Tomek, I.: Two modifications of CNN. IEEE Trans. Syst. Man Cybern. **6**, 769–772 (1976)
43. Tong, H.L., Ng, H., Yap, T.V.T., Ahmad, W.S.H.M.W., Fauzi, M.F.A.: Evaluation of feature extraction and selection techniques for the classification of wood defect images. J. Eng. Appl. Sci. **12**(3), 602–608 (2017)
44. Wegner, D.M., Abell, J.A., Wincek, M.A.: Automated stochastic method for feature discovery and use of the same in a repeatable process. US Patent App. 14/997,854 (2017)
45. Wilson, D.L.: Asymptotic properties of nearest neighbor rules using edited data. IEEE Trans. Syst. Man Cybern. **3**, 408–421 (1972)
46. Wu, D., Jennings, C., Terpenny, J., Gao, R.X., Kumara, S.: A comparative study on machine learning algorithms for smart manufacturing: tool wear prediction using random forests. J. Manuf. Sci. Eng. **139**(7), 071018 (2017)
47. Xia, Y., Liu, C., Liu, N.: Cost-sensitive boosted tree for loan evaluation in peer-to-peer lending. Electr. Commer. Res. Appl. **24**, 30–49 (2017)
48. Yao, X., Moon, S.K., Bi, G.: A hybrid machine learning approach for additive manufacturing design feature recommendation. Rapid Prototyping J. **23**(6), 983–997 (2017)
49. Zhang, L., Cao, Y., Yang, F., Zhao, Q.: Machine learning and visual computing. Appl. Comput. Intell. Soft Comput. **2017** (2017)
50. Zhou, Q., Zhou, H., Li, T.: Cost-sensitive feature selection using random forest: selecting low-cost subsets of informative features. Knowl. Based Syst. **95**, 1–11 (2016)

A Hybrid Optimization Algorithm for Standardization of Maintenance Plans

Eduardo Gilabert[1(✉)], Egoitz Konde[1], Aitor Arnaiz[1], and Basilio Sierra[2]

[1] IK4-TEKNIKER Intelligent Information Systems Unit, Inaki Goenaga 5, 20600 Eibar, Spain
{eduardo.gilabert,egotiz.konde, aitor.arnaiz}@tekniker.es

[2] Department of Computer Science and Artificial Intelligence, University of the Basque Country (UPV/EHU), Manuel Lardizabal Ibilbidea, 1, 20018 San Sebastian, Spain
b.sierra@ehu.es

Abstract. This paper presents an algorithm to solve the combinatorial optimization problem in the definition of preventive maintenance plans. This problem is not easy to solve, since tasks performed on assets can be redundant or not required. The objective is a more accurate definition of plans to reduce maintenance costs, where different optimization algorithms can be used. Taking a single optimization algorithm approach could serve to find an optimal solution depending of the use case, but it does not find reliable results in a generic way. The new hybrid approach with 4 different algorithms shows that better results are obtained than the use of the individual optimization algorithms.

Keywords: Preventive maintenance · Optimization algorithms Planning

1 Introduction

The prescriptive analytics considers not only the business data, also how the decisions affect the costs and benefits accounts, and what restrictions and considerations should be considered in the actions that are going to be carried out. This automatically generates realistic action policies that have a direct impact on benefits. In many cases, the decision process is not systematic. The decision to be made is framed in a context defined by a series of restrictions. These restrictions set the conditions for a decision to be valid, but there are many valid decisions in the same context. Usually when making these decisions, there is implicit a way of assessing the quality of one decision over another, an objective function that allows to distinguish between a better valid solution and another, also valid, but which is worse (Fig. 1).

On the other hand, maintenance research and development is gaining importance worldwide because of the irruption of technologies that enable advanced maintenance solutions, including prescriptive analytics techniques. Maintenance management involves different concepts and activities (Garg and Deshmukh 2006): Part of the

J. Del Ser et al. (Eds.): IDC 2018, SCI 798, pp. 135–144, 2018.
https://doi.org/10.1007/978-3-319-99626-4_12

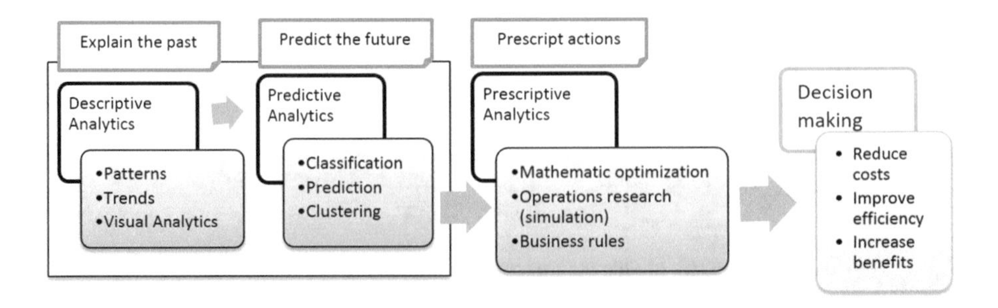

Fig. 1. Prescriptive analytics techniques and benefits.

literature deals with (a) development and selection of optimum maintenance strategies (Bevilacqua and Braglia 2000; Christer 1999; Gilabert et al. 2015; Koren and Ulsoy 2002; Scarf 1997); other major areas of research describe (b) the modelling and programming of the machine maintenance (Duffuaa et al. 1999) and (c) the theory of reliability, replacement and the determination of frequency of inspection (Jardine et al. 2001). Finally, other researchers concentrate on (d) simulations to measure a maintenance system (Banks 2005; Barros et al. 2003). In this paper our objective is to focus on a problem related to maintenance strategies that is scarcely tackled so far in literature: It is related to the development of standardized maintenance plans related to PM tasks, in complex maintenance scenarios with different fleet characteristics as well as maintenance resources. These scenarios appear in areas such as fleet transport systems (buses, trains), infrastructures, or elevators. Here, Preventive Maintenance (PM) tasks should be identified and developed to manage the failure. PMs are value-added tasks conducted using the least labor, downtime and materials to complete the tasks. Therefore, fleet maintenance requires standardized job plans. And the definition of plans concerning the PM tasks to be done by maintenance crew is a crucial problem prior to resource planning. The development of these plans is not trivial. Product maintenance specifications can include up to one hundred different actions, to be executed with different frequencies, and with some differences depending of product versions. On the other hand, not all maintenance crew is able to execute all actions (some of them may be better reserved to senior specialists) and there may be even different contracts regarding the frequency of inspections depending on the SLA (service level of agreement contracted) or the country legislation, to name only some examples. These characteristics fits as a challenge for automated decision-making trough expert optimization models. Decision making tools help users to improve the ability of solving problems and the capabilities of production systems (Medina-Oliva et al. 2015; Molenaar 2001). Optimization is also linked to decision-making, minimizing the maintenance costs or the ecological impact (Jiang et al. 2018). In this case, the standardization of maintenance plans cannot be solved usually by accurate optimization algorithms, which are those that guarantee finding an optimal solution. What disables the application of accurate optimization algorithms to these problems is the computation time. This fact has a consequence in the scientific community: it is the born of meta-heuristics techniques. Despite not finding the optimal solution, these techniques ensure that there is a good solution with the advantage that this search could be carried out in a reasonable execution time. Therefore, depending on the needs, there can be multiple optimization criteria simultaneously. It is possible to have different optimization systems or include commitments for calculations with partially connected criteria.

The rest of this paper is organized as follow. Section 2 introducing the problem of standardization of maintenance plans. Section 2 shows the proposed hybrid architecture and the specific algorithm inside. Section 4 presents the experimentation of the solution. Finally, the conclusions are given by integrating also future research directions.

2 Standardization of Maintenance Plans

The optimization methods addressed to tackle this problem that appears in literature are varied: Some of the optimization criteria are related to smoothing the preventive maintenance work load of the crew (Ben-Daya et al. 2009), other related to short-term (Dedopoulos and Shah 1995) and long term (Garg and Deshmukh 2006) scheduling of planned maintenance work. Hybrid optimization has been also used in maintenance for determining periodic inspections (Phan and Zhu 2015). Instead, this paper is centered in optimizing the group of tasks to be done by the maintenance crew in a fleet, taking into account that having too much different task plans could be difficult to be managed by the staff. A maintenance plans is defined as a set of multiple tasks of short duration. For instance, in case of a car could be (a) check tire pressure, (b) change oil and filters, and many other tasks with a defined duration and frequency. The definition and realization of these tasks depend on the configuration of the assets, applying also the corresponding legislations. Theoretically the optimal solution would consist on defining a different maintenance plan for each specific asset. In any case, that solution is not feasible, since the operators in charge of performing maintenance tasks cannot have a specific plan for each asset. A more efficient solution is to define a single plan (or a reduced set of plans) because the worker can perform the tasks efficiently. For this reason, maintenance plans must be standardized for all assets in a system, even if this implies the execution of tasks that are not strictly necessary on certain assets. Several maintenance factors or policies have an influence and must be considered when making the definition of the plans. These considerations suppose a problem of combinatorial optimization. The optimization problem must take into account the following factors, considering that it applies to a set of assets of a similar nature:

- Asset configuration: Each asset has a set of tasks to execute depending on its configuration.
- Tasks and frequency: Each task has a minimum frequency which depends on different legislations, safety criteria or customer requests.
- Client contract: The asset belongs to a client, that depending on the contract can have certain additional tasks (e.g. cleaning) or tasks that are made more frequently.
- Plans: The tasks to be executed on an asset are grouped in different plans, due to their different frequency requirements.

- Visits: The maintenance of an asset requires N plans carried out in V visits and by a certain operator skill. Obtaining the number of visits associated with each plan is another optimization problem.
- Worker profiles: the worker perform some tasks depending on their skills.
- Maximum plan duration.
- Balance parameter: the maximum difference of minutes between plans.

3 The Hybrid Algorithm

The hybrid algorithm tries to solve two different optimization problems. On the one hand, searching the best combination between plans and visits. Input data only provides the total number of plans and visits related to the asset. For instance, the requirements could be 2 plans in 6 visits, and the algorithm must give as a result that Plan 1 should be done in 2 visits and Plan 2 in 4 visits (or maybe other combination as 1-5 or 3-3). On the other hand, the hybrid algorithm searches the best standardization of tasks in the plans/visits to perform, according to the input requirements. Fig. 2 depicts the algorithm architecture used in the hybrid optimization. During the optimization, the tasks are distributed in the maintenance plans, which have previously been assigned a certain number of visits. This process is done in 4 different algorithms, obtaining as a result the best plan provided after the execution of:

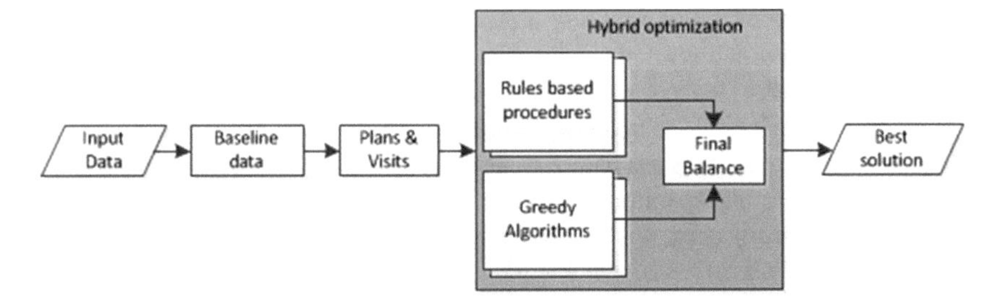

Fig. 2. Hybrid optimization algorithm architecture.

1. Rules procedure minimizing time.
2. Rules procedure balancing plans.
3. Greedy algorithm minimizing time.
4. Greedy algorithm balancing plans.

The main reason of using a hybrid approach is due to 2 different objectives are managed at the same time:

- The minimization of the total time required to perform the plan on the asset, that is, the sum of all tasks in visits, in every plan.

$$min \sum_{i=1}^{i=n} \sum_{j=1}^{j=m} t_j v_i \tag{1}$$

where n is the number of plans, m the number of tasks related to plan i, t_j the duration of task j, and v the number of visits of plan i.

– The minimization among the differences in duration of plans (maximum balance)

$$\min\left(\left(\max_i \sum_{j=1}^{j=m} t_j v_i\right) - \left(\min_i \sum_{j=1}^{j=m} t_j v_i\right)\right) \tag{2}$$

Next main parts of the algorithm are described in detail.

3.1 Baseline Data

During the first step the input information must be provided to perform the optimization. In detail:

- Main asset configuration. The concept of a plan standardization is not to provide the specific configuration, but it should be defined in an upper level (e.g. in case of cars, they could be gas or diesel, involving different maintenance tasks).
- Country legislation. Usually countries specify different frequency inspections related to maintenance tasks.
- Client contract type, the levels of SLA. For instance: Low cost, Normal and VIP.
- Worker profiles, in relation with their skills. For instance, Technician and Specialist.
- For each worker profile, number of maintenance plans and total visits by year.
- Maximum duration of a maintenance plan, in minutes.
- Maximum difference among plan durations in minutes (B), in order to provide balanced plans.

According to the previous inputs, the baseline data is obtained. This data consists on a set of tasks and their compulsory visits for each task. The Optimal Time is defined as the sum of these task durations multiplied by their visits.

$$OT = \sum_{i=1}^{i=m} t_i v_i \tag{3}$$

where m is the total number of tasks, t_i is the duration of the task and v_i the minimum number of visits of t_i. This Optimal Time will be used for validation purposes. The evaluation criterion is defined as the difference between the Total Time and the Optimal Time.

3.2 Plans vs Visits

As introduced before, obtaining the number of visits associated with each plan is another optimization problem. The maintenance of an asset requires N plans carried out in V visits and by a certain operator skill. This is calculated by another instance of the hybrid optimization algorithm. However, this execution is performed in a small number of iterations (<10). The target of Plans & Visits process consists in selecting the most appropriate plans/visits relation so that the search is more limited, and the search algorithm starts from a defined configuration. The step in this process starts with the calculation of all possible combinations of visits and plans by operator profiles. For example, for an expert profile, 3 plans and 6 visits (Expert-3-6), the number of visits of each plan could be: 1-1-4; 1-2-3; 2-2-2. For each possible combination, the hybrid algorithm with 10 iterations is executed. The best result obtained (E.g. 1-2-3), is the configuration used in the next step.

3.3 Rules Procedures

This first procedure follows this sequence:

- Tasks are ordered from largest to smallest duration.
- Sequentially, each task is assigned to a plan, using this algorithm:
  - Selection of possible candidate plans: do not include the task previously and meet the operator profile requirement.
  - Select the plan where the number of times matches the number of visits of the task, or in its absence the smallest absolute difference.
  - In case of a tie, the plan with the minor number of visits is selected.

The second procedure follows a similar sequence, except in case of tie in last step: the plan with the minor duration is selected in this case. This change produces a more balanced result.

3.4 Greedy Algorithms

The greedy algorithms are based on covering the complete space of possible solutions to the problem. We have selected the backtracking algorithm. This technique is the direct application of the search method known as first in depth. The algorithm consists on:

1. Select an option among the possible ones.
2. For each selection, consider every possible option recursively, in a loop (search in depth).
3. Return the best solution found of all evaluated solutions.

In our case, the first greedy algorithm performs this sequence: (1) If a task can be carried out by several profiles, all the possibilities are considered; (2) Combinatorial search of assignment of all possible tasks to each visit is done. That is, a visit could be considered as an empty box where the different combination of tasks is inserted.

The second greedy algorithm is similar except it works with sorted visits: in each search iteration, the visits are sorted by their absolute difference between the number of visits of the task and the number of visits of plan, prioritizing a maximum balance among the plans.

3.5 Final Balance

The Final Balance consists of 2 procedures which works with the found solution, trying to improve it. The final balance follows this sequence:

1. A greedy algorithm that tries to remove the tasks of the plans that exceed the maximum allowed or the longest in case of not having it. It moves them to the plans of shorter duration.
2. If the flattening criterion (B) is not yet met, another greedy algorithm adds unrealized tasks to the plans of shorted duration. This Final Balance is applied in the execution of the 4 algorithms implemented.

4 Use Case

The hybrid algorithm has been validated using truck forklift fleet job standardization. In this use case are considered two of forklifts in a fleet: electrical and diesel. Also, the maintenance crew is categorized into specialists and technicians. Due to a user requirement, a short execution time is needed, and the execution of greedy instances has been limited to 1000 iterations, enabling an execution time minor than 2 min. Moreover, two different experiment have been done. In the first experiment, the country evaluated has a lot of diversity in maintenance plans, and the maintenance company needs to standardize the plans to solve the problem of diversity. The frequency of maintenance is set to 2 months, by one specialist. The optimization has been executed setting 3 plans at 3 min of maximum difference in duration (B = 3). The Optimal Time (OT) is 82.57 min. Table 1 provides results when the greedy algorithm is isolated from previous rules procedures.

Table 1. Greedy algorithm isolated in Specialist-2-6

Skill	Plan	Visits	Duration (min)
Specialist	#1	1	19.14
Specialist	#2	2	18.46
Specialist	#3	3	18.07

Annual total time: 110.27 min.
Difference OT: +27.70 min.

However, when using the hybrid algorithm, the annual total time for maintenance is almost 2 min shorter.

The usage of this hybrid optimization algorithm improves the results comparing the usage of just a greedy algorithm, and even the computational time is improved. Actually, the standardization of plans using the final result in Table 2, supposes an increase of 7% in annual total time maintenance, and therefore a similar increase in maintenance costs, but the company consider this as an indirect cost required that allows to save money in learning, reduce human mistakes caused by having too much different plans and simplify maintenance planning for maintenance crew load balancing.

Table 2. Hybrid algorithm result in Specialist-2-6

Skill	Plan	Visits	Duration (min)
Specialist	#1	1	18.10
Specialist	#2	2	18.71
Specialist	#3	3	17.60

Annual total time: 108.32 min.
Difference OT: +25.75 min.

In the second experiment considers the situation where specialist executes the maintenance every month during a year, setting only 2 different plans. In an annual maintenance, the first plan is executed every 6 months and the second one the rest of months. OT is 261.88 min. The initial results are shown in Table 3.

Table 3. Specialist-2-12

Skill	Plan	Visits	Duration (min)
Specialist	#1	10	18.65
Specialist	#2	2	42.7

Annual total time: 271.91 min.
Difference OT: +10.03 min.

In this case the objective is to save money using two types of workers (specialist and technician). In this case the specialist executes 2 plans every 6 months, and the technician executes 1 plan in 10 visits, setting a maximum difference among plan durations of 1 min by profile (B = 1). Table 4 shows how the expert skill improves the time while keeping the basic tasks duration, related to previous situation in Table 3. This result means an improvement of 10 min by truck. Assuming a truck fleet of 100 units and considering that technician cost/hour is 40% less than using the specialist worker, the savings are close to 32% in this case.

Table 4. Specialist -2-12; Technician -1-10; B = 1

Skill	Plan	Visits	Duration (min)
Specialist	#1	1	38.85
Specialist	#2	1	38.53
Technician	#3	10	18.65

Annual total time: 261.88 min.
Difference OT: 0 min.

5 Conclusion

In an optimization problem, a hybrid approach improves the results obtained in comparison with the use of one single algorithm. This paper presents a new algorithm to solve the combinatorial optimization problem in standardization of maintenance plans, using a hybrid approach with 4 different algorithms. The usage of this algorithm allows to reduce costs or to enable an easier later planning. This algorithm is completely scalable in number of tasks, plans, visits and operator profiles. The hybrid combinations have been implemented in a multi-stage process. However, after this work we considered that it is possible to take advantage of the combination of algorithms in other ways, for instance, using a hierarchical approach. Next step will consider an Estimation of Distribution Algorithm (EDA) on the top, that will be fed by the optimal results found by 4 different kind of optimization algorithm: Rule-based, greedy, evolutionary and simulated annealing. The future objective will be to improve this algorithm expanding both the optimization typologies and the way to combine their results.

References

Banks, J.: Discrete-Event System Simulation. Pearson Education, New Delhi (2005)

Barros, A., Grall, A., Berenguer, C.: A maintenance Policy Optimized with imperfect monitoring and/or partial monitoring. In: Annual Reliability and Maintainability Symposium, pp. 406–411. IEEE (2003)

Ben-Daya, M., Ait-Kadi, D., Duffuaa, S.O., Knezevic, J., Raouf, A.: Handbook of Maintenance Management and Engineering, vol. 7. Springer, London (2009)

Bevilacqua, M., Braglia, M.: The analytic hierarchy process applied to maintenance strategy selection. Reliab. Eng. Syst. Safe **2000**(70), 71–83 (2000)

Christer, A.H.: Developments in delay time analysis for modeling plant maintenance. J. Oper. Res. Soc. **50**, 1120–1137 (1999)

Duffuaa, S.O., Raouf, A., Campbell, J.D.: Planning and Control of Maintenance Systems: Modeling and Analysis. Wiley, New York (1999)

Garg, A., Deshmukh, S.G.: Maintenance management: literature review and directions. J. Qual. Maint. Eng. **12**(3), 205–238 (2006)

Gilabert, E., Fernandez, S., Arnaiz, A., Konde, E.: Simulation of predictive maintenance strategies for cost-effectiveness analysis. Proc. Inst. Mech. Eng., Part B: J. Eng. Manuf. **231** (13), 2242–2250 (2015). 0954405415578594

Dedopoulos, I.T., Shah, N.: Optimal short-term scheduling of maintenance and production for multipurpose plants. Ind. Eng. Chem. Res. **34**(1), 192–201 (1995)

Jiang, A., Dong, N., Tam, K.L., Lyu, C.: Development and optimization of a condition-based maintenance policy with sustainability requirements for production system. Math. Probl. Eng. **2018**, 19 pages (2018)

Koren, Y., Ulsoy, A.G.: Vision, principles and impact of reconfigurable manufacturing systems. Powertrain Int. **5**(3), 14–21 (2002)

Jardine, A.K.S., Banjevic, D., Wiseman, M., Buck, S., Joseph, T.: Optimizing a mine haul truck wheel motors' condition monitoring program Use of proportional hazards modeling. J. Qual. Maint. Eng. **7**(4), 286–302 (2001)

Medina-Oliva, G., Weber, P., Iung, B.: Industrial system knowledge formalization to aid decision making in maintenance strategies assessment. Eng. Appl. Artif. Intell. **37**, 343–360 (2015)
Moubray, J.M.: Reliability-Centered Maintenance, 2nd edn. Industrial Press, New York (1997)
Phan, D.T., Zhu, Y.: Multi-stage optimization for periodic inspection planning of geo-distributed infrastructure systems. Eur. J. Oper. Res. **245**(3), 797–804 (2015)
Scarf, P.: On the application of mathematical models in maintenance. Eur. J. Oper. Res. **99**(3), 493–506 (1997)

Labelling Drifts in a Fault Detection System for Wind Turbine Maintenance

Iñigo Martinez[1(✉)], Elisabeth Viles[2], and Iñaki Cabrejas[1]

[1] NEM Solutions, 20009 San Sebastian, Spain
{imartinez,icabrejas}@nemsolutions.com
[2] University of Navarra - Tecnun, 20018 San Sebastian, Spain
eviles@tecnun.es

Abstract. A failure detection system is the first step towards predictive maintenance strategies. A popular data-driven method to detect incipient failures and anomalies is the training of normal behaviour models by applying a machine learning technique like feed-forward neural networks (FFNN) or extreme learning machines (ELM). However, the performance of any of these modelling techniques can be deteriorated by the unexpected rise of non-stationarities in the dynamic environment in which industrial assets operate. This unpredictable statistical change in the measured variable is known as concept drift. In this article a wind turbine maintenance case is presented, where non-stationarities of various kinds can happen unexpectedly. Such concept drift events are desired to be detected by means of statistical detectors and window-based approaches. However, in real complex systems, concept drifts are not as clear and evident as in artificially generated datasets. In order to evaluate the effectiveness of current drift detectors and also to design an appropriate novel technique for this specific industrial application, it is essential to dispose beforehand of a characterization of the existent drifts. Under the lack of information in this regard, a methodology for labelling concept drift events in the lifetime of wind turbines is proposed. This methodology will facilitate the creation of a drift database that will serve both as a training ground for concept drift detectors and as a valuable information to enhance the knowledge about maintenance of complex systems.

Keywords: Failure detection · Predictive maintenance
Concept drift · Supervised learning · Neural networks
Extreme learning machine · Wind turbine · Expert labelling

1 Introduction

The prevailing competitive marketplace demands companies from asset-intensive industries to create cost-efficient processes, both in manufacturing and in maintenance. Unforeseen breakdowns not only can cause expensive downtime, but

J. Del Ser et al. (Eds.): IDC 2018, SCI 798, pp. 145–156, 2018.
https://doi.org/10.1007/978-3-319-99626-4_13

also safety and environmental detriments that may lead to injuries or fatalities, as well as enormous legal expenses. In this sense, the ability to forecast machinery failure is vital for reducing maintenance costs, operation downtime and safety hazards. It is therefore essential to develop failure detection techniques to monitor the health of a system, which are encompassed in Condition Based Maintenance (CBM) strategies [1,2] and more recently, in prognostics and health management (PHM) [3–5].

A failure management system consists of several interrelated modules [6]. At first, failure prone situations are identified, using a continuous measure that judges the current situation as more or less failure prone; this is known as the failure detection stage. Then, after failure detection, the diagnosis stage is invoked, in order to find out where the error is located and where its root cause may be. Once the diagnosis is completed, maintenance actions are scheduled.

This article focuses in the failure detection module, which provides real-time monitoring of industrial processes by forecasting machinery health based on condition data and predicting possible incipient failures. Several industrial sectors have adopted these type of systems to improve their maintenance processes and to manage their assets more effectively. In particular, we are interested on applications to the maintenance of wind turbines, where various strategies have already been developed [7,8].

Technical approaches for building models in incipient failure detection systems can be categorized broadly into data-driven, model-based, and hybrid approaches. More details regarding the mentioned techniques can be found at [9]. This study deals with a pure data-driven approach, where information coming in real-time from different sensors is taken into account, and detection of possible anomalies is provided after learning the normal and expected behaviour from the industrial assets.

By using data sources, different strategies can be followed to build normal behaviour models, such as stochastic models, machine learning algorithms, Bayesian and fuzzy classifiers, time series prediction or pattern recognition. In the analyzed failure detection system a state-of-the-art neural network, Extreme Learning Machine (ELM), has been applied, due to its ability to easily model dynamic non-linear behaviours [10], and also because of its wide use in the prognostics of industrial systems and wind turbines [11–16].

Whatever the selected model may be, an initial training batch is necessary for learning relationships between variables. Taking a feed-forward neural network as example, the network is trained with an historical data set, a fixed and static information about the past events of the system. By comparing the expected behaviour with its real and current functioning, both a normal behaviour deviation degree as well as an estimation certainty degree are obtained. These are used to recognize an anomaly, which afterwards can be related to a known failure mode [17]. These failure detection strategies are based on the hypothesis that any asset should behave similarly under resembling conditions, concluding that a deviation from the model should speak for a symptom of dysfunction.

However, the performance of any of these modelling techniques can be deteriorated by the unexpected rise of non-stationarities in the dynamic environment in which industrial assets operate. Unpredictable statistical changes in the measured variable are desired to be detected by means of statistical detectors and window-based approaches. In order to evaluate the effectiveness of current detectors and also to design an appropriate novel technique for an specific wind turbine maintenance application, it is essential to dispose beforehand of a characterization of the existent drifts. Under the lack of information in this regard, in this article a methodology for labelling concept drift events in the lifetime of wind turbines is proposed.

Having introduced the necessity for predictive maintenance and failure detection systems on Sect. 1, the rest of the article is structured as follows: the irruption of non-stationary data in failure detection systems is explained on Sect. 2, along with the main adaptation techniques to this issue. On Sect. 3, an expert-based platform for drift labelling is described. Finally, Sect. 4 presents the main conclusions and paths for future work and improvement.

2 Non-stationary Data in Traditional Machine Learning

One of the problems that arise from the presented failure detection systems is that data is expected to be independent and identically distributed (i.i.d) along all the lifetime of the asset [18]. Unfortunately, in real industry applications, this assumption does not often hold true. In fact, due to the non stationarity in the monitored variables of the assets, the normality models need to be checked periodically and actions need to be taken to adapt to the new normality [19].

Regarding failure detection systems, the non stationarity can be due to multiple causes, such as sensors recalibration, replacement of a component during a corrective maintenance intervention, the wearing of some mechanical component, etc. The causes of the non-stationary data are critical to decide whether to take actions or not. Yet, model adaptation should only occur if the detection quality is seriously affected. An illustrative example can help understand this issue: the wearing of components or the lost of efficiency represent events that are desired to detect by studying deviations from the normality model, implying that the model should not be updated. Unfortunately, the representation of these events vary enormously from asset to asset and from time to time. This high variability implies some difficulties to find patterns that help to differentiate between a need-to-update situation and a maintain-model one.

Even though failure prediction systems are very vulnerable to non-stationarities, at the moment this issue being solved with human supervision. However, companies are monitoring more and more assets each year, and they are starting to reach up to a point where a manual checking is unfeasible. As an example, the recreation of the normal wind turbine operation can consist of about 10 models of subsystems or critical parts of the wind turbine. For 1.000 wind turbines, an architecture of 10,000 normal behaviour models would be managed. Moreover, taking into account current trends, with more wind farms being installed worldwide each year, it is necessary to include an automated system that guarantees the normality of each model for the current situation. This automated system would reduce as much as possible the human intervention in the system, thus allowing the scalability of the detection technology.

2.1 Concept Drift Theory

Current artificial intelligence and machine learning based applications live under the assumption that the systems they predict are static and stationary [18], neglecting that in reality, the world and the data generating processes are often dynamic and non-stationary. Unexpected changes in the environment or perturbed, incorrect and missing data can statistically detach the measured variable from the monitored feature, inducing to wrong decisions. Moreover, the spreading of online deployments with learned models gives increasing urgency to the development of efficient and effective mechanisms to address learning in the context of non-stationary distributions, or as it is commonly called concept drift [18–23]. Hence, concept drift can be defined as the unpredictable statistical change in measured variables, making the static and time-invariant model useless.

A concept is statistically defined on the literature [18] as the joint probability distribution of predictors (independent) variables and response (dependent) variables at a given period t: $Concept = P_t(X,Y)$, where X denotes a random variable over vectors of predictor values, and Y represents a random variable over the output or response. A concept drift occurs whenever a pair of periods of time t and u depict different joint distributions: $P_t(X,Y) \neq P_u(X,Y)$. According to Bayesian theory, a change in the joint probability distribution can occur due to a change either in the prior probability distribution $P_t(Y)$, in the class conditional probability distribution $P_t(X|Y)$, or in the posterior probability distribution $P_t(Y|X)$.

There is the implicit assumption that drift occurs over discrete periods of time, bounded before and after by stable periods without drift [20]. Under this assumption, the speed of drift can be quantified [18] as sudden if there is a sharp boundary, as gradual if the transition is smooth, and reoccurring if drifts repeat over time.

A quantitative characterization of drifts can be found in [20]. These measurements would help to create a taxonomy of drifts for a given application. Here the most relevant measures of drift are summarized:

Magnitude Distance between the concepts at the start t and end u of the period of drift. The metric used is usually the Hellinger Distance or the Total Variation Distance.

$$Magnitude_{t,u} = D(t,u) \tag{1}$$

For example, the Hellinger Distance H is a metric that measures the difference between two probability distributions P and Q, and can be defined as:

$$H_{P,Q} = \frac{1}{\sqrt{2}} ||P - Q||^2 \tag{2}$$

Duration The elapsed time over which a period of drift occurs. This measurement is critical for the differentiation of a fault an a concept drift.

$$Duration_{t,u} = u - t \tag{3}$$

Path Length Length of the path that the drift traverses during a period of drift, i.e. the sum of the drift magnitude between pairs of consecutive points.

$$PathLen_{t,u} = \lim_{n \to \infty} \sum_{k=0}^{n-1} D\Big(t + \frac{k}{n}(u - t), t + \frac{k+1}{n}(u - t)\Big) \tag{4}$$

2.2 Adaptation Techniques

Under the presence of concept drift, passive or active solutions are available in order to prevent a loss of performance in the detection quality or in the prediction accuracy. On one hand, passive strategies continuously adapt the model without the need of detecting a change. They aspire to maintain an up-to-date model at all times, either by retraining the model on the most recently observed samples, or by enforcing an ensemble of classifiers [24]. As the stability-plasticity dilemma defines, a single-classifier model is not able to retain existing knowledge and at the same time learn new information, so ensembles of classifiers are usually put to work [25]. Ensembles continuously update the weights of the fusion rule or create and remove models from the pool, being more accurate than single-classifier models, and easily incorporating new data or forgetting irrelevant knowledge.

However, in a failure detection system, the prediction accuracy is expected to decrease when a failure appears, and thus the objective is not to improve the performance, but to preserve the detection quality. In addition, passive adaptation techniques allow the normality model to learn from anomalies and failures as well, thus expanding the decision boundary and reducing the detectability of the system. Consequently, passive solutions should be discarded for failure detection applications, and active solutions should be preferable.

Active solutions rely on triggering mechanisms, that is, detection methods that indicate whether a drift has occurred or not based on a change in the statistics of the data-generating process. Change-detection tests [26–31] can be triggered either by monitoring the distribution of unlabeled observations or by a change in the prediction performance of the model. Whatever the monitored signal may be, these methods explicitly localize the change point in time, and invoke the substitution of the model with a new one, trained with recent data, that maintains the prediction accuracy and the overall detectability. In this article the attention is onto the active detection of concept drifts, whereas posterior actions, such as the retraining of the model are not studied.

3 Proposed Methodology: Drift Labelling

Failure detection technologies that take into consideration deviations from normal behaviour models will suffer from the concept drift problem as time passes on. This maintainability problem of failure detection technologies is in fact a reality in industrial applications. Appropriate adaptation techniques to tackle concept drifts can only be designed if there is a previous knowledge of the types of drifts that can appear. Therefore, this research rises from the current lack of knowledge regarding drift detection in real industrial assets, where a taxonomy of drifts is hard to discern and theoretical drift detectors have unknown performances.

In order to address the problem of normality changes in the performance of complex systems this article proposes a methodology for the identification, record keeping, and posterior classification of drifts in the specific case study of wind turbines maintenance. The result is the development of a platform that allows experts to create a reference dataset of behaviour changes in the normal performance of wind turbines. This will serve as training ground for the posterior classification of such changes, and also to test the effectiveness of those detectors in drift situations. At the same time, this will enhance the knowledge about the operation of the wind turbines, and as a consequence, the improvement of the applied failure detection systems.

3.1 Platform Overview

The process starts by collecting data from the wind turbines. In this regard, the operating and environmental conditions of virtually all wind turbines in operation today are recorded by the turbines' supervisory control and data acquisition (SCADA) system in 10-minute intervals [32]. The number of signals available to the turbine operator varies considerably between different manufacturers as well as between generations of turbines by the same manufacturer. The recreation of the normal wind turbine operation can consist of 10 models of critical parts, each of them monitoring a known variable. In this article the wind turbine's power model is illustrated as example (Fig. 1), where the ambient temperature, the wind speed and the wind turbulence act as predictors. These variables are available in almost all SCADA systems.

As was stated on Sect. 2.1, drifts can take diverse forms (sudden, gradual...) and appear on any variable (inputs, output). The proposed method tries to identify changes in the posterior probability distribution $P_t(Y|X)$, that is, in the output-inputs relationship, since those changes directly affect the performance of the failure detector. Therefore, the normal behaviour model residual—difference between the actual power and the predicted one—will be shown to the expert.

In order to automatize the labelling, an interactive web platform has been developed using the Shiny package in R. This platform shows the model residual as a time series, and experts are allowed to label any drift or event by dragging the desired period on the chart. The selected period is then classified, based on

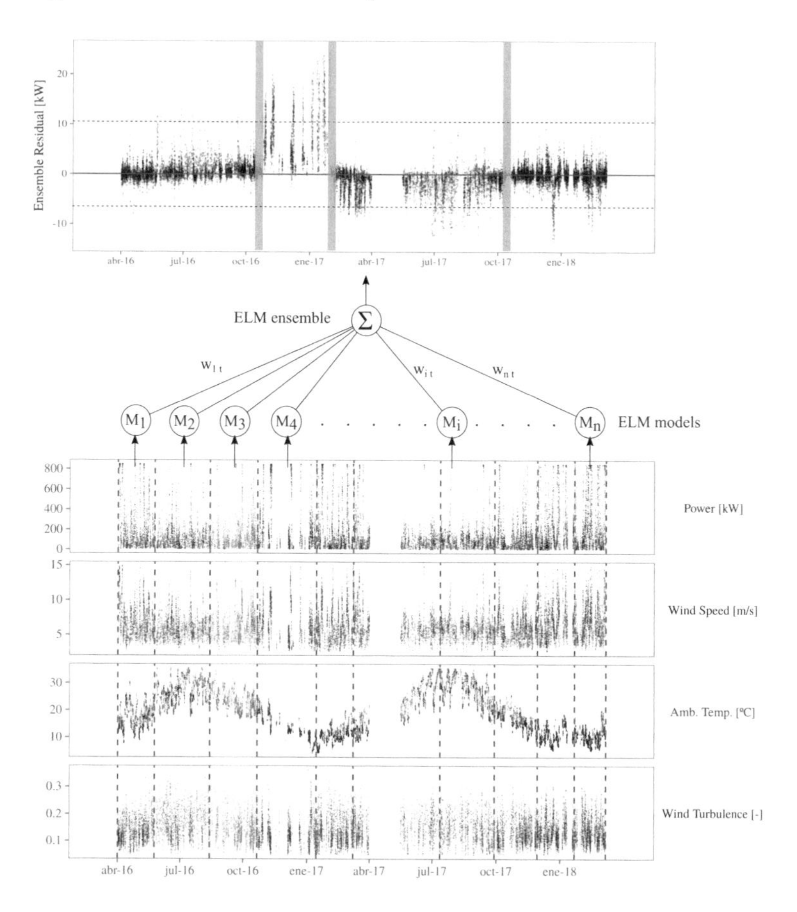

Fig. 1. Each ELM model is trained with independent batches and combined onto the ensemble

the severity of the drift and the possible causes (sensor mis-calibration, maintenance action, power limitation, etc.). Each time a period is labeled, it will be visualized on the chart and registered on the database, along with the expert-user and model information.

Due to the fact that this platform uses visual inspection to label periods of drift, the results can be subjected to the expert's own criteria. In subsequent work, expert's opinion will be complemented with information coming from the asset's maintainers, such as maintenance work orders. In order to evaluate appropriately the effectiveness of the labelling system and reduce this source of uncertainty, redundancy between multiple experts is proposed. Apart from allowing multiple users to label the same cases, each user will indicate a qualitative measure of confidence about their labels.

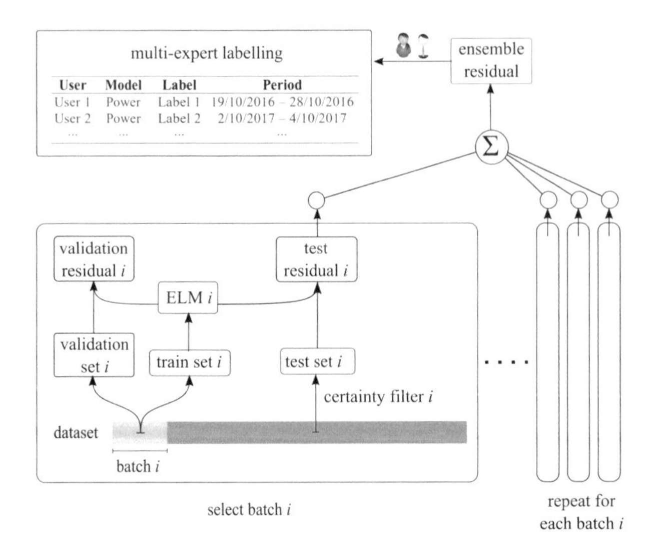

Fig. 2. Overview of the multi-expert platform. Experts label periods of drift on the ensemble residual. Each ELM model contributes to the ensemble, and is trained with an independent batch.

3.2 ELM Ensemble

Normal behaviour models are created by training a single layer extreme learning machine (ELM), which has the same topology as a single layer FFNN, with the difference that the hidden layer weights are not tuned, but randomly assigned. This slight modification makes ELM training extremely fast, since only output weights need to be optimized, which is worked out by a simple ridge regression [33]. ELM have already been used to model the real operation of wind turbines, showing great generalization properties [13–16].

In order to make the model predictions more robust and trustworthy, an ensemble of models [34] is used rather than an individual model. Each model from the ensemble is trained with a different batch of data, which will avoid two issues:

- On one side, the training data can contain failures and behaviour changes, which should not be learned. Rather than selecting a fault-free training set, the generalization capability of the weighted combination of individual models will automatically not learn anomalies and features that rarely appear, but the general behaviour, also guaranteeing a low validation error.
- On the other side, when a subset of the whole data is used as training set, it is possible that it may not completely represent the entire numerical space of the variables. A model should only predict the response when there is enough certainty about the inputs, otherwise the estimation may not be reliable. An example of such situation can be seen with the ambient temperature, where a

clear trend occurs along the year. The entire year should be selected in order to cover all temperature values and make certain predictions.

The residual calculation process has been illustrated on Fig. 1. Each individual ELM model is trained with a fixed size batch. A random sample of the batch is used for validation, whereas the remaining trains the model. The rest of the dataset is first filtered by certainty to the training set, and then is passed to the model to predict the response. The certainty filter makes sure the inputs are known to the model. This process is repeated for each batch, as has been represented on Fig. 2, thus obtaining the ensemble of models, and the combination of residuals, which is shown to the experts.

3.3 Preliminary Results

A comparison of the labelled periods of drift with the triggering of some active detectors has been included in Table 1. A group of 4 experts labelled events from 98 wind turbines ranging 3 years of data. Overall, the experiment showed that even though detectors match the manually labelled periods, captured by a high sensitivity, low precision results put into evidence that detectors trigger when other type of drifts happen or when a noisy signal undermines the performance of the detectors.

Table 1. Precision and sensitivity of other detectors on manually labelled periods, which are assumed to be the true condition. The bold numbers indicate the maximum of each metric.

	ADWIN	CUSUM	GMA	HDDM_A	HDDM_W	PH	SEED	SeqDrift1	SeqDrift2	STEPD
Precision	0.422	0.408	**0.571**	0.304	0.432	0.413	0.370	0.412	0.519	0.261
Sensitivity	0.711	**0.816**	0.316	0.737	0.500	0.684	0.789	0.737	0.737	**0.816**

The detectors were implemented using MOA, an open-source framework [35] for dealing with massive evolving data streams. The optimal parameters for each detector were extracted from the article by Gonçalves et al. [26]. The definition for the tested detectors can be found in [18,26–31].

4 Conclusions and Further Work

In this article, a methodology for the identification, record keeping, and posterior classification of drifts in a wind turbine maintenance case has been presented. In real complex systems, normality changes are not as clear as in artificial datasets, so it is difficult to implement active adaptation techniques, such as drift detectors. By following the proposed method, a database of behaviour changes can be created with the help of experts, serving both as a training ground for concept drift detectors and as a valuable information to enhance the knowledge of complex systems.

In addition, with this novel method, different types of drifts – sudden, gradual, recurrent – can be classified with a certain degree of objectivity. From the obtained labels, drift features such as magnitude, duration and path length can be estimated as well.

As was stated on Sect. 1, the problem of concept drift can be found in any failure prediction system that uses deviations from a normal behaviour model to identify incipient failures. The methodology explained on Sect. 3 has only been applied to a wind farm maintenance case. This methodology should be easily extended to other complex systems where no information about the possible behaviour changes exists.

After obtaining the characterization of drifts, the effectiveness of current drift detectors has been evaluated for this specific industrial application. A future line of research will focus on the design of novel drift detection techniques, to be tested in real-world environments with the purpose of improving the quality of the current drift detectors, tested in Sect. 3.3. In this way, an ensemble of detectors [36–39] could provide robustness in real life applications, where the representation of drift events vary enormously from asset to asset and from time to time. Just like individual classifiers are limited by the stability-plasticity dilemma, a single detector is not able to discern several types of drifts. In fact, most concept drift detectors found on the literature are based on assumptions about the distribution of the variables and try to identify changes whenever their stated hypothesis do not hold true.

Acknowledgements. This research has been supported by NEM Solutions, a technology-based company focused that provides intelligent maintenance of complex systems to O&M businesses.

References

1. Jardine, A.K., Lin, D., Banjevic, D.: A review on machinery diagnostics and prognostics implementing condition-based maintenance (2006). https://doi.org/10.1016/j.ymssp.2005.09.012
2. Heng, A., Zhang, S., Tan, A.C.C., Mathew, J.: Mech. Syst. Signal Process. **23**(3), 724 (2009). https://doi.org/10.1016/j.ymssp.2008.06.009
3. Zio, E., Kadry, S.: Diagnostics and prognostics of engineering systems: methods and techniques, pp. 333–356 (2012). https://doi.org/10.4018/978-1-4666-2095-7.ch017⟩
4. Vichare, N.M., Pecht, M.G.: IEEE Trans. Compon. Packag. Technol. **29**(1), 222 (2006). https://doi.org/10.1109/TCAPT.2006.870387
5. Cheng, S., Azarian, M.H., Pecht, M.G.: Sensor systems for prognostics and health management (2010). https://doi.org/10.3390/s100605774
6. Salfner, F., Lenk, M., Malek, M.: ACM Comput. Surv. **42**(3), 1 (2010). https://doi.org/10.1145/1670679.1670680
7. Yang, W., Court, R., Jiang, J.: Renew. Energy **53**, 365 (2013). https://doi.org/10.1016/j.renene.2012.11.030
8. Sheng, S., Veers, P.: Machinery Failure Prevention Technology (MFPT): The Applied Systems Health Management Conference 2011, vol. 2, p. 5, October 2011

9. Al-Turki, U.M., Ayar, T., Yilbas, B.S., Sahin, A.Z.: SpringerBriefs in Applied Sciences and Technology, pp. i–iv (2014). https://doi.org/10.1007/978-3-319-06290-7
10. Kubat, M.: Knowl. Eng. Rev. **13**(4), S0269888998214044 (1999). https://doi.org/10.1017/S0269888998214044
11. Liu, Z., Gao, W., Wan, Y.H., Muljadi, E.: IEEE Energy Conversion Congress and Exposition (ECCE) (August), 3154 (2012). https://doi.org/10.1109/ECCE.2012.6342351
12. Pelletier, F., Masson, C., Tahan, A.: Renew. Energy **89**, 207 (2016). https://doi.org/10.1016/j.renene.2015.11.065
13. Qian, P., Ma, X., Wang, Y.: Autom. Comput. (ICAC) **11** (2015). https://doi.org/10.1109/IConAC.2015.7313974
14. Qian, P., Ma, X., Zhang, D.: Energies **10**(10), 1583 (2017). https://doi.org/10.3390/en10101583
15. Saavedra-Moreno, B., Salcedo-Sanz, S., Carro-Calvo, L., Gascón-Moreno, J., Jiménez-Fernández, S., Prieto, L.: J. Wind Eng. Ind. Aerodyn. **116**, 49 (2013). https://doi.org/10.1016/j.jweia.2013.03.005
16. Wan, C., Xu, Z., Pinson, P., Dong, Z.Y., Wong, K.P.: IEEE Trans. Power Syst. **29**(3), 1033 (2014). https://doi.org/10.1109/TPWRS.2013.2287871
17. Garcia, M.C., Sanz-Bobi, M.A., del Pico, J.: Comput. Ind. **57**(6), 552 (2006). https://doi.org/10.1016/j.compind.2006.02.011
18. Gama, J., Žliobaitė, I., Bifet, A., Pechenizkiy, M., Bouchachia, A.: ACM Comput. Surv. **46**(4), 1 (2014). https://doi.org/10.1145/2523813
19. Žliobaite, I.: International Conference on Machine Learning, pp. 1009–1017 (2010). https://doi.org/10.1002/sam
20. Webb, G.I., Hyde, R., Cao, H., Nguyen, H.L., Petitjean, F.: Data Mining Knowl. Discov. **30**(4), 964 (2016). https://doi.org/10.1007/s10618-015-0448-4
21. Tsymbal, A.: Computer Science Department, Trinity College Dublin **4**(C), 2004 (2004). http://citeseerx.ist.psu.edu/viewdoc/summary?doi=10.1.1.58.9085
22. Hoens, T.R., Polikar, R., Chawla, N.V.: Prog. Artif. Intell. **1**(1), 89 (2012). https://doi.org/10.1007/s13748-011-0008-0
23. Ditzler, G., Roveri, M., Alippi, C., Polikar, R.: Learning in nonstationary environments: a survey (2015). https://doi.org/10.1109/MCI.2015.2471196
24. Krawczyk, B., Minku, L.L., Gama, J., Stefanowski, J., Woźniak, M.: Inf. Fusion **37**, 132 (2017). https://doi.org/10.1016/j.inffus.2017.02.004
25. Mouret, J.B., Tonelli, P.: Stud. Comput. Intell. **557**, 251 (2015). https://doi.org/10.1007/978-3-642-55337-0_9
26. Gonçalves, P.M., De Carvalho Santos, S.G.T., Barros, R.S.M., Vieira, D.C.L.: A comparative study on concept drift detectors (2014). https://doi.org/10.1016/j.eswa.2014.07.019
27. Sobolewski, P., Woźniak, M.: Adv. Intell. Syst. Comput. **226**, 329 (2013). https://doi.org/10.1007/978-3-319-00969-8_32
28. Sebastião, R., Gama, J.: 14th Portuguese Conference on Artificial Intelligence, pp. 353–364 (2009). http://citeseerx.ist.psu.edu/viewdoc/summary?doi=10.1.1.233.1180
29. Santos, S., Barros, R., Gonçalves, P.: Proceedings - International Conference on Tools with Artificial Intelligence, ICTAI, vol. 2016, pp. 1077–1084, January 2016. https://doi.org/10.1109/ICTAI.2015.153
30. Pears, R., Sakthithasan, S., Koh, Y.S.: Mach. Learn. **97**(3), 259 (2014). https://doi.org/10.1007/s10994-013-5433-9

31. Ross, G.J., Adams, N.M., Tasoulis, D.K., Hand, D.J.: Pattern Recogn. Lett. **33**(2), 191 (2012). https://doi.org/10.1016/j.patrec.2011.08.019
32. Bangalore, P., Patriksson, M.: Renew. Energy **115**, 521 (2018). https://doi.org/10.1016/j.renene.2017.08.073
33. Huang, G.B., et al.: Neurocomputing **70**(1–3), 489 (2006). https://doi.org/10.1016/j.neucom.2005.12.126
34. Lan, Y., Soh, Y.C., Huang, G.B.: Ensemble of online sequential extreme learning machine (2009). https://doi.org/10.1016/j.neucom.2009.02.013
35. Bifet, A., Holmes, G., Kirkby, R., Pfahringer, B.: J. Mach. Learn. Res. **11**, 1601 (2010). http://portal.acm.org/citation.cfm?id=1859903
36. Maciel, B.I.F., Santos, S.G.T.C., Barros, R.S.M.: Proceedings - International Conference on Tools with Artificial Intelligence, ICTAI, vol. 2016, pp. 1061–1068, January 2016. https://doi.org/10.1109/ICTAI.2015.151
37. Sobolewski, P., Woźniak, M.: J. Univ. Comput. Sci. **19**(4), 462 (2013)
38. Woźniak, M., Ksieniewicz, P., Kasprzak, A., Puchała, K., Ryba, P.: Advances in Intelligent Systems and Computing, vol. 525, pp. 27–34 (2017). https://doi.org/10.1007/978-3-319-47274-4_3
39. Du, L., Song, Q., Zhu, L., Zhu, X.: Comput. J. **58**(3), 457 (2015). https://doi.org/10.1093/comjnl/bxu050

Time Series Forecasting in Turning Processes Using ARIMA Model

Alberto Jimenez-Cortadi[1(✉)], Fernando Boto[1], Itziar Irigoien[2], Basilio Sierra[2], and German Rodriguez[1]

[1] Tecnalia, Industry and Transport Division, Paseo Mikeletegui 7, 20009 San Sebastian, Spain
alberto.jimenezcortadi@tecnalia.com
[2] University of Basque Country (UPV/EHU), Manuel Lardizabal Ibilbidea 1, 20018 San Sebastian, Spain

Abstract. A prediction model which is able to predict the tool life and the cutting edge replacement is tackled. The study is based on the spindle load during a turning process in order to optimize productivity and the cost of the turning processes. The methodology proposed to address the problem encompasses several steps. The main ones include filtering the signal, modeling of the normal behavior and forecasting. The forecasting approach is carried out by an Autoregressive Integrated Moving Average (*ARIMA*) model. Results are compared with a robust *ARIMA* model and show that the previous preprocessing steps are necessary to obtain greater accuracy in predicting future values of this specific process.

Keywords: Robust statistics · Process normality detection
Time series forecasting · ARIMA models

1 Introduction

The maintenance of machining processes is one of the most studied topics in manufacturing industry, due to its relevance for the process behavior and the economic impact on the production plans. Two different maintenance methods are applied in most of the industries; On the one hand, preventive maintenance, based on the theoretical life of the tool, which is commonly instigated by a periodic alarm. On the other hand, predictive maintenance uses strategies such as inferring the remaining useful life (*RUL*) of the tool, which mostly depends on the wear measurements. It is critically important to assess the RUL of an asset while in use since it has impacts on the planning of maintenance activities, spare parts provision, operational performance, and the profitability of the owner of an asset [17].

The objective of predictive maintenance is to predict when equipment failure may occur and to prevent a failure by performing maintenance. Ideally, this approach enables the system to have the lowest possible maintenance frequency. Many techniques can be used in order to achieve this aim, which heavily

J. Del Ser et al. (Eds.): IDC 2018, SCI 798, pp. 157–166, 2018.
https://doi.org/10.1007/978-3-319-99626-4_14

depends on the process studied. These techniques must achieve two main goals: being effective at predicting failures and providing enough time for upcoming maintenance.

This study is applied to a machining process in the automotive sector. The workpiece is a bearing conic ring. The material machined was Steel 100Cr6. The machining operation was cylindrical turning with two toolposts, one to mechanize the internal surface of the part and another one for the external surface.

The experiments were conducted on a CNC turning centre with two toolposts which can be moved into two axes independently with longitudinal and cross feed movement. The cutting tool holder used was a C3-MTJNR-22040-16 equipped with a TNMG160408-PF insert for the external surface and a special tool holder equipped with a WNMG060408-PM insert for the inner diameter. During the machining process the rotating revolutions were set to 1800 rpm.

This work has the aim of improving tool life in a particular turning process. This will enable the machine to have a lower number of tool changes which means that more pieces can be machined each year. This goal can be obtained by the prediction of the spindle load charge. Robust ARIMA methods are able to preprocess the signal before obtaining the prediction. In this study, a new preprocessing method is developed with expert process knowledge and ARIMA model is used for prediction. The developed method is compared with robust model and results shows that the preprocessing is necessary since an increase of the accuracy of the robust method forecasting is obtained.

The paper is organized as follows. Section 2 shows an analysis of the state of the art about remaining useful life RUL and different approaches generally used. Section 3 presents the methodology applied, whose results are in Sect. 4. Finally, conclusions and future work are exposed in Sect. 5.

2 Related Work

Extracting theoretical and empirical knowledge of an industrial process is one of the most important issues in manufacturing. The machining industry is interested in modeling their machines in order to have more efficient maintenance planning for unexpected behaviors. Data based modeling is a step forward when improving maintenance strategies applied to the machine in terms of predicting remaining useful life RUL, and knowing if the tooling process will maintain its function over time [10]. A wide variety of models have been developed for time series prediction, such as support vector machines(SVM) and their regression version Support Vector Regressors (SVR) [12], finite element models and also Autoregressive integrated moving average ($ARIMA$).

Among these, the $ARIMA$ model has turned out to be one of the most popular time series forecasting models. Yuan et al. [19] applied this method in order to predict China's primary energy consumption and compares it with a Gray differential model (GM). ARIMA models are also used in applications such as prediction of electricity prices [8] and gas demand [3]. The $ARIMA$ forms a general class of linear models that have historically seen wide use in modeling and forecasting time series [4]. ARIMA models are derived from the more common Autoregressive Moving Average (ARMA) model, which models a time series using two parts, an autoregressive (AR) part and a Moving Average (MA) part. However, since ARMA models can only be used to model stationary processes ARIMA models are often employed, as they can be used to model non-stationary time series signals. Applications and studies in RUL with ARIMA models can be seen in some works in the literature. For example in [13,16] evaluate the performance of different methods, including ARIMA, to predict the RUL of Lithium-ion Batteries is evaluated. Wu et al. [18] proposes an improved ARIMA approach to predict the future machine status, the prediction of the vibration characteristic in rotating machinery.

In this work, we analyze the ARIMA approach by comparing a preprocessing methodology with a robust ARIMA method.

3 Methodology

In this section the methodology used for experiments is explained. First of all we need to monitorize the process, where the acquired data is obtained with specific hardware for machining processes, which provides several measurements obtained during the turning process. These signals are obtained with a frequency of 5 Hz, which may be increased in future developments. In this work spindle load is studied as this is the main variable regarding the wear and the tool life of the process according to expert knowledge.

The used signal is obtained as a collection of observations made sequentially through time, a time series (see Fig. 1). These time series are divided into smaller ones that are associated to a tool and when the tool is changed the series ends, which are denominated manufacturing sequence. Each tool machines multiple pieces until the tool is changed due to a preventive maintenance rule. Once a sequence is finished, that tool is removed and another tool is taken for the next manufacturing sequence. The signal corresponding to a piece is cropped by experience with another signal acquired in the process, so we are able to indicate the beginning and the end of each piece as can be seen in Fig. 2.

In this work, each sequence is studied separately applying the methodology explained in the following paragraphs: characterization, outliers detection, filtering, abnormal behavior detection and forecasting.

3.1 Characterization of the Signal

This is a matter of representing or characterizing a signal fragment defined as a time series T, by means of a vector of values F. This vector can have a temporal relation or not, depending on the technique that is used.

$$T = (t_1, ...t_n) \Rightarrow F = (f_1...f_P) \qquad where\ n >> P\,,$$

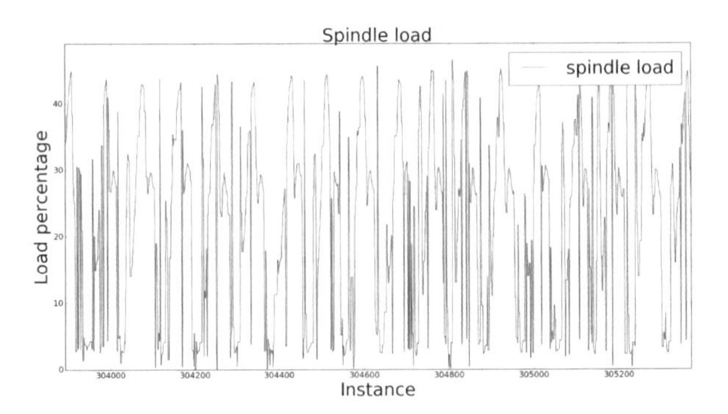

Fig. 1. Spindle load time series associated to a tool including 16 pieces.

That way, the initial length of the series T, n, is notably reduced to length P. Several time-series representations have been proposed as exposed in [2]. As a starting point of this work, we are going to use a simple but effective feature according to expert knowledge and it is explained int he following lines:

As the pieces are machined with a similar effort from the tool as can be appreciated in Fig. 2, in the case studied here the maximum value applied by the spindle is selected as the most important value; these maximum values are ordered according to the manufacturing sequence providing a new time series. These time series are the ones we will analyze in the rest of this work to infer the RUL of the tool.

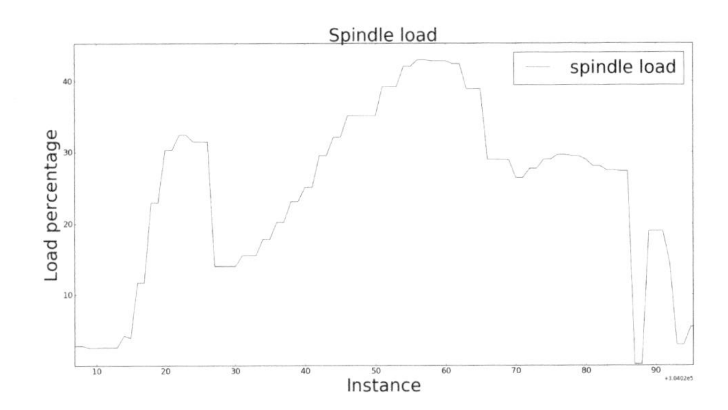

Fig. 2. Spindle load charge on a piece, extracted from the time series.

3.2 Outlier Detection

Let F be the time series signal obtained from the characterization and p the number of piece corresponding the value $F(p)$. Low acquisition frequency implies to have some not well measured values. This produces some outliers in the time series, which do not represent the monitorizing process well. Those values, known as outliers, must be deleted from the signals. To achieve this aim [14] proposed *three sigma edit rule* method which has been applied by [9] in his work on instability detection.

This simple robust method allows a median value to be obtained of each force component $Me = Median\{F\}$ and the corresponding absolute standard deviations, which are given as:

$$|F(p) - Me_F|, \quad p = 1, \ldots, P;$$

where P is the number of machined pieces for a series.

The median value of these deviations is calculated and divided by 0.6745 as indicated in [14]:

$$MADN(p) = \frac{Median\{|F(p) - Me_F|\}}{0.6745}$$

The robust three-sigma edit rule (see Maronna [14]) establishes that the observation $F(p)$ is an outlier if:

$$|F(p) - Me_F| > r$$

where r is a threshold value. Under the normality assumption, often $r = 3$ is set (hence the "three-sigma rule") and it is the value set in this work. Observations beyond the threshold are considered as outliers. In order to compare different series, detected outliers (machined pieces) are removed from all the original series.

3.3 Filtering

In order to visualize any trend in the time series, a filter to smooth the trend is applied. There are lot of filters in the bibliography that enables us to achieve this goal. In this study, a comparison between 2 of them is made as was suggested in [11].

On the one hand, Savitzky-Golay [6] is applied, which uses a polynomial approximation of the nearest data, where polynomial degree g and the number of points q are parameters that must be set up by experience. Let F be the time series which is filtered. On each point p, a new value is determined by the function:

$$F(p) = a_0 + a_1(p - p_0) + a_2(p - p_0)(p - p_1) + \ldots + a_g(p - p_0)(p - p_1) \ldots (p - p_{g-1})$$

where $a_0, \ldots, a_g$ values are determined by the g nearest points $(g-1)/2$ previous ones, $(g-1)/2$ ones after and the pth value. Note that g is an odd number and $q > g$.

On the other hand, Wiener filter [5] is studied, which is linear least square error filter, based on statistical estimations of the signal. The Wiener filter works in the frequency domain and requires knowledge of the variances of the signals.

3.4 Abnormal Behavior Detection

To achieve this objective, first, those series which are considered as abnormal must be detected and removed. Due to the fact that the current maintenance of the process is based on preventive maintenance with a simple rule, there are series that must be removed as they do not fulfill the criteria laid by the rule.

Second, series which are extremely different from the rest are deleted in order to keep the normal ones. This is obtained through the application of an iterative K-Means, where if a cluster C_1 is small compared to C_2, that is 10 times smaller or more, all the signals in C_1 are removed and the process is repeated until the clusters have a similar shape. This concludes with an amount of series that are similar. Based on those series, a mean series, which represents the normal behavior of the process is calculated.

3.5 Forecasting

Forecasting is always a goal in a huge variety of fields. In machining processes, predicting breakdowns enables the process to be more efficient. To this aim, a wide variety of techniques have been used. One of the most important and widely used time series forecasting models is the autoregressive integrated moving average $ARIMA$ model [7], which is used in this study.

In the case of a time series model, the effect of outliers is more serious than in a regression model because the presence of an outlier at time t affects not only that period but also subsequent periods. Robust ARIMA estimates based on a recursive robust filter replaces the observations suspected of being outliers with cleaned values.

For the forecasting of the spindle two approaches are compared; First, an ARIMA model for forecasting is developed in [15], which is combined with preprocessing approach explained in previous sections: outlier detection (Sect. 3.2), filtering (Sect. 3.3) and abnormal detection (Sect. 3.4). Second, a robust ARIMA method is used [1]. This robust method is applied to raw data without any preprocessing.

4 Results

As previously explained, for each machined piece a representative value is extracted, which in this case is the maximum load during the process. Figure 2 shows which value is extracted from the signal. This feature extraction is made for all the machined pieces of the series, and ordering those values sequentially a new time series is obtained (see Fig. 3). This series is shown to have some outliers, which are caused by the data acquisition frequency, and must be removed

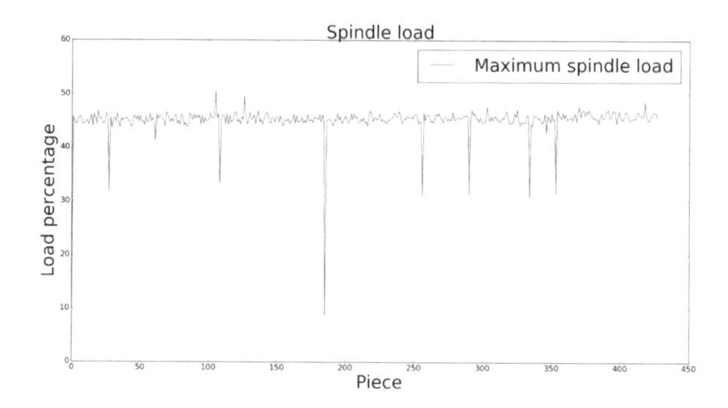

Fig. 3. One time series spindle load before outlier removal.

in order to obtain an appropriate signal. This removal is obtained with the three-sigma edit rule method.

To reduce signal noise and to detect sequence trends, a filter is applied to the signal. In this study, two different filters are applied, which are shown in Fig. 4. It is appreciated that *Savitzky-Golay* filter achieves the aim of detecting sequence trends better, thus this is the method selected for the results analysis. The series are now as shown in Fig. 5.

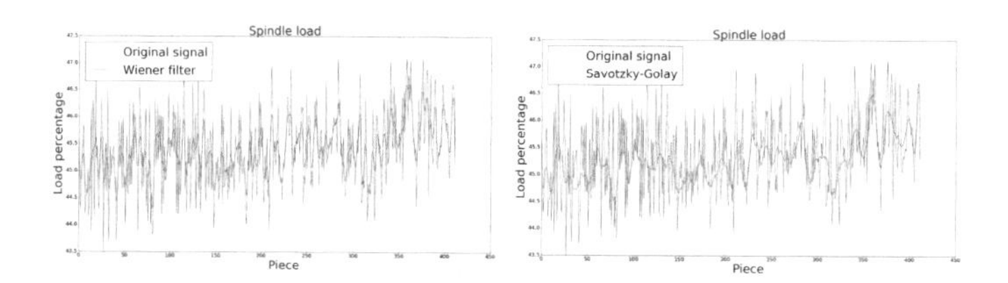

Fig. 4. Both *Wiener*(left) and *Savitzky-Golay*(right) filters applied to one signal.

At this point, in order to obtain a sequence which explains the process normality, some of the series must be removed. First, series which are less than 500 pieces are removed, which is the number of pieces for preventive maintenance in the machine nowadays (see Fig. 5). Next, a K-Means clustering is applied with two clusters in order to remove abnormal series such as the purple one shown in Fig. 5.

With all the 22 remaining series, a mean value is calculated, which is considered as a normality model and which is used for predicting spindle load values. This signal is used only for the *ARIMA* model, while robust *ARIMA* is used with raw data.

Signal is divided into training and tested values, where 96% of the data are extracted for the training one. In Table 1, obtained results are exposed, where it can be seen that the prediction with *ARIMA* model with the preprocessing

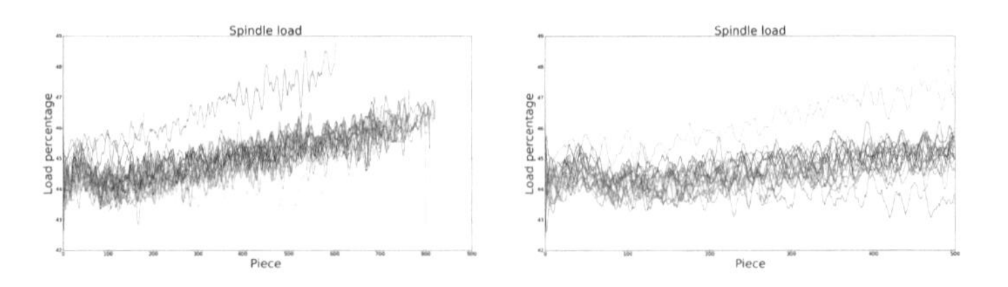

Fig. 5. All the series filtered (left). The series with more than 500 pieces machined (right) (Color figure online).

alternative is better. This suggests that the preprocessing performed on this process is necessary.

Table 1. Differences between real data and predicted value.

	t_1	t_2	t_3	t_4	t_5	t_6	t_7	t_8	t_9	t_{10}
Robust method	12.68	15.23	14.28	14.18	15.40	16.14	15.51	13.61	14.79	14.85
ARIMA model	2,20	4,66	3,62	3,83	4,96	5,76	4,99	3,10	4,33	4,38

Finally, based on the whole signal, future predicted values are calculated. Results (see Table 2) show that the spindle load is still controlled and therefore more pieces could be machined.

Table 2. Forecasting spindle values values.

	t_1	t_2	t_3	t_4	t_5	t_6	t_7	t_8	t_9	t_{10}
ARIMA model	45.24	45.27	45.32	45.21	45.40	45.25	45.35	45.33	45.32	45.35

5 Conclusions and Future Work

In this paper a methodology to predict spindle load values is carried out. Within this approach a new preprocessing method based on the expert knowledge of the process and *ARIMA* forecasting are developed. The results are compared with a robust *ARIMA* method which a priori could work in the presence of outliers and without previous preprocessing. Nevertheless it is shown that the preprocessing included in the approach increases the accuracy obtained, appearing the preprocessing as necessary. Forecasting values for ARIMA model has shown that machine could extend its *RUL*.

In order to obtain a better abnormal behavior detection to model the process, new strategies will be developed with more variables captured by the hardware. Obtaining a different characterization of each piece and increasing the acquired frequency for the spindle load signal is also proposed for future work.

References

1. Bianco, A.M., Martinez, E., Ben, M.G., Yohai, V.: Robust procedures for regression models with ARIMA errors. In: COMPSTAT, pp. 27–38. Springer, Heidelberg (1996)
2. Boto, F., Lizuain, Z., Cortadi, A.J.: Intelligent maintenance for industrial processes, a case study on cold stamping. In: Proceeding of International Joint Conference SOCO 2017-CISIS 2017-ICEUTE 2017 León, Spain, pp. 157–166. Springer, Heidelberg, 6–8 September 2017
3. Buchanan, W.K., Hodges, P., Theis, J.: Which way the natural gas price: an attempt to predict the direction of natural gas spot price movements using trader positions. Energy Econ. **23**(3), 279–293 (2001)
4. Butler, S.: Prognostic algorithms for condition monitoring and remaining useful life estimation. National University of Ireland, Maynooth (2012)
5. Chen, J., Benesty, J., Huang, Y., Doclo, S.: New insights into the noise reduction wiener filter. IEEE Trans. Audio Speech Lang. Process. **14**(4), 1218–1234 (2006)
6. Chen, J., Jönsson, P., Tamura, M., Gu, Z., Matsushita, B., Eklundh, L.: A simple method for reconstructing a high-quality ndvi time-series data set based on the savitzky-golay filter. Remote Sens. Environ. **91**(3), 332–344 (2004)
7. Chen, J., Zeng, G.-Q., Zhou, W., Du, W., Lu, K.-D.: Wind speed forecasting using nonlinear-learning ensemble of deep learning time series prediction and extremal optimization. Energy Convers. Manag. **165**, 681–695 (2018)
8. Contreras, J., Espinola, R., Nogales, F.J., Conejo, A.J.: Arima models to predict next-day electricity prices. IEEE Trans. Power Syst. **18**(3), 1014–1020 (2003)
9. Cortadi, A.J., Boto, F., Irigoien, I., Sierra, B., Suarez, A.: Instability detection on a radial turning process for superalloys. In: Proceeding of International Joint Conference SOCO 2017-CISIS 2017-ICEUTE 2017 León, Spain, pp. 247–255. Springer, Heidelberg, 6–8 September 2017
10. Galar, D., Palo, M., Van Horenbeek, A., Pintelon, L.: Integration of disparate data sources to perform maintenance prognosis and optimal decision making. Insight-Non-Destr. Test. Cond. Monit. **54**(8), 440–445 (2012)
11. Jimenez Cortadi, A.: Deteccion de anomalias para maquinas de estampacion en frio mediante tecnicas de data mining (2015)
12. Kaytez, F., Taplamacioglu, M.C., Cam, E., Hardalac, F.: Forecasting electricity consumption: a comparison of regression analysis, neural networks and least squares support vector machines. Int. J. Electr. Power Energy Syst. **67**, 431–438 (2015)
13. Laayouj, N., Jamouli, H.: Remaining useful life prediction of lithium-ion battery degradation for a hybrid electric vehicle. Global Adv. Res. J. Eng. Technol. **4**(2), 016–023 (2015)
14. Maronna, R., Martin, R.D., Yohai, V.: Robust statistics. Wiley, Chichester (2006). ISBN
15. McKinney, W., Perktold, J., Seabold, S.: Time series analysis in python with statsmodels. Jarrodmillman. Com, pp. 96–102 (2011)
16. Saha, B., Goebel, K., Christophersen, J.: Comparison of prognostic algorithms for estimating remaining useful life of batteries. Trans. Inst. Meas. Control. **31**(3–4), 293–308 (2009)
17. Shin, J.-H., Jun, H.-B.: On condition based maintenance policy. J. Comput. Des. Eng. **2**(2), 119–127 (2015)

18. Wu, W., Hu, J., Zhang, J.: Prognostics of machine health condition using an improved ARIMA-based prediction method. In: 2nd IEEE Conference on Industrial Electronics and Applications, 2007 ICIEA 2007, pp. 1062–1067. IEEE (2007)
19. Yuan, C., Liu, S., Fang, Z.: Comparison of china's primary energy consumption forecasting by using ARIMA (the autoregressive integrated moving average) model and gm (1, 1) model. Energy **100**, 384–390 (2016)

A New Distributed Self-repairing Strategy for Transient Fault Cell in Embryonics Circuit

Zhai Zhang(✉), Yao Qiu, and Xiaoliang Yuan

Nanjing University of Aeronautics and Astronautics,
Sub-box 269 of Main Post Box 159, No. 169 Sheng Tai West Road,
Jiang Ning District, Nanjing, Jiangsu, People's Republic of China
wolnyzhang@nuaa.edu.cn

Abstract. The embryonics circuit with cell array structure has the prominent characteristics of distributed self-controlling and self-repairing. Distributed self-repairing strategy is a key element in designing the embryonics circuit. However, all existing strategies of embryonics circuit mainly aim at the permanent faults, and lack of the transient faults. It would be a huge waste of hardware if a cell was permanently eliminated due to a local transient fault, and the waste will result in seriously low hardware utilization in those environments dominated by transient faults. In this paper, a new distributed self-repairing strategy named fault-cell reutilization self-repairing strategy (FCRSS) is proposed, where the cells with transient fault could be reused. Two mechanisms of elimination and reconfiguration are mixed together. Those transient fault-cells can be reconfigured to achieve fault-cell reutilization. Then, methods to design of all the modules are described in details. Lastly, circuit simulation and reliability analysis results prove that the FCRSS can increase hardware utilization rate and system reliability.

Keywords: Distributed self-repairing strategy · Transient fault
Fault-cell reutilization · Reliability analysis

1 Introduction

Fault-tolerant based on reconfigurable hardware is a feasible and effective method to improve reliability [1,2]. Embryonics circuit is a novel hardware with the characters of self-testing and self-repairing, and all processes are autonomously distributed and have adaptability [3–6]. Self-repairing strategy in embryonics circuit designing is one of the most important factors, which controls the process of cell elimination and reconfiguration. An appropriate self-repairing strategy can greatly increase the utilization rate of hardware and improve the reliability of electronic system.

At present, research in self-repairing strategy of embryonics circuit mainly focuses on the realization ways of fault cells elimination. Tyrrell et al. proposed the row/column elimination strategy with a very simple cell structure

J. Del Ser et al. (Eds.): IDC 2018, SCI 798, pp. 167–177, 2018.
https://doi.org/10.1007/978-3-319-99626-4_15

and working process. However, the consumption of hardware is tremendous for the entire row/column cells elimination for only one fault in any cell. [7,8] Mange et al. put forward the single cell elimination strategy, which requires cell memorizers backup large numbers of configuration information (genes) [9,10]. It is an extremely redundancy of hardware. Szasz proposed a self-repairing strategy based on a fixed structure of 9-cell array, but with a large number of redundant cells. [11,12] Samie et al. proposed a self-repairing strategy based on prokaryotic cell array with the less configuration information and longer reconfiguration time. [13–15] CAI Jinyan et al. came up with a cyclic removal self-repairing strategy based on bus structure, but it would cost a lot of time [16,17]. All these findings remove the fault cell directly regardless of the type of faults.

The traditional self-repairing strategies hold that all fault cells should be eliminated no matter what kind of fault type they belong to. However single-event upsets (SEU) fault and single-event transients (SET) fault are accounting for around 90% in space environment [14,18,19]. It certainly will bring the result of low utilization rate of hardware for repairing the transient fault cell which could be healed just by reconfiguration.

This paper propose a fault-cell reutilization self-repairing strategy, which adds a fault-cell reconfiguration stage during the period of replacement instead of the direct elimination in traditional strategies. The circuit design method of each module inside cells is expounded. The effectiveness of the new self-repairing strategy is verified by the analysis of reliability and consumption in comparison with the traditional cell elimination self-repairing strategy.

2 Self-repairing Strategies of Embryonics Hardware

A typical architecture of the embryonics hardware, as shown in Fig. 1, is a two-dimensional cell array, all cells are same in internal, cells connect with the left and right cells in the same row directly, and the upper and lower cells are connected through the AR module. Each cell is composed of four modules: Controller, Function circuit, I/O routing switch and Configuration register. Controller controls all operations of the cell. Function circuit is the processing block. The I/O routing switch is responsible for connecting and transferring data with surrounding cells. The Configuration register deposits all configuration information of cells.

2.1 Cell Elimination Self-repairing Strategy (CESS)

The CESS is widely used in embryonics circuit. It works as follows: when a working cell was detected failure (shown in Fig. 2a cell_01), the cell elimination in row will be triggered, functions of working cells are shifted to the cells on the right direction begin with the fault cell, which is shown in Fig. 2b. If spare cells are more than fault cells, the array can be repaired without spare rows, otherwise the row elimination will be triggered as Fig. 2d [5,9,10].

The CESS has two shortcomings. One is the hardware waste when remove the whole cell for inside partial fault. The other one is that it considers all faults as

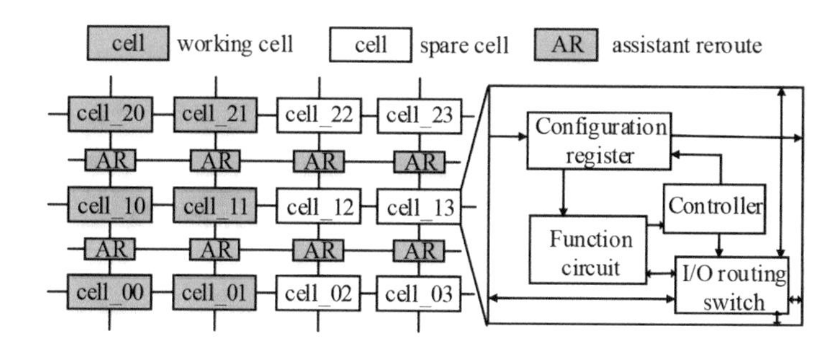

Fig. 1. Two-dimensional cell array and internal modules

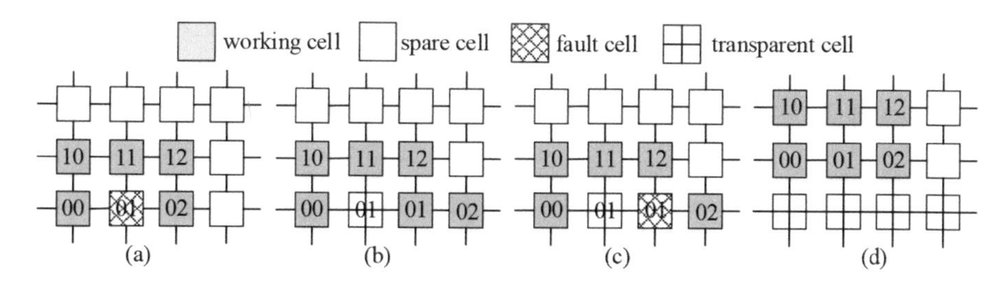

Fig. 2. Principle of cell elimination self-repairing strategy

fixed permanent faults. For the transient fault cell, removal without reutilization will result in great increase of hardware consumption.

2.2 Fault-Cell Reutilization Self-repairing Strategy (FCRSS)

The new strategy of FCRSS combines with two mechanisms of elimination and reconfiguration. The whole procedure can be divided into three stages: fault-cell detection, fault-cell elimination and fault-cell reconfiguration.

Fault-cell detection stage: Dual-modular redundancy (DMR) structure was used to detect functional fault of modules in configuration register and look-up table (LUT).

Fault-cell elimination stage: when a cell was detected in failure by itself, the fault cell is configured to be transparent immediately, and its function will be replaced by the right cell, which is same as the CESS strategy.

Fault-cell reconfiguration stage: the new strategy FCRSS regards the eliminated cells which in transparent as spare cells, which can be reconfigured with the configuration information backup in the left neighbor cell.

As shown in Fig. 3, the FCRSS is introduced with an example by the process of three transient faults in the middle row.

1. Figure 3a shows a normal working cell array.
2. Working cell_12 suffers a fault, and the right neighbor spare cell replaces its function (Fig. 3b). Then, the working cell_12 switches to transparent state and becomes transparent cell_12 (Fig. 3c).

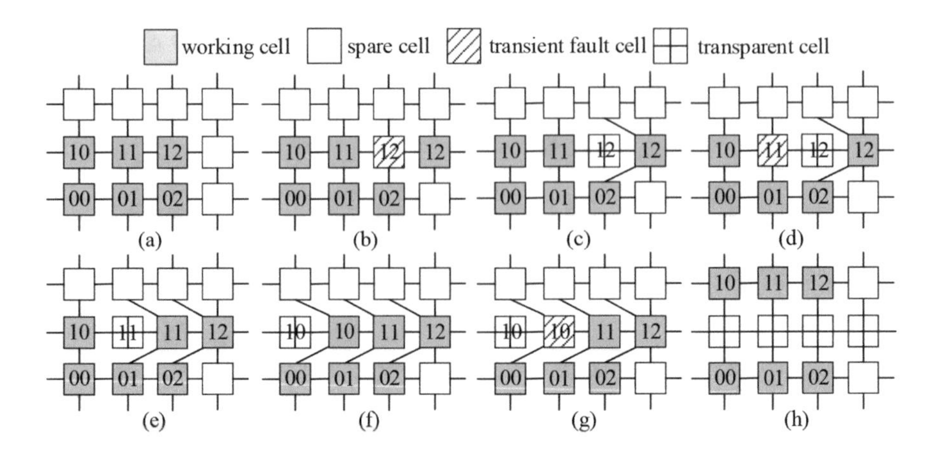

Fig. 3. The process of fault-cell reutilization self-repairing strategy

3. A transient fault occurs in working cell_11 in Fig. 3d. At that time, the transparent cell_12 will replace the function of cell_11 as a spare cell(control by the fault signal of left cell), then the cell array works fault-free. The prerequisite for the successful replacement of cell_11s function by transparent cell_12 is that the transparent cell_12 suffered a transient fault that can be repaired by reconfiguration. The self-detection circuit in the transparent cell can offer fault signals and be removed again if the transparent cell cant be repaired by reconfiguration.
4. As shown in Fig. 3e, the cell array still works normally with two transient faults in middle row. Then, if cell_10 is suffered faults, its function is shifted to the transparent cell_11 in right direction, which is shown in Fig. 3f.
5. This array just allows the function of left cells be replaced by the right cells. When there are two fault cells on the left, such as Fig. 3g, the array will not be repaired by spare cells, and then row elimination will be triggered.

In strategy CESS, the array like Fig. 3a can only be repaired once inside one row, while the new strategy FCRSS can be repaired up to 4 times. With the increasing of the number of spare cells, the number of fault cells repaired in a row by the new self-repairing strategy will increase as well.

3 Circuit Modules

The Controller module should be innovative designed in FCRSS. Circuits of Function circuit and I/O routing switch are same as those in the traditional self-repairing strategies [20]. To achieve the goal of reusing, Configuration register of the right neighbor cell should have the ability of cyclic backup the configuration information of the left cell. Assistant reroute (AR) module controls the connections between cells in different rows.

3.1 Controller

Controller in charge of the function of generating the shifting signal of the Configuration register and the switching signal of the I/O routing switch. The Controllers state transition diagram, as shown in Fig. 4, contains three states: working state, transparent state and reconfiguration state.

In the initial phase, there are only working cells and spare cells in cell array, and working cells are in the working state. When set is "1", cells are self-tested. When a fault was self-detected inside a working cell (self_fault = "1"), the fault cell switches to transparent state and sends shift signal to the right cell. Cell in the transparent state switches to the reconfiguration state after receiving the shift signal of the left cell (front_shift = "1") and reconfigure itself with the backup configuration information. The cell comes into working state and the self-repairing of the cell array is completed after the reconfiguration has completed (self_fault = "0"). Otherwise, the cell will be marked as transparent again (self_fault = "1") and sent shift signal to the right cell.

Figure 5 shows the timing diagram of Controller. Reclk, front_shift and self_fault are the input signals of the Controller, and shift, bypass and cell_state are the output signals. The specific meanings of each signal are: reclk is the reconfiguration clock, front_shift is the shift signal of the left neighbor cell, and self_fault is the output to the OR gate of all detector inside the cell, which is a signal flag of cell fault. Cell_state signal shows the cell status, it is "1" when the cell in working state, otherwise is "0". Shift is the output signal of the left cell to trigger reconfiguration of right cell, which direct connects the front_shift of the right cell. Bypass is control signal for the I/O routing switch and Assistant reroute module.

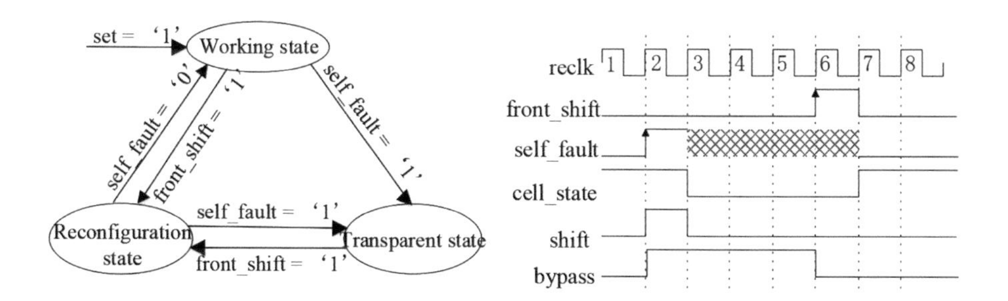

Fig. 4. State transition diagram of Controller

Fig. 5. Timing diagram of Controller

The working process of Fig. 5 is as follows:

1. Firstly, at reclk 1, the cell is in working state.
2. At reclk 2, the cell suffers a fault, so the fault signal self_fault is changed to "1" and bypass is set to "1" at the same time, the cell enters transparent state, and the shift signal is set to "1".

3. One clock later, the array is self-repaired already and the cell is in transparent state. So cell_state and shift turns into "0". The self_fault signal is ineffective because the self-testing circuit cant output to the self_fault in the transparent state cells.
4. At reclk 6, front_shift signal turns to "1" means a fault was detected in left cells, then the bypass signal is set to "0" for cell substitution, and the cell switch to the reconfiguration state.
5. After one clock, reconfiguration process is over, cell_state is set to "1", and the cell will work correctly.

3.2 Configuration Register (CR)

Configuration register saves all the configuration information for Function circuit and I/O routing switch. The new self-repairing strategy requires Configuration register hold the configuration information of the adjacent left cell and update the backup configuration information. Figure 6 is a schematic diagram of configuration register with cyclic backup structure. The bold letters like mxx represent the configuration information of native cell, and the non-bold letters express the backup configuration information of left cell. The letters mxx represents the configuration information of cell_xx.

The connection used to bypass the fault configuration register is connected only when the cell is transparent, otherwise it is expressed by dashed line. If the CR of native cell fails, this cell switches to transparent state, the CR chain is directly connected to the left cell with the right cell. At the same time, the backup configuration information of all the cells on the right side of the fault cell is shifted into the working CR. In next clock, the new CR chain moves the working configuration information of the left cell into the backup CR of the adjacent right cell.

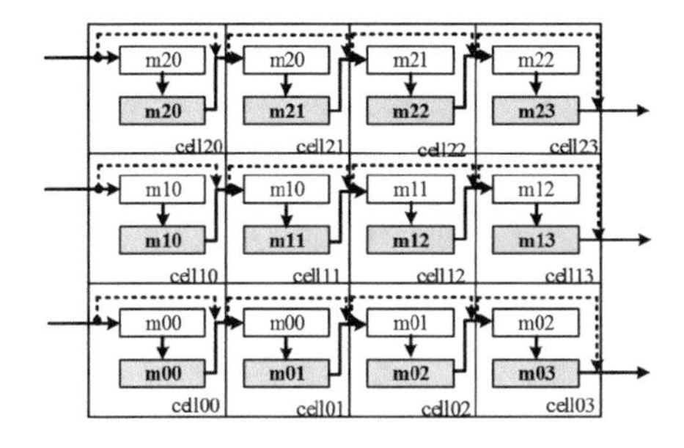

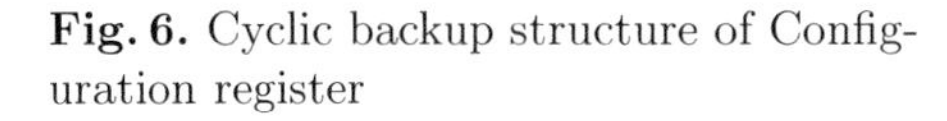
Fig. 6. Cyclic backup structure of Configuration register

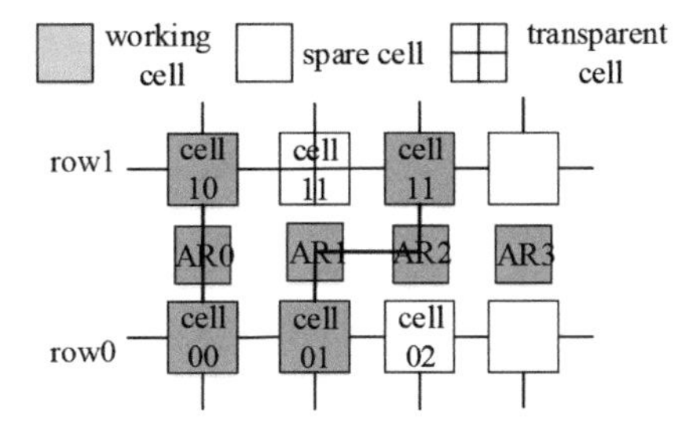

Fig. 7. Connection diagram of Assistant reroute module

3.3 Assistant Reroute Module

The AR module is a connection module between rows. Signal bypass in every cell controls the switching between the up and down rows. Figure 7 shows a connection diagram of AR module. There are four cells in working, cell00 connects with cell10 by AR0, while the bypass of transparent cell_11 controls the connection of cell01 and new working cell_11 by AR1 and AR2.

4 Simulation and Reliability Analysis

A 4-bit binary adder circuit was realized on the electronic cell array, which is composed of 12 electronic cells with array structure 3×4. Xilinx ISE is used to synthesize the circuit and ISim for function simulation. The same modules was used to implement CESS to compare with FCRSS.

4.1 I/O Simulation Results

Figure 8 is the simulation results of the example, where x[3:0] and y[3:0] are two addends, s[3:0] is sum, and c is carry. In this simulation, external injection signals are injected as faults into the cell array. In Fig. 8, two transient fault signals (cell_11_fault and cell_10_fault), which both last one reconfiguration clock, were injected, The meanings of signals in the sequence chart are as follows: cell_xx_fault is the fault signal of cell_xx, cell_xx_configbits is the native configuration information of cell_xx.

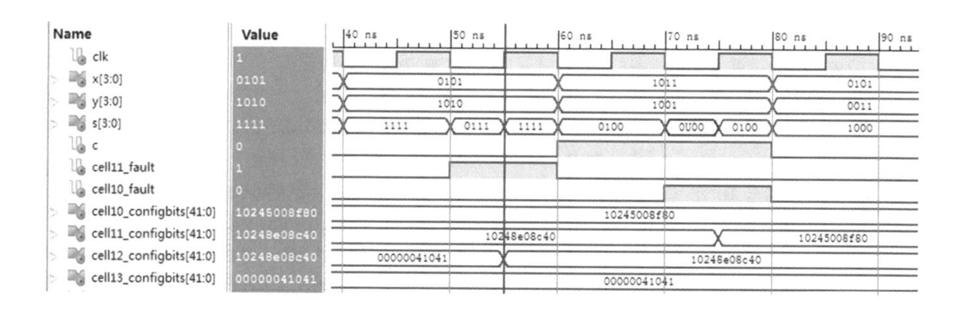

Fig. 8. Simulation sequence chart of self-repairing process for transient faults

Figure 8 shows the sequence chart of self-repairing process for two transient faults. The specific process is as follows:

1. At 50 ns, the working cell_11 is injected with a transient fault signal, the corresponding cell array is shown in Fig. 9a. In this situation, signal cell_11_fault is “1” and s[3:0] output the erroneous result during 50–55 ns. At 55 ns, the right spare cell_12 uses the backup configuration information to configure itself and becomes the new working cell_11, as shown in Fig. 8, cell_12_configbits becomes cell_11_configbits. At 60 ns, cell_11 enters into transparent state and the cell array is self-repaired, which is shown in Fig. 9b.

2. At 70 ns, the working cell_10 is injected with a transient fault signal. In Fig. 8, cell_10_fault is set to "1" and s[3:0] output the erroneous result during 70–75 ns. The fault cell_11 has become transparent cell_11, and then the cell_11 would be configured as working cell10 to complete the array repairing at 75 ns, as shown in Fig. 9d.

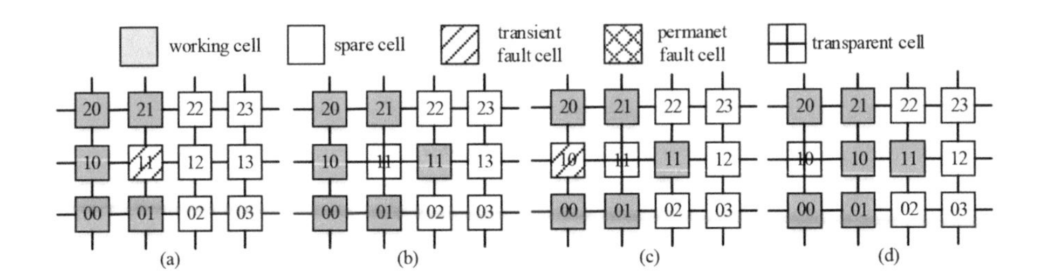

Fig. 9. Self-repairing process corresponding to self-repairing strategy for transient faults

4.2 Hardware Consumption and Time Consumption

Taking the 3×4 cell array as an example, the Spartan3 XC3S400 chip is selected to use. Compared with the traditional cell elimination strategy, the hardware resource consumption of the slices, flip-flops and LUTs of the new strategy FCRSS with the 3×4 array only increases about 10% of the hardware for designing the new Controller and some cell status signals.

The new strategy FCRSS spends one reconfiguration clock when spare-cell replace fault-cell with transient fault cell, which is the same as traditional cell elimination strategy. If the transparent cell is a permanent fault cell, which cant be repaired by reconfiguration, the time consumption will increase by one reconfiguration clock. Therefore, the self-repairing time is two reconfiguration clocks because the permanent fault cell would still be detected in failure after it has been reconfigured and then trigger the new elimination. So, the more permanent fault cells, the higher time consumption.

4.3 Reliability Analysis

For a cell array of size $N \times M$ cells, it will be considered working if a sub-array of size $n \times m$ is faultless. Where N, n are the rows of total array and the working cell array respectively, the M and m are the corresponding numbers of columns (where $M \geqslant m$, $N \geqslant n$). To analyze the reliability of the new strategy FCRSS, several definitions are proposed as follows:

- Defining Q is the in-row repairable number, which indicates the number of cells that can be repaired inside the same row. In strategy CESS, $Q = M - m$, while in FCRSS, Q will be far greater than this value for fault-cell reutilization.

- Defining L is the proportion of working cells in the row, so $L = m/M$.
- Defining S as the proportion of transient faults, which represents the proportion of transient faults to total faults (sum of transient faults and permanent faults).
- Defining M_e as the equivalent number of row cells in the new strategy. Q is equivalent to the numbers of spare cells in calculation of reliability, and number of working cells m remains unchanged, so $M_e = m + Q$.

In the new strategy FCRSS, Q is a variable. Therefore, the reliability model should be modified when calculating $MTTF$ [3]:

The row reliability $R_r(t)$ can be calculated in Eq. 1.

$$R_r(t) = \sum_{i=m}^{M_e} \binom{M_e}{i} e^{-\lambda it}(1 - e^{-\lambda it})^{M_e - i} \tag{1}$$

The reliability $R_a(t)$ and $MTTF$ of the array are shown in Eqs. 2 and 3

$$R_a(t) = \sum_{j=n}^{N} \binom{N}{j} R_r(t)^j (1 - R_r(t))^{N-j} \tag{2}$$

$$MTTF = \int_0^\infty R_a(t)dt = \int_0^\infty \sum_{i=n}^{N} \binom{N}{i} R_r(t)^i (1 - R_r(t))^{N-i} \tag{3}$$

where λ is failure rate, it is a constant, and the dimension is $10^{-6}/h$.

Take $M = N = 100$, $m = n = 80$, $\lambda = 1 \times 10^{-6}/h$, this paper analyzes the change trend of Q firstly, then the array reliability and $MTTF$.

Figure 10 shows the array reliability curves with different S. It can be seen from the figure that the failure time of the array is extended as the increasing of S gradually, and the array reliability increases gradually too. Moreover, both of them are greater than the values in cell elimination strategy where S is 0.

Figure 11 shows the variation curve of $MTTF$ with different proportions of transient faults S. It can be seen from the figure that the $MTTF$ of the two self-repairing strategies are the same when S is 0. With the increasing of the proportion of transient faults, the $MTTF$ increases gradually of new self-repairing strategy FCRSS, and the larger the S, the faster the $MTTF$ increases. When the S is greater than 0, the FCRSSs $MTTF$ is superior to the cell elimination self-repairing strategys. The larger the S is, the more obvious the $MTTF$ increasing is.

Although the self-repairing strategy FCRSS consumes more clocks for the permanent fault cells, but with the increasing of S, the reconfiguration time will decreases gradually, and when $S = 1$, it will down to the same value in the CESS. The large proportion of S means most of the cell faults are transient fault, and the extra reconfiguration time consumed by the new self-repairing strategy will be reduced, while the reliability of the array will rise with the increase of S. So, the FCRSS in this paper can only pay a small amount of reconfiguration time to win much higher hardware utilization rate and reliability in those environment with high transient failure.

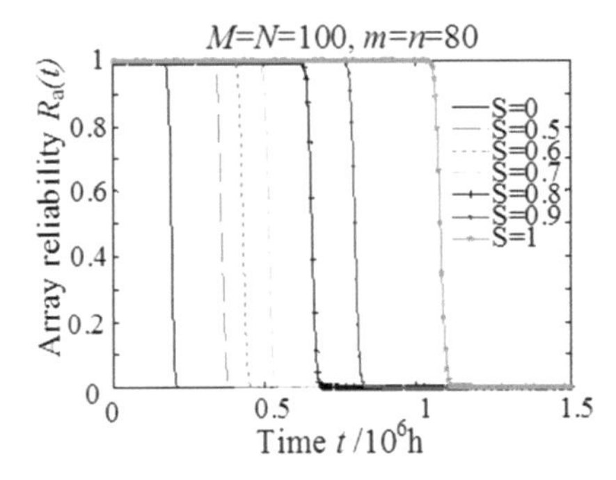

Fig. 10. Array reliability variation changes with different S

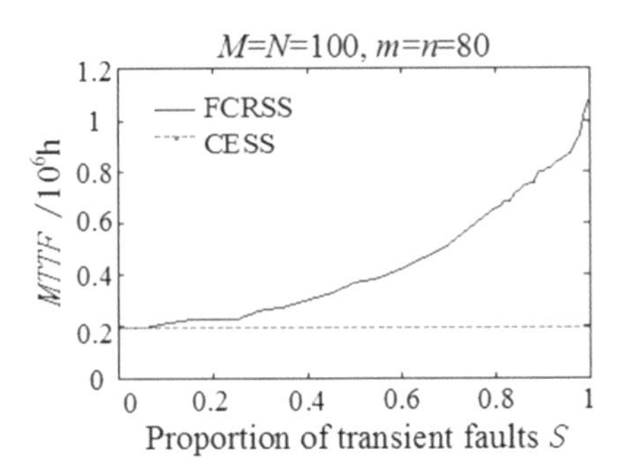

Fig. 11. $MTTF$ variation changes with S in different strategies

5 Conclusions

This paper describes the principle of fault-cell reutilizing self-healing strategy and gives the circuit design method of the cell modules. The effectiveness of the new self-healing strategy is verified by the circuit design and simulation. The simulation results show that the cell array can implement the logic function correctly and reuse the transient fault cells.

The advantages of high hardware utilization rate and system reliability of FCRSS was concluded through the analysis results of self-repairing capability and reliability, and the advantages will be more significant with the higher proportion of transient faults.

Acknowledgements. This study was co-supported by the National Natural Science Foundation of China (No. 61202001 and 61402226) and the Fundamental Research Funds for the Central Universities of NUAA (No. NS2018026 and NS2012024).

References

1. Mange, D., Sanchez, E., Stauffer, A., et al.: Embryonics: a new methodology for designing field-programmable gate arrays with self-repair and self-replicating properties. In: IEEE Transactions on Very Large Scale Integration (VLSI) Systems (1998)
2. Kretzschmar, U., Gomez-Cornejo, J., Astarloa, A., Bidarte, U., Del Ser, J.: Synchronization of faulty processors in coarse-grained TMR protected partially reconfigurable FPGA designs. Reliab. Eng. Syst. Saf. **151**, 1–9 (2016)
3. Stauffer, A., Mange, D., Rossier, J.: Bio-inspired self-organizing cellular systems. Biosystems **94**(1–2), 164–169 (2008)
4. Zhang, Z., Wang, Y.R.: Method to reliability improvement of chip self-repairing hardware by array layout reformation. Acta Aeronaut. Astronaut. Sin. **35**(12), 3392–3402 (2014). [Chinese]
5. Zhang, Z., Wang, Y.R., Yang, S.S., Yao, R., Cui, J.: The research of self-repairing digital circuit based on embryonic cellular array. Neural Comput. Applic **17**(2), 145–151 (2008)

6. Zhang, Z., Wang, Y.R.: Cell granularity optimization method of embryonics hardware in application design process. Acta Aeronaut. Astronaut. Sin. **37**(11), 3502–3511 (2016). [Chinese]
7. Canham, R.O., Tyrrell, A.M.: A multilayered immune system for hardware fault tolerance within an embryonic array. Artif. Immune Syst. **1**(1), 3–11 (2002)
8. Canham, R.O., Tyrrell, A.M.: A hardware artificial immune system and embryonic array for fault tolerant systems. Genet. Program. Evolvable Mach. **4**(4), 359–382 (2003)
9. Ortega, S.C., Mange, D., Smith, S.: Embryonics: a bio-inspired cellular architecture with fault-tolerant properties. Genet. Program. Evolvable Mach. **1**(3), 187–215 (2000)
10. Tempesti, G., Mange, D., Mudry, P.A.: Self-replicating hardware for reliability: the embryonics project. ACM J. Emerg. Technol. Comput. Syst. **3**(2), 1–21 (2007)
11. Szasz, C., Virgil CX, F., Husi, G.: Embryonic systems implementation with FPGA-based artificial cell network hardware architectures. Asian J. Control. **12**(2), 208–15 (2010)
12. Husi, G., Szasz, C., Chindris, V.: Artificial immune system implementation upon embryonic machine for hardware fault-tolerant industrial control applications. Glob. J. Comput. Sci. Technol. **10**(4), 60–6 (2010)
13. Samie, M., Dragffy, G., Tyrrell, A.M.: Novel bio-inspired approach for fault-tolerant VLSI systems. IEEE Trans. Very Large Scale Integr. Syst. **21**(10), 1878–1891 (2013)
14. Bremner, P., Liu, Y., Samie, M.: SABRE: a bio-inspired fault-tolerant electronic architecture. Bioinspiration Biomim.S **8**(1), 1–16 (2013)
15. Wang, N.T.: Research of self-repairing technique based on prokaryotic bio-inspired array. National University of Defence Technology, Changsha (2011)
16. Wang, T., Cai, J.Y., Meng, Y.F.: Design of bus-based embryonic array and selection method for mounts of spare cells. Comput. Eng. Appl. **53**(8), 44–49 (2017). [Chinese]
17. Wang, T., Cai, J.Y., Meng, Y.F.: A novel embryonics electronic cell array structure based on functional decomposition and circular removal self-repair mechanism. Adv. Mech. Eng. **9**(9), 1–16 (2017)
18. Normand, E.: Single-event effects in avionics. IEEE Trans. Nucl. Sci. **43**(2), 461–474 (2002)
19. Yao, R., Chen, Q.Q., Li, Z.W.: Multi-objective evolutionary design of selective triple modular redundancy systems against SEUs. Chin. J. Aeronaut. **28**(3), 804–13 (2015)
20. Zhang, Y.: Designed and implementation of embryonic circuit oriented to self-repair on chip. Nanjing University of Aeronautics and Astronautics, Nanjing (2008)

Mobility and Smart Cities

Solving the Open-Path Asymmetric Green Traveling Salesman Problem in a Realistic Urban Environment

Eneko Osaba[1(✉)], Javier Del Ser[1,2,3], Andres Iglesias[4,5], Miren Nekane Bilbao[2], Iztok Fister Jr.[6], Iztok Fister[6], and Akemi Galvez[4,5]

[1] TECNALIA Research & Innovation, Derio, Spain
eneko.osaba@tecnalia.com
[2] University of the Basque Country (UPV/EHU), Bilbao, Spain
[3] Basque Center for Applied Mathematics (BCAM), Bilbao, Spain
[4] Universidad de Cantabria, Santander, Spain
[5] Toho University, Funabashi, Japan
[6] University of Maribor, Maribor, Slovenia

Abstract. In this paper, a driving route planning system for multi-point routes is designed and developed. The routing problem has modeled as an Open-Path and Asymmetric Green Traveling Salesman Problem (OAG-TSP). The main objective of the proposed OAG-TSP is to find a route between a fixed origin and destination, visiting a group of intermediate points exactly once, minimizing the CO_2 emitted by the car and the total distance traveled. Thus, the developed transportation problem is a complex and multi-attribute variant of the well-known TSP. For its efficient solving, three classic meta-heuristics have been used: Simulated Annealing, Tabu Search and Variable Neighborhood Search. These approaches have been chosen for its easy adaptation and rapid execution times, something appreciated in this kind of real-world systems. The system developed has been built in a realistic simulation environment, using the open source framework Open Trip Planner. Additionally, three heterogeneous scenarios have been studied in three different cities of the Basque Country (Spain): Bilbao, Gazteiz and Donostia. Obtained results conclude that the most promising technique for solving this problem is the Simulated Annealing. The statistical significance of these findings is confirmed by the results of a Friedman's non-parametric test.

Keywords: Route planning · Traveling Salesman Problem
Emission reduction · Simulated Annealing · Tabu Search
Variable Neighborhood Search

1 Introduction

Route planning has gained a crucial importance in mobility management in last years. Traveling long and short distances daily has become a habit for many

J. Del Ser et al. (Eds.): IDC 2018, SCI 798, pp. 181–191, 2018.
https://doi.org/10.1007/978-3-319-99626-4_16

people world-wide. In this sense, two different route planning modes can be highlighted: multi-modal and mono-modal transportation. The first one aims at providing the traveler feasible routes between origin and destination using diverse public and private transportation modes. The second mode consists on routes performed by a single transport type. This second kind of routes are the focus of this study. More concretely, and as will be detailed later, routes performed by car, which are the most common ones for medium-length journeys. Furthermore, routes conducted by car are also the most usual ones for long-length paths, when the scheduling of public transportation is not the most suitable one. For these reasons, mono-modal transportation is widely referenced by the users. In this context, the most common mono-modal routes are the ones performed by car, bike or walking, developed for different purposes. Even for planning these kind of routes users take advantage of several route planners that are available in the market or in the web.

In recent years, multiple route planning systems available on the Internet have been developed for a wide range of transportation modes. Furthermore, most of these planners are open and freely available, popularizing their use in multiple contexts. These tools, accessible thought different platforms such as smartphones, tablets and computers, grant flexibility to the users for designing routes in any place and at any time. Furthermore, almost all these platforms for the building of paths composed by intermediate points, but not with the enough flexibility. Tools such as Google Maps[1] allow for the introduction of intermediate points, but they are visited in order of appearance, where not optimization is needed. In this sense, there are plenty of situations in which an optimized design of multi-point routes for this kind of platforms would be highly profitable for the end user. Some examples of these situation could be the transportation on demand [1] or the goods transportation. The lack of literature exploring this interesting novel aspect has motivated us to conduct the present study.

Concretely, the main contribution of this work is the development of a driving route planning system for multi-point routes, modeling the problem as an Open-Path Asymmetric Green Traveling Salesman Problem (OAG-TSP). The main objective of the developed algorithm—system—etc. is to find the optimal route between two different points, visiting a set of intermediate points exactly once, minimizing the distance traveled and the CO_2 emitted by the car. Furthermore, for the OAG-TSP solving three different classic meta-heuristics have been used: Simulated Annealing (SA, [2]), Tabu Search (TS, [3]) and Variable Neighborhood Search (VNS, [4]).

The system developed in this study has been deployed in a realistic simulation environment, using the Open Trip Planner (OTP, [5]) open platform. In overall, three different real-world scenarios comprise the conducted experimentation. These scenarios have been placed in the three most important cities of the Basque Country (Spain): Gazteiz, Bilbao and Donostia. Additionally, three different configurations have been used for each scenario, resulting on an experimentation composed by nine separated tests. For these experiments, two

[1] https://www.google.es/maps/.

different real data sources have been used: the Open Street Map of the whole Basque Country, and its Digital Elevation Model (DEM).

1.1 Background and Research Contribution

As can be deduced from its name, the proposed OAG-TSP is a complex variant of the well-known Traveling Salesman Problem (TSP, [6]). The TSP is one of the more studied benchmarking problems in artificial intelligence and operations research. The main reasons for its popularity and importance is its easy formulation and its complex resolution, which suppose a challenge for the scientific community. In this sense, the TSP is the focus of lots studies annually [7,8]. Furthermore, the focus of the current scientific community is directed to the multi-attribute or rich variant of the TSP [9]. These kinds of problems are special cases of the well-known conventional TSP, with the distinction of having multiple constraints and complex formulations. These problems have a great scientific interest. While they maintain their complexity, their social interest is also high, as their applicability to real-world situations is greater than the conventional version. Recent examples of these multi-attribute routing problems can be found in [10,11] or [12]. The OAG-TSP developed for this research can be placed in this category.

The originality of this paper rests on two different concepts: the first one is the application of a green multi-point route optimization algorithm into a car route planning open platform. This way, users can plan their driving paths using intermediate places and ordering them as an OAG-TSP. The other main originality is the modeling of the routing problem. This is the first work in the literature exploring the application of the well-known Vehicle Specific Power (VSP) concept into a vehicle routing problem, allowing in this way the reduction of pollutant gas emissions in TSP open path routes. Additionally, the application of such problem into a realistic environment supposes an added value for the conducted research [13]. This is so since it is infrequently done for TSP family problems, usually used as benchmarking problems.

The remaining of this work paper is structured as follows. In Sect. 2, the proposed OAG-TSP for facing the car route problem is described. In Sect. 3, the description of the deployed environment based on Open Trip Planner for facing the established requirements are described. Next, the experimentation performed is discussed in Sect. 4. Finally, this paper ends in Sect. 5 with conclusions and an outline of further research paths.

2 Problem Definition

As has been mentioned in the introduction, the routing problem addressed in this study has been modeled as an OAG-TSP problem. Considering this novel problem as a realistic variant of the TSP, it can be represented as a complete graph $\mathcal{G} \doteq (\mathcal{V}, \mathcal{A})$, where $\mathcal{V} \doteq \{v_1, v_2, \ldots, v_N\}$ denotes the group of nodes that represent the graph, which are the different points that the car should visit. On the other hand, $\mathcal{A} \doteq \{(v_i, v_j) : v_i, v_j \in \mathcal{V}\times\mathcal{V}, i,j \in \{1,\ldots,N\}\times\{1,\ldots,N\}, i \neq j\}$ is the set of edges connecting every pair of nodes in $\mathcal{V}$, which represent the streets of the cities. Furthermore, each edge has an associated cost $d_{ij} \in \mathbb{R}^+$, related to the traveling costs thorough this arc. Additionally, due to the asymmetric characteristic of the problem, $d_{ij} \neq d_{ji}$ in most of the cases. The reason for this asymmetry is related to the difference in traffic flow in both directions. In addition to this weight characteristic, each edge has an emission value $em_{ij} \in \mathbb{R}^+$ associated, which is based on the well-known VSP [14] measure. This metric follows the same philosophy as d_{ij} in terms of asymmetry, and the mathematical procedure for calculating its value is described later, in Subsect. 2.1.

With all this, the proposed OAG-TSP hinges on the discovery of a route that visits every node once and only once, starting and ending in two different defined nodes, and minimizing an objective function composed by the emission and the traveling costs. This optimization problem can be formally stated as

$$\text{minimize} \quad f(\mathbf{X}) = \sum_{i=1}^{N}\sum_{\substack{j=1\\ i\neq j}}^{N} w_d(\psi_d(i,j)) + w_{em}(\psi_{em}(i,j)) \tag{1a}$$

$$\text{subject to} \quad \sum\nolimits_{\substack{j=1\\ i\neq j}}^{N} x_{ij} = 1, \quad \forall j \in \{1,\ldots,N\}, \tag{1b}$$

$$\sum\nolimits_{\substack{i=1\\ i\neq j}}^{N} x_{ij} = 1, \quad \forall i \in \{1,\ldots,N\}, \tag{1c}$$

$$\sum\nolimits_{\substack{i\in\mathcal{S}\\ j\in\mathcal{S}\\ i\neq j}} x_{ij} \geq 1, \quad \forall \mathcal{S} \subset \mathcal{V}, \tag{1d}$$

$$w_d + w_{em} = 1,\ 0 < w_d, w_{em} < 1 \tag{1e}$$

where $\mathbf{X} \doteq [x_{ij}]$ is a $N \times N$ binary matrix whose entry $x_{ij} \in \{0,1\}$ takes value 1 if edge (i,j) is used in the solution. With all this notation in mind, the objective function represented in Expression (1a) is the sum of costs and emissions associated to all the arcs contained in the solution. Additionally, each of these separated values are weighted using two parameters w_d and w_{em}. Clauses (1b) and (1c) guarantee that each node is visited only once. Finally, (1d) ensures the absence of sub-tours, and forces that any subset of nodes S has to be abandoned at least one time. This constraint is crucial for avoiding cycles along the route.

2.1 Quantifying CO_2 Emission

As has been pointed before, the VSP formula has been used as base for the CO_2 emission calculation. This concept is a formalism used in the evaluation of vehicle emission, and it was first developed by Jimenez in 1998 [15] at the Massachusetts Institute of Technology.

Briefly explained, the VSP is the summation of the load resulting from acceleration, hill climbing, rolling resistance and aerodynamic drag, all of them divided by the mass of the corresponding vehicle. This measure is usually represented in kilowatts per tonne, and it depicts the instantaneous power demanded by the vehicle divided by its mass. In general terms, this formula can be used to calculate the vehicle CO_2 emissions whether it is combined with remote-sensing and dynamometer measurements.

In the early 2000s, the United States Environmental Protection Agency (EPA) adopted the VSP to calculate of vehicle pollutant gas emissions. As can be read in the technical report published by the EPA in [16], the concept of VSP has been considered in a wide variety of research and applied studies in different ways, showing its potential and usefulness as metric for characterizing vehicle emissions. With the aim of facilitating this calculation, a generalization of the formula was developed for light duty vehicles:

$$VSP = v[1.1a + 9.81(\text{atan}(\sin(grade))) + 0.132] + 0.000302v^3, \qquad (2)$$

where VSP is given in $kW/Metric\ Ton$, v in m/s, a in m/s^2, and grade is the road slope measured in percentage (%). In any case, as can be read in [16] the VSP cannot be directly applied for calculating reliable emissions. For this reason, the North Carolina State University (NCSU) proposed a *binning* approach in which operational bins were defined based on speed, acceleration, and power demand, and estimates within each modal bin were refined using ordinary least squares (OLS) regression analysis [17]. In this sense, the Table 1 is used for calculating the resultant VSP mode based on a VSP value. Furthermore, once the VSP is established in a specific VSP Mode, the corresponding emitted CO_2 grams per second can be deduced using the column represented as $em_{i,j}$ in the same table. In this sense, for every second a light vehicle spends in a street, the CO_2 value related to its VSP Mode is added. Finally, as will be mentioned later, the Digital Elevation Model has been used for calculating the elevation and the grades of each street of the cities.

Table 1. VSP Bin definitions developed by NCSU.

VSP Mode	VSP [kW/Metric Ton]	$em_{i,j}$	VSP mode	VSP [kW/Metric Ton]	$em_{i,j}$
1	VSP < –2	2.25	8	13 ≤ VSP < 16	6.50
2	–2 ≤ VSP < 0	2.00	9	16 ≤ VSP < 19	7.00
3	0 ≤ VSP < 1	1.50	10	19 ≤ VSP < 23	7.25
4	1 ≤ VSP < 4	3.50	11	23 ≤ VSP < 28	8.00
5	4 ≤ VSP < 7	4.75	12	28 ≤ VSP < 33	9.00
6	7 ≤ VSP < 10	5.50	13	33 ≤ VSP < 39	9.50
7	10 ≤ VSP < 13	6.00	14	39 ≤ VSP	10.0

Summarizing, the $\psi_{em}(i, j)$ associated to each street is calculated using the following formula: $\psi_{em}(i, j) = seconds\_in\_street * em_{i,j}$. In other words, for calculating the CO_2 emitted by a car traversing a street, the seconds that the vehicle needs to go through this street is multiplied by the associated VSP.

3 Description of the Simulation Environment

As has been highlighted along this paper, the developed routing system has been deployed using OTP. Briefly explained, OTP is an open source framework for mono and multi-modal journey planning. It is based on the client-server model, providing a map-based web interface, which has been used for the tests carried out during this research. Furthermore, OTP provides a REST API service for third-party applications, and it gives the opportunity of working with different open data standards. These are the main advantages that have encourage us for using this framework:

- It is open source, which facilitates its modification and adaptation to the proposed scenarios.
- It efficiently works with OSM and GeoTIFF, providing the structure to automatically build the street network and its elevations. As it is described in the following Sect. 4, these two standards are the ones used for this work.
- It is well documented, and it has an active community working behind. This fact facilitates the understanding of the framework.

For this research, the last version developed for the OTP has been used, which has been written in JAVA programming language. Additionally, the one written in JAVA programming language has been used. For this reason, all the modifications done for the correct adaptation of our conditions have been conducted using this language. The basic version of the OTP is not able to fulfill all the characteristics of the proposed study requisites, for this reason, both the source code and the client have been modified to meet the main objectives.

Because they fall outside the scope of this paper, the implementation details are not described in this paper. In any case, the main modification carried out is the adding of the meta-heuristic algorithms for solving the proposed OAG-TSP. Specifically, the local search algorithms that have been used are the SA, TS and VNS. We have chosen these three methods for the experimentation for various reasons. First of all, all these algorithms are well-known in the literature, and they have shown their efficiency for solving this kind of problem in multiple times [18,19]. Furthermore, they are easy to implement and to adapt to the proposed problem, and they do not need a high computational effort for their running. This last reason has been the most important one. This is so since the route building by the OTP is a complex process requiring lots of calculations, leading to high execution times in case we use complex meta-heuristics. Some further modifications on the OTP code are related with the addition of the VSP calculation, and the consideration of intermediate point for the route calculation.

4 Experiments and Results

In this section, the experimentation carried out in this research is described. In overall, three different scenarios have been considered. These scenarios correspond to the three most important cities of the Basque Country (Spain):

San Sebastian, Vitoria-Gazteiz and Bilbao. Each of these cities are recognized by different characteristics in terms of geographic profile and street distribution, making this experimentation heterogeneous enough to be representative. Furthermore, each instance has been generated using a fixed origin and destination, and 18 randomly chosen intermediate points.

For conducting the experimentation correctly, two open data sources have been used on the deployed platform:

- *Digital Elevation Model*: This data source is used with the intention of setting the elevations of the streets of the whole Basque Country. OTP uses these information for assigning the corresponding elevation to the entire street network, and it is employed for calculating the CO_2 emission of each route. Furthermore, this source is directly used by OTP in GeoTIFF format, and it has been openly taken from *SRTM Tile Grabber*[2] database.
- *OSM map file*: In order to build the complete street network for properly building the routes, the corresponding map file is needed. This file has been openly obtained from *Planet OSM* platform[3], using *BBBike*[4]. Through this open tool, OSM maps have been downloaded in Protocolbuffer Binary Format (PBF), containing all the nodes, ways and relations necessary to build the map. These OSM files are directly consumed by OTP, which automatically constructs the full road network.

Additionally, it is interesting to mention that all the tests conducted in this work have been run on an Intel Core i7-7600 computer, with 2.90 GHz and a RAM of 16 GB. Furthermore, each scenario and configuration has been run 10 times for each algorithm. It is interesting to point that three different configurations have been tested for each scenario:

- Green alternative (G-Alt): In this case, the minimization of the CO_2 issued by the car has been prioritized. For this reason, the objective function used has been the following one: $w_d = 0.2$, $w_{em} = 0.8$.
- Short alternative (S-Alt): For this alternative, the minimization of the length of the route is prioritized over the emission, using the following objective function: $w_d = 0.8$, $w_{em} = 0.2$.
- Balanced alternative (B-Alt): For this last case, both objectives are considered using the same weights: $w_d = 0.5$, $w_{em} = 0.5$.

Regarding the parameterization of the algorithms, and following the good practices listed in [20], we have used the same functions and configuration for all the three cases with the aim of performing a fair comparison. The initial solutions of the used SA, TS and VNS are generated randomly, maintaining the position of the first and last node along the whole execution, and corresponding to the origin and the destination inserted by the user. Regarding the ending criterion, a maximum number of 200 iterations has been set after an empirical performance

[2] http://dwtkns.com/srtm/.
[3] http://planet.osm.org.
[4] http://download.bbbike.org/osm.

analysis. In this sense, it should be mentioned again that the calculation of a route by the OTP is a demanding process. For this reason, the maximum number of iterations should not be high, avoiding the increasing of the algorithm runtime. In relation to the successor functions used, TS and SA used the well-known swapping operator, while the VNS uses also the insertion function [21]. Furthermore, the initial temperature of SA has been set in 0.95, and the cooling factor in 0.01. On the other hand, the size of tabu list of the TS has been established in $problem\_size \times 2$, where $problem\_size$ is the number of nodes in the scenario. Table 2 summarizes the parameterization for each meta-heuristic.

Table 2. Parametrization of all developed methods.

SA		TS		VNS	
Parameter	Value	Parameter	Value	Parameter	Value
# evaluations	200	N. of evaluations	200	N. of evaluations	200
Successor function	Swapping	Successor function	Swapping	Successor function	Insertion & Swapping
Cooling factor	0.01	Size of tabu list	$problem\_size \times 2$	Probability of choosing each function	0.5
Temperature	0.95	Memory type	Short term [3]		

In the following Table 3, the results obtained by the SA, TS and VNS are show for each scenario and configuration. As can be observed on this table, SA is the algorithm that has reached the best results. Specifically, SA has obtained best outcomes in 7 out of 9 of the cases. Besides that, and following the guidelines in [22], one statistical test has been carried out to resolve the statistical relevance of the results. Thus, the Friedman's non-parametric test for multiple comparison allows proving if there are significant differences in the results obtained by all reported methods. This way, in the last row of Table 3, we have displayed the mean ranking returned by this nonparametric test for each of the compared algorithms and scenarios (the lower the rank, the better the performance). Additionally, the Friedman statistic obtained is 8.667. The confidence interval has been set in 99%, being 5.991 the critical point in a χ^2 distribution with 2° of freedom. Since $8.667 > 5.991$, it can be concluded that there are significant differences among the results.

Additionally, and in order to graphically show the best results obtained by each technique, the best green routes obtained for each considered scenario are depicted in Fig. 1. Being specific, the first route corresponds to Gazteiz, and it has been obtained by the VNS. The second of the maps shows the best route found in Bilbao, reached by the SA. Finally, the third path correspond to Donostia, and it has been got also by the SA.

Table 3. Results obtained by the SA, TS and VNS for each scenario and configuration, and average rankings returned by the Friedman's non-parametric test.

GASTEIZ			
	SA	TS	VNS
G-Alt	64281.7 ± 3504.9	66914.5 ± 4343.6	**63749.5** ± 4930.3
S-Alt	**59150.3** ± 5728.3	63315.4 ± 3887.8	61157.7 ± 5826.9
B-Alt	**58471.6** ± 4556.8	62381.6 ± 4188.5	61644.5 ± 4248.2
BILBAO			
	SA	TS	VNS
G-Alt	**45862.0** ± 3144.7	48466.8 ± 4963.9	48869.2 ± 3500.1
S-Alt	**46346.6** ± 4841.8	49594.5 ± 3536.0	51318.5 ± 4095.2
B-Alt	**45155.9** ± 2576.5	48443.5 ± 3351.9	49684.1 ± 2814.6
DONOSTI			
G-Alt	**35244.9** ± 3431.5	38142.1 ± 2550.4	37291.8 ± 2739.4
S-Alt	34969.6 ± 2648.6	37673.5 ± 2807.6	**34207.6** ± 1886.3
B-Alt	**33103.1** ± 1263.1	36911.6 ± 2229.8	37313.2 ± 2948.6
Friedman's non-parametric test			
Rank	**1.4444**	1.7778	2.7778

Fig. 1. Best routes found for each scenario. Top left: Gazteiz. Top right: Bilbao. Bottom: Donostia.

5 Conclusions and Further Work

In this work, a driving route planning system for multi-point routes is designed and developed, modeling the transportation problem as an Open-Path and Asymmetric Green Traveling Salesman Problem. The main goal of the OAG-TSP is to find a promising route between two different points, visiting a set of intermediate points exactly once, minimizing the CO_2 emitted by the driver and the distance traveled. In this sense, the developed problem is a complex and multi-attribute variant of the well-known TSP. For the OAG-TSP solving three different methods have been used: the SA, the TS and the VNS, which have been chosen for its easy adaptation and rapid execution times. The system developed has been built in a realistic simulation environment, using the OTP. Furthermore, three different scenarios have been studied in three different cities of the Basque Country (Spain): Bilbao, Gazteiz and Donostia.

Several future work lines have been planned for the short term. The first one is the development of further tests in different cities of Spain, using more algorithms. Furthermore, additional research will be done regarding the transportation problem, applying further realistic restrictions and conditions.

Acknowledgements. E. Osaba and J. Del Ser would like to thank the Basque Government for its funding support through the EMAITEK program.

References

1. Cordeau, J.F., Laporte, G., Potvin, J.Y., Savelsbergh, M.W.: Transportation on demand. Handb.S Oper. Res. Manag. Sci. **14**, 429–466 (2007)
2. Van Laarhoven, P.J., Aarts, E.H.: Simulated annealing. In: Simulated Annealing: Theory and Applications, pp. 7–15. Springer, Netherland (1987)
3. Glover, F., Laguna, M.: Tabu search. In: Handbook of Combinatorial Optimization, pp. 3261–3362. Springer, Boston (2013)
4. Mladenović, N., Hansen, P.: Variable neighborhood search. Comput. Oper. Res. **24**(11), 1097–1100 (1997)
5. Open Trip Planner. https://github.com/opentripplanner. Accessed 30 Nov 2017
6. Hoffman, K.L., Padberg, M., Rinaldi, G.: Traveling salesman problem. In: Encyclopedia of Operations Research and Management Science, pp. 1573–1578. Springer, Boston (2013)
7. Malaguti, E., Martello, S., Santini, A.: The traveling salesman problem with pickups, deliveries, and draft limits. Omega **74**, 50–58 (2018)
8. Elgesem, A.S., Skogen, E.S., Wang, X., Fagerholt, K.: A traveling salesman problem with pickups and deliveries and stochastic travel times: an application from chemical shipping. Eur. J. Oper. Res. (2018)
9. Vidal, T., Crainic, T.G., Gendreau, M., Prins, C.: Heuristics for multi-attribute vehicle routing problems: a survey and synthesis. Eur. J. Oper. Res. **231**(1), 1–21 (2013)
10. Veenstra, M., Roodbergen, K.J., Vis, I.F., Coelho, L.C.: The pickup and delivery traveling salesman problem with handling costs. Eur. J. Oper. Res. **257**(1), 118–132 (2017)

11. Arnesen, M.J., Gjestvang, M., Wang, X., Fagerholt, K., Thun, K., Rakke, J.G.: A traveling salesman problem with pickups and deliveries, time windows and draft limits: case study from chemical shipping. Comput. Oper. Res. **77**, 20–31 (2017)
12. Osaba, E., Yang, X.S., Fister, I., Del Ser, J., Lopez-Garcia, P., Vazquez-Pardavila, A.J.: A discrete and improved bat algorithm for solving a medical goods distribution problem with pharmacological waste collection. Swarm and Evolutionary Computation (2018)
13. Osaba, E., Del Ser, J., Bilbao, M.N., Lopez-Garcia, P., Nebro, A.J.: Multi-objective design of time-constrained bike routes using bio-inspired meta-heuristics. In: Proceedings of 8th International Conference on Bioinspired Optimization Methods and their Applications (2018 in press)
14. Jimenez, J.L., McClintock, P., McRae, G., Nelson, D.D., Zahniser, M.S.: Vehicle specific power: a useful parameter for remote sensing and emission studies. In: Ninth CRC On-Road Vehicle Emissions Workshop, San Diego, CA (1999)
15. Jimenez-Palacios, J.L.: Understanding and quantifying motor vehicle emissions with vehicle specific power and tildas remote sensing. Massachusetts Institute of Technology (1998)
16. Hart, C., Koupal, J., Giannelli, R.: Epa's onboard analysis shootout: overview and results (no. epa/420/r-02/026). Technical report (2002)
17. Koupal, J., Hart, C., Brzezinski, D., Giannelli, R., Bailey, C.: Draft emission analysis plan for moves GHG. Technical report, US Environmental Protection Agency (2002)
18. Li, H., Alidaee, B.: Tabu search for solving the black-and-white travelling salesman problem. J. Oper. Res. Soc. **67**(8), 1061–1079 (2016)
19. Todosijević, R., Mjirda, A., Mladenović, M., Hanafi, S., Gendron, B.: A general variable neighborhood search variants for the travelling salesman problem with draft limits. Optim. Lett. **11**(6), 1047–1056 (2017)
20. Osaba, E., Carballedo, R., Diaz, F., Onieva, E., Masegosa, A., Perallos, A.: Good practice proposal for the implementation, presentation, and comparison of metaheuristics for solving routing problems. Neurocomputing **271**, 2–8 (2018)
21. Osaba, E., Díaz, F.: Comparison of a memetic algorithm and a tabu search algorithm for the traveling salesman problem. In: Federated Conference on Computer Science and Information Systems (FedCSIS), pp. 131–136. IEEE (2012)
22. Derrac, J., García, S., Molina, D., Herrera, F.: A practical tutorial on the use of nonparametric statistical tests as a methodology for comparing evolutionary and swarm intelligence algorithms. Swarm Evol. Comput. **1**(1), 3–18 (2011)

Road Traffic Forecasting Using NeuCube and Dynamic Evolving Spiking Neural Networks

Ibai Laña[1(✉)], Elisa Capecci[2], Javier Del Ser[3], Jesus L. Lobo[1], and Nikola Kasabov[2]

[1] TECNALIA, 48160 Derio, Spain
{ibai.lana,javier.delser,jesus.lopez}@tecnalia.com
[2] KEDRI - Auckland University of Technology (AUT), 1010 Auckland, New Zealand
[3] TECNALIA, University of the Basque Country (UPV/EHU) and Basque Center for Applied Mathematics (BCAM), Bizkaia, Spain
{ecapecci,nkasabov}@aut.ac.nz

Abstract. This paper presents a new approach for spatio-temporal road traffic forecasting that relies on the adoption of the NeuCube architecture based on spiking neural networks. The NeuCube platform was originally conceived and designed to process electroencephalographic (EEG) signals considering their temporal component and their spatial source within the brain. Its neural representation allows for a visual analysis of connectivity among different locations, and also provides a prediction tool harnessing the predictive learning capabilities of dynamic evolving Spiking Neural Networks (deSNNs). Taking advantage of the NeuCube features, this work focuses on the potential of spatially-aware traffic variable forecasts, as well as on the exploration of the spatio-temporal relationships among different sensor locations within a traffic network. Its performance, assessed over real traffic data collected in 51 locations in the center of Madrid (Spain), is superior to that of other machine learning techniques in terms of forecasting accuracy. Moreover, we discuss on the interactions and relationships among sensors of the network provided by Neucube, which may provide valuable insights on the traffic dynamics of the city under study towards enhancing its management.

Keywords: Traffic forecasting · Spiking neural networks · NeuCube

1 Introduction

The prediction of road traffic variables such as flow, occupancy or speed has been a matter of extensive research in the last decades. Anticipating the future traffic state is of great interest for institutions that manage and regulate traffic and users of road networks [27]. Thereby, models for traffic forecasting have been developing over years, initially based on time-series analysis [2,18], and

J. Del Ser et al. (Eds.): IDC 2018, SCI 798, pp. 192–203, 2018.
https://doi.org/10.1007/978-3-319-99626-4_17

thereafter gravitating towards non-parametric and machine learning models such as k-nearest neighbors (kNN), Artificial Neural Networks (ANN), Support Vector Machines (SVM), Bayesian Networks and Fuzzy Logic, among others alike [26]. Likewise, the target of predictions has drifted from freeway to urban contexts, where traffic dynamics are more complex and are affected by congestion, signals, interactions with close nodes and a variety of unexpected events [25]. Hence, building accurate predictive models for urban traffic becomes more practically feasible when available data are abundant and more granular. Moreover, when traffic data are provided by a network of sensors (also referred to as *detectors*), correlations among them allow for more complex traffic models [22]. Aligned with this, it is increasingly common to find works that prospect for road network interrelations that help obtaining predictive models for their individual nodes, although network-wide predictions are still considered a challenge of paramount relevance for future Intelligent Transportation Systems (ITS) [16,28].

Urban network-wide forecasts have been approached by means of diverse techniques in recent works. Among them, [4] proposed a spatio-temporal version of the kNN model with improved results up to 12 step forecasting. A spatio-temporal random effects model is proposed by [29] to obtain predictions in many locations simultaneously, improving the outcomes of ARIMA models. Another approach in [33] is based on Bayesian Networks, which are able to model the relations between predicted variable and surrounding sources of traffic. In [1], predictions are obtained for a whole network, considering the ratio of traffic flow that propagates from a specific link to its adjacent links. The availability of large amounts of data is exploited in [31], where authors propose an ensemble of predictors to obtain online forecasts.

As stated above, urban spatio-temporal traffic forecasting has been widely addressed by researchers in recent years. Surprisingly, despite their renowned capability to process time distributed signals and unveil spatio-temporal relationship from data, Spiking Neural Networks (SNNs) remain unexplored to date for this particular application. SNNs have been used to forecast variables out of the domain of brain signals, in diverse areas such as electrical loads, financial data or grain yield [15,20,32]. However, they are hardly found to obtain traffic predictions. Specifically, only the recent work in [24] formulates a traffic-related pattern recognition problem and uses it as a use case for the application of a SNN-based architecture called NeuCube. It is this architecture which allows for a spatial mapping of a road traffic network and its understanding.

NeuCube represents a computational framework for the creation of machine learning models inspired in the brain functioning. This framework has been implemented in the Knowledge Engineering and Discovery Research Institute (KEDRI) of the Auckland University of Technology (New Zealand). The NeuCube framework, as presented in [13] and schematically depicted in Fig. 1, is the result of the advancements made in the evolving SNN (eSNN) model proposed in [12,30], which was part of the Evolving Connectionist Systems (ECOS). ECOS are multi-modular, adaptive and knowledge representation computer systems based on neural networks; following their principles, the first implementation of a NeuCube-based model was developed for brain data modeling and analysis using neuromorphic, brain-like SNNs as information processing system [7]. Now, the NeuCube is used as a multi-modular generic system, which can be adapted to different applications across a wide variety of domains, such as environmental/ecological modeling [3], global/personalized brain data modeling [6] or neurogenetic models [8].

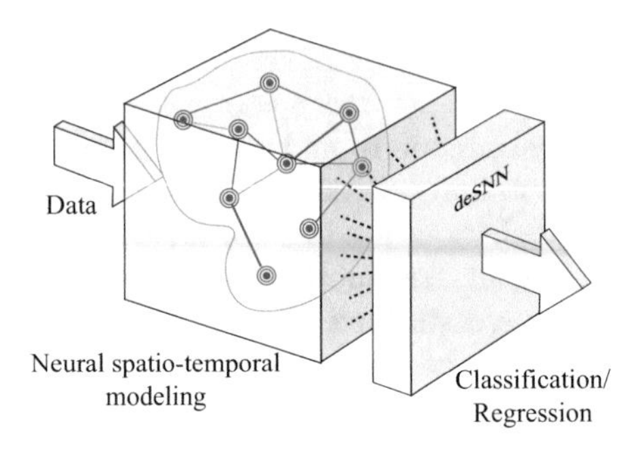

Fig. 1. NeuCube architecture. Input variables are mapped into the network where correlations of data are found. In a second stage, those correlations are used to define the inputs of a SNN model, used to perform the regression.

In this manuscript NeuCube is used for modeling and regression of spatio-temporal traffic data captured over the city of Madrid (Spain). Specifically, by mapping the data features into the SNN architecture, we aim at studying the spatial correlation of traffic generated in these locations. The spatio-temporal insights provided by the Neucube architecture are shown to shed light on correlation patterns between different traffic loops installed over Madrid, paving the way for future research lines focused on exploiting this information for traffic prediction purposes.

2 Materials and Methods

This research work is based on real traffic volume observations obtained from the Madrid Open Data Portal [19]. The Madrid City Council provides through this website access to traffic data around the whole city, publishing 15-min aggregates and live 5-min aggregates of flow, occupancy and speed data in more than 3600 measuring stations (loops). Loops have been selected under the following criteria: (a) placement in a business district for highly changing traffic profiles; (b) availability of data, as failures are common in many sensors, yielding large series of missing data [17]; and (c) at least 50 sensors in the nearby network to assess spatio-temporal correlations. The area depicted in Fig. 2 has been found to meet these 3 criteria.

Data have been captured in 51 locations of the business center of the city for the first half of year 2017. These data contain minor missing data portions that have been imputed according to simpler methods described in [17]. This previous work also presented an imputing strategy based on using predictions for a certain location relying in data obtained from other locations. This same technique is employed in this work. Traffic flow detected in one of the locations will be used as target variable, while the traffic at the other locations will act as input data; hence, the developed model will allow to predict traffic in a certain location with information available from other locations, depending on their spatio-temporal correlations.

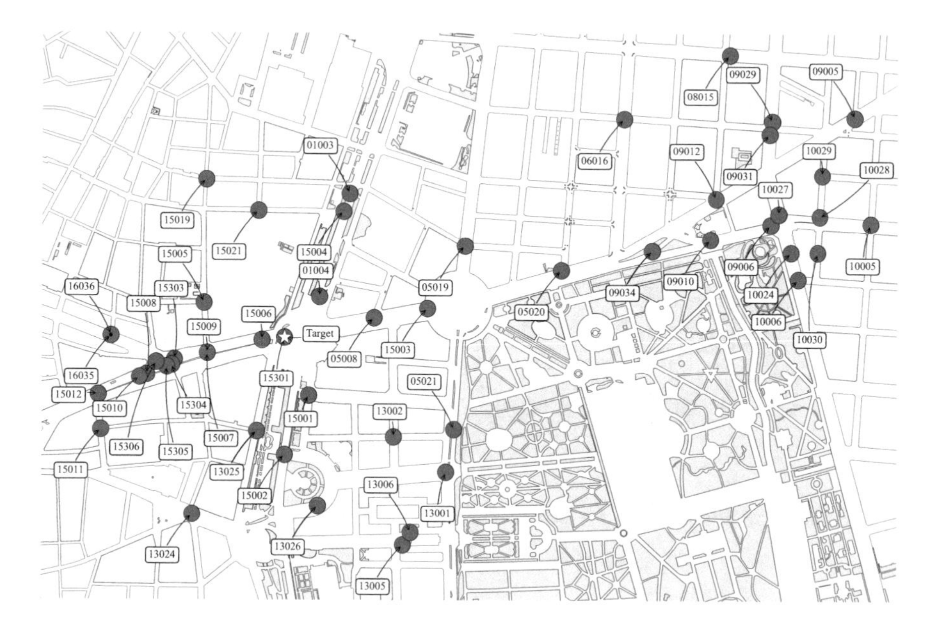

Fig. 2. Location of 51 sensors in Madrid city center. The detector marked with a $\star$ symbol represents the forecast target location.

After missing values imputation, data have been preprocessed by encoding them to obtain short-term predictions in one of the locations using the information provided by surrounding sensors. The NeuCube platform allows for a spatio-temporal prediction by encoding brain signals as data matrices: each column or feature represents a sensor of the experimental setup, and each row contains the readings of sensors in different moments. Once signals have been processed and propagated through the network, NeuCube yields the interrelations of sensors by showing their level of connectivity. Traffic sensors have been mapped to a two-dimensional grid, and data have been encoded according to Fig. 3.

As NeuCube is a generic system, we tailored it for our application by following this scheme:

- Real-valued data vectors were encoded into spike sequences using the adaptive threshold-based (ATB) encoding algorithm proposed in [24].

- Input variables were mapped into the network, which is populated by leaky integrate-and-fire (LIF) neurons [10] separated by a small-world (SW) connectivity distance.
- Unsupervised learning of the spatio-temporal spike sequences was performed using spike-timing-dependent plasticity (STDP) rules [21] to capture the spatio-temporal correlation of data.
- Supervised learning was performed by using dynamic evolving spiking neural networks (deSNN, [14]), which assign output data to the corresponding regression variables.
- Optimization of the numerous parameters of the cube to improve the regression accuracy.
- The connectivity and spiking activity generated in the cube during learning is recorded and the resulting activity analyzed for a better understanding of data and the processes that generated them.

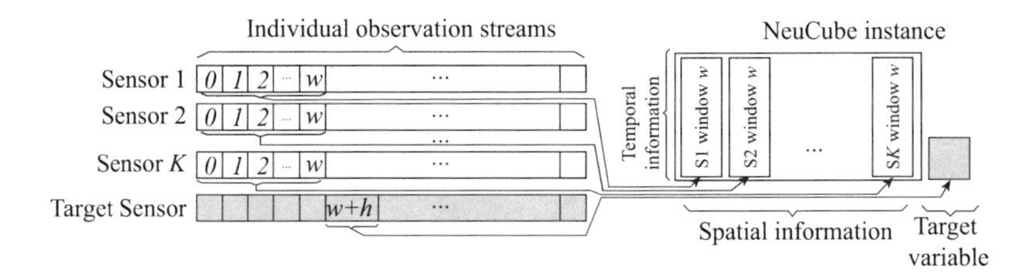

Fig. 3. Data transformation for NeuCube instance generation from streams of loop data.

As per [24], the ATB encoding algorithm is implemented as a bi-directional threshold that transform each vector of time-series data into spike trains according to

$$ATB = \mu + s\ \sigma, \tag{1}$$

where μ is the mean of the differential signal with respect to time, $x(t) = \{x_{t_2} - x_{t_1},\ x_{t_3} - x_{t_2}, \ldots,\ x_{t_n} - x_{t_{n-1}}\}$, calculated by using all samples; σ is its standard deviation; and s is a scale parameter of σ. This bi-directional threshold considers only significant changes in the signal gradient and generates two types of spike trains: one that corresponds to the signal increment (a positive spike train) and one that corresponds to the signal decay (a negative spike train). Positive and negative spike trains are fed into the cube using two different input nodes, which possess identical mapping coordinates.

All the input nodes are mapped into specific (x, y, z) coordinates of the cube. As we are dealing with a two-dimensional space (the map of the city of Madrid), the number of neurons in the cube and their locations were calculated accordingly. More specifically, every neuron of the cube represents a point into a grid of $90\,(x) \times 60\,(y)$ positions (physical locations of the sensors) so that the proportions in the distances among loops are maintained. According to this mapping, the number of neurons in the cube was set to 5400, and then all neurons placed in spaces corresponding to buildings or parks were eliminated, as traffic cannot propagate through them. Each traffic sensor is associated to an input neuron, a point in this grid mapped to the sensor coordinates. This allows studying the possible spatio-temporal correlation between the locations of the traffic sensors and the timing of the data.

2.1 Initialization and Unsupervised Learning

The generated spikes are presented into a cube of LIF neurons. The cube is stochastically initialized to assign synaptic connections to neurons. This is made according to the SW connectivity distance that makes neighboring nodes highly and strongly interconnected. This influences not only the initialization, but also the learning process in the architecture. Every connected neuron is associated with a weight that is calculated as the product of a random number $[-0.1, +0.1]$ times the inverse of the Euclidean distance $d(i, j)$ between two neurons, according to their (x, y, z) coordinates ($z = 0$ in this case). These neurons are called pre- and a post- synaptic neurons, as per the direction of the incoming signal. As in the mammalian brain, the number of inhibitory synapses is found to be about 20–30% [5], also in the cube 20% of these weights are set to be negative (or inhibitory), while 80% are set to be positive (or excitatory). Another parameter – the distance threshold D_{thr} – impacts on the value of $d(i, j)$. This parameter is calculated as:

$$D_{thr} = \max\ (d(i,j))\ p, \tag{2}$$

where p is the parameter of the SW connectivity.

The cube is trained in an unsupervised way to modify the initially set connection weights, by applying Hebbian-like learning rules [11]. Following the STDP protocol, the strength of a synapse between two neurons is modified according to the intensity and timing of their activation. This permits to learn sequential temporal relations from the data, and it is implemented as follows:

$$w_j(t) = \begin{cases} w_j(t-1) \pm \alpha/\Delta t & \mathrm{t_j} \neq \mathrm{t_i}, \\ w_j(t-1) & \mathrm{t_j} = \mathrm{t_i}, \end{cases} \tag{3}$$

where α is the rate of the STDP learning algorithm, and Δ_t indicates the time since the last spike was emitted by the post-synaptic neuron j. If a pre-synaptic neuron i fires before a post-synaptic neuron j, then its connection weight $w_{j,i}$ increases. Otherwise, it decreases.

2.2 Supervised Learning and Regression Outputs

The deSNN is used as a supervised learning algorithm to assign regression outputs to the respective variables. This algorithm is used as it combines the rank-order (RO) learning rule [23] with the STDP [21] temporal learning. According to this algorithm, data used during unsupervised training is propagated one more time in the cube to train the output model. Then, each training sample is associated to an output neuron, which is connected to each and every neuron of the cube. Initially, connection weights between input and output neurons are zero, and then are modified according to the RO rule as:

$$w_{i,j} = mod^{order(i,j)}, \tag{4}$$

where mod indicates a modulation factor and $order(i,j)$ the order of the first incoming spike. Subsequently, the weights are modified according to the spike driven synaptic plasticity (SDSP) learning rule [9]:

$$w_{i,j}(t) = \begin{cases} w_{i,j}(t-1) + drift & \mathrm{S_j(t) = 1}, \\ w_{i,j}(t-1) - drift & \mathrm{S_j(t) = 0}, \end{cases} \tag{5}$$

where $drift$ indicates a parameter used to modify the weights, and $S_i(t)$ is the occurrence of the spikes arriving from neuron i at a time t after the first one was emitted. Every generated output neuron is trained to recognize the spikes generated by a corresponding labeled input into regression variables. If data do not activate existing output neurons, new neurons are generated to match the data along with their new connections.

All the above make NeuCube a suitable framework for revealing complex spatio-temporal patterns concealed in the data, setting the rationale why this architecture has been chosen for analyzing spatio-temporal traffic data.

3 Experiments and Results

NeuCube platform provides a complete environment to analyze spatio-temporal relationships. Once traffic data are encoded as described in Fig. 3, they are fed to the NeuCube and transformed into spikes that propagate through its neurons according to the rules described in the previous section. Once the training phase of the NeuCube is finished, it yields a map of the interaction between the input neurons depicted as a chord diagram in Fig. 4.

The width of the lines between each pair of nodes displays the relative amount of synaptic connections that have been produced between those nodes, after encoding the input data into spikes and feeding them the trained cube. These interactions represent the relationship between traffic profiles in both points. For a connection to be established between two neurons (locations in the map), it is required that the synapses propagate through all the neurons that are in their way, which have also been represented in the cube set up. Relationships are found in local clusters, for instance in the area where 09xxx and 10xxx loops are placed, and also among loops 15xxx and 13xxx, due to the propagation and small-world parameters. A strong correlation is needed for two distant nodes to be connected, hence stronger connections are found between closer loops, such as 15301 and 15006 or 15008, 15306 and 15010, which are placed next to each other. Other strong interactions such as those between 09005 and 10028 or 15010 and 13025 are found for more distant loops, suggesting a highly related traffic profile. Weaker interactions also allow for a traffic behavior analysis.

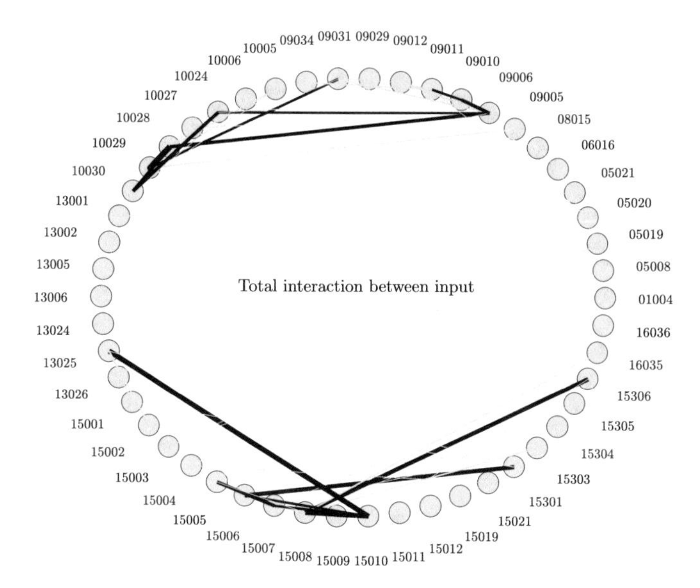

Fig. 4. Chord diagram representing the spatio-temporal interaction among different inputs to the network of neurons.

Beyond a visual analysis of the network connectivity, interactions found in the NeuCube provide the weight configuration for its last layer, a deSNN regression algorithm, rarely used in the traffic prediction domain. This configuration adapted to the real spatio-temporal relations is expected to boost the performance of the regression algorithm. In order to quantitatively assess the accuracy scored by the regression model, a benchmark with other commonly used machine learning methods is implemented.

For all these cases input data are the same, encoded as 1000 feature instances. For each measured value of the target sensor, a window of $w = 20$ past values is taken for each of the other 50 sensors, prior to the target observation at $h = 1$ time steps (prediction horizon). This yields a sample consisting of a value containing 1000 values and a target value, as described in Fig. 5. With this setup, data are divided into a train set and a test set, containing 75% and 25% of the available instances respectively. Train set is then used to tune the NeuCube operation parameters: STDP rate, Refractory Time, Mod, Drift and Similarity. To this end we resort to a grid search efficiently traversed by a genetic optimization tool, which is provided off-the-shelf by the NeuCube platform. When the optimal parameter set is found, NeuCube and deSNN are trained with train set and tested with the remaining 25%.

The methods selected to compare against the deSNN cover a range of machine learning approaches frequently used in traffic forecasting problems and are listed as follows: k-nearest neighbors (KNN), decission tree regression (DTR), random forest regression (RFR), AdaBoost regression (ADAR), gradient boost regression (GBR), ridge regression (RID), stochastic gradient descent regression (SGD),

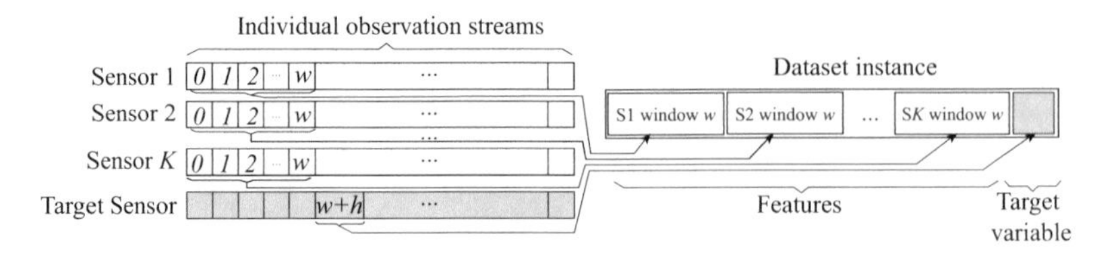

Fig. 5. Data transformation scheme for machine learning algorithms dataset generation from streams of loop data.

support vector machine (SVR), and extreme learning machine (ELM). All of them have been run through a 10-fold cross validation process in order to tune their parameters, using the same 75% of train data as with NeuCube.

Table 1. Comparative results of all the proposed methods.

Metric	KNN	DTR	RFR	ADAR	GBR	RID	SGD	SVR	ELM	deSNN
R^2	0.89	0.91	0.93	0.91	0.92	0.91	0.92	0.91	0.90	0.95
$RMSE$	171	155	138	153	144	150	147	150	157	133

Predictive results of deSNN model (in terms of R^2 and Root Mean Square Error) are compared to the rest of regression models in Table 1. Most of them perform reasonably well, considering that test data correspond to a different period of the year than training data. As the training phase only covers around 4 months of data, some of the situations produced in the test set (due to seasonality) have never been observed before by the model. However, deSNN yields a remarkable improved performance. Furthermore, the connectivity information obtained in the initial NeuCube stage allows not only for a better understanding of the interactions between nodes, but also for a more efficient predictive algorithm that disregards (through trainable weights) the input data that are not useful for its purpose.

4 Conclusions and Future Work

Unveiling complex spatio-temporal traffic correlations in urban contexts has acquired momentum in recent years, as a way to tackle the prediction of highly fluctuating traffic in metropolitan areas. In this regard, this manuscript has examined the applicability of the NeuCube spatio-temporal processing architecture to the spatio-temporal analysis of traffic data. The platform has been successfully used for classification and regression in areas not related to its original purpose, but has never been applied within the road traffic domain. The architecture of NeuCube propagates the input data through a network of nodes that represent the spatial disposition of deployed road sensors, also considering their temporal distribution. Interactions of different intensity are found among the sensors assisting the initial configuration of the deSNN prediction algorithm. With these tuned inputs, the deSNN prediction model at the core of the NeuCube platform provides forecasts able to outperform traditional supervised learning models. The reported performance gains are substantial when compared to other methods. Moreover, the use of this platform provides an insightful mechanism to observe and analyze traffic correlations among different measuring locations. These relationships allows building efficient network models, using only information of relevant sensors.

Although NeuCube is able to model interactions among a large number of locations, its current architecture restricts its predictive capabilities to one sensor at a time. Further research efforts should be conducted towards endowing NeuCube with the capability to simultaneously predict all the considered locations, thereby providing combined forecasts for a whole network.

Acknowledgements. This work was supported by the EU project Pacific Atlantic Network for Technical Higher Education and Research - PANTHER (grant number 2013-5659/004-001 EMA2), and by the Basque Government (EMAITEK program).

References

1. Abadi, A., Rajabioun, T., Ioannou, P.A.: Traffic flow prediction for road transportation networks with limited traffic data. IEEE Trans. Intell. Transp. Syst. **16**(2), 653–662 (2015)
2. Ahmed, M.S, Cook, A.R.: Analysis of freeway traffic time-series data by using box-jenkins techniques, vol. 722 (1979)
3. Bose, P., Kasabov, N.K., Bruzzone, L., Hartono, R.N.: Spiking neural networks for crop yield estimation based on spatiotemporal analysis of image time series. IEEE Trans. Geosci. Remote. Sens. **54**(11), 6563–6573 (2016)
4. Cai, P., Wang, Y., Lu, G., Chen, P., Ding, C., Sun, J.: A spatiotemporal correlative k-nearest neighbor model for short-term traffic multistep forecasting. Transp. Res. Part C: Emerg. Technol. **62**, 21–34 (2016)
5. Capano, V., Herrmann, H.J., de Arcangelis, L.: Optimal percentage of inhibitory synapses in multi-task learning. Sci. Rep. **5**, 9895 (2015)
6. Capecci, E., Kasabov, N., Wang, G.Y.: Analysis of connectivity in neucube spiking neural network models trained on eeg data for the understanding of functional changes in the brain: a case study on opiate dependence treatment. Neural Netw. **68**, 62–77 (2015)
7. Chen, Y., Hu, J., Kasabov, N., Hou, Z.G., Cheng, L.: Neucuberehab: a pilot study for eeg classification in rehabilitation practice based on spiking neural networks. Neural Inf. Process. **8228**, 70–77 (2013)
8. Espinosa-Ramos, J.I., Capecci, E., Kasabov, N.: A computational model of neuroreceptor dependent plasticity (NRDP) based on spiking neural networks. IEEE Trans. Cogn. Dev. Syst. **99**, 1–1 (2017)
9. Fusi, S.: Spike-driven synaptic plasticity for learning correlated patterns of asynchronous activity. In: International Conference on Artificial Neural Networks, pp. 241–247. Springer, Heidelberg (2002)

10. Gerstner, W.: Plausible Neural Networks for Biological Modelling. Kluwer Academic Publishers, Dordrecht (2001). vol What's different with spiking neurons?
11. Hebb, D.O.: The Organization of Behavior: A Neuropsychological Approach. Wiley, New York (1949)
12. Kasabov, N.: Evolving Connectionist Systems: The Knowledge Engineering Approach. Springer, Heidelberg (2007)
13. Kasabov, N.: Evolving spatio-temporal data machines based on the neucube neuromorphic framework: design methodology and selected applications. Neural Netw. **78**, 1–14 (2016)
14. Kasabov, N., Dhoble, K., Nuntalid, N., Indiveri, G.: Dynamic evolving spiking neural networks for on-line spatio- and spectro-temporal pattern recognition. Neural Netw. **41**, 188–201 (2013)
15. Kulkarni, S., Simon, S.P., Sundareswaran, K.: A spiking neural network (SNN) forecast engine for short-term electrical load forecasting. Appl. Soft Comput. **13**(8), 3628–3635 (2013)
16. Laña, I., Del Ser, J., Vélez, M., Vlahogianni, E.I.: Road traffic forecasting: recent advances and new challenges. IEEE Intell. Transp. Syst. Mag. **10**, 93–109 (2018a)
17. Laña, I., Olabarrieta, I.I., Del Ser, J., Vélez, M.: On the imputation of missing data for road traffic forecasting: new insights and novel techniques accepted pending to be published. Transp. Res. Part C: Emerg. Technol. **90**, 18–33 (2018b)
18. Levin, M., Tsao, Y.: On forecasting freeway occupancies and volumes. Transp. Res. Rec. **773**, 47–49 (1980)
19. Madrid Open Data Portal (2018). http://datos.madrid.es. Accessed 31 Jan 2018
20. Reid, D., Hussain, A.J., Tawfik, H.: Spiking neural networks for financial data prediction. In: The 2013 International Joint Conference on Neural Networks (IJCNN), pp. 1–10. IEEE (2013)
21. Song, M.K.D., Abbott, L.F.: Competitive hebbian learning through spike-timing-dependent synaptic plasticity. Nat. Neurosci. **3**(9), 919–926 (2000)
22. Stathopoulos, A., Karlaftis, M.G.: A multivariate state space approach for urban traffic flow modeling and prediction. Transp. Res. Part C: Emerg. Technol. **11**(2), 121–135 (2003)
23. Thorpe, S.J., Gautrais, J.: Rank order coding. In: Computational Neuroscience, pp. 113–118. Springer, Heidelberg (1998)
24. Tu, E., Kasabov, N., Yang, J.: Mapping temporal variables into the neucube for improved pattern recognition, predictive modeling, and understanding of stream data. IEEE Trans. Neural Netw. Learn. Syst. **28**(6), 1305–1317 (2017)
25. Van Arem, B., Kirby, H.R., Van Der Vlist, M.J.M., Whittaker, J.C.: Recent advances and applications in the field of short-term traffic forecasting. Int. J. Forecast. **13**(1), 1–12 (1997)
26. Van Hinsbergen, C., Van Lint, J., Sanders, F.: Short term traffic prediction models. In: World Congress on Intelligent Transport Systems (2007)
27. Vlahogianni, E.I., Golias, J.C., Karlaftis, M.G.: Short-term traffic forecasting: overview of objectives and methods. Transp. Rev. **24**(5), 533–557 (2004)
28. Vlahogianni, E.I., Karlaftis, M.G., Golias, J.C.: Short-term traffic forecasting: where we are and where we're going. Transp. Res. Part C: Emerg. Technol. **43**, 3–19 (2014)
29. Wu, Y., Chen, F., Lu, C., Yang, S.: Urban traffic flow prediction using a spatio-temporal random effects model. J. Intell. Transp. Syst. Technol. Plan. Oper. **2450**, 1–12 (2015)
30. Wysoski, S.G., Benuskova, L., Kasabov, N.: Evolving spiking neural networks for audiovisual information processing. Neural Netw. **23**(7), 819–835 (2010)

31. Xu, J., Deng, D., Demiryurek, U., Shahabi, C., Van Der Schaar, M.: Mining the situation: spatio-temporal traffic prediction with big data. IEEE J. Sel. Top. Signal Process. **9**(4), 702–715 (2015)
32. Yang, L., Zhongjian, T.: Prediction of grain yield based on spiking neural networks model. In: 2011 IEEE 3rd International Conference on Communication Software and Networks (ICCSN), pp. 171–174. IEEE (2011)
33. Zhu, Z., Peng, B., Xiong, C., Zhang, L.: Short-term traffic flow prediction with linear conditional Gaussian Bayesian network. J. Adv. Transp. 1–13 (2016)

A Preliminary Study on Automatic Algorithm Selection for Short-Term Traffic Forecasting

Juan S. Angarita-Zapata[1,2](✉), Isaac Triguero[3](✉), and Antonio D. Masegosa[1,2,4](✉)

[1] Faculty of Engineering, University of Deusto, Av. Universidades, 24, 48007 Bilbao, Spain
js.angarita@deusto.es

[2] DeustoTech-Fundación Deusto, Deusto Foundation, Av. Universidades, 24, 48007 Bilbao, Spain
ad.masegosa@deusto.es

[3] Automated Scheduling, Optimisation and Planning (ASAP), School of Computer Science, University of Nottingham, Jubilee Campus, Wollaton Road, Nottingham NG8 1BB, UK
Isaac.Triguero@nottingham.ac.uk

[4] IKERBASQUE, Basque Foundation for Science, 48011 Bilbao, Spain

Abstract. Despite the broad range of Machine Learning (ML) algorithms, there are no clear baselines to find the best method and its configuration given a Short-Term Traffic Forecasting (STTF) problem. In ML, this is known as the Model Selection Problem (MSP). Although Automatic Algorithm Selection (AAS) has proved success dealing with MSP in other areas, it has hardly been explored in STTF. This paper deepens into the benefits of AAS in this field. To this end, we have used Auto-WEKA, a well-known AAS method, and compared it to the general approach (which consists of selecting the best of a set of algorithms) over a multi-class imbalanced classification STTF problem. Experimental results show AAS as a promising methodology in this area and allow important conclusions to be drawn on how to improve the performance of ASS methods when dealing with STTF.

1 Introduction

Urban development and population growth have risen the levels of congestion that in turn generate socio-economic and environmental issues. One contemporary policy to deal with congestion is the development of STTF systems. STTF is the prediction of near future traffic measures for fixed locations, road segments, or entire links [15]; which in consequence allows users to plan ahead their movements along the roads.

The recent emergence of telecommunications technologies integrated to transportation infrastructure generates vast volumes of traffic data. Such unprecedented data availability and growing computational capacities have increased

J. Del Ser et al. (Eds.): IDC 2018, SCI 798, pp. 204–214, 2018.
https://doi.org/10.1007/978-3-319-99626-4_18

the use of ML to approach STTF. The main strength of ML, with respect to traffic theory models, is its ability to predict short-term traffic using current and historical data without the need of knowing theoretical traffic mechanisms.

The literature on STTF reports a great variety of ML algorithms applications as Neural Networks (NNs), Support Vector Machines (SVMs), k-Nearest Neighbors (k-NN) or Random Forest (RF) [5,15]. Nevertheless, given the broad range of ML methods there are no baselines to select the most appropriate algorithm and its best hyper-parameter setting given the characteristics of an STTF problem. In ML, this is known as the MSP, and AAS has been one of the most successful approaches to address it so far. It aims at automatically finding the ML algorithm and hyper-parameters configuration pair which maximize a performance measure on given data, using an optimization strategy that minimizes a predefined loss function.

Although AAS methods have approached the MSP with high performance in other research areas [7], to the best of authors' knowledge only the work in [14] has tackled the MSP in STTF. The proposed AAS method aims at predicting average speed in a time horizon of 5 min using a time series regression approach. In this research, we conduct a preliminary study to keep exploring the benefits of AAS in STTF focusing on a classification STTF problem in multiple time horizons and using a different AAS method, Auto-WEKA. To this end, we compare the AAS method versus the general approach in STTF, which consists of selecting the best of a set of commonly used ML algorithms. Concretely, we compare Auto-WEKA with four state-of-the-art ML algorithms (NN, SVM, k-NN and RF) in the task of forecasting traffic Level of Service (LoS) using real data.

The rest of this paper is structured as follow. Section 2 presents related work in ML and AAS algorithms applied to STTF. Section 3 shows the methodology of this research. Then, Sect. 4 exposes results followed by the Conclusions in Sect. 5.

2 Related Work

This section reviews literature related to ML and AAS in the context of STTF. We start by describing how STTF can be addressed from a ML perspective. Then, ML methods for STTF are discussed, and finally we review existing AAS methods.

Short-Term Traffic Forecasting from a Machine Learning perspective

In recent years, STTF is being influenced by the great availability of data provided by Intelligent Transportation Systems. Some technologies, such as Automatic Vehicle Identification, Electronic Tolls, and GPS, collect individual traffic data related to each vehicle on the road; meanwhile, others collect macroscopic traffic measures (averages of many vehicles) as Vehicle Detection Stations (VDS). These technologies and contemporary computational advances have caused a leap in the way STTF is approached switching from a traffic theory-based perspective to a data-driven one, with a special focus on ML. In this work, we center on ML applied to VDS data, because it is the most common type of data available and used in literature [9].

From a ML perspective, STTF is approached by building a model from traffic historical data to make predictions on new and unseen data. Depending on the type of input and output (predicted) data, different ML approaches can be used. When the traffic measure to be forecasted is continuous (e.g. speed or flow) it should be dealt as a regression problem. In the case that input variables are ordered by time, the approach is time series regression, which requires defining the prior time steps and the number of lagged variables to predict the forecasted traffic measure. On the other hand, when the predicted value is discrete, the prediction should be addressed as a classification problem (e.g. Traffic LoS).

ML algorithms for STTF using VDS data

ML methods applied to STTF can be categorized into single or hybrid. The first type corresponds to adaptations of existing ML algorithms and in turn, they can be classified as parametric and non-parametric. The parametric category assumes the relationship between the explanatory and response variables as known; meanwhile, the non-parametric ones are able to model nonlinear relationships without requiring the mentioned assumptions. Commonly non-parametric algorithms used are NNs, SVMs, k-NN, and RF [15]. As mentioned before, the other approach of ML algorithms is hybridization. Within it, two or more algorithms, from ML or even other areas, are combined to find synergies that improve their isolated performance. Some recent examples are [8], where authors integrate a Boltzmann Machine with Recurrent NNs, and [6], where Genetic Algorithms are integrated with Fuzzy Systems.

Nevertheless, despite the great variety of ML methods, dealing with the MSP in STTF is not a trivial task, as mentioned before. The general approach to tackle the MSP in STTF consists of testing a set of algorithms with multiple hyper-parameter combinations and select the best one. This requires expert knowledge and a lot of human effort. Nowadays, AAS has received a lot of attention in ML because of its promising results in dealing with the MSP with low human intervention.

Automatic ML Algorithm Selection in STTF

As stated above, AAS method deals with MSP as an optimization problem whose objective consists of finding the ML algorithm, from a pre-defined base of algorithms, and its hyper-parameter configuration that maximizes an accuracy measure on a given ML problem. The first method in tackling simultaneously the selection of algorithm and hyper-parameters in ML was Auto-WEKA [13]. It uses Bayesian optimization to search for the best pair (algorithm, hyper-parameter setting) and its base of algorithms are the 39 ML methods implemented in WEKA (a well-known open-source ML software that contains algorithms for data analysis and predictive modelling). Subsequently, Komer et al. [3] and Feurer et al. [1] developed Hyperopt-sklearn and Auto-sklearn, respectively. They automatically select ML algorithms and hyper-parameter values from scikit-learn[1]. In the case of [3], the AAS method uses Hyperopt Python library for the optimization process, concretely a Bayesian optimization method as Auto-WEKA. Meanwhile, Auto-sklearn stores the best combination of ML algorithm and hyper-parameters that have been found for each previous ML problem and using meta-learning chooses a starting point for a sequential optimization process. More recently, Sparks et al. [12] proposed a method that supports distributed computing for AAS, and Sabharwal et al. [10] developed a cost-sensitive training data allocation method that assesses a pair (algorithm, hyper-parameters setting) on a small random sample of the data-set and gradually expands it over time to re-evaluate it when the combination is promising. For this research, we select Auto-WEKA because of the wider variety of the base of algorithms in comparison with the others approaches reviewed. Furthermore, unlike the methods presented by Sparks et al. and Sabharwal et al. that only consider a pre-defined set of hyper-parameters combinations, Auto-WEKA has no limitations in the hyper-parameter space to be explored.

To the best authors' knowledge, only one work has tackled the MSP in STTF [14]. In this work, Vlahogianni proposed a meta-modelling technique that, based on surrogate modelling and a genetic algorithm with an island model, optimizes both the algorithm selection and the hyper-parameter setting. The AAS task is performed from an algorithms base of three ML methods (NN, SVM and Radial Base Function) that forecast average speed in a time horizon of 5 min using a time series regression approach. The main differences between this work and Vlahogianni's one lay on the addressed problem and the AAS method. Regarding the problem, we predict traffic LoS along multiple time horizons using a classification approach; and with respect to the method, we use an AAS method that considers a much broader base of algorithms, which is an important aspect in the MSP as we will discuss later.

3 Methodology

This research seeks to keep exploring the benefits AAS can bring to STTF. To accomplish this, we compare to what extent the results of AAS differ from the general approach in STTF, in which a set of Reference Algorithms (RAs) is tested over the forecasting problem in hand and the one with best performance metrics is chosen. We select Auto-WEKA as AAS method for the reasons explained above; and NN, SVM, $k-NN$, and RF as the RAs that represent the general approach. These algorithms are the most commonly used one in recent STTF literature. Due to space limitations, the details of Auto-WEKA are omitted. The interested reader is referred to [13] for further details. The following part of the section is devoted to exposing how short-term forecasting of traffic LoS can be

[1] Scikit-learn is a Python library of ML algorithms: http://scikit-learn.org.

Table 1. Speed and Density data sets

Data-sets	# Instances	# Attributes	# Instances per class	Imbalance Ratios
SD_5m_5, SD_5m_15, SD_5m_30, SD_5m_45, SD_5m_60	9979	12	A = 4534, B = 3681, C = 867, D = 891, E = 6	A/B=1.2; A/C=5.2; A/D=5.1; A/E=755.7
SD_1h_60, SD_1h_120	2159	7	A = 984, B = 790, C = 268, D = 117	A/B=1.2; A/C=3.7; A/D=8.4
DD_5m_5, DD_5m_15, DD_5m_30, DD_5m_45, DD_5m_60	9979	14	A = 2194, B = 559, C = 1075, D = 3541, E = 261	D/A=1.61; D/B=6.3; D/C=3.3; D/E=1.3
DD_1h_60, DD_1h_120h	2158	9	A = 471, B = 125, C = 267, D = 822, E = 473	D/A=1.7; D/B=6.6; D/C=3.07; D/E=1.7

approached as a classification problem; as well as to describe the data-sets and the experimental set-up used.

Short-Term Forecasting of Traffic Service Quality as a Classification Problem

In this work, STTF is focused on predicting the quality of traffic service, at a specific location, through a categorical measure named LoS. For freeways, LoS is used to categorize the quality levels of traffic, through letters from A to E in a gradual way[2], based on performance measures such as speed, density, and volume/capacity [11].

In this sense, we are approaching the forecasting of traffic service quality from a classification approach, concretely, as a multi-class classification problem. Based on speed and density data (calculated as flow/speed), which are continuous variables, we estimate how congested will be the road at the detector location, in different short-term time horizons through the LoS measure. This categorical measure is estimated from two univariate approaches, which means that speed and density are independently used to predict LoS using the intervals defined for them in [11].

It is important to clarify that the forecasting of LoS could be also addressed as a regression or time series problem predicting either speed or density and then discretizing the results to obtain a categorical interval of LoS. However, we chose the classification approach to explore a different problem to the one published in [14], to deepen into the benefits of AAS in a different area of STTF.

Data-sets and Experimental Set-Up

Data used in this work is provided by PeMS[3]. According to recent literature, this data source is highly used in the area of STTF because of its high quality data, availability of various traffic measures and its public accessibility. The route selected for our experiments is the California Interstate I405-N. Particularly, we focus on the detector VDS 771826 located at the post-mile 0.11 on this freeway.

[2] Category A indicates light to moderate traffic, whereas a category E means extended delays.
[3] http://pems.dot.ca.gov.

Traffic measures collected from the detector are speed and flow in aggregation times of 5 min and 1 h for both measures.

From this data, we generate 14 data-sets: seven using speed and seven using density as traffic data, respectively. Time horizons in which LoS is predicted are 5, 15, 30, 45, 60, and 120 min using data granularity of 5 min or 1 h depending on the case (granularity means how often the traffic measures are taken and aggregated). To better identify the data-sets, they are named following the next structure: $TrafficData\_Granularity\_TimeHorizon$.

Attributes of data-sets with 5 min granularity are *Day of the week; Hour of the day; Minute of the Hour; Quarter hour of the day; traffic measure at past 5, 15, 30, 45, and 60 min; current traffic measure; and current LoS*, where traffic measure could be average speed or density depending on the respective data-set. In the case of data-sets with 1 h granularity, these are *Day of the week; Hour of the day; traffic measure at past 1 and 2 h; current traffic measure; and current LoS*; again, the traffic measure could be average speed or density. In addition, Table 1 presents the number of instances and attributes of each data-set, together with the Imbalance Ratios (IRs) calculated dividing the number of instances of the majority class over the number of instances of each of the rest of classes. IR values show that the generated data-sets are imbalanced, although with different degrees.

For the experimentation with Auto-WEKA, three execution times were considered: 15, 150, and 300 min. These correspond to the time that the method can take to find the best ML algorithm and its hyper-parameter configuration for a given data-set. Furthermore, five repetitions with different initial seeds were carried out for each execution time. In the case of the RAs, we test them using WEKA. The process of evaluating every RA over a data-set was done with 20 repetitions with different initial seeds, and using the default hyper-parameter setting offered by WEKA. We have not performed any optimization or extra-adjustment of the RAs' hyper-parameters because our aim is to compare the performance of AAS versus RAs using the same human effort for both of them in order to make a fairer comparison.

4 Results

This section presents the results obtained with the experimental set-up proposed in the previous section. We evaluated the performance of the AAS method and the RA using the metric *G-measure (mGM)* that is applied for multi-class imbalanced data in classification problems [4]. Its calculation is expressed as $mGM = \sqrt[M]{\prod_{i=1}^{M} precision_i \cdot recall_i}$ wherein $G - measure$ on $i - th$ class is estimated as $GM_i = \sqrt{precision_i \cdot recall_i}$, and M is the total number of classes.

Table 2 shows the mean and standard deviation (between brackets) of the mGM values obtained by both the RAs and Auto-WEKA over all repetitions for each data-set. mGM values in bold indicate the best result in every data-set achieved from either any of the RAs or any of the Auto-WEKA's execution times. As it can be seen, in general, the AAS method performs better than

$k - NN$, NN and SVM; but worse than RF that is the best RA along most of the data-sets. Nevertheless, the improvement of RF w.r.t Auto-WEKA is small in most cases, ranging from 0.02 to 0.097, being even negative in three cases ($DD\_1h\_120, SD\_1h\_60, SD\_1h\_60$) where the AAS method obtains better mGM values than RF. This result is interesting because, in order to get to the conclusion that RF is the best RA, the user should run all RAs over all data-sets and compare their performance. However, according to these results, running Auto-WEKA only once, and therefore employing less human effort, the user can expect results very similar to the best RA and even better in some cases. Regarding data-sets characteristics, we can see that they do influence the differences between results of Auto-WEKA and RF. Concretely, for data-sets with a granularity of 5 min, for both traffic and density data-sets, the divergences between these two methods are greater for long-term time (in favour of RF). If we take into account the granularity of the data-sets, Auto-WEKA works especially well on those with 1 h granularity improving RF in all cases except DD_1h_120.

Table 2. Mean mGM values and their standard deviations (in brackets) obtained for density and speed data-sets by RA and AAS method.

	Reference algorithms				Auto-WEKA		
Data-sets	k-NN	NN	RF	SVM	15mET	150mET	300mET
DD_5m_5	0.485 (0.02)	0.577 (0.03)	**0.585 (0.02)**	0.523 (0.02)	0.564 (0.02)	0.563 (0.02)	0.563 (0.02)
DD_5m_15	0.47 (0.02)	0.531 (0.03)	**0.559 (0.01)**	0.465 (0.02)	0.495 (0.02)	0.485 (0.01)	0.529 (0.02)
DD_5m_30	0.433 (0.01)	0.472 (0.03)	**0.535 (0.01)**	0.339 (0.02)	0.462 (0.02)	0.451 (0.02)	0.463 (0.04)
DD_5m_45	0.424 (0.02)	0.460 (0.03)	**0.530 (0.01)**	0.332 (0.03)	0.460 (0.02)	0.478 (0.03)	0.479 (0.03)
DD_5m_60	0.436 (0.01)	0.441 (0.03)	**0.531 (0.01)**	0.356 (0.01)	0.427 (0.03)	0.455 (0.06)	0.481 (0.05)
DD_1h_60	0.601 (0.02)	0.568 (0.05)	**0.679 (0.02)**	0.476 (0.13)	0.622 (0.01)	0.637 (0.03)	0.646 (0.03)
DD_1h_120	0.498 (0.03)	0.445 (0.04)	0.576 (0.03)	0.36 (0.03)	0.543 (0.02)	0.557 (0.02)	**0.578 (0.01)**
SD_5m_5	0.671 (0.04)	0.773 (0.01)	**0.775 (0.02)**	0.756 (0.01)	0.746 (0.03)	0.763 (0.03)	0.764 (0.03)
SD_5m_15	0.587 (0.04)	0.557 (0.03)	**0.610 (0.02)**	0.499 (0.02)	0.572 (0.01)	0.571 (0.02)	0.572 (0.02)
SD_5m_30	0.528 (0.02)	0.394 (0.06)	**0.566 (0.02)**	0.322 (0.02)	0.453 (0.07)	0.469 (0.07)	0.452 (0.05)
SD_5m_45	0.454 (0.05)	0.356 (0.06)	**0.545 (0.01)**	0.572 (0.07)	0.408 (0.0)	0.471 (0.08)	0.453 (0.06)
SD_5m_60	0.467 (0.02)	0.368 (0.10)	**0.550 (0.01)**	0.538 (0.01)	0.461 (0.0)	0.469 (0.02)	0.482 (0.03)
SD_1h_60	0.424 (0.03)	0.473 (0.03)	0.482 (0.04)	0.40 (0.03)	0.532 (0.02)	**0.553 (0.03)**	0.462 (0.03)
SD_1h_120	0.286 (0.03)	0.322 (0.08)	0.366 (0.03)	0.259 (0.13)	0.372 (0.03)	**0.422 (0.03)**	0.399 (0.03)

Another interesting aspect is the relation between the execution time and the performance of the models provided by Auto-WEKA. For density data-sets, longer execution times contribute to obtaining better results although the improvements are very small. In the case of speed data-sets, results improve when the Auto-WEKA's execution time increases from 15 to 150 min, but they are worse when we pass from 150 to 300 min. Through some analyses, we observed that this worsening is due to the over-fitting produced by the hyper-parameters selected by Auto-WEKA. This result indicates that it is necessary to introduce mechanisms in the AAS method to deal with over-fitting, especially when execution times are high.

Table 3. Average Rankings of Friedman Test and Adjusted p-Values obtained through Holm post-hoc test

	Friedmans's avg. ranking			Holm's adj. p-values		
Algorithms	ET_{15min}	ET_{150min}	ET_{300min}	ET_{15min}	ET_{150min}	ET_{300min}
k-NN	3.5	3.7143	3.6429	**3.93e–4**	**8.6e–5**	**1.45e–4**
NN	3.1786	3.2857	3.2143	**2.026e–3**	**1.056e–3**	**1.636e–3**
RF	1.2143	1.2143	1.2143	-	-	-
SVM	4.2857	4.3571	4.3571	**1e–6**	**1e–6**	**1e–6**
Auto-WEKA	2.8214	2.4286	2.5714	**7.161e–3**	**4.2165e–2**	**2.3151e–2**

Another important aspect is the low performance of Auto-WEKA and the RAs, with mGM values below 0.7 in all data-sets, except for SD_5m_5. In analyses that are not included here because of space limitations, we corroborated that this behavior is mainly due to the high imbalance of the data-sets, especially in density ones, and the poor accuracy and recall of the methods when predicting the minority classes. This situation shows us that in the design of AAS methods for STTF problems, it is necessary to include mechanisms that allow addressing imbalance either by data pre-processing techniques or by adjusting the hyper-parameters of the ML methods.

To assess whether the differences in performance observed among the RAs and Auto-WEKA variants are significant or not, we made use of non-parametric statistical tests. Two statistical tests have been applied, following the guidelines proposed in [2]. First, the Friedman's test for multiple comparisons has been applied to check whether there are significant differences among the studied methods. Given that the p-value returned by these tests is 0, the null hypothesis can be rejected in all cases. The mean ranking returned by this test is displayed in Table 3, confirming the better global results of RF against both the rest of RAs and Auto-WEKA, and also the better global results of Auto-WEKA versus $k - NN$, NN and SVM.

Holm post-hoc test has also been applied using RF as control algorithm (because it is the method that achieved the best overall performance) to assess the significance of the differences with respect to the other RAs and Auto-WEKA. Table 3 also presents the adjusted p-values returned by this test. In order to highlight significant differences, those p-values lower than 0.05 are in bold. Looking the Table 3 there are important differences in the test's outcomes. It can be said that RF results improve significantly all the RAs including Auto-WEKA with their three execution times, although their p-values are greater than the ones of the other RAs.

To finalize with this section, we analyze the classifiers selected by Auto-WEKA over all data-sets. Table 4 summarizes how many times an algorithm is selected to forecast congestion along the data-sets. It is important to clarify that Auto-WEKA has a base of 39 algorithms and the ones that were not suggested for the data-sets evaluated are not included in Table 4. As each data-set was evaluated with three Auto-WEKA's running times along five repetitions in each of them, one algorithm can be chosen at most 15 times per data-set. In Table 4, RF is in general the most chosen algorithm except for DD_5m_5, where it is the second most selected algorithm behind *Logistic*, and for SD_1h_60 and SD_1h_120 where RF is not chose any time. In those data-sets Auto-WEKA improves RF because is able to find alternative methods that work better than RF, especially $DecisionTable$ and LMT. Another interesting fact is that the RAs *k-NN* and SVM (named by Auto-WEKA as *IBk* and SMO, respectively) are only chosen once and four times, in that order, and NN is not suggested any time despite of being three of the most widely used algorithms in literature. Moreover, in the case of the base of algorithms used in [14], only SVM is

Table 4. Classifiers selected by Auto-WEKA and absolute frequency in which they were suggested for density and speed data-sets

Classifiers	*DD_5m_5*	*DD_5m_15*	*DD_5m_30*	*DD_5m_45*	*DD_5m_60*	*DD_1h_60*	*DD_1h_120*	***Subtotal DD***	*SD_5m_5*	*SD_5m_15*	*SD_5m_30*	*SD_5m_45*	*SD_5m_60*	*SD_1h_60*	*SD_1h_120*	***Subtotal SD***	***Total DD+SD***
BayesNet	0	0	0	0	0	0	0	0	0	0	0	0	0	2	0	2	2
Logistic	9	0	0	0	0	0	0	9	2	0	0	0	0	0	0	2	11
SMO	0	0	1	1	0	0	0	2	0	2	0	0	0	0	0	2	4
IBk	0	0	0	0	0	0	0	0	0	1	0	0	0	0	0	1	1
Kstar	0	1	0	0	0	0	0	1	0	3	4	3	0	0	0	10	11
LWL	0	0	0	0	0	0	0	0	0	1	0	0	0	1	3	5	5
AdaBoostM1	0	1	0	0	0	0	0	1	0	0	1	0	0	1	0	2	3
Bagging	0	2	1	1	1	2	3	10	3	0	0	0	0	2	2	7	17
RandomCommittee	0	0	0	0	1	2	1	4	1	0	0	0	2	3	1	7	11
RandomSubSpace	2	1	2	1	3	0	0	9	2	0	0	0	0	0	0	2	11
Vote	0	1	0	0	0	1	1	3	0	3	1	0	1	1	1	7	10
DecisionTable	0	1	0	0	0	0	1	2	0	0	0	0	0	0	6	6	8
J48	0	0	0	0	2	0	0	2	0	0	0	0	0	1	1	2	4
LMT	0	0	3	2	0	2	2	9	2	0	0	0	0	4	1	7	16
RandomForest	4	8	8	10	8	8	7	53	5	5	9	12	12	0	0	43	96

selected by Auto-WEKA, which in our opinion stands out the importance of a broad algorithm base to perform AAS in STTF.

5 Conclusions

In this paper, we have focused on deepening into the benefits of AAS in the field of STTF. To accomplish this, we have compared to what extent the results of AAS differs from the general approach in STTF. We used Auto-WEKA as AAS method and *NN*, *SVM*, *k-NN* and *RF* as RAs representatives of the general approach. Concretely, our comparisons were made based on a multi-class imbalanced problem, consisting on the prediction of traffic LoS, at a fixed location, over different time horizons ahead. From the results we drawn interesting conclusions: the AAS method improves three out of the four RAs, and obtain similar results to *RF*, the best RA; with a lower human effort, the user can expect similar o even better results than the best RA; higher execution times for Auto-WEKA not always leads to better results due to over-fitting issues; and the performance shown by the RAs and Auto-WEKA was poor in general because they shown problems to deal with imbalance data.

Further research lines that we aim to explore in the future are: (I) including mechanisms, within the base of algorithms used by the AAS method, to deal with imbalanced data-sets and over-fitting; (II) testing different optimization strategies for finding the best pair (algorithm and hyper-parameter setting); (III) integrating more data preprocessing techniques to the AAS process; and (IV) approaching the forecasting of LoS also as a time series regression problem to explore what algorithms are more suitable depending on the modelling approach chosen.

Acknowledgements. This project has received funding from the European Union's Horizon 2020 research and innovation programme under grant agreement No. 636220 and the Marie Sklodoska-Curie grant agreement No. 665959. This work has been also supported by the research projects TIN2014-56042-JIN from the Spanish Ministry of Economy and Competitiveness.

References

1. Feurer, M., Klein, A., Eggensperger, K., Springenberg, J., Blum, M., Hutter, F.: Efficient and robust automated machine learning. Adv. Neural Inf. Process. Syst. **28**, 2962–2970 (2015)
2. Garcia, S., Fernandez, A., Luengo, J., Herrera, F.: Advanced nonparametric tests for multiple comparisons in the design of experiments in computational intelligence and data mining: experimental analysis of power. Inf. Sci. **180**(10), 2044–2064 (2010)
3. Komer, B., Bergstra, J., Eliasmith, C.: Hyperopt-sklearn: automatic hyperparameter configuration for scikit-learn. In: Proceedings of SciPy, pp. 33–39 (2014)
4. Krawczyk, B., McInnes, B.T., Cano, A.: Sentiment classification from multi-class imbalanced twitter data using binarization. In: Hybrid Artificial Intelligent Systems, pp. 26–37 (2017)

5. Lana, I., Ser, J.D., Velez, M., Vlahogianni, E.I.: Road traffic forecasting: recent advances and new challenges. IEEE Intell. Transp. Syst. Mag. **10**(2), 93–109 (2018)
6. Lopez-Garcia, P., Onieva, E., Osaba, E., Masegosa, A.D., Perallos, A.: A hybrid method for short-term traffic congestion forecasting using genetic algorithms and cross entropy. IEEE Trans. Intell. Transp. Syst. **17**(2), 557–569 (2016)
7. Luo, G.: A review of automatic selection methods for machine learning algorithms and hyper-parameter values. Netw. Model. Anal. Health Inform. Bioinform. **5**(1), 5–18 (2016)
8. Ma, X., Yu, H., Wang, Y., Wang, Y.: Large-scale transportation network congestion evolution prediction using deep learning theory. PloS one **10**(3), e0119,044 (2015)
9. Oh, S., Byon, Y.J., Jang, K., Yeo, H.: Short-term travel-time prediction on highway: a review of the data-driven approach. Transp. Rev. **35**(1), 4–32 (2015)
10. Sabharwal, A., Samulowitz, H., Tesauro, G.: Selecting near-optimal learners via incremental data allocation. In: Proceedings of the Thirtieth AAAI Conference on Artificial Intelligence, AAAI 2016, pp. 2007–2015 (2016)
11. Skycomp, I.B.M.: Major High- way Performance Ratings and Bottleneck Inventory. Maryland State Highway Administration, the Baltimore Metropolitan Council and Maryland Transportation Authority, State of Maryland (2009)
12. Sparks, E.R., Talwalkar, A., Haas, D., Franklin, M.J., Jordan, M.I., Kraska, T.: Automating model search for large scale machine learning. In: Proceedings of SoCC 2015, pp. 368–380 (2015)
13. Thornton, C., Hutter, F., Hoos, H.H., Leyton-Brown, K.: Auto-WEKA. In: Proceedings of the 19th ACM SIGKDD International Conference on Knowledge Discovery and Data Mining - KDD 2013, pp. 847–855 (2013)
14. Vlahogianni, E.I.: Optimization of traffic forecasting: intelligent surrogate modeling. Transp. Res. Part C: Emerg. Technol. **55**, 14–23 (2015)
15. Vlahogianni, E.I., Karlaftis, M.G., Golias, J.C.: Short-term traffic forecasting: where we are and where we're going. Transp. Res. Part C: Emerg. Technol. **43**, 3–19 (2014)

Solving an Eco-efficient Vehicle Routing Problem for Waste Collection with GRASP

Airam Expósito-Márquez, Christopher Expósito-Izquierdo(✉), Julio Brito-Santana, and José A. Moreno-Pérez

Departamento de Ingeniería Informática y de Sistemas, Universidad de La Laguna, 38271 La Laguna, Spain
{aexposim,cexposit,jbrito,jamoreno}@ull.es

Abstract. We address in this work the optimization of real waste collection in the island of La Palma (Canary Islands, Spain). The waste containers are of two types: paper-carton and plastic packaging. The optimization criterion in the problem is to collect those containers with the highest fill level in such a way that the environmental impact is minimized. In order to solve this optimization problem we firstly estimate the fill level of the containers by exploiting historic data and later we use a meta-heuristic procedure to design the collection routes. The computational experiments reveal the optimization technique is effective and efficient due to the fact that it allows to improve the current collection process according with several eco-efficient indicators.

Keywords: Waste solid problem
Vehicle Routing Problem · GRASP · Green logistics · Eco-efficient

1 Introduction

The production of solid waste is growing in today's society, becoming a major environmental problem [8]. The concern about the environment and sustainable development leads to devote resources to the policies and actions of planning and management of solid waste. Indeed, waste management is a critical issue for most of regions and cities around the world [11]. This a key challenge for the sustainable development, especially in island regions with small dimensions. An efficient waste management reduces negative effects in the environment and guarantees the safety and quality of life for citizens.

The present work is part of a large technological transfer project aimed at analyzing and improving the integral management of waste at the Island of La Palma, one of the seven main Canary Islands (Spain). In particular, this project is focused on the collection, transfer, and transport of paper-carton and plastic waste and seeks to analyze the criteria, factors and strategies for collection, route planning, collection schedule and the appropriate number of required resources

J. Del Ser et al. (Eds.): IDC 2018, SCI 798, pp. 215–224, 2018.
https://doi.org/10.1007/978-3-319-99626-4_19

for waste collection around the island. Specifically, in this work we focus our attention on the waste collection activity (*i.e.*, the process of collecting waste from the existing containers and emptying their content in the dump).

Integrating environmental considerations within the waste management in the activities of transport of materials is part of the so-called Green Logistics [12]. Green Logistic is a term concerned about the economic impact of logistics policies on the enterprises carrying them out and also effects on daily life, such as the effects of pollution on the environment [7]. Among these operations, transportation has the most harmful effects on the environment, such as CO_2 emissions and energy consumption. In this context, road freight transportation is a major contributor to carbon dioxide equivalent emissions and fuel consumption. Green vehicle routing is a branch of green logistics which refers to vehicle routing problems where externalities of using vehicles, such as carbon dioxide-equivalents emissions, are explicitly taken into account so that they are reduced through better planning [3].

The Vehicle Routing Problem (VRP) consists of determining the least cost routes for a fleet of vehicles to satisfy the demands of a set of customers, subject to temporary or length constraints. The VRP was introduced more than 50 years ago by Dantzig and Ramser [5]. This optimization problem and its variants, with additional features, have been of the most widely studied so far [4]. The common objective is to minimize the total cost traveled by all the vehicles, but this objective can be oriented in terms of eco-efficiency or reduction of negative externalities. It is necessary to find trade-off between the economic cost and the environmental cost in routing problems. With green orientation, the VRP can be classified in: (i) the Green VRP (GVRP), (ii) the pollution routing problem (PRP), and (iii) the VRP in reverse logistics (VRPRL). GVRP deals with the minimization of energy, PRP deals with the minimization of greenhouse gas emissions, whereas VRPRL deals with the distribution aspects of reverse logistics [10].

In scientific literature there are many recent works about GVRP, some of them are cited below. In [13] a bi-objective Fuel efficient Green Vehicle Routing Problem (F-GVRP) with varying speed constraint is discussed. The problem is solved using Particle Swarm Optimization. A practical solution approach for the GVRP that uses a mixed-integer linear formulation is proposed in [1] with a new formulation that offers compactness and flexibility. The paper in [2] addresses a novel model for the multi-trip Green Capacitated Arc Routing Problem (G-CARP) with the aim of minimizing total cost including the cost of generation and emission of greenhouse gases. A Hybrid Genetic Algorithm (HGA) is developed to solve the problem.

Waste collecting and recycling is green activity in itself. Moreover, from a green logistics perspective, there is also a requirement of conducting the transport activities in a sustainable manner. The eco-efficient routing problem options which reduce environmental impact are investigated. This work incorporates negative environmental externalities of transport as a measure of efficiency in route planning. We propose a waste collection model, which strategy is to pick up high priority containers, in a daily basis. High priority containers are those that are full above the threshold of 2/3 of their capacity therefore have to be picked up immediately to avoid that the waste are deposited on the outside or do not separate. The objective is to collect paper-carton and plastic packaging containers with the highest fill level in such a way that the environmental impact is minimized. Lowering fuel costs can be achieved through lower kilometers traveled, and therefore achieving shorter routes is consistent with our objective. The model under which the applied problem cited in this paper is addresed can be considered as a multi-depot routing problem whose starting and ending route locations are known.

To solve this problem we use a solution algorithm based on the framework of Greedy Randomized Adaptive Search Procedure (GRASP) [6]. GRASP is a relevant meta-heuristic that is appropriate for solving real and large instances of routing problems. GRASP has been successfully applied to a wide variety of combinatorial optimization problems including route planning [14].

The remainder of the present paper is organized as follows. Section 2 introduces the optimization problem to solve. Then, the algorithmic proposal to address the problem is described in Sect. 3. The computational experiments and the obtained results are presented in Sect. 4. Finally, the main conclusions and several lines for further research are provided in Sect. 5.

2 Eco-efficient Vehicle Routing Problem

In this section we provide a detailed description of the Eco-efficient Vehicle Routing Problem (Ee-VRP). The waste collection problem consists of routing a fleet of vehicles to collect waste while maximizing the fill level of containers. This optimization problem is focused on the collection of paper-carton and plastic packaging containers. The routes to be designed are associated with a known type of recyclable waste and have a limited duration, which is relative to the working day of the vehicle drivers. The problem here addressed can be considered as a multi-depot routing problem whose starting and ending route locations are known. Specifically, there are three start and end plants for the routes, whose locations are in Breña Alta, Mazo, and Los Llanos, municipalities of La Palma. The planning of the routes to be carried out considers an extended planning horizon of several working days.

The feasible solutions of the optimization problem are composed of a route to be carried out by each available vehicle in the company each day within the planning horizon. In this regard, each route departs from the starting depot and finishes in the end plant. In addition, each route is a sequence of waste collection points to be visited by the corresponding vehicle in such a way that the containers located in each point are all collected. The strategic objective of this project is to increase the recycling of waste in the long term, encouraging the citizen to recycle by transferring non-recyclable waste to recyclable waste. According to the recycling company, full or overflowing containers discourage citizens to recycle. For this reason, this is one of the key elements to avoid in the optimization problem. Indeed, this project proposes to give collection priority to those waste collection points whose percentage of fill is high, avoiding a possible fill or overflowing thereof.

The Ee-VRP is defined by means of a complete directed graph $G = (V, A)$, where $V = \{v_0, v_1, ..., v_{n-1}\}$ is the set of n locations and $A = \{(v_i, v_j) : v_i, v_j \in V, i \neq j\}$ is the arc set. In this case, $V = P \cup E$. Set P is defined as $P = \{p_0, p_1, ..., p_{m-1}\}$, where m is the number of waste collection points. The set E with three elements is defined as $E = \{e_0, e_1, e_2\}$, each element $e_i \in E$ corresponds to a start and end points of the route.

Each $p_i \in P$ corresponds to a waste collection point and has an associated fill rate $q_i \geq 0$, a service duration $s_i \geq 0$ and a fill level $f_i(k)$ where k is the number of days the waste collection point has been receiving waste. Additionally, each $p_i \in P$ has a set of recyclable waste container set $C_i = \{c_i^0, c_i^1, ..., c_i^{l-1}\}$ where each element $c \in C_i$ corresponds to a recyclable waste container. Accordingly, the set C which contains all containers that belong to waste collection points is defined as $C = \bigcup_{i=1}^{m} C_i$. Hence, each service duration s_i of the waste collection point i is defined as $s_i = \sum_{j=0}^{l-1} s_j^{'}$ where $s_j^{'}$ is the service duration of collecting the container c_j.

The planning horizon considers t days, also referred to as set $T = \{1, 2, ..., t\}$. The routes at which are based r vehicles that have unlimited capacity and maximum working time D. This route time limit mentioned above is associated with the working day of the truck drivers who perform the collection of recycled waste. For each arc $(v_i, v_j) \in A$ there is a known the travel time $t_{ij} \geq 0$ and the distance $d_{ij} \geq 0$. The problem then consists in designing r feasible vehicle routes for each of the t days and kind of recyclable waste.

3 Solution Approach

In order to solve the optimization problem described in the previous section, hereafter we propose a solution approach composed of the following stages carried out one after the other:

- Determining the fill level of the containers located on the scenario under analysis.
- Determining the most efficient routes to collect the containers by means of the fleet of vehicles.

3.1 Determining Fill Level

One of the most highlighted issues when planning the collection process of the company is to determine what are the most attractive containers to collect by the available vehicles. These must be selected in such a way that the filling rate could be as high as possible. In this case, the attractiveness measure is computed from their fill level. Unfortunately, due to the lack of sensorization associated with the containers, their fill level must be estimated on the basis of the data reported by the company.

The fill level of those containers belonging to a given waste collection point is obtained as a function whose argument is the number of days since the last collection. It is therefore assumed that each container $i \in C$ is filled lineally over days. That is:

$$f_i(k) = q_i \cdot k + b_i, \quad (1)$$

where $k \geq 0$ is the number of days since the last collection, q_i is the filling rate, and b_i is the level of the container after collected. It is worth mentioning that the filling rate of each container is computed as the average level of that container when collected previously by the company. Furthermore, the containers can be full, and thus no additional waste is allowed. $f_i(k) > 1$ indicates the container $i \in C$ is overflowed after k days since its last collection.

3.2 Designing Routes

Once the fill level of each container is estimated, the routes to be followed by the existing vehicles can be designed with the aimed of optimizing the percentage of filling during the collection process. For this purpose, a GRASP is proposed. It is a well-known meta-heuristic that combines a semi-greedy algorithm with a local search method in a multistart procedure [15].

The pseudocode of the proposed GRASP to solve the Ee-VRP is depicted in Algorithm 1:

```
1  for each route r ∈ routes do
2      Initialize route r ← (start, finish);
3      while improvement do
4          Initialize the set Θ of candidate elements;
5          while Θ is not empty and constraints are satisfied do
6              Evaluate the candidate elements in Θ;
7              Build the restricted candidate list (RCL);
8              Select an element p from the RCL at random;
9              Add p to route r;
10             Update the set of candidate elements;
11             Reevaluate the candidate elements in C;
12         end
13         Apply improvement strategy to r
14     end
15 end
```

Algorithm 1. Pseudocode of the proposed Greedy Randomized Adaptive Search Procedure aimed at solving the E-VRP

The goal of the previous optimization technique is to report the set of routes to be followed by the fleet of vehicles in such a way that the percentage of filling is maximized. With this goal in mind the routes are designed one after the other from the first day up to the last day within the planning horizon (lines 1–15). Each route is here defined as a sequence of waste collection points to be visited by the corresponding vehicle. The first step in the design process is to build a route including the starting and final positions for the vehicle and with no waste collection points (line 2). Then, one waste collection point is added to the route under construction at each step until no more waste collection points are available or the problem constraints are not satisfied (lines 5–12).

The attractiveness of each potential waste collection point when added to the current route in a specific position is computed according to the impact on the objective function value. However, due to the temporal constraints of the routes, some waste collection points cannot be included without exceeding this bound. For this reason, these waste collection points are ruled out from the selection process. An ordered list composed of the best candidates to be added to the route is built (line 7). The cardinality of this set is set by the user. Lastly, one of the candidates, p, is selected at random from the restricted candidate list (line 8) and added to the route in the corresponding position (line 9).

It should be noted that the semi-greedy building process provides feasible solutions of the optimization problem under analysis. However, the quality of the routes could be enhanced by applying some improvement strategy. In this case, the Lin-Kernighan heuristic proposed for the Traveling Salesman Problem is used [9]. Applying this technique allows to reduce the distance traveled by the vehicle when visiting the sequence of waste collection points. This potential improvement enables to add new waste collection points to the route under construction (lines 3–14). Finally, the process is finished when all the routes are designed and any improvement can be obtained.

4 Computational Experiments and Results

This section describes the computational experiments carried out in the case study and the results obtained from them.

This work focuses on a real case study in La Palma, Canary Islands. Hence, a portion of the data which the computational experiments have been done, are real data. The input data will refer mainly to the location of waste collection points and the plants in which deliver the collected waste, the filling level of the containers and the travel time and distance spent in the transfer between the points and the operation of the vehicles.

The containers are distributed across the island in groups termed collection points. All the containers are homogeneous and has a capacity of 3.000 liters. There are 774 containers around the island, located in a total of 338 collection points. 375 of them are dedicated to contain paper and cardboard, whereas the remaining 399 containers are dedicated to plastic packaging.

Regarding distances and travel times between collection points, a part of this data was provided by the company. Specifically, the part related to the movements between collection points that the company has carried out and are stored in its database. In order to compensate for the lack of data regarding distances and travel times between collection points that the company did not have in its database, for each pair of collection points, travel times and distances were extracted through web services from Google Maps API. Once the data was extracted from Google Maps, a correction factor was applied to the travel times that allows a better fitting of the approximation provided by Google Maps to the characteristics of the company vehicles that perform the recyclable waste collection service.

The experiments presented along this section were carried out on a personal computer equipped with an Intel Core 2 Duo E8500 3.16 GHz and 4 GB of RAM. Additionally, the GRASP was programmed by using the Java SE 8 language. In this case, around 1 min of computational time is required by the technique to solve the scenario under analysis. However, it is worth mentioning that this factor is not particularly relevant for the recycling company due to the fact that the design of routes would be eventually performed during the night before each Monday.

Table 1 summarizes the computational results obtained by the optimization technique introduced in Sect. 3 and the scenario lived by the recycling company during the working days of the first week of October 2017. This week has been specified by the company to analyze the proposal due to the fact that it has no holidays in all the municipalities of the island and relevant events that perturb the recycling behavior.

The table includes several indicators related to the logistic management and environmental impact of the recycling process. These indicators are divided into three groups and are used widely by the company across the country to assess the recycling processes in the different regions. In this case, the former group is composed of those indicators related to the number, duration, and distance of the routes to carry out. The second group of indicators are related to the state of the number and fill level of the containers. Lastly, the third group of indicators are related to the impact of the recycling process in terms of environmental impact. Some of these indicators are the amount of waste to recycle, the fuel consumption, and the distance and time required to collect one kilogram of recyclable waste.

As can be seen in the computational comparison showed in Table 1, using the GRASP presented in Sect. 3 allows to improve most of the indicators under analysis. Only one route per vehicle is designed each day with a maximum duration of 6.5 h. This was a suggestion of the recycling company in order to analyze the feasibility of collecting the containers within a working day of a driver by vehicle. The remaining 1.5 h are dedicated to potential contingencies (*e.g.*, traffic jams, changes in the location of containers, etc.) and rest periods. The goal behind this suggestion is to study how to reduce labor costs in the middle term. Nowadays, the company is using more than 10 routes during a week (13 routes in the comparative scenario), which require a large number of working hours. In this case, 128.9 h instead of only 65 h proposed through the GRASP. However, this reduction in the working hours demands to increase the distance traveled by the vehicles. The reason is found in the selection strategy of waste collection points used by the GRASP. In particular, only those containers with the highest fill level are collected, without taking into account the distance to achieve them.

Table 1. Summary of the comparison carried out between the proposed algorithmic approach and the current scenario during the first week of October 2017

Indicator	Proposed scenario	Current scenario
Routes	10	13
Time (h)	65	128.9
Time per route (h)	6.5	9.915
Distance (km)	1383.924	1162.534
Distance per route (km)	138.392	89.426
Containers	788	1215
Container by day	157.6	243
Overflowing containers	115	Unknown
Fill level (%)	78.995	61.113
Kilograms (kg)	15792.94	18563.074
Kilograms by hour (kg/h)	243.134	144.011
Consumption (l)	624.225	1238.729
Consumption by kilogram (l/kg)	0.040	0,067
Distance by kilogram (km/kg)	0,088	0,063
Time by kilogram (m/kg)	0,247	0.427

Moreover, in spite of the reduction in the number of routes and time, the fill level of the containers is appreciably better in the proposal (78.995% in comparison with 61.113%). Simultaneously, the productivity of the vehicles during the collection process can be increased according to the reported routes. In this case, the amount of recyclable waste collected in the proposed routes is nearly equivalent to the current, but the consumption of the vehicles and the time required to collect the waste is substantially reduced. This reduction is especially relevant to the recycling company due to the fact that it gives rise to a lower environmental impact.

Finally, Fig. 1 seeks to illustrate a few illustrative examples of the routes obtained by means of the GRASP introduced in Sect. 3 for one day of the planning horizon in comparison with those routes that were carried out by the recycling company. As can be checked, the proposed routes are larger and collect a lower number of containers across the island. However, determining what are the most attractive containers to collect allows to improve the amount of recyclable waste in each collection at the expense of increasing the distance traveled by the vehicles.

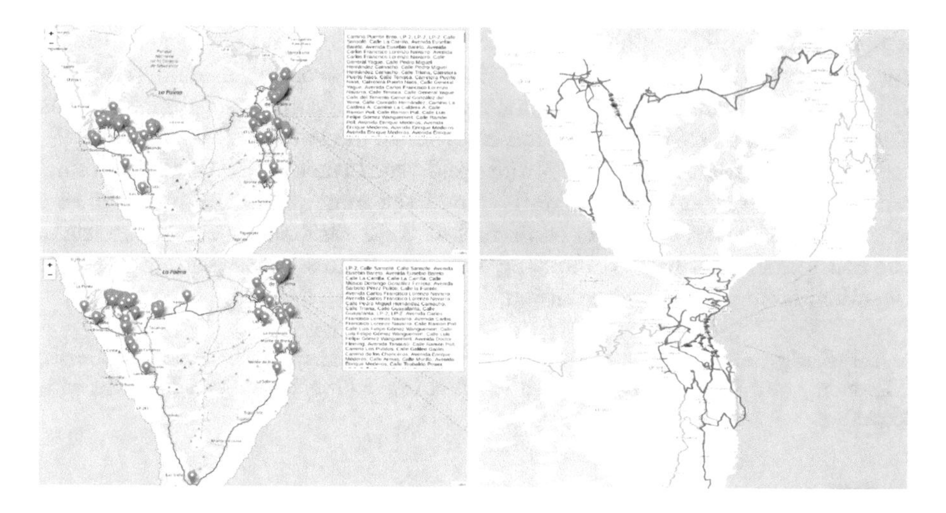

Fig. 1. Comparison of routes reported by the GRASP (left figures) and those carried out by the company during 5th October, 2017 (right figures) by the two available vehicles

5 Conclusions and Further Works

This work proposes a planning of the routes for collecting waste of paper-carton and plastic packaging containers at La Palma, Canary Islands. The applied techniques are mainly a probabilistic model to estimate the fill level of the containers based on the historical data, a mathematical model of the optimization problem for the daily routes to collect the containers with a high level of filling, a meta-heuristic based on GRASP for the daily construction of routes that improve their performance, and a simulation to validate the quality and robustness of the solutions provided. The computational results for paper-carton and plastic waste collection shown that propose planning reduces both total traveling distances, operational hours of vehicles and fuel consume in comparison with those of practical scenarios. The proposed model could be used for scenarios having similar components and vehicles characteristics. Those results are significant to practitioners and local policy makers.

Future work will introduces constraints about trucks maximum capacity. Specifically, capacity of 4500 kg. in the case of paper-carton waste and 3200 kg. in the case of plastic package waste. in the case of containers. This constraint requires that, if the truck capacity reaches the maximum along a route, the truck returns to the corresponding plant. Then two situations can be given, either the same vehicle continues the route after passing through the plant or another truck completes the route if it was previously going through that point. Another line of future work is related to the inclusion of time windows in containers.

The quality of the waste collection service is another of the future lines of work. In connection with this, all containers must be picked up at least once a week and it is not possible to pass in front of container and not pick it up. The new constraints mentioned above are in relation to avoiding misunderstandings and complaints from citizens and municipal governments regarding minimum service conditions and non-collection or overflow of containers.

The introduction of new constraints and requirements related to fuel refueling and consumption optimization is one of the most promising future lines of work. Along the routes, trucks can refuel at 2 gas stations on the whole island. The consideration of limited refueling along the routes in the planning of waste collection, is an interesting extension.

Acknowledgments. This work has been partially funded by the Spanish Ministry of Economy and Competitiveness with FEDER funds (TIN2015-70226-R) and also by Ecoembes.

References

1. Leggieri, V., Haouari, M.: A practical solution approach for the green vehicle routing problem. Transp. Res. Part E: Logist. Transp. Rev. **104**, 97 – 112 (2017)
2. Tirkolaee, E.B., Hosseinabadi, A.A.R., Soltani, M., Kumar, A., Wang, J.: A hybrid genetic algorithm for multi-trip green capacitated arc routing problem in the scope of urban services, vol. 10 (2018)
3. Bektaş, T., Demir, E., Laporte, G.: Green vehicle routing. In: Green Transportation Logistics, pp. 243–265. Springer, Switzerland (2016)
4. Braekers, K., Ramaekers, K., Van Nieuwenhuyse, I.: The vehicle routing problem: state of the art classification and review. Comput. Ind. Eng. **99**, 300–313 (2016)
5. Dantzig, G.B., Ramser, J.H.: The truck dispatching problem. Manage. Sci. **6**(1), 80–91 (1959)
6. Feo, T.A., Resende, M.G.C.: Greedy randomized adaptive search procedures. J. Glob. Optim. **6**(2), 109–133 (1995)
7. Giusti, L.: A review of waste management practices and their impact on human health. Waste Manage. **29**(8), 2227–2239 (2009)
8. Guerrero, L.A., Maas, G., Hogland, W.: Solid waste management challenges for cities in developing countries. Waste Manage. **33**(1), 220–232 (2013)
9. Helsgaun, K.: An effective implementation of the lin-kernighan traveling salesman heuristic. Eur. J. Oper. Res. **126**(1), 106–130 (2000)
10. Lin, C., Choy, K.L., Ho, G.T.S., Chung, S.H., Lam, H.Y.: Survey of green vehicle routing problem: past and future trends. Expert. Syst. Appl. **41**(4), 1118–1138 (2014)
11. Marshall, R.E., Farahbakhsh, K.: Systems approaches to integrated solid waste management in developing countries. Waste Manage. **33**(4), 988–1003 (2013)
12. McKinnon, A., Browne, M., Whiteing, A., Piecyk, M.: Green Logistics: Improving the Environmental Sustainability of Logistics. Kogan Page Publishers (2015)
13. Poonthalir, G., Nadarajan, R.: A fuel efficient green vehicle routing problem with varying speed constraint (F-GVRP). Expert. Syst. Appl. **100**, 131–144 (2018)
14. Resende, M.G.C., Ribeiro, C.C.: Greedy randomized adaptive search procedures: advances, hybridizations, and applications. In: Handbook of Metaheuristics, pp. 283–319. Springer, Heidelberg (2010)
15. Resende, M.G.C., Ribeiro, C.C.: GRASP: The basic heuristic, pp. 95–112. Springer, New York (2016)

Fostering Agent Cooperation in AmI: A Context-Aware Mechanism for Dealing with Multiple Intentions

Arthur Casals[1,3(✉)], Assia Belbachir[2], Amal El-Fallah Seghrouchni[3], and Anarosa Alves Franco Brandão[1]

[1] Escola Politécnica - Universidade de São Paulo, São Paulo, SP, Brazil
{arthur.casals,anarosa.brandao}@usp.br
[2] IPSA, Paris, France
assia.belbachir@ipsa.fr
[3] Sorbonne Université, LIP6- CNRS UMR 7606, Paris, France
amal.elfallah@lip6.fr

Abstract. Ambient Intelligent (AmI) environments dynamically provide contextual information to intelligent agents that interact with them. In such environments, could these agents cooperate to improve their goal achievement, considering multiple intentions from several agents? With multiple agents, cooperation will depend on each agent's own intentions. Agents adapt to dynamic changes in the environment using context-aware planning mechanisms such as the Contextual Planning System (CPS), which proposes an optimal plan for a single agent based on the current context. In this paper we present the Collective CPS (CCPS), an opportunistic cooperative planning mechanism for multiple agents in AmI environments. CCPS allows agents to partially delegate their own plans or to collaborate with other agents' plans during their execution, while retaining individual planning capabilities. A working scenario is shown for a realistic AmI environment, such as a Smart Campus.

1 Introduction

Ambient Intelligence (AmI) is a reference to electronic environments in which electronic devices or systems can perceive and respond to the presence of people, while also being able to communicate with each other [1]. In such environments, cooperation and coordination among different systems may exist in different forms, and for different purposes. This paper presents an approach to cooperatively deal with multiple intentions for a specific type of intelligent agent that interacts with AmI environments, denominated ambient agent (AA). The cooperative model in place is opportunistic, i.e., cooperation may exist if there are favorable conditions for it to occur. Since AmI environments provide dynamic contextual information, the AA's reasoning process must adapt to environment changes as efficiently as possible, while still making it possible for the AA to achieve its goals.

J. Del Ser et al. (Eds.): IDC 2018, SCI 798, pp. 225–234, 2018.
https://doi.org/10.1007/978-3-319-99626-4_20

Adopting Multiagent Systems (MAS) to deal with contextual information in AmI applications is interesting because of its very nature, where different entities autonomously interact in a dynamic and uncertain environment [3,15]. MAS whose agents follow the BDI model are specially tailored for AmI due to its inherent use of contextual information in the form of *beliefs*. These agents, however, may have different and independent goals, and still need cooperation to perform specific tasks. This can be illustrated in a real-world situation: if a person has an appointment but cannot be there in time, getting a ride from someone will make this goal achievable. Other similar AmI-related situations include cooperative collision warnings in vehicular ad-hoc networks [8] and opportunistic problem-solving using ontology negotiation [2].

While extensive research has been conducted in the field of intelligent agent cooperative planning [7,10,13,18], most of the existing work relies on central coordination or communication mechanisms. Such restrictions can make it difficult to use the existing research in AmI scenarios where agents possess individual goals, but cooperation without previous commitment is still beneficial. With that in mind, we propose a cooperative planning mechanism that can be used by an AA to delegate specific plans to other agents. The objective is to provide cooperation in each of the agents' planning processes without having to commit to common goals, or make use of central planning engines of any kind. In terms of scientific contribution, this work also makes it possible to deal with multiple intentions while maintaining the collaboration mechanism among intelligent agents strictly opportunistic. In addition, we also improve the BDI agent reasoning mechanism by using a preemptive planning mechanism in conjunction with contextual planning. Thus, the proposed mechanism makes possible for different agents to cooperate among themselves to achieve goals otherwise unachievable by delegating the execution of specific plans. It is important to notice that the cooperation mechanism does not assume benevolent agents, nor it is based in existing commitments of any sort: AAs only cooperate among themselves if it is on their best interests.

This paper is organized as follows: Sect. 2 introduces relevant aspects of planning and cooperation used along this paper, with a description of the contextual planning system (CPS [6]). Section 3 describes the Collective CPS (CCPS) by presenting its formal structure and related algorithms. Section 4 presents, as proof of concept, an implementation of CCPS in a real-world based scenario, and describes the experiments performed. Finally, Sect. 5 concludes this paper and presents some perspectives for future work.

2 Background

In this section we present an overview of planning and cooperation in the domain of BDI agents, followed by the CPS, which is the contextual planning mechanism used as a basis for the present work.

2.1 Planning and Cooperation

The planning process used by a BDI agent involves choosing a set of actions to be executed and it is triggered whenever new information is perceived. Re-planning over new information allows the agent to adapt its plans accordingly and re-evaluate its goals at each interaction, so it can adopt the best course of action at the time.

Environmental changes can occur independently from the agent (such as the natural passage of time), or as a consequence of an executed action. One agent's actions can impact the perceptions of other agents in the same environment. If agents have different goals, one agent's actions may cause another agent's goals to become unachievable. Using coordination mechanisms can mitigate these impacts [9], and eventually increasing the overall planning efficacy or efficiency in face of a dynamic environment [12,14]. It may also be the case that coordinated planning is a requirement for solving the problem at hand.

While there exist solutions aimed at solving situations that require cooperative planning processes [7,10,18], they usually consider joint goals situations, either relying on central coordination systems or being limited by the problem domain. AA could benefit from cooperative planning in situations where individual efficiency is affected by environmental conditions, ultimately executing certain tasks more efficiently. If multiple agents were to cooperate, context-dependent tasks could be delegated among them, and otherwise unachievable goals could be successfully achieved. Our approach aims at exploring situations where contextual-dependent tasks may be delegated by one AA as part of its planning process, giving the coordination process a contextual dynamic characteristic. We will detail this approach in the following paragraphs, from the formalism involved to the planning mechanism.

2.2 Contextual Planning System (CPS)

In the original BDI-agent model, an interpreter manages the agent states related to its beliefs (B), desires (D), and intentions (I) in order to achieve goals through planning. Let $\mathbb{A} = \{A_1, A_2, ..., A_n\}$ be the set of agents and $A_i \in \mathbb{A}$.

Chaouche *et al.* [6] proposed a model to deal with multiple intentions for an agent A_i. It represents an agent plan $\tilde{P}(A_i)$ as a tree structure composed of a set of intention plans ($\hat{P}_j$) and elementary plans ($P_{l,k}$). $\tilde{P}(A_i)$ is composed of multiple intention plans $\hat{P}_j$, and each intention plan corresponds to the achievement of an agent's intention.

An intention plan $\hat{P}_j$ can be an alternate of multiple elementary plans: $\hat{P}_j = \{P_{j,0}, ..., P_{j,q}\}$. Alternating elementary plans allows achieving an specific intention in different manners, each one expressed by an elementary plan.

Elementary plans $P_{j,k}$ are described by *AgLOTOS expressions* [4], referring to behaviour expressions $E_{i,k}$ expressed as an ordered finite set of observable actions a_i to be executed by the agent. Any AgLOTOS expression is associated with contextual information related to the (current) BDI state of an agent. These elementary plans are obtained from a library of plans (*LibP*).

Using this model, Chaouche *et al.* [5] proposed a predictive planning mechanism that uses the contextual information to verify which among the actions known by the agent are feasible. "Feasibility" of an action is determined by any conditions the agent must met to execute it. For instance, moving from point A to point B is only feasible if agent is at point A. Once this verification occurs, CPS selects an execution path that maximizes the number of satisfied intentions [5], producing an optimal plan $\tilde{P}(A_i)$. In the next section we will explain how this plan is used by the Collective Contextual Planning System (CCPS) in conjunction with its cooperative mechanism.

3 Collective Contextual Planning System (CCPS)

The Collective Contextual Planning System (CCPS) structure is designed to process plans from multiple CPS-capable agents. The novelty here, in comparison with the CPS mechanism, is that when an agent plan cannot be executed for any reason, the agent can ask for help and delegates part of this plan to other agents, avoiding re-planning. By doing this, the agent eliminates the problem while maintaining the consistency of its goals. At the same time, all the parallelism and concurrence taken into consideration by the original CPS mechanism are maintained.

In the first step of the CCPS mechanism, an agent has an initial plan resulting from the CPS mechanism execution and verifies if it can be executed without the help of any other agent. If that is not the case, the agent detects which actions are preventing the plan from being executed. These actions are listed in a help request message that is sent to other agents. After receiving this message, each agent decides about offering help for each action and reply its answer. The requester agent and decides about accepting help and inform the helper agent its decision. In the case where no other agent is able to provide help, the agent tries to replace the intention associated with the plan that cannot be executed with a different one that could provide similar results. If no such intention is found, then the original intention is removed.

The idea behind this mechanism is similar to the contract net protocol [16], designed for distributed problem-solving. Aspects related to connection and contract negotiation, however, are simplified - either by considering that it is provided by the agent's environment or by taking the opportunistic aspect into account. The concepts involved in CCPS will be formalized and explained in the next paragraphs.

3.1 Formal Framework

In this section, we introduce some of the definitions necessary to formally describe the CCPS and its algorithm.

- Environment: $Env(t)$ is a set of logical propositions representing contextual information that can be perceived by any agent at time t.

- Action: An action a represents the finest granular activity performed by the agent in the environment. If an agent A_i moves from point X to point Y, there is an associated action $move(X, Y)$. Each action a is subject to three sets of logical propositions (as in STRIPS [11]): preconditions list ($Pre(a)$); delete list ($Del(a)$); and post-conditions list ($Post(a)$). An action a if *feasible* if all of its preconditions $Pre(a)$ are satisfied by the conditions of $Env(t)$. After an action a is executed, its conditions can be changed by $Del(a)$ or $Post(a)$- also altering $Env(t)$.
- Intention: Each agent $A_j \in \mathbb{A}$ has a set of intentions, and each intention is associated with a weight factor that serves as a comparator index, used to determine *order of importance* between two intentions. Thus, the intention function over the known agents is defined as $I : \mathbb{A} \rightarrow \mathbb{I}x\mathbb{W}$, and the set of intentions for an agent is expressed by:

  $$I(A_j) = \{(i_0, w_0), (i_1, w_1), ..., (i_n, w_n)\}, A_j \in \mathbb{A}, i_0, ..., i_n \in \mathbb{I}, w_0, ..., w_n \in \mathbb{N}^*$$

  Two intentions are comparable according to their associated weight factors: given $(i_p, w_p), (i_q, w_q) \in I(A_j), 0 \leq p, q \leq n$, i_p is at least as important as i_q if $w_q \leq w_p$.
- Plan: Plan definition follows exactly the one from CPS, presented at Sect. 2.2.
- Plan Feasibility: Being a composite and partially ordered set of actions, the feasibility of an elementary plan $P_{i,j}$ can be given by the feasibility of its actions: $P_{j,k}$ is feasible in $Env(t)$ if and only if all of its actions are feasible in $Env(t)$. An intention plan $\hat{P}_j$ is feasible if and only if at least one of its elementary plans is feasible. In the case of an agent plan $\tilde{P}(A_i)$, we determine that it is completely feasible if and only if all of its intention plans are feasible.

There might be situations where it is necessary to consider partial feasibility of an agent plan $\tilde{P}(A_i)$. Suppose that $\tilde{P}(A_i)$ contains a set of intention plans $\{\hat{P}_1, ..., \hat{P}_q\}$ referring to $I_j = \{(i_0, w_0), (i_1, w_1), ..., (i_j, w_j)\}, I_j \subseteq I(A_i)$. In this case, if only $\hat{P}_j$ is unfeasible and w_j is relatively much lesser than the other intentions' weights, it could make sense for the plan $\tilde{P}(A_i)$ to be executed anyway, while the intention I_j is still unfeasible. However, if we have a situation where w_j is relatively much greater than the other intentions' weights, we should not want to proceed with $\tilde{P}(A_i)$ at all, and re-planning becomes necessary.

Let us also consider the following notations: the average weight of all intentions associated with all intention plans in $\tilde{P}(A_i)$ is given by $\overline{w_I}$, and the average weight of all intentions associated with all *feasible* intention plans in $\tilde{P}(A_i)$ is given by $\overline{w_F}$. With that in mind, we define the concept of partial feasibility for an agent plan: $\tilde{P}(A_i) = \{\hat{P}_1, ..., \hat{P}_j\}$ is partially feasible if $\overline{w_F} \geq \overline{w_I}$. In the rest of this work, we will consider an agent plan $\tilde{P}(A_i)$ unfeasible if it is not partially feasible. In the same manner, a plan $\tilde{P}(A_i)$ will be considered feasible if it is at least partially feasible.

It is also important to formalize a few aspects related to agent communication - in particular, the aspects involving the exchanged messages between agents, and their possible effects on the CCPS mechanism. When an agent A_j requests help to other agents, this is done through a message M containing at least one unfeasible action selected from $\tilde{P}(A_j)$. This message also contains *temporal limits* and a *deadline*. Temporal limits are associated with actions, and are formed by two different limits: a beginning (when an action should start), and an end (when an action should be finished). The deadline, on the other hand, is associated with the message and it is the time limit for the message to be answered. Once the deadline expires, the agent won't expect any replies to that particular message.

The concepts of complementary and equivalent intentions are also important to the CCPS mechanism since they are related to the cooperative process. Suppose that there are no other agents present in the environment, or no other agent can provide help. In this case, the agent asking for help tries to find any *complementary and equivalent intentions* that may exist within its original set of intentions. Complementarity and equivalence are defined considering that different intentions can have similar or equal results when their plans are executed. We consider that two intentions are complementary if at least one precondition of one of the intentions corresponds to a post-condition of the other. Additionally, two intentions are equivalent if at least one precondition of these intentions is identical, and if all the post-conditions of both intentions are the same.

4 Experiments

To verify the proper functioning of the CCPS structure, we implemented a proof-of-concept (PoC) version of the mechanism and tested it in conjunction with an application scenario. The CCPS PoC mechanism used the algorithm previously described and was implemented in Java. All interactions between the agents were simulated with the use of the JADE framework[1]. This implementation was evaluated according to a formal experimentation process appropriate for case studies in the Software Engineering domain [19]. The study protocol used in this evaluation was: (*i*) Using the concept of opportunistic cooperation as a basis, create a scenario with predictable outcomes; (*ii*) Defining the situations in which the scenario would be executed; (*iii*) Defining the theoretical outcome of the scenario for each of the situations; (*iv*) Running the scenario for each of the situations, using the PoC implementation; (*v*) Analyzing the results, comparing the obtained data with the expected outcomes for each situation; and (*vi*) Discussing the results, pointing out any relevant issues or points of attention regarding the evaluation and its elements.

The objective of the evaluation was to observe if the CCPS mechanism could properly identify and act on situations involving *opportunistic cooperation*. Hence, we decided to study a specific scenario considering (*i*) isolated agents, with no communication in-between, and (*ii*) agents that are able to communicate and delegate individual tasks to each other. We used the CCPS formalization as a frame of reference to determine which would be the behavior in both situations, so that we could compare the outcome of the experiments with a theoretical baseline. By running the same scenario under different contextual

[1] http://jade.tilab.com.

conditions (simulating a dynamic environment), we wanted to compare how the use of the newly implemented opportunistic cooperation mechanism would affect the performance of each agent in terms of reaching individual goals. Our expectation was that the use of the CCPS mechanism would allow the agents to reach their individual goals more frequently when exposed to different environment conditions.

4.1 Scenario

To evaluate the CCPS mechanism in the domain of Ambient Intelligence (AmI), we chose to model a scenario related to Smart Cities, as described by Streitz [17]. According to this author, one of the goals of designing smart cities should be "enabling people to experience everyday life and work". Achieving this goal could be done through the use of location-based services and communication infrastructure used to connect different citizens among themselves and the city.

Having these concepts in mind, we based our scenario in a situation involving four different agents: Alice, Bob, Claire, and Damien. All agents play different roles within the campus of a University. Some of the agents could cooperate to achieve their individual intentions, such as retrieving a specific resource (book) or getting to a specific location (parking) in a specific time (taking the bus respecting the bus schedule). Cooperation among the agents was passively stimulated with the use of contextual restrictions, such as means of transportation, the agents' initial locations, temporal restrictions (duration of different actions, and time limits in which they should be performed), and so on. The scenario location was modeled after a university campus, and possesses different locations: a laboratory, three different offices, a parking lot, a bus stop, and the residence of all agents. Each of the agents possess the following routines (used to model their intentions): (*i*) *Morning preparation* (actions that take place before each agent leaves their residence); (*ii*) *Going to the university*; (*iii*) *Going to work*; and (*iv*) *Going back home*. Also, individual conditions were attributed to the agents (Alice and Claire have cars, but Bob and Damien take the bus; Alice and Bob work in the same office; and a few others).

4.2 Situations and Expected Outcomes

The scenario described above was executed in three different environment conditions (contexts), considering starting time (enough time to achieve time-sensitive intentions), inherently unachievable intentions, and the presence of equivalent intentions. Considering the aforementioned experiment situations, the expected results in the next paragraphs refer to the situations where communication between the agents is possible. The situation where the agents cannot communicate with each other was used as a reference to measure the efficacy of the implementation.

In the first context (1), the scenario started at 8 a.m., and each agent was given a set of intentions to be evaluated and eventually achieved along the day (e.g., going to work, working, coming back home). These intentions did not necessarily require cooperation of agents, and the temporal restrictions related to all actions involved were set in a manner that cooperation was not necessary at any part of the planning process. In this context, we expected that none of the agents would ask for any help.

In the second context (2), the starting time was kept at 8 a.m., but a few unsatisfiable intentions were also included (e.g., asking a book to the agents when none of them have it), with no equivalent or complementary intentions that could be used. In the case of the satisfiable intentions, in some cases we included equivalent intentions to reach the same goal (going back home by bus or by a ride in another agent's car). In this context, we expected that the agents with unsatisfiable intentions to ask the other agents for help; however, since they were designed to be absolutely unfeasible, we expected the agents to ask for help and - in the absence of any replies - to try and find any equivalent intentions that could be used. Ultimately, we expected that the agents to discard the unfeasible intentions and re-plan using the remaining intentions.

In the third context (3), we started the scenario at 9 a.m., but included unfeasible intentions with equivalent feasible intentions. This context was supposed (*i*) to allow opportunistic cooperation to effectively take place and (*ii*) to allow the agents to properly use the equivalent intention mechanism. In this context, we expected the agents to use more time in the planning process, but also to achieve more goals (compared to the non-communicating situation).

4.3 Results

Once properly coded and parametrized within the implementation, the scenario described in Sect. 4.1 was executed, and the results were compared with the expected outcomes for each context.

In the first experiment (context 1), the agents behaved as expected. Both the non-communicating and the communicating agents were able to generate achievable plans for all of their intentions using the same amount of time. Our conclusion was that the verification involved in checking if cooperation was needed had no impact in the computational resources required.

In the second experiment (context 2), the communicating agents spent considerable additional time (about 200% more) trying to ask for help and trying to find an equivalent intention before reaching the same plan as the non-communicating agents. However, the outcome was still consistent with the expected results.

In the third experiment (context 3), the communicating agents also took considerably more time to reach a final plan, but in this case their plans also had much more achievable intentions than the non-communicating agents. In some cases, where all intentions were dependent on the achievability of a single one (duration and temporal limit for reaching specific locations), some non-communicating agents had considerable less intentions included in their plans, since the others were simply not achievable (as expected). On the other hand, the plan created by the communicating agents, while feasible, took a considerable amount of time to be generated.

5 Conclusion and Future Work

In this paper we formally presented CCPS, a cooperative contextual planning mechanism to be used by AA in AmI scenarios. CCPS was designed reusing an existing contextual planning mechanism (CPS) in conjunction with cooperation based on delegation. The resulting structure makes it possible for an agent to achieve goals otherwise unachievable through the cooperation with other agents running in the same environment. We were able to properly evaluate the CCPS mechanism by simulating a specific scenario in different contexts, using a proof-of-concept implementation and comparing the obtained results with the theorized expected situations. While we were able to confirm the proper functioning of the mechanism, there are a few considerations that must be taken into account.

The first one is related to efficiency: we observed that taking advantage of opportunistic cooperation situations also drastically increased the planning process time. In a real-world situation, this could negatively impact the goal achievement success rates of the agents using the CCPS mechanism. This situation is not completely unexpected, however, and we expect to address in future research.

Another consideration related to the experiments performed was related to the communication process. When the CCPS mechanism finds an unfeasible intention, the communication process that takes place is simple, but not optimal. I a worst-case scenario, the environment might be flooded by help requests. We also intend to study this limitation by testing different constraint mechanisms.

Finally, there are multiple aspects yet to be explored in the proposed CCPS mechanism. Different parts of its algorithm can be detailed studied, and concepts such as complementary and equivalent intentions can be refined. Implementation can also be extended to support multiple scenario variations in runtime, and more complex scenarios can be used to verify how the CCPS mechanism performs under different conditions. We intend to explore all of these aspects in future work.

Acknowledgements. Arthur Casals is supported by CNPq, grant no. 142126/2017-9.

References

1. Aarts, E., Wichert, R.: Ambient intelligence. In: Technology Guide, pp. 244–249. Springer (2009)
2. Bailin, S.C., Truszkowski, W.: Cooperation between intelligent information agents. In: International Workshop on Cooperative Information Agents, pp. 223–228. Springer (2001)
3. Baldoni, M., Müller, J.P., Nunes, I., Zalila-Wenkstern, R. (eds.) Engineering Multi-Agent Systems - 4th International Workshop, EMAS 2016. Lecture Notes in Computer Science, Singapore, Singapore, 9–10 May 2016, Revised, Selected, and Invited Papers, vol. 10093. Springer (2016)
4. Chaouche, A.-C., El Fallah Seghrouchni, A., Ilié, J.-M., Saïdouni, D.E.: A Higher-order agent model with contextual management for ambient systems. In: Transactions on Computational Collective Intelligence XVI, LNCS, vol. 8780, pp. 146–169. Springer, Berlin, Heidelberg (2014)

5. Chaouche, A.-C., El Fallah-Seghrouchni, A., Ilié, J.-M., Saidouni, D.E.: From intentions to plans: a contextual planning guidance. In: Intelligent Distributed Computing VIII, pp. 403–413. Springer International Publishing (2015)
6. Chaouche, A.-C., El Fallah-Seghrouchni, A., Ilié, J.-M., Saïdouni, D.-E.: Learning from situated experiences for a contextual planning guidance. J. Ambient Intell. Humanized Comput. **7**(4), 555–566 (2016)
7. Di Febbraro, A., Sacco, N., Saeednia, M.: An agent-based framework for cooperative planning of intermodal freight transport chains. Transp. Res. Part C Emerg. Technol. **64**, 72–85 (2016)
8. Dua, A., Bawa, S.G., Kumar, N.G.: Efficient data dissemination in vehicular ad hoc networks. Ph.D. thesis (2016)
9. Durfee, E.H.: Scaling up agent coordination strategies. Computer **34**(7), 39–46 (2001)
10. Engesser, T., Bolander, T., Mattmüller, R., Nebel, B.: Cooperative epistemic multi-agent planning for implicit coordination. arXiv preprint arXiv:1703.02196 (2017)
11. Fikes, R.E., Nilsson, N.J.: Strips: a new approach to the application of theorem proving to problem solving. Artif. Intell. **2**(3–4), 189–208 (1971)
12. Jung, D., Zelinsky, A.: An architecture for distributed cooperative planning in a behaviour-based multi-robot system. Robot. Auton. Syst. **26**(2–3), 149–174 (1999)
13. Nigon, J., Gleizes, M.-P., Migeon, F.: Self-adaptive model generation for ambient systems. Procedia Comput. Sci. **83**, 675–679 (2016)
14. Osawa, E.: A metalevel coordination strategy for reactive cooperative planning. ICMAS **95**, 297–303 (1995)
15. Piette, F., Caval, C., Dinont, C., El Fallah-Seghrouchni, A., Taillibert, P.: A multi-agent solution for the deployment of distributed applications in ambient systems. In: Engineering Multi-Agent Systems - 4th International Workshop, EMAS 2016, Singapore, Singapore, 9–10 May 2016, Revised, Selected, and Invited Papers, pp. 156–175 (2016)
16. Smith, R.G.: The contract net protocol: high-level communication and control in a distributed problem solver. IEEE Trans. Comput. **12**, 1104–1113 (1980)
17. Streitz, N.: Citizen-centred design for humane and sociable hybrid cities. Hybrid City, pp. 17–20 (2015)
18. Torreno, A., Onaindia, E., Sapena, O.: Fmap: distributed cooperative multi-agent planning. Appl. Intell. **41**(2), 606–626 (2014)
19. Wohlin, C., Runeson, P., Höst, M., Ohlsson, M.C., Regnell, B., Wesslén, A.: Experimentation in Software Engineering. Springer Science & Business Media, Heidelberg (2012)

Robotics and Video Games

Distributed Formation Tracking of Multi Robots with Trajectory Estimation

Ali Alouache(✉) and Qinghe Wu

Beijing Institute of Technology, Haidian District, Beijing 100081, China
{alouacheali,qinghew}@bit.edu.cn

Abstract. This paper investigates distributed formation tracking of multi robots with virtual robot as reference trajectory subject to communication failure. The objective is to propose a control approach which improves the performances of the formation in term of stability and robustness. Suppose fixed and directed communication topology, the control law is developed for each robot using extended consensus algorithm with a time varying reference trajectory. Meanwhile, polynomial regression method is implemented for estimating the trajectory of the virtual robot to overcome communication failure. At the end, Matlab simulations are carried out and the comparative results demonstrate the effectiveness of the proposed approach.

Keywords: Mobile robot · Formation control · Graph theory
Communication failure · Polynomial regression · Stability

1 Introduction

Recently, coordination control of multi robotic systems has gained a lot of research interests due to the substantial technological advances such as intelligent robots, network communication and automatic control [1–3]. Formation control is a basic and important research of coordinated control problems in multi robotic systems for accomplishing complex and sophisticated tasks [4,5].

Control methods for formation of mobile robots can generally be partitioned into three class approaches as follows: behaviour based method [6–8], virtual-structure method [9–11] and leader-follower method [12–14].

Communication of multi robots systems plays an important role in practice, and its several issues that always exist were recently raised by many researchers in the literature. Cooperative fault detection algorithms to detect the inter-vehicle communication failures in a network of multiple vehicle systems are proposed in [15]. The management of formation flight control under communication failure is proposed in [16]. Robust adaptive sliding mode approach to the asymptotic consensus problem for a class of multi agent systems with time-varying additive actuator faults and communications perturbation is proposed in [17]. A hybrid technique combined particle swarm optimization and consensus protocol

J. Del Ser et al. (Eds.): IDC 2018, SCI 798, pp. 237–246, 2018.
https://doi.org/10.1007/978-3-319-99626-4_21

is proposed in [18] for motion coordination of autonomous mobile groups to deal with communication disconnection.

Communication exchanged between the robots may fail very often in real time applications due many reasons like noises or external disturbances for instance. Consequently, communication failure causes degradation in the performances of the formation. In [19] communication failure is investigated for single integrator multi agent system based on fixed communication topology. In a previous work [20] we proposed performance comparison of two approaches for formation control of multiple nonholonomic wheeled mobile robots without considering communication failure. In [21] we proposed genetic algorithms (GA) to optimize $l-\varphi$ controller and to overcome the problem of communication failure for formation tracking of multi robots. In [22] we proposed least square estimation for formation control of single integrator multi agent system with a time varying reference state subject to communication disturbance.

Based on this background, in this paper we investigate communication failure for distributed formation control of multi robots with virtual robot as reference trajectory. Suppose fixed and directed communication topology based on graph theory, then we discuss the consensus algorithm with a time varying reference state and the design of the control law for each robot based on extended consensus algorithm such that the formation tracks a time varying reference trajectory. Our main contribution in this paper is to propose a modified control law based on polynomial regression method for estimating the reference trajectory of the virtual robot to deal with communication failure, and improving the performance of the formation in term of stability and robustness.

The rest of the paper is structured as follows. Section 2 reviews some preliminaries about graph theory. Section 3 presents consensus algorithms with time varying reference state. Section 4 presents formation control of multi robots based on extended consensus algorithm. Section 5 discusses communication failure and the proposed approach based on polynomial estimation. Section 6 presents simulation analysis. Finally, some conclusions and future works are presented in Sect. 7.

2 Graph Theory

Consider a formation of multi robots that is interconnected and can share communication information among the robots. A directed graph is a pair (V_p, ϵ_p), where the set $V_p = 1, ..., p$ is a finite nonempty node set and $\epsilon_p \subseteq V_p \times V_p$ is an edge set of ordered pairs of nodes, called edges. The edge (i, j) in the edge set of a directed graph denotes that vehicle j can receive information from vehicle i but not necessarily vice versa. Undirected graph can be viewed as a special case of a directed graph, where an edge (i, j) in the undirected graph corresponds to edges (i, j) and (j, i) in the directed graph. A directed tree is a directed graph in which every node has exactly one parent expect for one node, called the root, which has no parent and which has a directed path to every other node. In undirected graphs, a tree is a graph in which every pair of nodes is connected by exactly one undirected path.

The adjacency matrix $A_p = a_{ij} \in R^{p\times p}$ of a directed graph (V_p, ϵ_p) is defined such that a_{ij} is a positive weight if $(j, i) \in \epsilon_p$, and $ij = 0$ if $(j, i) \notin \epsilon_p$. Self-edges are not allowed unless otherwise indicated. The adjacency matrix of an undirected graph is defined analogously except that $a_{ij} = a_{ji}$ for all $i \neq j$ because $(j, i) \in \epsilon_p$ implies $(i, j) \in \epsilon_p$. Define the matrix $L_p = [l_{ij}] \in R^{p\times p}$ as $l_{ii} = \sum_{(j=1),(j\neq i)}^{n} = a_{ij}$, $l_{ij} = -a_{ij}, i \neq j$. Note that if $(j, i) \notin \epsilon_p$ then $l_{ij} = -aij = 0$. The matrix L_p satisfies $l_{ij} \leq 0, \sum_{(j=1),(j\neq i)}^{n} l_{ij} = 0$, $i = 1, ..., p$. For an undirected graph, L_p is symmetrical and it is called the Laplacian matrix.

3 Consensus Algorithm with Time Varying Reference State

Consider an interconnected system composed of n agents with single integrator dynamics given as following.

$$\dot{\xi}_i = u_i \tag{1}$$

where $i = 0, ..., n$, or in matrix form, then Eq. (1) may be rewritten as follows.

$$\dot{\xi} = [-L_{n(t)} \otimes I_m]\xi \tag{2}$$

where $\xi = [\xi_1^T, ..., \xi_n^T]^T$. $\xi_i \in R^m$ and $u_i \in R^m$ are respectively the information state and the information control input of the ith vehicle. $L_{n(t)} \in R^n$ is the non-symmetrical Laplacian matrix at time t and $\otimes$ denotes the Kronecker product. I_m denotes the m dimensional identity matrix.

Suppose a multi agent system which consists of n identical vehicles as described above with an additional vehicle labelled $n + 1$, which acts as the unique virtual leader of the team. Therefore the $(n + 1)^{th}$ vehicle is named as the leader and the vehicles $1, , n$ are named the followers.

The vehicle $n + 1$ has the information state as $\xi_{n+1} = \xi^r \in R^m$ where ξ^r denotes the time varying reference state. The consensus reference state satisfies the following.

$$\dot{\xi}^r = (t, \xi^r) \tag{3}$$

where $(.,.)$ is bounded, piecewise continuous in t and locally Lipschitz in ξ^r. The consensus tracking problem with a time varying reference state is solved iff $\xi_i(t) \to \xi^r(t), i = 1, ..., n$ as $t \to \infty$. Assume the team of the vehicles as an interaction topology with $n + 1$ nonempty nodes. Therefore according to the graph theory tutorial given in the Sect. 2, we can get the corresponding directed graph G_{n+1} for the multi agent system i.e. $G_{n+1} = (V_{n+1}, \epsilon_{n+1})$, the adjacency matrix $_{+1}$, and the matrix l_{n+1} which is the non-symmetrical Laplacian matrix associated with the graph G_{n+1}. The consensus tracking problem with a time-varying reference state is solved with the following algorithm [23].

$$u_i = \frac{1}{\eta_i}\sum_{j=1}^{n} a_{ij}(t)[\dot{\xi}_j - \gamma(\xi_i - \xi_j)] + \frac{1}{\eta_i}\sum_{j=1}^{n} a_{i(n+1)}(t)[\dot{\xi}^r - \gamma(\xi_i - \xi^r)] \tag{4}$$

where $i = 1, ..., n, j = 1, ..., n+1$, $a_i(t)$ is the (i, j) entry of the adjacency matrix $A_{n+1}(t)$ at time t. γ is a positive constant scalar, and $\eta_{i(t)} = \sum_{j=1}^{n+1} a_{ij}(t)$.

Theorem 1. [23]. The formation achieves consensus with time varying reference state based on the control law (4) if and only if communication graph has spanning tree.

4 Controller Design for Formation of Multi Robots Based on Extended Consensus Algorithm

Suppose a formation that is composed of multiple $n+1$ nonholonomic wheeled mobile robots on the X-Y plane as shown in Fig. 1. The kinematics of the ith mobile robot is described as follows.

$$\dot{x}_i = V_i cos(t); \dot{y}_i = V_i sin(t); \dot{\theta}_i = w_i \tag{5}$$

where (x_i, y_i) and θ_i denote respectively the cartesian and orientation of the ith wheeled mobile robot with respect to the inertial frame at time (t). (V_i, w_i) denote the linear and angular velocity of the wheeled mobile robot.

Let the $(n+1)^{th}$ robot denotes the virtual leader of state $\xi^r(t)$ which is available only for the subgroup leaders of the formation.

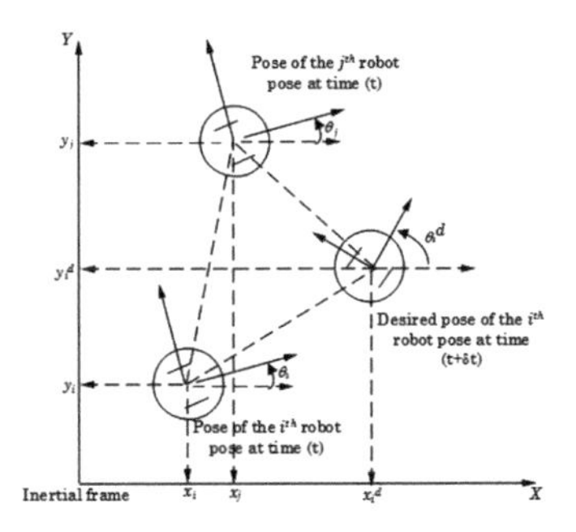

Fig. 1. Modelling of the multi robots formation based on virtual robot

Remark 1. The motion of the virtual leader is independent from the other robots and it has no incoming information from the other robots.

Suppose that the ith robot denotes the subgroup leader and its desired trajectory at time $(t+\delta t)$ is given as follows $r_i^d = (x_i^d, y_i^d, \theta_i^d)^T$, where δt denotes the sampling time. Due to the nonholonomic constraint of the wheeled mobile robot, then the desired orientation θ_i^d of the ith robot must satisfies the following equation.

$$\dot{x}_i^d sin(\theta_i^d) - \dot{y}_i^d cos(\theta_i^d) = 0 \tag{6}$$

The desired control inputs $u_i^d = (V_i^d, w_i^d)$ of the ith wheeled mobile robot at time $(t+\delta t)$ are defined by the following equations.

$$\begin{cases} V_i^d = \sqrt{(\dot{x}_i^d)^2 + (\dot{y}_i^d)^2} \\ w_i^d = \frac{(\dot{x}_i^d)(\ddot{y}_i^d) - (\ddot{x}_i^d \dot{y}_i^d)}{(\dot{x}_i^d)^2 + (\dot{y}_i^d)^2} \end{cases} \tag{7}$$

Under condition that both $\dot{x}_i^d$ and $\dot{y}_i^d$ dont equal zero.

The coordinates of the jth follower robot shown in Fig. 1 are defined by the following variables.

$r_j = (x_j, y_j)^T$ denotes the actual pose of the jth robot with respect to the inertial frame at time (t). $r_j = (x_j^d, y_j^d)^T$ denotes the desired pose of the jth follower robot at time $(t + \delta t)$. $r_j^{d_i} = (x_j^{d_i}, y_j^{d_i})^T$ denotes the desired deviation of the jth follower robot relative to the ith subgroup leader robot.

According to the pose of the ith robot subgroup leader at time t and time $(t + \delta t)$ respectively, therefore the desired coordinates of r_j^d for the jth follower robot are defined as follows.

$$\begin{bmatrix} x_j^d \\ y_j^d \end{bmatrix} = \begin{bmatrix} x_i \\ y_i \end{bmatrix} + \begin{bmatrix} cos\theta_i & -\ sin\theta_i \\ sin\theta_i & cos\theta_i \end{bmatrix} \begin{bmatrix} x_{ji}^d \\ y_{ji}^d \end{bmatrix} \tag{8}$$

Therefore, the control law for the jth robot u_j^d at time $(t + \delta t)$ is defined based on extended consensus algorithm as follows [24].

$$u_j^d = \dot{r}_j^d - a_i(r_j - r_j^d) - \sum_{n}^{j=1} a_{ij}((r_j - r_j^d) - (r_i - r_i^d)) \tag{9}$$

where a_i is a positive scalar. The variable a_{ij} is the (i, j) entry of the adjacency matrix.

If each jth follower robot can track its desired trajectory $r_j^d(x_j^d, y_j^d)$ accurately, then the desired formation can be maintained accurately based the control law (9).

Remark 2. The formation achieves consensus with a time varying reference state if and only if the communication graph has a spanning tree.

Remark 3. According to Theorem 1 and Remark 2, then only directed communication graphs with spanning tree are considered in this work.

5 Communication Failure

In this section polynomial regression method is proposed to deal with communication failure. The problem of communication failure occurs when the communication is missing between the virtual leader and the subgroupleaders due to external disturbances or errors of the sensors.

We suppose that communication failure between the robots occurs during a finite interval of time and the state of the virtual leader ξ^r is available again after the interval of communication failure.

We propose an estimation labeled as $\tilde{\xi}^r$ for the reference state ξ^r, therefore we need to approximate ξ^r as a polynomial of finite order such that the residual error between the reference state ξ^r and its corresponding estimation $\tilde{\xi}^r$ is minimum.

The algorithm (4) based on the estimated trajectory $\tilde{\xi}^r$ can be rewritten as follows.

$$u_i = \frac{1}{\eta_i}\sum_{j=1}^{n} a_{ij}(t)[\dot{\xi}_j - \gamma(\xi_i - \xi_j)]$$
$$+\frac{1}{\eta_i}\sum_{j=1}^{n} a_{i(n+1)}(t)[\lambda[\dot{\xi}^r - \gamma(\xi_i - \xi^r)] + (1-\lambda)[\dot{\tilde{\xi}}^r - \gamma(\xi_i - \tilde{\xi}^r)]] \quad (10)$$

where the variable λ is defined as follows. 1 if communication is valid and 0 if communication is failed.

For both cases of λ, the multi agent system achieves consensus with a time varying reference state if communication topology is directed and has spanning tree. Suppose that the estimated trajectory $\tilde{\xi}^r$ can be expressed as following.

$$\tilde{\xi}^r = a_0 + a_1 t + a_2 t^2 + ... + a_n t^n \quad (11)$$

Minimizing the error $E(t)$ of (11) means that its derivative with respect to the parameters $(a_i, i = 0, ..., n)$ equal zero. The derivative of the error $E(t)$ of (11) with respect to parameters $(a_i, i = 0, ..., n)$ yields the following equation.

$$\frac{dE}{da_i} = 0 \quad (12)$$

Equation (12) yields the following.

$$\begin{cases} \frac{dE}{da_0} = 0 \\ \frac{dE}{da_1} = 0 \\ \vdots \\ \frac{dE}{da_n} = 0 \end{cases} \quad (13)$$

To estimate the polynomial $\theta = [a_0, a_1, ..., a_n]$, then it is necessary to solve the following system.

$$\begin{bmatrix} n & \sum_{i=1}^{n} p_i & \cdots & \sum_{i=1}^{n} p_i^n \\ \sum_{i=1}^{n} p_i & \sum_{i=1}^{n} p_i^2 & \cdots & \sum_{i=1}^{n} p_i^{n+1} \\ \vdots & \vdots & \vdots & \vdots \\ \sum_{i=1}^{n} p_i^{n+1} & \sum_{i=1}^{n} p_i^{n+2} & \cdots & \sum_{i=1}^{n} p_i^{2n} \end{bmatrix} \begin{bmatrix} a_0 \\ a_1 \\ \vdots \\ a_n \end{bmatrix} = \begin{bmatrix} \sum_{i=1}^{n} p_i \\ \sum_{i=1}^{n} p_i q_i \\ \vdots \\ \sum_{i=1}^{n} p_i^n q_i \end{bmatrix} \quad (14)$$

where $i = 1, ..., m$; $n \leq m + 1$ and (p_1, q_1),(p_2, q_2),...,(p_m, q_m) being the data collected for the ith robot along directed axis.

6 Simulation Analysis

Section 6.1 presents kinematic linearization of the mobile robot and Sect. 6.2 demonstrates simulation results.

6.1 Kinematic Linearization

The kinematic model of the wheeled mobile robot is nonlinear, therefore it is necessary to linearize it into single integrator dynamics. We need to define a head point named h_i which is located at distance d_i from the ith robots center. The coordinates of the head point are given as follows.

$$\begin{bmatrix} h_{xi} \\ h_{yi} \end{bmatrix} = \begin{bmatrix} x_i \\ y_i \end{bmatrix} + d_i \begin{bmatrix} cos\theta_i \\ sin\theta_i \end{bmatrix} \tag{15}$$

The time derivative of equation (20) is given as follows.

$$\begin{bmatrix} \dot{h}_{xi} \\ \dot{h}_{yi} \end{bmatrix} = \begin{bmatrix} cos\theta_i & -d_i sin\theta_i \\ sin\theta_i & -d_i cos\theta_i \end{bmatrix} \begin{bmatrix} v_i \\ w_i \end{bmatrix} \tag{16}$$

Let the following.

$$\begin{bmatrix} \dot{h}_{xi} \\ \dot{h}_{yi} \end{bmatrix} = \begin{bmatrix} u_{xi} \\ u_{yi} \end{bmatrix} \tag{17}$$

6.2 Simulation Results

Consider a formation of three multiple wheeled mobile robots moving on the X-Y plane with one leader denoted L and two follower robots denoted F1 and F2 respectively. The reference trajectory of the virtual leader ξ^r(t) is available only for the subgroup leader(L) and it is given by the following equations.

$$\begin{cases} x(t) = t \\ y(t) = cos(x) \end{cases} \tag{18}$$

The coordinates of three mobile robots at the initial time are given as follows.

Leader: (0, 0); Follower 1: (0, 5); Follower 2: (0, 10).

Consider communication topology as shown in Fig. 2, hence the graph is strongly connected with spanning tree. The communication failure between the virtual leader and the followers occurs in time interval [20 s, 25 s].

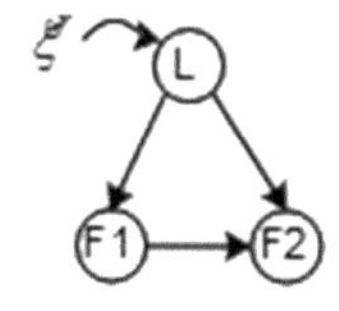

Fig. 2. Communication topology of the formation

The results of applying algorithm (10) without estimation can be seen in Fig. 3(a) and (b) respectively. Therefore, the formation fails to achieve consensus

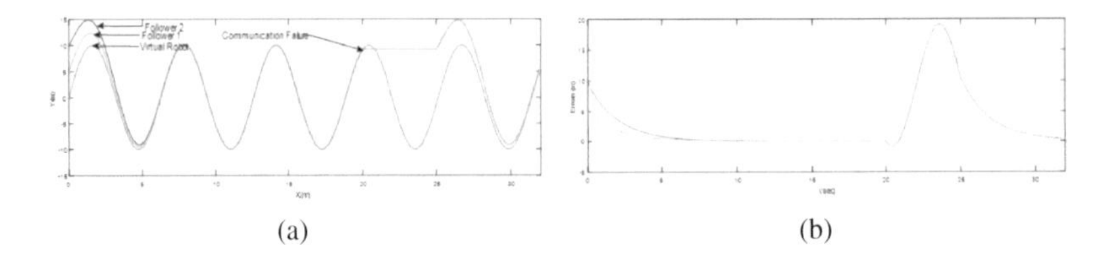

(a) (b)

Fig. 3. Results of formation tracking without estimation

at the instant of communication failure (t = 20 s) as shown in Fig. 3(a), also large tracking errors are resulted due to communication failure as shown in Fig. 3(b).

The solution is discussed in Sect. 5 by using polynomial regression method for estimating the reference trajectory. Hence, the results of algorithm (10) with estimation are shown in Fig. 4(a) and (b) respectively.

It can be noted that the formation effectively tracks the reference trajectory as shown in Fig. 4(a). Also, compared to the results of Fig. 3(b), hence the tracking errors due to communication failure are neglected as shown in Fig. 4(b).

These results prove the effectiveness of the proposed approach to remedy communication failure between the robots of the formation.

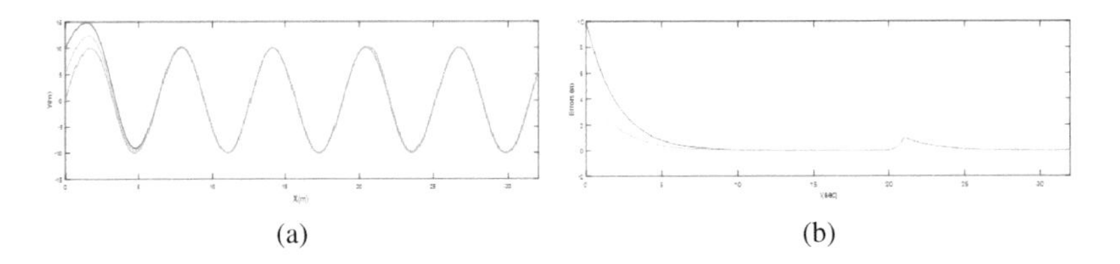

(a) (b)

Fig. 4. Results of formation tracking with estimation using ploynomial regression

7 Conclusions

This paper proposes stable approach for distributed formation tracking of multi robots subject to communication failure. Suppose virtual robot as reference trajectory and fixed communication topology. Polynomial regression method is proposed for estimation of the reference trajectory to remedy communication failure and to improve the stability and robustness of the formation. The comparative results demonstrated the effectiveness of the proposed approach where the trajectory of the formation is stable and the errors due to communication failure are neglected based on trajectory estimation. In the future works, we attempt to investigate variable communication topology with noisy data because this paper deals with fixed communication topology. Also obstacle avoidance based on potential function with the design of the controller in practical applications.

Acknowledgements. This work was partially supported by the National Natural Science Foundation of China under Grant 61321002.

References

1. Cao, Y., Yu, W., Ren, W., Chen, G.: An overview of recent progress in the study of distributed multi-agent coordination. IEEE Trans. Ind. Inform. **9**(1), 427–438 (2013)
2. Oh, K., Park, M., Ahn, H.: A survey of multiagent formation control. Automatica **53**, 424–440 (2015)
3. Qin, J., Ma, Q., Shi, Y., Wang, L.: Recent advances in consensus of multi-agent systems: a brief survey. IEEE Trans. Ind. Electron. **64**(6), 4972–4983 (2017)
4. Xue, D., Yao, J., Wang, J., Guo, Y., Han, X.: Formation control of multi-agent systems with stochastic switching topology and time-varying communication delays. IET Control Theory Appl. **7**(13), 1689–1698 (2013)
5. Sun, X., Cassandras, C.G.: Optimal dynamic formation control of multi-agent systems in environments with obstacles. In: 2015 54th IEEE Conference on Decision and Control (CDC), Osaka, pp. 2359–2364 (2015)
6. Liu, Z., Kubota, N.: Hybrid learning approach based on multi-objective behavior coordination for multiple robots. In: 2007 International Conference on Mechatronics and Automation, Harbin, pp. 204–209 (2007)
7. Yang, F., Liu, S., Dong, D.: Robot behavior and service-based motion behavior structure design in formation control. Robot **34**(1), 120–128 (2012)
8. Mendiburu, F.J., Morais, M.R.A., Lima, A.M.N.: Behavior coordination in multi-robot systems. In: 2016 IEEE International Conference on Automatica (ICA-ACCA), Curico, pp. 1–7 (2016)
9. Jawhar, G., Hasan, M., Maarouf, S.: Formation path following control of unicycle-type mobile robots. Robot. Auton. Syst. **58**(5), 727–736 (2010)
10. Hasan, M., Jawhar, G., Maarouf, S.: Nonlinear coordination control for a group of mobile robots using a virtual structure. Mechatronics **21**(7), 1147–1155 (2011)
11. Chen, L.E.I., Baoli, M.A.: A nonlinear formation control of wheeled mobile robots with virtual structure approach. In: Proceedings of the 34th Chinese Control Conference, 28–30 July, Hangzhou, China (2015)
12. Shao, J., Xie, G., Wang, L.: Leader-following formation control of multiple mobile vehicles. IET Control Theory Appl. **1**(2), 545–552 (2007)
13. Luca, C., Fabio, M., Domenico, P.: Leader-follower formation control of nonholonomic mobile robots with input constraints. Automatica **44**(5), 1343–1349 (2008)
14. Yang, L., Cao, Z., Tan, M.: Dynamic formation control for multiple robots in uncertain environments. Robot **32**(2), 283–288 (2010)
15. Izadi, H.A., Gordon, B.W., Zhang, Y.: Cooperative communication failure detection for multiple vehicle systems. In: AIAA Guidance, Navigation, and Control Conference, 2–5 August 2010, Toronto, Ontario, Canada (2010)
16. Shin, J., Jin Kim, H., Kim, S., Yoon, Y.: Formation flight control under communication failure. In: Proceedings of the 1st International Conference on Robot Communication and Coordination (RoboComm 2007), Piscataway, NJ, USA, Article 47, 4 pages. IEEE Press (2007)
17. Jin, X.Z., Yuan, Z.H.: Robust adaptive sliding mode consensus of multiagent systems with perturbed communications and actuators. Math. Probl. Eng. **2013**, 9 pages (2013)
18. Grandi, R., Falconi, R., Melchiorri, C.: Coordination and control of autonomous mobile robot groups using a hybrid technique based on particle swarm optimization and consensus. In: Proceeding of the IEEE International Conference on Robotics and Biomimetics (ROBIO) Shenzhen, China, December 2013

19. Li, H., Wu, Q., Sabir, D.: Consensus tracking algorithms with estimation for multi-agent system, CCDC (2015)
20. Alouache, A., Wu, Q.: Performance comparison of consensus protocol and approach for formation control of multiple nonholonomic wheeled mobile robots. J. Mechatron. Electr. Power Veh. Technol. **8**, 22–32 (2017)
21. Alouache, A., Wu, Q.: Tracking control of multiple mobile robot trajectory by genetic algorithms. Electrotech. Electron. Autom. (EEA) **65**(4), 155–161 (2017). ISSN 1582-5175
22. Alouache, A., Wu, Q.: Consensus based least square estimation for single integrator multi agent with a time varying reference state. J. Electron. Sci. Technol. (2018, Accepted for publication)
23. Ren, W.: Multi-vehicle consensus with a time-varying reference state. Syst. Control. Lett. **56**(7–8), 471–483 (2007)
24. Wei, R., Nathan, S.: Distributed coordination architecture for multirobot formation control. Robot. Auton. Syst. **56**(4), 324–333 (2008)

Applying Evolutionary Computation Operators for Automatic Human Motion Generation in Computer Animation and Video Games

Luis de la Vega-Hazas[1], Francisco Calatayud[1], and Andrés Iglesias[2,3](✉)

[1] BINARYBOX STUDIOS, Calle Juan XXIII, 1, 39001 Santander, Spain
[2] Universidad de Cantabria, Avda. de los Castros, s/n, 39005 Santander, Spain
iglesias@unican.es
[3] Toho University, 2-2-1 Miyama, Funabashi 274-8510, Japan

Abstract. This paper presents an evolutionary computation scheme for automatic human motion generation in computer animation and video games. Given a set of identical physics-driven skeletons seated on the ground as an initial pose (similar for all skeletons), the method applies forces on selected bones seeking for a final stable pose with all skeletons standing. Such forces are initially random but then modulated by a set of evolutionary operators (selection, reproduction, and mutation) to make the digital characters learn to stand up by themselves. An illustrative example is discussed in detail to show the performance of this approach. This method can readily be extended to other skeleton configurations and other interesting motions with little modification. Our approach represents a significant first step towards automatic generation of motion routines by applying evolutionary operators.

Keywords: Artificial intelligence · Evolutionary computation
Computer animation · Automatic motion generation · Skeletal model
Virtual actors

1 Introduction

Nowadays, artificial intelligence (AI) techniques are increasingly used in computer animation and video games as powerful tools to improve the performance of current development frameworks, simplify and speed up the graphical animation and game creation pipelines by task automation and support for digital production, and enhance the quality and realism of the final product. Sophisticated AI and machine learning techniques are currently being applied for behavioral animation of the NPCs (non-player characters) in video games [1–5]. Crowd simulation in computer movies can now be generated automatically taking advantage of swarm intelligence and other AI techniques [7]. New AI-assisted software tools are now available for automatic generation of terrains, buildings, cities, characters, and assets. Expert systems are applied to provide goal-directed

J. Del Ser et al. (Eds.): IDC 2018, SCI 798, pp. 247–258, 2018.
https://doi.org/10.1007/978-3-319-99626-4_22

specifications for cameras and lights. Machine learning and reasoning can be applied to the automatic generation of cinematic scripts for computer movies. And the list of applications is continuously growing and expanding to meet the current needs of computer animation and video games industries [10].

Human animation is one of the fields that can mostly benefit from recent advances in AI. Roughly, three approaches are mainly applied in the field: manual generation, physics-based, and data-driven. Human motion in computer animation is typically carried out by professional animators by using controllers and other software tools for the manipulation of joints, constraints, deformers and so on. This process relies on a number of shape and motion parameters, which are determined by the animator in a rather manual way. As expected, the process becomes error-prone and time-consuming, and demands a lot of expertise from the animators for realistic and believable motion. Physics-based approaches capture the natural aspect of a particular motion by applying a set of physical laws describing its dynamics. These methods are slow and very demanding in terms of CPU and memory storage, as they require to solve huge systems of equations. Also, in many cases the obtained solutions lead to unnatural trajectories and unrealistic motions. In data-driven techniques (e.g., motion capture), the motion of real human actors is tracked and recorded with cameras or other devices. This technology is very popular in the entertainment industry for movies and video games to get more realistic human movements (take, for instance, the acclaimed blockbuster *Avatar*). However, it requires specialized hardware for motion capture, which is costly and difficult to use. Furthermore, it is based on the records of specific actors, so it cannot be extended to any human character.

In conclusion, neither of the previous approaches is actually well suited for automatic motion generation. This is the motivation of our current work in the field. In particular, in this paper we raise the following questions:

Can a digital character learn to make a particular motion autonomously (i.e., without human intervention)? If so, how?

Obviously, these questions are too ambitious and general to be answered here. In this paper we focus on a particular motion: to stand up. We look for a motion pattern where the digital character is lying on the floor in a seated pose (initial pose) and at the end is standing, getting firmly into a stable upright position on his/her feet (final pose). Stated as such, this is a typical problem in inverse kinematics. It is also a difficult one, since it is highly multimodal: there might be several (even infinite) solutions for the motion pattern from the initial to the final pose. As a starting point, in this paper we try to elucidate whether it is possible to obtain one of such solutions *automatically and autonomously.* To this purpose, we rely on *evolutionary computation* (EC), a subfield of artificial intelligence focused on algorithms for global optimization inspired by biological evolution. Generally speaking, EC algorithms are population-based metaheuristic methods inspired by the principles and mechanisms of biological evolution, such as selection, mutation, recombination, crossover, and so on. A great advantage of such methods is that they do not make any assumption on the problem to be solved such as the functional structure of the objective function or the underlying fitness landscape. This feature make these methods ideal tools for dealing with problems subjected to uncertainty, noise, and with little (or none) information about the problem, which is actually our case here.

Evolutionary computation has already been applied to computer graphics starting with the seminal work by Karl Sims in the 80s [8,9], mostly based on genetic algorithms. More recently, these ideas have been applied to evolutionary design through evolutionary operators in [11]. Our work takes the inspiration from these sources. However, these approaches are targeted at computer design rather than computer animation and hence, they do not address the issue of motion generation. On the other hand, genetic algorithms have been used for optimizing the character gaze behavior of virtual characters in [6]. Evolutionary computation is also applied in the *Darwin's Nightmare* game to drive the exploration of a large combinatorial space defining the behavior and appearance of enemy crafts. The focus on these works is *not* on behavioral animation rather than on motion animation, as it happens in this paper. The main originality of our approach consists of applying the principles of evolutionary computation directly to a population of skeletons through virtual forces acting on selected bones. In this sense, the genotypes of our evolutionary scheme are the vectors of applied forces rather than the motion routines or the poses themselves.

1.1 Aims and Structure of the Paper

This paper presents an evolutionary computation scheme for automatic human motion generation in computer animation. Our approach applies forces on a given population of identical physics-driven human skeletons. Such forces are modulated by a set of evolutionary operators carefully chosen so as to make the digital characters learn to stand up by themselves.

The structure of this paper is as follows: in Sect. 2 we describe briefly the skeletal model used in this paper. Then, our evolutionary computation approach is presented through an in-depth discussion of its main evolutionary operators in Sect. 3. Our experimental results are reported in Sect. 4. The paper closes with the conclusions and some ideas for future work in the field.

2 The Skeletal Model

In this paper we consider a collection of η human virtual actors $\{\mathscr{H}_i\}_{i=1,\ldots,\eta}$. The physical structure of each virtual character $\mathscr{H}_i$ is described by a skeletal model $\mathscr{S}_i$, which provides the geometric rigid structure of the virtual body much like its real-world counterpart actually does. The skeleton $\mathscr{S}_i$ consists of two components: a collection of bones and a collection of joints. For each $\mathscr{S}_i$ the set of bones, denoted as $\mathscr{B}_i$, consists of λ_i bones $\mathscr{B}_i = \{\mathbf{b}^i_j\}_{j=1,\ldots,\lambda_i}$, while the set of joints $\mathscr{T}_i$ consists of all joints $\{\mathscr{T}^i_{k,l}\}_{k,l}$ connecting bones $\mathbf{b}^i_k$ and $\mathbf{b}^i_l$. The joints provide a natural representation for the constraints of different parts of the virtual body by a proper selection of the kind of deformations allowed and their degrees of freedom (DOFs). In this sense, they play a quite similar role to the real-world human body joints in which they are originally based on.

To form a skeleton, the bones must be connected. A natural way to do so is to use a hierarchy, where each bone belongs to a parent and can have one or several children connected to it. For human body representation, we use a hierarchical tree in which the primary bone, called the root bone and represented onwards as $\mathbf{b}_1^i$, is located in a central part of the body, generally the spine. All other bones $\mathbf{b}_j^i$ $(j > 1)$ are children of this root bone, either directly (first level) or indirectly (higher levels) via other intermediate bones $\mathbf{b}_k^i$, which are children of the root bone (and possibly of other bones as well) and parents of this bone (and possibly of others too). In this way, for any given bone $\mathbf{b}_j^i$ we can define two types of sequences of bones: the forward (or outer) sequences, denoted by $\{\phi_p(\mathbf{b}_j^i)\}_p$ and given by all bones that are children of $\mathbf{b}_j^i$ and connected by joints continuously, and the backward (or inner) sequences, denoted by $\{\varphi_q(\mathbf{b}_j^i)\}_q$ and given by all bones connecting $\mathbf{b}_j^i$ with the root bone through joints in a continuous way. These sequences are important because affecting a bone $\mathbf{b}_j^i$ (for instance, by applying a force F) also affects all of its children, i.e. all bones in any forward sequence $\phi_p(\mathbf{b}_j^i)$, while the bones in the backward sequence $\varphi_q(\mathbf{b}_j^i)$ are not affected by F. Although this is opposed to what happens in real life, it simplifies the model and reduces its computational load.

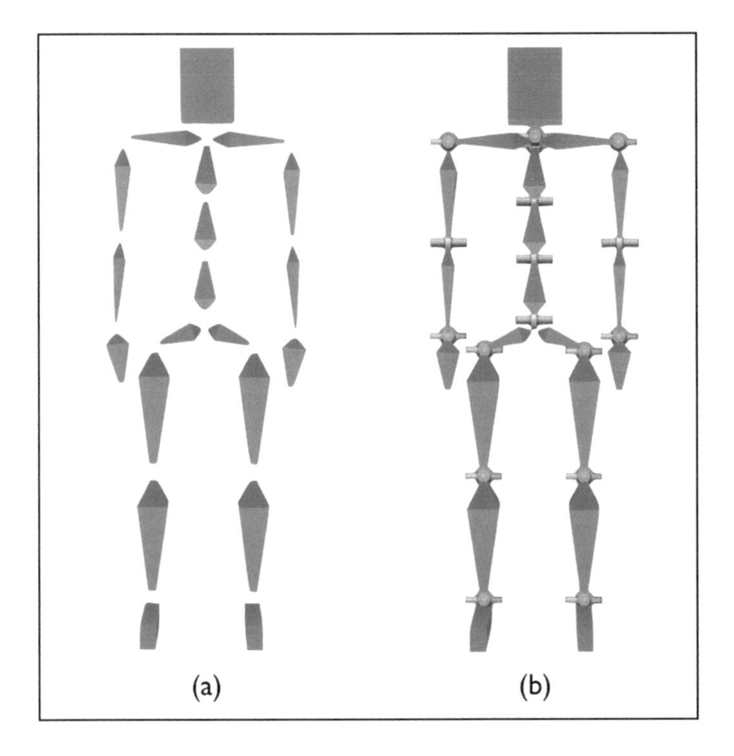

Fig. 1. Skeletal model used in this paper: (left) bones; (right) joints.

Table 1. List of bones and joints of the skeleton used in this paper.

Bones		*Joints*	
1. Spine1	11. L-Foot	1. Spine1	11. R-Shoulder
2. Spine2	12. R-Clavicle	2. Spine2	12. R-Elbow
3. Spine3	13. R-UpperArm	3. Spine3	13. R-Wrist
4. L-Clavicle	14. R-LowerArm	4. Neck	14. R-Hip
5. L-UpperArm	15. R-Hand	5. L-Shoulder	15. R-Knee
6. L-LowerArm	16. R-Hip	6. L-Elbow	16. R-Ankle
7. L-Hand	17. R-UpperLeg	7. L-Wrist	
8. L-Hip	18. R-LowerLeg	8. L-Hip	
9. L-UpperLeg	19. R-Foot	9. L-Knee	
10. L-LowerLeg	20. Head	10. L-Ankle	

Figure 1 shows the prototypical skeletal model for human actors used in this paper. The figure is organized in two parts, displaying the set of bones (left) and the set of joints (right) and labelled as (a) and (b), respectively. As the reader can see, we consider a very basic and simple (yet good enough for the purposes of this paper) skeleton, comprised of 20 interconnected bones and 16 joints, fully reported in Table 1. The initial 'L' and 'R' letters followed with a hyphen stand for left and right, respectively. Figure 2 shows the top-down hierarchical tree of all bones of our skeleton model. It consists of a undirected graph where nodes of the tree represent the bones, which are connected by edges. As shown in the figures, the tree starts with the root bone at the top of the hierarchy. From it,

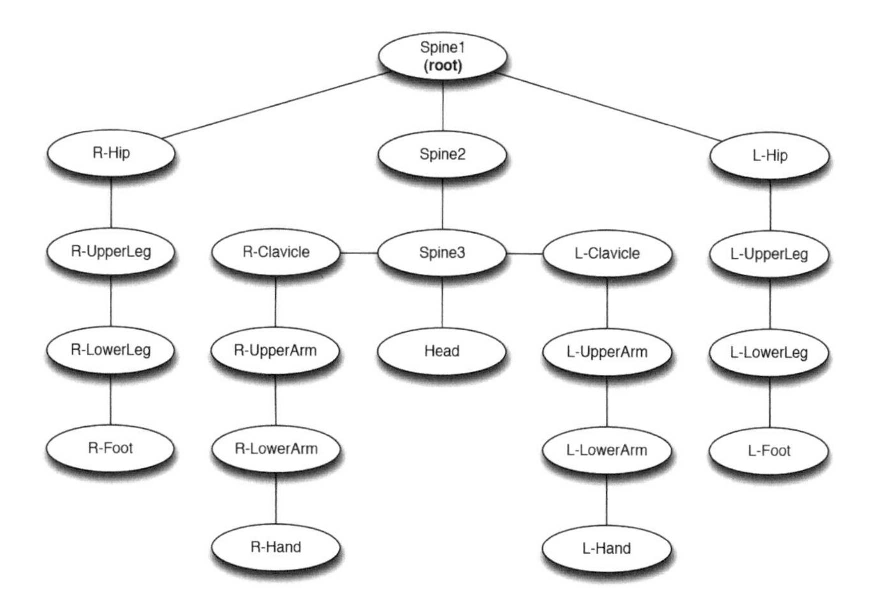

Fig. 2. Skeleton represented by a top-down hierarchical tree.

there are five different sequences of connected bones going down until terminal nodes (feet, hands and head) are found.

3 The Method

Our method is a population-based evolutionary computation approach that operates on an initial population of skeletons $\{\mathscr{S}_i\}_{i=1,\ldots,\eta}$ all lying on the floor and then moving to get seated. This initial movement is identical for all skeletons and has been pre-processed and then stored, so it is not affected by our approach. The method starts when all skeletons are seated on the floor at exactly the same pose and works in an iterative fashion. At initial time of each generation g, a force vector ${}^{[g]}\mathscr{F} = \left\{{}^{[g]}\mathscr{F}_1, \ldots, {}^{[g]}\mathscr{F}_\eta\right\}$ is applied, where ${}^{[g]}(.)$ is used to indicate the generation and each force ${}^{[g]}\mathscr{F}_i$ is applied on skeleton ${}^{[g]}\mathscr{S}_i$. At its turn, ${}^{[g]}\mathscr{F}_i = \left\{{}^{[g]}\mathscr{F}_{i,j}\right\}_{j=1,\ldots,\lambda_i}$, where force ${}^{[g]}\mathscr{F}_{i,j}(t)$ is applied on bone $\mathbf{b}^i_j$ at instance time τ of generation g. This force takes the form of an impulsive force consisting of a burst of χ forces $\{\xi^\kappa_{i,j}\}_{\kappa=1,\ldots,\chi}$ applied sequentially at times $\tau + \kappa \Delta t$ as:

$$ {}^{[g]}\mathscr{F}_{i,j}(t) = \sum_{\kappa=1}^{\chi} {}^{[g]}\xi^\kappa_{i,j} \delta_{\mathbf{T}}(\tau + \kappa \Delta t) \tag{1} $$

where $\delta(.)$ is the Dirac delta function and $\mathbf{T}$ is the vector of time impulses of the force. For simplicity, in this paper we assume that $\lambda_i = \lambda$, $\forall i$. We also assume that all forces in our method are exerted vertically upwards except for gravity, which works in the opposite direction. As a result of the action of these forces, the skeletons move for a while until reaching a stable position (in our case, until all skeletons keep fully static for 5 s). Once the steady-state is attained, the new skeleton configurations are evaluated and ranked according to a given fitness function. In this work, the fitness function $\mathbf{F}$ is defined as the sum of three terms: ${}^{[g]}\mathbf{F}(\mathscr{S}_i) = {}^{[g]}\mathbf{F}_1(\mathscr{S}_i) + {}^{[g]}\mathbf{F}_2(\mathscr{S}_i) + {}^{[g]}\mathbf{F}_3(\mathscr{S}_i)$, where each term evaluates a specific feature of the skeleton configuration. Assuming that the body is resting but stretched, for a skeleton to be standing it is required that:

(1) the feet are firmly standing on the ground;
(2) the hip should lie on the imaginary vertical axis Y that goes upwards from the feet to the head and at a distance from the ground given by the length of the skeleton from the foot base to the hip;
(3) the axes of the head and the body are aligned so the center of the head lies on the vertical axis Y and with the same orientation, and the distance from the ground to the top of the head corresponds to the body height.

Functions $\mathbf{F}_i$ are associated with the three conditions, respectively. We remark that conditions (1)–(3) are ideal and, hence, very difficult to replicate accurately. They are also unreasonable, as we expect different characters to stand up differently, similar to how human beings actually do. For these reasons, we allow a threshold error given by three imaginary boxes $\{B_i\}_i$. B_1 is a 2D box on the ground marking the available area for feet placement, providing some flexibility on constraint (1) by allowing the feet to be placed freely within this area as long as they are stepping on the ground. B_2 and B_3 are volumetric boxes intended to provide some flexibility on the position and stance by allowing lateral and vertical displacements of the hip and the head respectively, provided that such bones still keep inside their respective boxes. Mathematically, these functions are defined as:

$$^{[g]}\mathbf{F}_1^i(\mathscr{S}_i) = \mathscr{D}_H\left(^{[g]}\mathbf{b}_{11}^i \cup {}^{[g]}\mathbf{b}_{19}^i, B_1\right) \tag{2}$$

$$^{[g]}\mathbf{F}_2^i(\mathscr{S}_i) = d_2\left(\gamma\left(^{[g]}\mathbf{b}_8^i \cup {}^{[g]}\mathbf{b}_{16}^i\right), \gamma(B_2)\right) \tag{3}$$

$$^{[g]}\mathbf{F}_3^i(\mathscr{S}_i) = d_2\left(\gamma\left(^{[g]}\mathbf{b}_{20}^i\right), \gamma(B_3)\right) \tag{4}$$

where $\mathscr{D}_H$ is the Hausdorff distance between sets, d_2 is the Euclidean distance, γ computes the 3D geometrical center of a set, and $\cup$ is the set union operator.

The ranked skeletons are sorted in increasing order and stored in a list L, whose first and last elements correspond to the best and worst values in current generation g, denoted as $^{[g]}S^*$ and $^{[g]}S_*$ respectively. Improvement over time is achieved by applying three evolutionary operators: selection, reproduction, and mutation. Similar to genetic algorithms and other evolutionary methods, the selection operator, denoted as $\odot$, is used to promote the best individuals according to the Darwinian survival of the fittest. In our method, we consider *elitism* so that the best solutions from the current generation are transferred directly to the next generation without further modification. We also consider a monoparental reproduction operator denoted as $\bowtie$ and based on *cloning*, where the best individual is cloned (i.e., copied identically) so that further operators are applied on the clones while preserving the original individual for elitism. The clones undergo mutation under the action of an mutation operator $\otimes$ that applies Gaussian white noise additive perturbations on $^{[g]}\mathscr{F}_{i,j}$. This procedure is repeated iteratively until $^{[g]}\mathbf{F}(\mathscr{S}_i) = 0, \forall i$, meaning that all skeletons in our population hold our constraints according to Eqs. (3)–(4).

4 Experimental Results

Figures 4 and 5 shows an illustrative example of our experimental results. In this experiment, we consider a set of 10 skeletons $\{\mathscr{S}_k\}_{k=1,\ldots,10}$, placed in a row and numbered from left to right. In this example the method is executed for 17 iterations, shown in sequence from top to bottom and in Fig. 3. For each iteration, we show the final configuration of the skeletons along with the three boxes B_i described in previous section. As mentioned above, our starting point consists of the skeletons seated on the ground (initial pose). Then, we consider an initial population of forces $^{[0]}\mathscr{F}_i$ applied on our skeletons. Since the skeletons are already seated, forces are applied only on the bones in the trunk (active bones) to capture better the real physics of the process. This means that $^{[0]}\mathscr{F}_i = \left\{^{[0]}\mathscr{F}_{i,j}\right\}_{j \in \mathscr{A}_B}$ where $\mathscr{A}_B$ is the list of indices for the active bones, given by: $\mathscr{A}_B = [1..7] \cup [12..15]$, where $[a..b]$ means all integer numbers between a and b (including

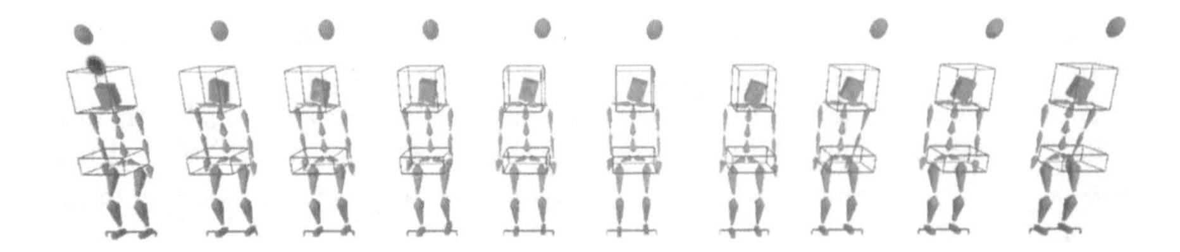

Fig. 3. Final results at iteration $g = 17$.

a, b if they are integers). The forces are initially chosen randomly according to a uniform distribution within a feasible interval $[\alpha_j, \beta_j]$ specific for each bone. Then, our method is applied iteratively so that the force values improve over time according to our fitness function. As shown in the figure, they still fail to get any skeleton upright for the 7 first iterations. At $g = 8$ the first standing skeleton is obtained for $i = 9$; it becomes the best ${}^{[8]}S^* = \odot\left(\{{}^{[8]}S_i\}_i\right)$, marked by a red balloon above its head. After ranking, the worst solution (corresponding to $i = 6$ and marked by a blue balloon above the head) is replaced by a clone of the best, $\bowtie({}^{[8]}S_9)$ and then mutated $\otimes\left[\bowtie({}^{[8]}S_9)\right]$. Then, the process is restarted again for $g = 9$ and so on. The full process after iteration $g = 7$ is summarized in Table 2. The table reports (in columns): the iteration number g, the index of the global best for this iteration $\bowtie({}^{[g]}S_*)$, the indices of the skeletons to be replaced by clones of the best and then mutated $\bowtie({}^{[g]}S_*)$, and the indices of the skeletons passed to the next iteration without further modification (except the best). Final results of this simulation example (reached for generation $g = 17$) are highlighted in bold. The corresponding graphical output for this iteration is depicted in Fig. 3. It shows that all skeletons are properly standing after 17 iterations.

Table 2. Results of iterations 9–17 for the example in Figs. 4 and 5.

g	${}^{[g]}S^*$	$\bowtie({}^{[g]}S_*)$	$\odot\left(\left\{{}^{[g]}S_i\right\}_i\right)$
7	9	6	1,2,3,4,5,7,8,10
8	7	6,2,	1,3,4,5,8,10
9	5	6,2,8	1,,3,4,7,9,10
10	3	6,2,8,10	1,4,5,7,9
11	3	6,2,8,10,9	1,4,5,7
12	7	6,2,8,10,9,4	1,3,5
13	4	6,2,8,10,9,4,1	3,5,7
14	1	6,2,8,10,9,4,1,3	5,7
15	1	6,2,8,10,9,4,1,3,5	7
16	1	6,2,8,10,9,4,1,3,5	7
17	**1**	**6,2,8,10,9,4,1,3,5**	**7**

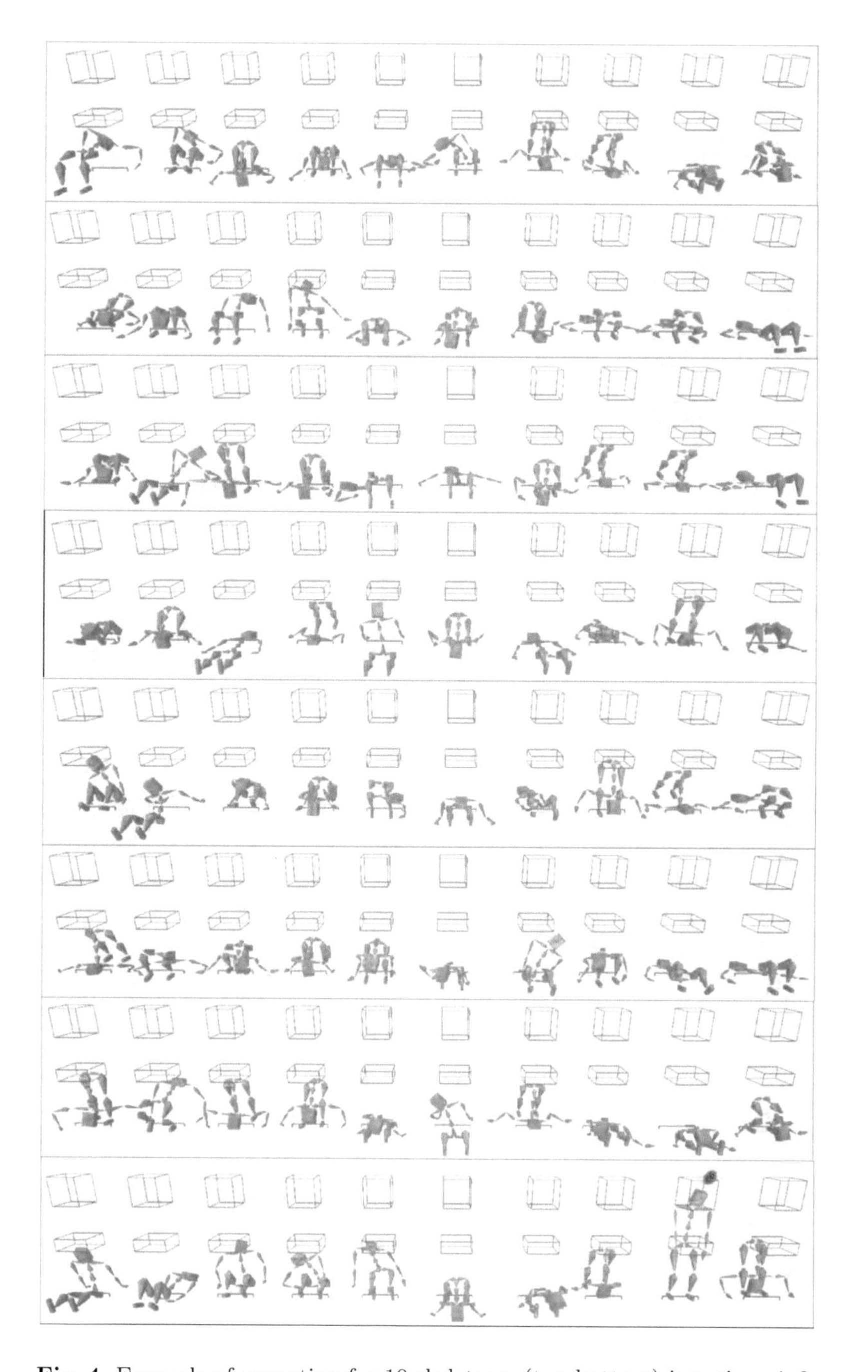

Fig. 4. Example of execution for 10 skeletons: (top-bottom) iterations 1–8

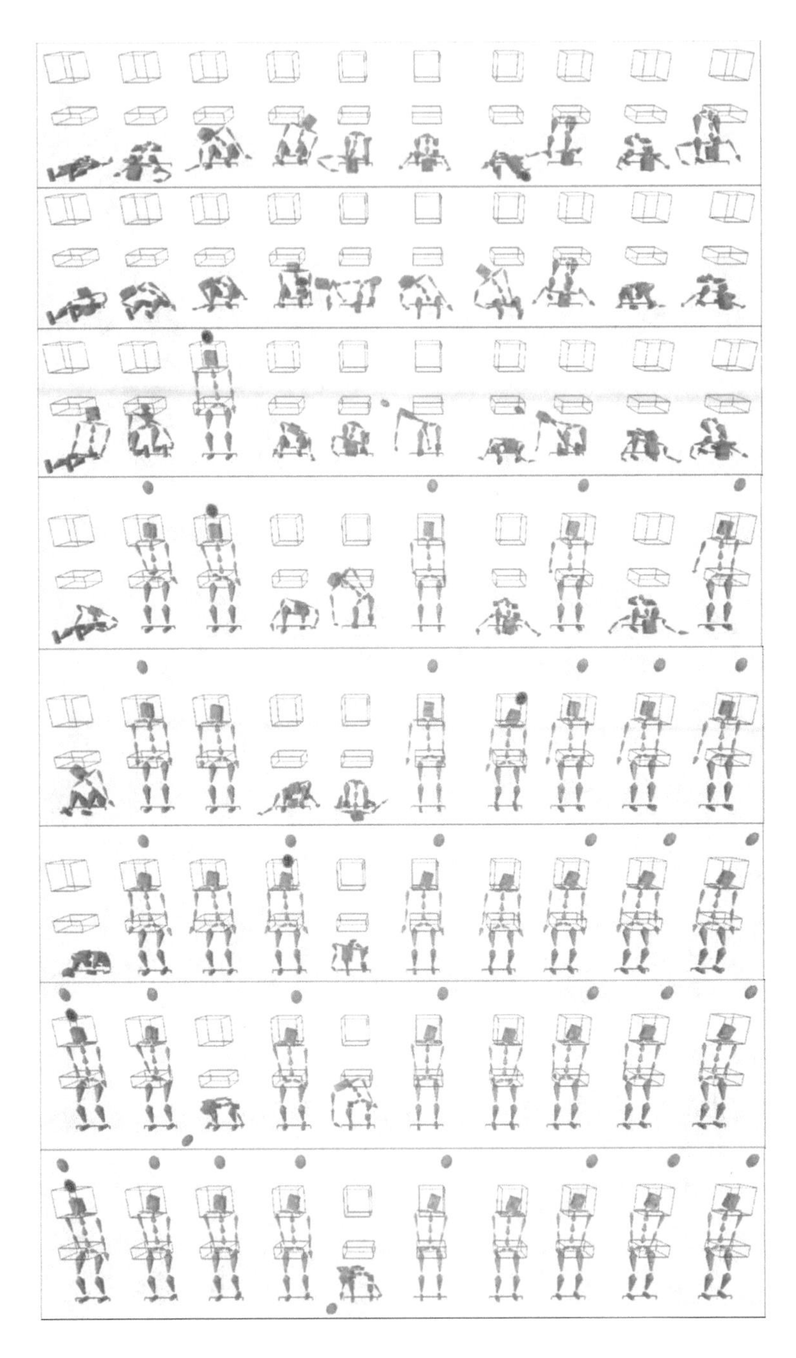

Fig. 5. Example of execution for 10 skeletons: (top-bottom) iterations 9–16

5 Conclusions and Future Work

In this paper we apply several evolutionary computation operators to the problem of automatic human motion generation in computer animation and video games. Given a set of identical physics-driven skeletons, the method applies forces on selected bones in order to obtain a final stable pose for all skeletons. Such forces are initially random but then modulated by a set of evolutionary operators (selection, reproduction, and mutation) to make the digital characters learn to move by themselves. An illustrative example for the standing motion routine is discussed to show the performance of this approach. This method can readily be extended to other skeleton configurations and other interesting motions with little modification. Our approach represents a significant first step towards automatic generation of motion routines by applying evolutionary operators.

The main conclusion of our work is that this evolutionary computation approach performs very well for the problem addressed in this paper. Although only one example is discussed in detail here, we performed several executions for this problem. All our simulations reached the target state in less than 50 iterations. This is a remarkable result, since the initial forces are totally random in all cases. This means that we can recreate a human motion routine without any knowledge about the forces to be applied or their intensity values.

Our computational work has been carried out in the popular game engine *Unreal Engine 4*, by *Epic Games*, by using a 3.8 GHz quad-core Intel Core i5, with 16GB of DDR3 memory, and a graphical card AMD RX580 with 8 GB VRAM. The source code has been created by the authors by using the visual scripting system *Blueprints*, with some extra code being programmed directly on C++. The CPU times of all our simulations are very good, as they run in real time for populations of skeletons up to $20 \sim 30$ individuals (excluding rendering).

Of course, this method also comes with some limitations. As shown in the discussed example, the convergence of the method is achieved at the expense of diversity. At every iteration the worst individual is replaced by a clone of the global best of the swarm; then, it undergoes mutation for diversity. For small populations, it is likely that this strategy leads to populations where almost all individuals are slightly modified copies of the global best after only a few iterations. This problem can be alleviated by introducing new random individuals in the population throughout the iterations while simultaneously applying elitism to keep the population size constant over the time. Also, it is sometimes difficult to determine which bones should be affected by the forces for a particular motion. Now, this task is performed manually, a tedious process that typically requires some expertise for acceptable results.

Regarding the future work, our scheme can be improved in many different ways. We plan to extend this approach to more difficult motion routines, likely requiring applying forces on large sets of bones for a proper and realistic motion pattern. We also wish to consider skeletons with a larger number of bones, particularly for motions involving the hands and feet, such as walking or grabbing an object. Finally, we plan to develop an automatic procedure for the proper determination of the bones that should be affected by the forces for a particular motion routine.

Acknowledgements. This research work has received funding from the project PDE-GIR of the European Union's Horizon 2020 research and innovation programme under the Marie Sklodowska-Curie grant agreement No 778035, the Spanish Ministry of Economy and Competitiveness (Computer Science National Program) under grant #TIN2017-89275-R of the Agencia Estatal de Investigación and European Funds FEDER (AEI/FEDER, UE), and the project #JU12, jointly supported by public body SODERCAN and European Funds FEDER (SODERCAN/FEDER UE).

References

1. Díaz, G., Iglesias, A.: Swarm intelligence scheme for pathfinding and action planning of non-player characters on a last-generation video game. In: Advances in Intelligent Systems and Computing, vol. 514, pp. 343–353 (2017)
2. Díaz, G., Iglesias, A.: Intelligent behavioral design of non-player characters in a FPS video game through PSO. In: Advances in Swarm Intelligence. Lecture Notes in Computer Science, vol. 10385, pp. 246–254 (2017)
3. Iglesias, A., Luengo, F.: Intelligent agents for virtual worlds. In: Proceedings of CW 2004, Tokyo, Japan, pp. 62–69. IEEE Computer Society Press, Los Alamitos (2004)
4. Iglesias, A., Luengo, F.: New goal selection scheme for behavioral animation of intelligent virtual agents. IEICE Trans. Inf. Syst. **E88–D**(5), 865–871 (2005)
5. Iglesias, A., Luengo, F.: AI framework for decision modeling in behavioral animation of virtual avatars. LNCS, vol. 4488, pp. 89–96 (2007)
6. Mori, H., Toyama, F., Shoji, K.: Optimization of character gaze behavior animation using an interactive genetic algorithm. Int. J. Asia Digit. Art Des. **21**(1–4), 25–31 (2017)
7. Schwab, B.: AI Game Engine Programming, 2nd edn. Course Technology, Boston (2009)
8. Sims, K.: Artificial evolution for computer graphics. In: ACM SIGGRAPH, pp. 319–328 (1991)
9. Sims, K.: Evolving virtual creatures. In: ACM SIGGRAPH, pp. 15–22 (1994)
10. Woodcock, S.: Game AI: the state of the industry 2000–2001: it's not just art, it's engineering. In: Game Developer, August 2001, pp. 36–44 (2001)
11. Xu, K., Zhang, H., Daniel Cohen-Or, D., Chen, B.: Fit and diverse: set evolution for inspiring 3D shape galleries. ACM Trans. Graph. **31**(4), Article No. 57 (2012)

A General-Purpose Hardware Robotic Platform for Swarm Robotics

Nureddin Moustafa[1], Akemi Gálvez[1,2], and Andrés Iglesias[1,2](✉)

[1] Universidad de Cantabria, Avda. de los Castros, s/n, 39005 Santander, Spain
{galveza,iglesias}@unican.es
[2] Toho University, 2-2-1 Miyama, Funabashi 274-8510, Japan

Abstract. Swarm intelligence is based on the recently-acquired notion that sophisticated behaviors can also be obtained from the cooperation of several simple individuals with a very limited intelligence but cooperating together through low-level interactions between them and with the environment using decentralized control and self-organization. Such interactions can lead to the emergence of intelligent behavior, unknown to the individual agents. One of the most remarkable applications of swarm intelligence is swarm robotics, where expensive and sophisticated robots can be replaced by a swarm of simple inexpensive micro-robots. In this context, this paper introduces a general-purpose hardware robotic platform suitable for swarm robotics. With a careful choice of its main components and its flexible and modular architecture, this robotic platform provides support to the most popular swarm intelligence algorithms by hardware. As an illustration, the paper considers four of the most popular swarm intelligence methods; then, it describes the most relevant hardware features of our approach to support such methods (and arguably many other swarm intelligence approaches as well) for swarm robotics.

Keywords: Swarm intelligence · Swarm robotics
General-purpose robot · Hardware robotic platform
Intelligent behaviors

1 Introduction

Swarm intelligence (SI) has been regarded as one of the most exciting new avenues of research in artificial intelligence (AI) during the last few decades. Unlike many other areas in AI, individuals in SI do not need to be actually *intelligent* in its most canonical sense. Instead, the intelligence in SI is typically obtained from the aggregation of very simple behavioral patterns by unsophisticated agents collaborating together to solve a complex problem. Amazingly, this kind of collective behavior had already been observed in some natural groups for centuries. Consider, for instance, the dynamics of colonies of social insects (ants, termites, bees, fireflies), where the group as a whole is able to construct complex nests and carry out many different sophisticated tasks unattainable

J. Del Ser et al. (Eds.): IDC 2018, SCI 798, pp. 259–271, 2018.
https://doi.org/10.1007/978-3-319-99626-4_23

for its individuals members. Another typical example is the behavior of a flock of birds when moving all together following a common tendency in their displacement. Other examples from nature include animal herding, fish schooling, and many others. Furthermore, these examples have been used as metaphors for popular SI methods, such as ant colony optimization (ACO) or particle swarm optimization (PSO). In SI methods, there is not a centralized intelligence controlling the swarm, taking decisions, and sending orders to the individual members about how to behave. In fact, such individual agents follow simple rules and have a very limited knowledge and intelligence. However, as a whole, the social group is capable of complex collective behaviors, which emerge from a small set of simple behavioral rules exploiting only low-level interactions between individuals and with the environment (stigmergy) using decentralized control and self-organization. The interested reader is referred to [5,8,21] for a comprehensive overview about the field of swarm intelligence, its history, main techniques, and applications. See also Sect. 2 for a brief description of some popular swarm intelligence methods.

A major reason for this interest on SI is its potential application in several fields. An illustrative example is *swarm robotics*, a field where swarms of simple self-organizing micro-robots are used to replace sophisticated and expensive robots to accomplish complex tasks [2,6,13,14,16,17]. As remarked in [1,3], swarm robotic systems offer several interesting advantages, such as:

- *Improved performance by parallelization*: swarm intelligence systems are very well suited for parallelization, because the swarm members can perform different actions at different locations simultaneously. This feature makes the swarm more flexible and efficient for complex tasks, as individual robots (or groups of them) can solve different parts of a complex task independently.
- *Task enablement*: groups of robots can do certain tasks that are impossible or very difficult for a single robot (e.g., collective transport of heavy items, dynamic target tracking, cooperative environment monitoring, and so on).
- *Scalability*: adding new robots to a swarm does not require reprogramming the whole swarm. Furthermore, because interactions between robots involve only neighboring individuals, the total number of interactions does not increase dramatically by adding new units, making the system highly scalable.
- *Distributed sensing and action*: a swarm of simple interconnected mobile robots deployed throughout a large search space possesses greater exploratory capacity and a wider range of sensing than a sophisticated robot. This makes the swarm much more effective in tasks such as exploration and navigation tasks (e.g., in disaster rescue missions).
- *Fault tolerance*: due to the decentralized and self-organized nature of the swarm, the failure of a single unit (or a small group of them) does not affect the completion of the given task.

All these advantages motivated a great interest in swarm robotics during the last two decades. The interested reader is referred to [12,15–17] and references therein for a brief description about previous work in the field.

1.1 Aims and Structure of This Paper

In this paper, we introduce a first prototype of a hardware robotic platform for swarm robotics designed to meet some important differential features:

- *It is affordable.* Since the robots must operate in swarms, it is important to keep their individual price as low as possible. In our proposal, we avoid expensive components, so that each robot costs about 50–60US$.
- *It is general-purpose.* Instead of a specialized goal-oriented robot, our design is general-purpose. This is possible thanks to its flexible design and modular architecture, avoiding fixed parts so that different components (e.g., sensors, holders, frames) can readily be added or removed to meet different goals.
- *It is suitable for swarm robotics.* In spite of its low-cost design, the robot CPU and memory are powerful enough to support some of the most popular swarm intelligence techniques running locally at software level. Furthermore, the robots of the swarm are highly interconnected with one another via standard communication interfaces such as *Wifi* and *Bluetooth* (in our case, they are built-in in our microprocessor single-board).

The structure of this paper is as follows: Sect. 2 provides a very brief description of some popular SI methods. Section 3 describes our design of the hardware robotic platform, including its architecture and main components, and the programming framework. Then, Sect. 4 discusses the main features of our proposal that provide support to the methods in Sect. 2 for swarm robotics applications. The paper closes with the conclusions and some future work in the field.

2 Some Popular Swarm Intelligence Algorithms

In this section, some of the most typical SI algorithms are briefly revisited. Since they are very popular and widely reported in the literature, we restrict our discussion to their main features without further detail, and refer the interested reader to our bibliography for a fully informed description. Relevant information about these methods will be used for our discussion in Sect. 4.

2.1 Particle Swarm Optimization

Particle Swarm Optimization (PSO) is a global stochastic optimization algorithm for dealing with problems where potential solutions (called *particles*) can be represented as vectors in a search space [7,8]. Particles are distributed over such space and provided with an initial velocity and the capacity to communicate with other neighbor particles, even the entire swarm. Particles "flow" through the solution space and are evaluated according to some fitness function after each instance. Particles evolution is regulated by two memory factors: memory of their own best position and knowledge of the global or their neighborhood's best. Particles of a swarm communicate good positions to each other and adjust their own position and velocity based on these good positions. As the swarm iterates, the fitness of the global best solution improves so the swarm eventually reaches the best solution. To this aim, each particle modifies its position P_i along the iterations by storing the coordinates P_i^b associated with the best solution (fitness) achieved so far. These values account for the *memory* of the best particle position. In addition, members of a swarm can communicate good positions to each other, so they can adjust their own position and velocity according to this information. To this purpose, we also collect the best global position P_g^b from the initial iteration. The evolution for each particle i is given by:

$$\begin{aligned} V_i(k+1) &= w\,V_i(k) + \gamma_1 R_1[P_g^b(k) - P_i(k)] + \gamma_2 R_2[P_i^b(k) - P_i(k)] \\ P_i(k+1) &= P_i(k) + V_i(k) \end{aligned} \tag{1}$$

where $P_i(k)$ and $V_i(k)$ are the position and the velocity of particle i at time k, respectively, w is called *inertia weight* and decide how much the old velocity will affect the new one and coefficients γ_1 and γ_2 are constant values called *learning factors*, which decide the degree of affection of P_g^b and P_i^b. This procedure is repeated several iterations until a termination condition is reached. The reader is referred to [4,5,7–9,21] for further details on this very popular algorithm.

2.2 Bat Algorithm

The *bat algorithm* (BA) is a bio-inspired SI method proposed in 2010 to solve optimization problems [23,24]. It is based on the behavior of microbats, which use a type of sonar called *echolocation*, with varying pulse rates of emission and loudness, to detect prey, avoid obstacles, and locate their roosting crevices in the dark. The idealization of the echolocation of microbats is as follows:

1. Bats use echolocation to sense distance and distinguish between food, prey and background barriers.
2. Each virtual bat flies randomly with a velocity $\mathbf{v}_i$ at position (solution) $\mathbf{x}_i$ with a fixed frequency f_{min}, varying wavelength λ and loudness A_0 to search for prey. As it searches and finds its prey, it changes the frequency of their pulses and adjust the rate of pulse emission r, depending on the proximity of the target.
3. It is assumed that the loudness will vary from an (initially large and positive) value A_0 to a minimum constant value A_{min}.

Under these idealized rules, the algorithm considers an initial population of bats, each representing a potential solution of the optimization problem and having a location $\mathbf{x}_i$ and velocity $\mathbf{v}_i$. The algorithm initializes these variables with suitable random values. Then, the pulse frequency, pulse rate, and loudness are computed for each individual bat. The swarm evolves in a discrete way over generations until the maximum number of generations is reached. For each g and each bat, new frequency, location and velocity are computed as:

$$f_i^g = f_{min}^g + \beta(f_{max}^g - f_{min}^g) \tag{2}$$

$$\mathbf{v}_i^g = \mathbf{v}_i^{g-1} + [\mathbf{x}_i^{g-1} - \mathbf{x}^*]\, f_i^g \tag{3}$$

$$\mathbf{x}_i^g = \mathbf{x}_i^{g-1} + \mathbf{v}_i^g \tag{4}$$

where $\beta \in [0,1]$ follows the random uniform distribution, and $\mathbf{x}^*$ represents the current global best location (solution), which is obtained through evaluation of the objective function at all bats and ranking of their fitness values. The best current solution and a local solution around it are probabilistically selected according to some given criteria. Then, search is intensified by a local random walk. For this local search, once a solution is selected among the current best solutions, it is perturbed locally through a random walk. If the new solution achieved is better than the previous best one, it is probabilistically accepted depending on the value of the loudness. In that case, the algorithm increases the pulse rate and decreases the loudness. The reader is referred to [10,11,23–25,28] for further details and in-depth analysis of the method and its implementation.

2.3 Firefly Algorithm

The *firefly algorithm* (FFA) is a SI algorithm introduced in 2009 for optimization problems [19]. It is based on the social flashing behavior of fireflies in nature. The key ingredients of the method are the variation of light intensity and formulation of attractiveness. In general, the attractiveness of an individual is assumed to be proportional to their brightness, which in turn is associated with the encoded objective function. In the firefly algorithm, there are three particular idealized rules, based on some of the major flashing characteristics of real fireflies:

1. All fireflies are unisex, so that one firefly will be attracted to other fireflies regardless of their sex;
2. The degree of attractiveness of a firefly is proportional to its brightness, which decreases as the distance from the other firefly increases due to the fact that the air absorbs light. For any two flashing fireflies, the less brighter one will move towards the brighter one. If there is not a brighter or more attractive firefly than a particular one, it will then move randomly;
3. The brightness or light intensity of a firefly is determined by the value of the objective function of a given problem. For optimization problems, the light intensity can simply be proportional to the value of the objective function.

For a full description of this method, the reader is kindly referred to [19–22].

2.4 Cukoo Search Algorithm

The *cuckoo search algorithm* (CSA) is a SI method proposed in 2009 [26] and inspired by the obligate interspecific brood-parasitism of some cuckoo species that lay their eggs in the nests of host birds of other species to escape from the parental investment in raising their offspring and to minimize the risk of egg loss to other species. In CSA, the eggs in the nest represent the pool of candidate solutions while the cuckoo egg represents a new coming solution. The method uses these new (potentially better) solutions associated with the parasitic cuckoo eggs to replace the current solution associated with the eggs in the nest. This replacement, carried out iteratively, will eventually lead to a very good solution. For computational reasons, CSA is also based on three idealized rules [27]:

1. Each cuckoo lays one egg at a time, and dumps it in a randomly chosen nest;
2. The best nests with high quality of eggs (solutions) will be carried over to the next generations;
3. The number of available host nests is fixed, and a host can discover an alien egg with a probability $p_a \in [0, 1]$. For simplicity, this assumption can be approximated by a fraction p_a of the n nests being replaced by new nests (with new random solutions at new locations).

More details on this algorithm can be found in [18,21,22,26,27].

3 Hardware Robotic Platform for Swarm Robotics

This section describes our hardware robotic platform for swarm robotics, including its architecture and main components, and the programming framework.

3.1 Hardware Architecture and Components

The main components of our robotic platform are shown in Fig. 1. They are:

1. the *chassis*: the robot is mounted on a rigid chassis, in orange in that figure. The chassis and other mechanical parts such as the holders have been generated by 3D printing from a PLA (polylactic acid) filament by using a domestic desktop 3D printer. The chassis hosts the battery with its board connectors and the electronics of our robotic unit.
2. the *wheels* and *servomotors*: robot movement is provided through two side wheels with power supplied by two continuous rotation servomotors, displayed in the picture. Unlike ordinary motors, servomotors can be individually controlled; they only require the angle of rotation for motion. Rotation is supported through an omni-directional ball caster wheel able to swivel in any direction.
3. a *battery*: a rechargeable power-efficient 3.7 V lithium-ion polymer battery is used in our implementation, with the advantage that it has higher specific energy than other lithium batteries.
4. a *boost step-up power*: for the battery to be voltage-compatible, we also include the module *Lithium 134n3p* charger, a built-in charge and discharge power MOS operating at an input voltage in the range 3.7 V–5.5 V with output voltage 5 V and providing charge and discharge management, temperature control and protection against over-temperature, output over-voltage, short circuit, heavy load over-charge and over-discharge.
5. a *single-board micro-computer*: in our implementation we use the popular micro-computer *Raspberry Pi Zero W*, one of the most affordable and cost-effective micro-computers in the market, with a price of 10US$. This micro-computer comes with a 32-bit *RISC ARMv6Z* architecture, featuring a *Broadcom BCM2835* system on a chip application processor. Its CPU is the *ARM1176JZF-S* core by ARM, running at 1 GHz. It also includes a graphical processor unit *Broadcom Video Core IV* running at 250 MHz, with support to Open GL and featuring a H.264/MPEG-4 AVC high-profile decoder

and encoder with support to 1080p (high-definition video mode). The system comes with 512 MB (shared with the GPU), 1 micro-USB (direct from the *BCM2835* chip), MIPI camera interface for video input, mini-HDMI at 1080p resolution and composite video for video output, 2 boards via the serial bus I^2S for audio input, a stereo audio through PWM on GPIO for audio output, and a MicroSDHC non-volatile memory card for data storage.

6. *built-in communication interface*: our single-board chip also provides support for communications via *Bluetooth 4.1* for very short distances (about a range of 10 m or less) , *802.11n* wireless LAN for wider areas (up to 100 m), and a FM receiver working in the range 65 MHz to 108 MHz FM bands, all through the *Cypress CYW43438* wireless chip. It also has an unpopulated HAT-compatible 40-pin GPIO header, and composite video and reset headers. These wireless communication options are used for communication and data exchange over short distances among the robots of the swarm and with a central server for tracking purposes.
7. a *stripboard*: used for further connectivity of all electronic components. It is located close to the board chip to gain access to all micro-computer pins.
8. an *ultrasound sensor*: in this work, we use the ultrasound sensor *HC-SR04*, manufactured by *ElecFreaks*. It is an ultrasonic sensor operating at 5 V DC that uses sonar to compute the distance to an object. Each *HC-SR04* module includes an ultrasonic transmitter, a receiver and a control circuit, with 4 pins for power, trigger (transmitter), echo (receiver), and ground. Each ultrasound pulse of our sensors operates at a constant frequency of 40 kHz, sending an 8 cycle burst of ultrasound pulses. The sensor captures its echo with signals lasting in the order of milliseconds. The accuracy range of the sensor is about 3 mm, with a traveling range of pulses between 2–500 cm. The ultrasound sensor is used for collision avoidance with static and dynamic objects (including other robots in the swarm) as well as with the boundaries of the physical 3D environment.
9. a LED RGB: a hand-made diode with RGB lights used to indicate different robot states, such as *active*, *idle*, *sleep*, and others.
10. a *magnetometer*: in this work, we use the triple-axis magnetometer board *HMC-5883L*. This user-friendly compass is a 3.3 V max chip with added circuitry to make it 5 V-safe logic and power, so that it can be connected to either 3 or 5 V microcontrollers. It uses I^2S serial bus for easy interface to communicate. Its internal functioning is based on the anisotropic magnetoresistive (AMR) technology by *Honeywell*, with AMR directional sensors having a full range of ± 8 gauss an a resolution of up to 2 milligauss. The magnetometer is used in this work for global spatial orientation of the robotic units of the swarm regardless their physical environment.
11. *infrared sensors*: used for collision avoidance, scene exploration and navigation throughout the 3D environment. In this work we consider a set of three infrared sensors deployed on a semi-ring holder located in the front of the robot to cover a wider exploration area. One of the IR sensors is located in the middle, and the other two near the corners in the front. They can detect

obstacles in the range of 30 cm in the sun light. They also come with a variable resistance to adapt the sensors to different short distances.

12. a *mini-camera*: used for image capture and navigation. Our model provides a resolution of 3280 × 2464 pixels, and support video capture at a frame rate of 30 FPS (frames per second) for a resolution of 1080p, 60 FPS at 720p and 90 FPS at 480p.
13. an *OLED micro-display*: a 0.96 in. *SDD1306* chipset with serial connection I^2C, and power consumption of 20 mA. It provides a resolution of 128 × 64 pixels and vision angle of 160° and is mainly used to display relevant information for tracking purposes and as user-interface with the board.

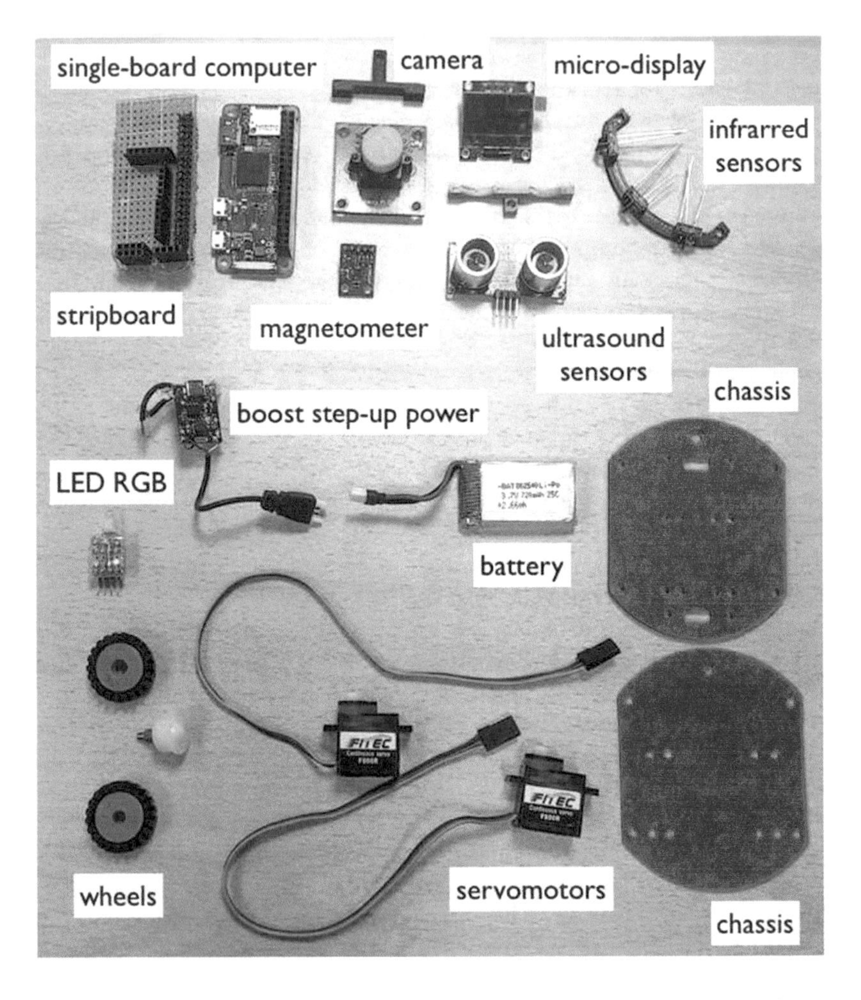

Fig. 1. Main components of our hardware robotic platform for swarm robotics.

All these components are connected to different I/O pins in a rather standard way (see Fig. 2). The detailed description of these connections is out of the scope of this paper and will be omitted here to keep the paper to a manageable size.

Fig. 2. Four views of the proposed general-purpose hardware robotic platform: (top-left) top view; (top-right) side view; (bottom-left) front view; (botom-right) rear view.

3.2 Programming Framework

There are several operating systems and programming frameworks that can be used for the *Raspberry Pi Zero* micro-computer. Among all possibilities for the operating system, we recommend to use *Raspbian*, a free operating system based on *Debian*, optimised for the *Raspberry Pi* hardware. *Raspbian* comes with over 35,000 packages and is the recommended operating system for normal use on a *Raspberry Pi*. Other popular options are *Snappy Ubuntu*, *Pidora*, *Arch Linux ARM*, *Gentoo Linux*, *FreeBSD*, or *RISC OS Pi*.

Regarding the programming framework, the choice is strongly dependent on the programming language used for coding. In our implementation, we propose to use *Phyton* running on *Wing IDE*. We remark that *Wing IDE* is not directly supported by the *Raspberry Pi*, although it is possible to set up *Wing IDE* on a computer connected to the *Raspberry Pi* to generate and debug *Python* code remotely. Other possibilities include *Kivy*, *GTK+*, *PyQt*, and *Glade*.

4 Support Features for Swarm Intelligence Algorithms

A major ingredient of all SI methods (not only those in this paper) is a population of unsophisticated individuals with the ability to compute their own positions according to some evolution equations and communicate such positions and other information to other members of the swarm. All these relevant features are incorporated in our proposal. We consider a swarm of individuals that are identical replicas of the hardware robotic platform described in previous section. This micro-robot includes all hardware components and features required to implement the previous methods for swarm robotics at full extent:

- Its *Raspberry Pi Zero W* micro-processor is powerful enough to perform all the required computations in real time for all methods described. Although this feature can arguably be accomplished with (cheaper) micro-controllers (e.g., *Arduino*), our proposal is much more powerful. Instead of running one program many times as micro-controllers do, we have a general-purpose full-fledged computer running on Linux and with the ability to run multiple programs in a complex way. Consequently, we have no limitations in terms of the SI algorithms to be implemented, the number of individuals in the swarm (even multiple swarms are easily supported) and the number and complexity of tasks the robotic units are assigned.
- All SI methods require *global positioning*, which is computed with the magnetometer and its built-in digital compass, so that the robot can determine its current position with suitable accuracy for many practical applications.
- All SI methods require *communication capabilities*. Built-in-board *Wifi* and *Bluetooth* interfaces are included in our robots, allowing them to communicate useful information (e.g., position) to other members of the swarm.
- Our robots support *collision avoidance*. This feature is not included in the SI methods, but it is a must for swarm robotics, where robots moving in a real environment can collide with static and dynamic objects. Our robotic units are equipped with several sensors (ultrasounds, infrared, camera), global positioning (magnetometer and digital compass), and communication features (*Wifi*, *Bluetooth*) to avoid collisions with obstacles and other robots.
- Different *sensors* can be used to cope with the specific features and needs required by each SI method, as follows:
  - The PSO method relies on positions, velocities and orientations. Changes in velocity and position can be computed through an accelerometer, while changes in orientation and rotational velocity can be measured through a gyroscope. These sensors are not currently embedded into the *Raspberry Pi* board, but they can easily be attached to it. In our case, we rely on the servomotors to compute distance and velocity.
  - The work in [15] proved that some ultrasound sensors are ideal for the bat algorithm. We propose to use the same sensor in this approach.
  - In this paper, we argue that infrared sensors are adequate for the firefly algorithm. Infrared sensors can be used to measure distances, as required by the algorithm. Furthermore, since these sensors can also determine

how "bright" the light is, we can go even further with the nature-based metaphor and use them to embed the concepts of brightness and attractiveness by hardware instead of computing them exclusively by software.
- Following the same rationale, a combination of some of our sensors (possibly including the mini-camera) can also be used for the CSA.

5 Conclusions and Future Work

This paper introduces a general-purpose hardware robotic platform suitable for swarm robotics. Our approach is based on a careful choice of its main hardware components (computing unit, sensors, communication interfaces) to support the most popular swarm intelligence algorithms by hardware. Our design has also a very flexible and modular architecture so that it can be adapted to different swarm intelligence methods and many applications with minimal (if any) modifications. As an illustration, four examples of popular swarm intelligence methods (particle swarm optimization, firefly algorithm, bat algorithm, and cuckoo search algorithm) are considered in this paper. The most important hardware features of our approach to support such methods (and arguably many other swarm intelligence approaches as well) for swarm robotics are discussed.

Future work includes the consideration of other swarm intelligence approaches with the (possible) addition of extra components to support them. On the other hand, we plan to apply our robotic platform to real-world problems that could benefit from the swarm robotics principles and techniques. Reducing the size while maintaining (or even increasing) the performance of our robotic platform is also part of our plans for future work in the field.

Acknowledgements. Research supported by: project PDE-GIR of the EU Horizon 2020 research and innovation program, Marie Sklodowska-Curie grant agreement No 778035; project #TIN2017-89275-R of Agencia Estatal de Investigación and EU Funds FEDER; and project #JU12, of SODERCAN and EU Funds FEDER.

References

1. Arkin, C.R.: Behavior-Based Robotics. MIT Press, Cambridge (1998)
2. Arvin, F., Murray, J.C., Licheng S., Zhang, C., Yue, S.: Development of an autonomous micro robot for swarm robotics. In: Proceedings of IEEE International Conference on Mechatronics and Automation – ICMA 2014, pp. 635–640 (2014)
3. Bonabeau, E., Dorigo, M., Theraulaz, G.: Swarm Intelligence: From Natural to Artificial Systems. Oxford University Press, New York (1999)
4. Delice, Y., Aydogan, E.K., Ozcan, U., Ilkay, M.S.: A modified particle swarm optimization algorithm to mixed-model two-sided assembly line balancing. J. Intell. Manuf. **28**(1), 23–36 (2017)
5. Engelbrecht, A.P.: Fundamentals of Computational Swarm Intelligence. Wiley, Chichester (2005)

6. Faigl, J., Krajnik, T., Chudoba, J., Preucil, L., Saska, M.: Low-cost embedded system for relative localization in robotic swarms. In: Proceedings of IEEE International Conference on Robotics and Automation - ICRA 2013, pp. 993–998 (2013)
7. Kennedy, J., Eberhart, R.C.: Particle swarm optimization. In: IEEE International Conference on Neural Networks, Perth, Australia, pp. 1942–1948 (1995)
8. Kennedy, J., Eberhart, R.C., Shi, Y.: Swarm Intelligence. Morgan Kaufmann Publishers, San Francisco (2001)
9. Nouiri, M., Bekrar, A., Jemai, A., Niar, S., Ammari, A.C.: An effective and distributed particle swarm optimization algorithm for flexible job-shop scheduling problem. J. Intell. Manuf. **29**(3), 603–615 (2018)
10. Osaba, E., Yang, X.S., Diaz, F., Lopez-Garcia, P., Carballedo, R.: An improved discrete bat algorithm for symmetric and asymmetric traveling salesman problems. Eng. Appl. Artif. Intell. **48**, 59–71 (2016)
11. Osaba, E., Yang, X.S., Fister Jr., I., Del Ser, J., Lopez-Garcia, P., Vazquez-Pardavila, A.J.: A discrete and improved bat algorithm for solving a medical goods distribution problem with pharmacological waste collection. Swarm Evol. Comput. (in Press)
12. Sahin, E.: Swarm robotics: from sources of inspiration to domains of application. In: Swarm Robotics. Lecture Notes in Computer Science, vol. 3342, pp. 10–20 (2005)
13. Suárez, P., Iglesias, A.: Bat algorithm for coordinated exploration in swarm robotics. Adv. Intell. Syst. Comput. **514**, 134–144 (2017)
14. Suárez, P., Gálvez, A., Iglesias, A.: Autonomous coordinated navigation of virtual swarm bots in dynamic indoor environments by bat algorithm. Lecture Notes in Computer Science, vol. 10386, pp. 176–184 (2017)
15. Suárez, P., Iglesias, A., Gálvez, A.: Make robots be bats: specializing robotic swarms to the bat algorithm. Swarm Evol. Comput. (in Press)
16. Tan, Y., Zheng, Z.Y.: Research advance in swarm robotics. Def. Technol. J. **9**(1), 18–39 (2013)
17. Wagner, I., Bruckstein, A.: Special issue on ant robotics. Ann. Math. Artif. Intell. **31**(1–4), 113–126 (2001)
18. Wang, H., Yang, D., Yu, Q., Tao, Y.: Integrating modified cuckoo algorithm and creditability evaluation for QoS-aware service composition. Knowl. Based Syst. **140**, 64–81 (2018)
19. Yang, X.S.: Firefly algorithms for multimodal optimization. Lectures Notes in Computer Science, vol. 5792, pp. 169–178 (2009)
20. Yang, X.S.: Firefly algorithm, stochastic test functions and design optimisation. Int. J. Bio Inspired Comput. **2**(2), 78–84 (2010)
21. Yang, X.S.: Nature-Inspired Metaheuristic Algorithms, 2nd edn. Luniver Press, Frome (2010)
22. Yang, X.S.: Engineering Optimization: An Introduction with Metaheuristic Applications. Wiley, New York (2010)
23. Yang, X.S.: A new metaheuristic bat-inspired algorithm. Studies in Computational Intelligence, vol. 284, pp. 65–74 (2010)
24. Yang, X.S.: Bat algorithm for multiobjective optimization. Int. J. Bio Inspired Comput. **3**(5), 267–274 (2011)
25. Yang, X.S.: Bat algorithm: literature review and applications. Int. J. Bio Inspired Comput. **5**(3), 141–149 (2013)
26. Yang, X.S., Deb, S.: Cuckoo search via Lévy flights. In: Proceedings of the World Congress on Nature & Biologically Inspired Computing, NaBIC, pp. 210–214. IEEE Press (2009)

27. Yang, X.S., Deb, S.: Engineering optimization by cuckoo search. Int. J. Math. Model. Numer. Optim. **1**(4), 330–343 (2010)
28. Yang, X.S., Gandomi, A.H.: Bat algorithm: a novel approach for global engineering optimization. Eng. Comput. **29**(5), 464–483 (2012)

Internet of Things

A Deep Learning Approach to Device-Free People Counting from WiFi Signals

Iker Sobron[1(✉)], Javier Del Ser[2,3], Iñaki Eizmendi[1], and Manuel Velez[1]

[1] University of the Basque Country (UPV/EHU), 48013 Bilbao, Spain
{iker.sobron,inaki.eizmendi,manuel.velez}@ehu.eus
[2] TECNALIA, University of the Basque Country (UPV/EHU), Bizkaia, Spain
javier.delser@tecnalia.com
[3] Basque Center for Applied Mathematics (BCAM), Bizkaia, Spain

Abstract. The last decade has witnessed a progressive interest shown by the community on inferring the presence of people from changes in the signals exchanged by deployed wireless devices. This non-invasive approach finds its rationale in manifold applications where the provision of counting devices to the people expected to traverse the scenario at hand is not affordable nor viable in the practical sense, such as intrusion detection in critical infrastructures. A trend in the literature has focused on modeling this paradigm as a supervised learning problem: a dataset with WiFi traces and their associated number of people is assumed to be available a priori, which permits to learn the pattern between traces and the number of people by a supervised learning algorithm. This paper advances over the state of the art by proposing a novel convolutional neural network that infers such a pattern over space (frequency) and time by rearranging the received I/Q information as a three-dimensional tensor. The proposed layered architecture incorporates further processing elements for a better generalization capability of the overall model. Results are obtained over real WiFi traces and compared to those recently reported over the same dataset for shallow learning models. The superior performance shown by the model proposed in this work paves the way towards exploring the applicability of the latest advances in Deep Learning to this specific case study.

Keywords: Device-free people counting · Internet of Things
Convolutional neural network · Deep Learning

1 Introduction

Non-invasive detection of human presence through the analysis of the propagation effects of radio-frequency signals has lately grasped a great deal of attention within the research community due to their straightforward utility in IoT applications such as intrusion detection and tracking [12,23], vital signs monitoring

J. Del Ser et al. (Eds.): IDC 2018, SCI 798, pp. 275–286, 2018.
https://doi.org/10.1007/978-3-319-99626-4_24

[1,14], elderly or children in care homes [19,29] and emotion identification [21], among many others [18]. Specifically, one person in motion close to a transmitter-receiver wireless communication link can perturb the propagated signal through the link in terms of refractions, reflections or fading, which may yield noticeable fluctuations in the received signal. Measurements of the channel quality information at the receiver can be exploited and analyzed for localization, tracking or gesture recognition.

Non-intrusive detection systems relying on this principle can exploit received signal strength indicators (RSSI) or channel state information (CSI) at the receiver as representative fingerprints of the wireless communication link. RSSI is generally employed due to its simplicity and low hardware requirements. However, their values are coarse information since the frequency-selective multipath information cannot be exploited. For that purpose, channel-related information (CSI) captured from the subcarriers of an orthogonal frequency-division multiplexing (OFDM) system is often alternatively employed [5]. Feature extraction combined with ML models have garnered most of the research activity in counting device-free people during the last decade. After retrieving a number of CSI traces from received packets for a predefined time window, feature extraction can be performed over time, frequency, Doppler-spectrum and other domains [29]. Since CSI fluctuations arise from a combined effect of several physical phenomena which depend on specific conditions of the environment, in general it is difficult to design a global set of predictors whose validity can be extrapolated to several different scenarios. In other words, feature engineering often produces a set of predictors whose predictive relevance is stringently circumscribed to the characteristics of the scenario at hand. The derivation of sufficiently general features for device-free people counting is a paradigm that remains unsolved to date.

In this context, the advent of Deep Learning architectures noted in the last few years has allowed to undertake predictive modeling problems on *raw* data by processing them through complex neural network architectures over space and time. This has been possible by virtue of new neuron models, which permit to preserve well-established learning algorithms for traditional neural networks (i.e., backpropagation) even if neurons differ significantly from the perceptron on which these predictive models were originally based. As such, Convolutional Neural Networks (CNN) combine locally neighboring regions of their inputs, making them specially suitable for image classification and analysis (e.g. segmentation). On the other hand, Recurrent Neural Networks (RNN) hinge on neuron models capable of detecting and remembering correlations over arbitrarily long input sequences, for which reason they are renowned predictive models for problems defined on time series data. Interestingly, these two broad families of Deep Learning models can be also combined into a single hybrid neural network encompassing convolutional and recurrent layers, of utmost utility when patterns to be captured span over space and time (e.g. video classification).

The portfolio of application scenarios where Deep Learning has shown an unprecedented performance in complex prediction tasks includes people counting over image data [32,33], and more recently, with WiFi-based CSI data. For instance, the authors in [27] apply Deep Learning to CSI-based indoor positioning using a fully connected neural network with four hidden layers. They also extract calibrated phase information from CSI data to train a network with three fully connected hidden layers for indoor positioning [28]. In contrast, [17] employs convolutional layers tailored for multichannel CSI data that extract meaningful features based on the findings of prior CSI studies for detecting events and states of indoor everyday objects. It is worth mentioning that in this latter work, Deep Learning models are utilized for automatically learning not only the pattern between features and the target class, but also the features themselves.

This paper joins this research line and contributes with a CNN proposal to infer the relationship between the time variability of CSI values over a time window and the number of people in the surroundings of the wireless communication link. To this end, raw I/Q (In-phase and Quadrature) data captured from the physical interface over time is rearranged and processed as if it were an image whose dimensions depend on both the amount of OFDM subcarriers and the duration of the time window. Our proposed approach can be regarded as a similar strategy to the use of single-dimensional CNN (also referred to as 1-CNN) to address predictive problems over time sequences wherein samples correlated with the target variable to be predicted are assumed to be close in time. The performance of the proposed model is assessed over a recently contributed public dataset for device-free people counting (`EHUCOUNT` [24]), which has so far been tackled by shallow machine learning models [25]. The obtained results, quantified in terms of nested cross-validated scores, show that the proposed CNN model outperforms significantly traditional supervised learning models. This promising research direction paves the way towards extending this study to hybrid CNN-RNN models, so as to capture correlations between raw signals and the number of people in the scene over arbitrarily long time intervals.

The rest of the manuscript is structured as follows: for the sake of completeness the literature related to device-free people counting is reviewed in Sect. 2, along with some preliminary material on CSI data for people counting. Next, details on the proposed CNN model are provided in Sect. 3. Section 4 presents the experimental setup and the utilized dataset, whereas Sect. 5 discusses on the obtained results. Finally, Sect. 6 concludes the paper by outlining future research lines.

2 Preliminaries and Related Work

In typical indoor scenarios, radio signals usually reach the receiver antenna after several reflections, scattering, and attenuation. As a result, each multi-path signal component is characterized by a different time delay, amplitude attenuation, and phase shift. Thus, the received signal can be expressed as

$$h(t) = \sum_{i=1}^{N} \alpha_i(t) e^{-j\phi_i(t)} \delta(\tau - \tau_i(t)), \tag{1}$$

where N denotes the number of multi-path signal components, and $\alpha_i(t)$, $\phi_i(t)$ and $\tau_i(t)$ represent the amplitude, phase, and time delay of the ith multi-path component, respectively. The Fast Fourier Transform (FFT) of $h(t)$ corresponds to the so-called *channel frequency response* or PHY layer CSI, which can be easily computed in OFDM systems through channel estimation processes. Most WiFi standards such as IEEE 802.11a/g/n/ac utilize OFDM waveforms. As a result, complex-valued CSI traces are necessarily estimated for channel equalization in the reception process. The complex-valued CSI of an OFDM symbol with K subcarriers sampled at time slot n is given by

$$\mathbf{H}(n) = [H_1(n), \ldots, H_K(n)], \tag{2}$$

where $H_i(n) \doteq |H_i(n)|e^{j\angle H_i(n)}$ with $|H_i(n)|$ and $\angle H_i(n)$ being the amplitude and phase of the CSI at the i-th subcarrier, respectively.

The presence of people around WiFi receivers shall modify the channel response at symbol time-varying amplitude, phase and time delay of the N paths in (1), thus leading to temporal variations in the amplitude and phase of the CSI samples in (2). Based on such fluctuations along the time domain different features can be extracted from CSI and fed to a ML model for estimating the current number of people present at the surrounds of the WiFi receivers. Such predictors can be computed in a very diverse fashion within a general processing flow. In [25], the authors reviewed different CSI-based features and ML models employed for human activity recognition in IoT environments, emphasizing on those gravitating on crowd counting and tracking problems. In this context, supervised schemes such as support vector machines (SVMs) [15, 20, 29, 30], Random Forest (RF) [31] or Logistic Regression [31] achieve in average terms notable performance scores at the expense of an important computational effort.

3 Proposed CNN Model for Device-Free People Counting

As anticipated in the introduction to this work, our proposed model relies on a multi-layer CNN architecture fed with a $K \times T \times 2$ tensor containing the data captured over T consecutive time instants and K OFDM subcarriers. In this data arrangement the number of *channels* in the tensor correspond to the I and Q components of the physical wireless signal captured in the receiver frontend (i.e. real and imaginary parts of $H_i(n)$ values). This tailored data structure permits to operate on the image in a similar fashion to image data, so that the CNN network described in what follows can learn correlation among neighboring regions in space and time.

Figure 1 depicts the specific CNN structure proposed in this work. The model essentially comprises a sequential concatenation of CNN layers: the first is composed by 5 different convolutional filters of size 3×3, whereas the second increases the number of filters to 10 while maintaining their dimensions. Nonlinear activation by a Rectifier Linear Unit (ReLu) and Max Pooling are placed in between both CNN layers. The former allows for a better gradient propagation during the training process of the model, whereas the second allows down-sampling the

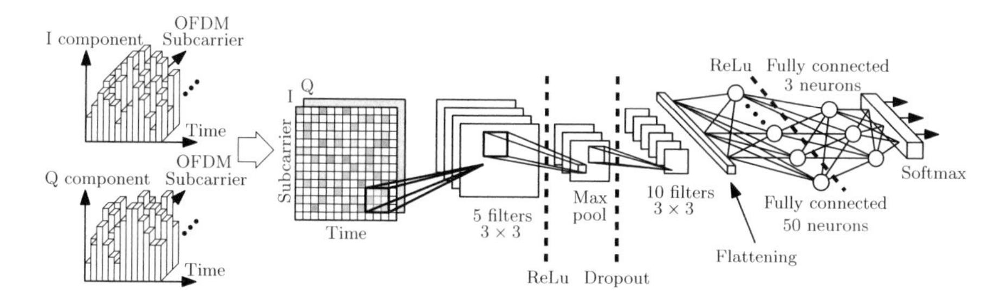

Fig. 1. Proposed CNN architecture particularized for a people counting scenario producing a supervised dataset with at most 2 passing people.

dimensionality of the output from the first CNN layer. In particular, each image is downscaled by a factor equal to $(2,2)$ (horizontal and vertical, respectively), with a stride equal to $(2,2)$. Overfitting is kept to a minimum by inserting a small amount of Dropout right after the pooling layer. The output of the second convolutional layer is serialized and input to a fully connected neural layer with 50 standard output neurons (linear perceptron). The output of this first layer is connected to a second fully connected layer through ReLu activation. The number of output neurons for this second layer is set equal to the number of possible people (target variable) in the dataset at hand. Softmax activation is applied to the output of this second neural layer so as to produce class probabilities for every predicted sample.

The overall model accounts to 253911 trainable parameters for the case with highest number of classes in the considered `EHUCOUNT` dataset (6 different labels), including connection weights, biases and filter coefficients. To iteratively adjust their value, gradient information of a loss function measured over the training data available for each scenario is backpropagated through the neural structure. This gradient information is used to iteratively adjust (*move*) such parameters in the proper direction indicated by the propagated gradient, for which a solver from the family of gradient descend optimization techniques is used. In particular we will employ an Adam solver with learning rate equal to 0.01. The choice of this optimization algorithm was made due to its good properties in terms of computational efficiency and easiness of implementation [13]. Since we are facing a multi-class classification problem, categorical cross-entropy is the selected loss function to be minimized through backpropagation, which is given by:

$$H(\mathbf{y},\widehat{\mathbf{y}}) = -\sum_{n=1}^{N_{tr}}\sum_{c=1}^{C}\mathbb{I}(y_n = c)\log P_n^c, \tag{3}$$

where $\mathbf{y}$ is the true class corresponding to the N_{tr}-sized training set, $\widehat{\mathbf{y}}$ their labels predicted by the developed model, P_n^c denotes the probability that the n-th tested example belongs to class c as told by the SoftMax activation layer, and $\mathbb{I}(\cdot)$ is an auxiliary function that takes value 1 if its argument is true (and 0 otherwise). For a more efficient training procedure, backward and forward passes are run over batches of 32 training examples, repeating the entire process of a total of 70 epochs.

Before proceeding with the experimental section of this work, it is important to note that the above design for the proposed CNN model has been pursued empirically by following good practices reported by the community in this regard. Recently the use of Evolutionary Computation has been postulated as a promising tool to automate the design and construction of complex Deep Learning architectures over massive datasets [8,16,26], with software already made available by the community. This research area can be regarded as a legacy of previous developments around the so-called Neuroevolution paradigm [9], yet at higher complexity levels due to the upsurge of new processing layers concurring in the Deep Learning realm. However, in this work we opted for a handcrafted deep learning architecture. Early trials with neuro-evolutionary packages resulted in Deep Learning architectures prone to overfitting which, after regularization, did not perform as good as the handcrafted approach described above. Nevertheless, automating the design process deserves further research efforts in the future, as clearly outlined in the section closing this manuscript.

4 Experimental Setup

In order to assess the performance of the proposed CNN architecture when learning from real WiFi data, a number of computational experiments have been carried out over the `EHUCOUNT` dataset [24], which provides full CSI from all $K = 53$ subcarriers of IEEE 802.11n signals recorded during 15 seconds over the 2.4 GHz band. The number of CSI traces per class and scenario ranges between 12000 and 15000, depending on synchronization issues at the signal decoding process.

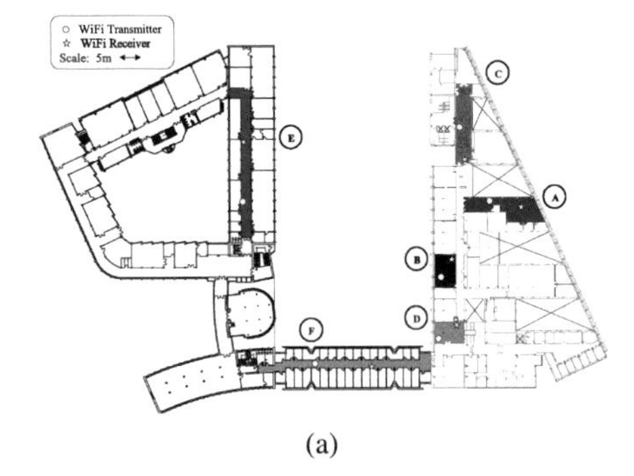

(a)

Scenario	People (#)
A (Office)	[0,1,2,3]
B (Lab)	[0,1,2,3,4,5]
C (Corridor)	[0,1,2,3,4]
D (Hall+Stairs)	[0,1,2,3,4,5]
E (Corridor)	[0,1,2,3,4]
F (Corridor)	[0,1,2,3,4,5]

(b)

Fig. 2. Physical location and number of people per scenarios (A to F) contained in the `EHUCOUNT` dataset.

Measurements have been carried out in six indoor scenarios where up to 5 people walked casually in the colored areas depicted in Fig. 2.a. In general, people tended to maintain the direction for a while before changing it according to the scenario topology. That is, one direction is maintained for a longer distance in corridors such as scenarios C, E and F. In contrast, people roamed around the rooms in scenarios A, B and D. In all cases, people were instructed to walk by the area of interest at a speed in the typical range for indoor walking dynamics (i.e. ≤ 3 km/h). Details about the number of people that took part in the measurements at each scenario are given in Table 2.b. More details about the dataset can be found in [25].

The CNN model has been compared with the following benchmark of *shallow* supervised learning models, which provide the highest scores as reported in [25]:

- Linear Discriminant Analysis (LDA), which seeks a linear combination of its input features so as to separate among instances belonging to different classes. To this end the classifier models the class conditional distribution of the data as a multivariate Gaussian distribution, whose mean and covariance matrices are estimated from the training data for every class [3].
- Random Forest (RF), an ensemble of fully grown decision tree classifiers that reduces variance in the prediction error by feeding each constituent tree classifier with a bootstrap and averaging (voting) their produced predictions [4].
- Gradient Boosting Classifier (GBC), another ensemble of decision tree learners whose learning algorithm relies on a sequential arrangement of such learners. In short, every classifier along the processing chain aims to improve the ensemble composed by the already trained models located in previous stages of the boosting chain [10, 11].
- Support Vector Machine (SVM, νSVM), which essentially hinge on computing a maximum-margin separating hyperplane over a highly-dimensional, nonlinear projection of the feature space in which the dataset is expected to be linearly separable [7, 22].

The above models are fed with a set of engineered features that conform to the state of the art in this field. For this purpose, we have selected a variety of features computed in time, frequency and Doppler spectrum domains over the provided CSI dataset. Time windows of 50 OFDM symbols have been used for the feature extraction in the time domain. As a result, a total of 374 features have been employed as shown in Table 1.

A nested cross-validation (CV) methodology has been adopted to compute different performance scores over each simulated case. Specifically, the measurements for every 2-tuple (scenario, model) are partitioned in $M_{outer} = 4$ equal parts or folds. Then, one of such M_{outer} parts is left apart for testing, and the parameters of the model are tuned by performing a stratified K_{inner}-fold cross-validation over the remaining $M_{outer} - 1$ folds, where M_{inner} is set to 4 in all cases. The parameter tuning is done based on a fine-grained value grid of the hyper-parameters of the model at hand, whose optimality is given by the average values of two scores computed over the M_{inner} folds as follows:

1. Accuracy, given by the ratio of correctly classified examples to the total number of validation/test samples.

2. Cohen's Kappa Coefficient, which measures the agreement between two datasets by accounting for the probability that data could coincide just by chance [6]. It is worth to note that this coefficient removes any interpretative bias due to an imbalanced class distribution in the datasets. A fair level of agreement between the predicted and the true number of people is given by values of this coefficient above 0.2, with 0.8 acknowledged as the minimum for declaring a *very good* agreement [2].

Table 1. Engineered features for the supervised learning models considered in the benchmark.

Features	Number of samples
Time-windowed amplitude-based eigenvalues	50
Time-windowed phase-based eigenvalues	50
Time-windowed mean of CSI amplitudes	53
Time-windowed variance of CSI amplitudes	53
Time-windowed skewness of CSI amplitudes	53
Time-windowed kurtosis of CSI amplitudes	53
Time-windowed inter-quartile range of CSI amplitudes	53
Doppler spectrum mean	1
Doppler spectrum variance	1
Doppler spectrum centroid	1
2nd order moment of Doppler spectrum	1
Doppler spectrum skewness	1
Doppler spectrum kurtosis	1
Time-frequency-windowed mean-std deviation ratio of CSI energy	1
Time-frequency-windowed skewness of CSI energy	1
Time-frequency-windowed kurtosis of CSI energy	1

For each model, a prediction is obtained employing the fold left for testing. The same model tuning and testing procedure is repeated by iterating on the remaining $M_{outer} - 1$ folds, after which an average score can be computed from the results obtained for every fold. Then the overall process is replicated for 20 repetitions by randomly reorganizing the measurements within the dataset so as to remove any influence of the partitioning on the performance of the models.

5 Results and Discussion

In Table 2, we show the mean and standard deviation statistics of the accuracy and Kappa scores computed over the 20 repetitions of the nested cross-validation method for the feature-based models and the CNN model in the different environments of the EHUCOUNT dataset. The table includes results obtained (1) with the entire set of features in Table 1 fed to every model, and (2) with a selected feature subset resulting from the application of a feature selection (FS) method prior to model training. A model-based approach is adopted for FS; that is, a RF model is first built over the training dataset with all features included, from which those accounting for a *relevance* greater than the median across all predictors are retained as the selected feature subset. The so-called relevance is computed as the average impurity decrease of the feature when it is utilized to grow a compounding tree of the RF model. The CNN model takes raw CSI traces without feature engineering in time windows of 50 OFDM symbols.

Table 2. Accuracy and Cohen's Kappa Coefficient (first and second row of every cell, respectively), without and with FS. Results are expressed as mean±std computed over nested CV with $M_{inner} = M_{outer} = 4$ folds and 20 repetitions.

			Model					
			LR	RF	GBC	μSVM	SVM	CNN
EHUCOUNT scenar-ios	A	NFS	0.68 ± 0.01	0.71 ± 0.01	0.71 ± 0.01	0.69 ± 0.01	0.69 ± 0.01	0.78 ± 0.01
			0.57 ± 0.01	0.61 ± 0.01	0.61 ± 0.01	0.58 ± 0.01	0.59 ± 0.01	
		FS	0.73 ± 0.01	0.70 ± 0.00	0.70 ± 0.01	0.75 ± 0.01	0.76 ± 0.01	0.72 ± 0.02
			0.64 ± 0.01	0.62 ± 0.01	0.60 ± 0.02	0.65 ± 0.01	0.65 ± 0.02	
	B	NFS	0.69 ± 0.01	0.72 ± 0.01	0.76 ± 0.00	0.71 ± 0.01	0.71 ± 0.01	0.79 ± 0.01
			0.62 ± 0.01	0.66 ± 0.01	0.71 ± 0.01	0.66 ± 0.01	0.66 ± 0.01	
		FS	0.74 ± 0.01	0.73 ± 0.00	0.76 ± 0.01	0.75 ± 0.01	0.75 ± 0.01	0.74 ± 0.01
			0.69 ± 0.01	0.68 ± 0.01	0.72 ± 0.01	0.68 ± 0.01	0.70 ± 0.01	
	C	NFS	0.63 ± 0.02	0.71 ± 0.00	0.71 ± 0.01	0.64 ± 0.01	0.64 ± 0.01	0.77 ± 0.01
			0.54 ± 0.01	0.65 ± 0.01	0.64 ± 0.01	0.55 ± 0.01	0.55 ± 0.01	
		FS	0.74 ± 0.01	0.73 ± 0.01	0.72 ± 0.01	0.72 ± 0.00	0.72 ± 0.02	0.71 ± 0.01
			0.67 ± 0.01	0.67 ± 0.01	0.65 ± 0.01	0.66 ± 0.01	0.69 ± 0.01	
	D	NFS	0.40 ± 0.01	0.55 ± 0.01	0.53 ± 0.01	0.48 ± 0.00	0.48 ± 0.00	0.59 ± 0.02
			0.28 ± 0.01	0.46 ± 0.01	0.44 ± 0.01	0.38 ± 0.01	0.38 ± 0.01	
		FS	0.46 ± 0.01	0.57 ± 0.01	0.54 ± 0.00	0.54 ± 0.01	0.54 ± 0.00	0.50 ± 0.02
			0.35 ± 0.01	0.48 ± 0.01	0.45 ± 0.01	0.43 ± 0.01	0.43 ± 0.01	
	E	NFS	0.72 ± 0.01	0.79 ± 0.00	0.81 ± 0.01	0.78 ± 0.01	0.80 ± 0.01	0.87 ± 0.02
			0.65 ± 0.01	0.74 ± 0.01	0.76 ± 0.01	0.73 ± 0.01	0.75 ± 0.01	
		FS	0.77 ± 0.01	0.80 ± 0.01	0.81 ± 0.01	0.80 ± 0.01	0.85 ± 0.01	0.83 ± 0.03
			0.72 ± 0.01	0.75 ± 0.01	0.76 ± 0.01	0.75 ± 0.01	0.79 ± 0.01	
	F	NFS	0.53 ± 0.01	0.61 ± 0.01	0.58 ± 0.01	0.58 ± 0.01	0.59 ± 0.01	0.77 ± 0.02
			0.44 ± 0.01	0.54 ± 0.01	0.50 ± 0.00	0.50 ± 0.01	0.51 ± 0.01	
		FS	0.58 ± 0.01	0.64 ± 0.01	0.60 ± 0.01	0.65 ± 0.02	0.66 ± 0.01	0.72 ± 0.04
			0.50 ± 0.01	0.57 ± 0.01	0.53 ± 0.01	0.56 ± 0.01	0.59 ± 0.01	

Taking a look at Table 2, we can firstly observe that performance indicators of the CNN model outperform those given by the feature-based ML models. In this sense, deep learning shows a great potential to cope with raw CSI data. The Kappa scores obtained with CNN for every scenario are above 0.5, which indicate a fairly good agreement between the true and predicted number of people. Additionally, except for D scenario, scores maintain an average Kappa around 0.7, which yield good prediction results for different scenarios (i.e. an accuracy close to 80%). Best scores are notably lower for D scenario, for which scores of the same model degrade down to an average accuracy of 59% and an average Kappa of 0.5.

If we compare the feature-based models with the DL model, we can see that FS provides closer performance indicators to those achieved with the CNN model. For instance, SVM with FS in the A scenario achieves an accuracy of 76% compared with 78% for CNN; 76% (GBC FS) vs 79% in B or 85% (SVM FS) vs 87% in the E scenario. This performance comparison with the feature-based models reinforces the physical characteristics of the scenario under study have more relevant influence on the ultimate performance of the model when feature engineering is employed.

6 Conclusions

This paper has elaborated on estimating the number of people traversing a given physical environment by analyzing the perturbations imprinted by the people themselves on the wireless signal emitted from a WiFi transmitter to a receiver. Specifically, we have addressed this problem from a modeling perspective based on a supervised set of prior examples captured over the scenario under analysis. Traditionally machine learning models have been used to learn the pattern between the WiFi signal captured at the receiver frontend and the actual number of people in the scenario. In this work we have proposed a novel Deep Learning architecture comprising two convolutional processing layers, Max Pooling and two fully connected neural layers, further incorporating elements aimed to avoid overfitting. The rationale for the designed architecture is to override any need for engineering predictors from the captured signals (in the form of e.g. amplitude and spectrum statistics), so that the model directly operates on the raw I/Q signals.

The proposed CNN scheme has been compared with feature-based ML models with real WiFi traces contained in the `EHUCOUNT` dataset, over which the following performance indicators are measured: accuracy and Cohen's Kappa coefficient. Nested cross-validation is used for the sake of a fair comparison between the models in the benchmark. Results are conclusive: the proposed CNN model outperforms shallow supervised learning models recently used in the literature with this dataset [25], even when FS is performed. These empirical findings underscore the great potential of Deep Learning models for device-free people counting, as they overcome the close dependence of traditional classifiers on the quality and predictive power of the features from which they learn.

Future research will be devoted to automating the construction of the Deep Learning model. Given the high number of structural degrees of freedom of this kind of models (not only their number of layers, but also their types and parameters), we plan to study different evolutionary approaches lately emerging in the community to optimize this structure. Furthermore, the transferability of the learned knowledge in one given scenario to other environments will be addressed given the renowned suitability of Deep Learning models to this paradigm by simply avoiding any parameter update (*layer freezing*) of the initial layers along the processing pipeline.

Acknowledgements. This work was supported in part by the Spanish Ministry of Economy and Competitiveness under project 5GnewBROS (TEC2015-66153-P MINECO/FEDER, EU) and by the Basque Government (IT683-13 and the EMAITEK program).

References

1. Abdelnasser, H., Harras, K.A., Youssef, M.: Ubibreathe: a ubiquitous non-invasive WiFi-based breathing estimator. In: 16th ACM International Symposium on Mobile Ad Hoc Networking and Computing, pp. 277–286. ACM (2015)
2. Altman, D.G.: Practical Statistics for Medical Research. Chapman and Hall/CRC, London (1990)
3. Balakrishnama, S., Ganapathiraju, A.: Linear discriminant analysis-a brief tutorial. Institute for Signal and information Processing **18**, 1–8 (1998)
4. Breiman, L.: Random forests. Mach. Learn. **45**(1), 5–32 (2001)
5. Cianca, E., De Sanctis, M., Di Domenico, S.: Radios as sensors. IEEE Internet Things J. **4**(2), 363–373 (2017)
6. Cohen, J.: A coefficient of agreement for nominal scales. Educ. Psychol. Meas. **20**(1), 37–46 (1960)
7. Cortes, C., Vapnik, V.: Support-vector networks. Mach. Learn. **20**(3), 273–297 (1995)
8. David, O.E., Greental, I.: Genetic algorithms for evolving deep neural networks. In: Companion Publication of the 2014 Annual Conference on Genetic and Evolutionary Computation, pp. 1451–1452. ACM (2014)
9. Floreano, D., Dürr, P., Mattiussi, C.: Neuroevolution: from architectures to learning. Evol. Intell. **1**(1), 47–62 (2008)
10. Friedman, J.H.: Greedy function approximation: a gradient boosting machine. Ann. Stat. **29**, 1189–1232 (2001)
11. Friedman, J.H.: Stochastic gradient boosting. Comput. Stat. Data Anal. **38**(4), 367–378 (2002)
12. Joshi, K.R., Bharadia, D., Kotaru, M., Katti, S.: Wideo: fine-grained device-free motion tracing using RF backscatter. In: NSDI, pp 189–204 (2015)
13. Kingma, D.P., Ba, J.: Adam: a method for stochastic optimization. arXiv preprint arXiv:14126980, published at ICLR 2015 (2014)
14. Liu, J., Wang, Y., Chen, Y., Yang, J., Chen, X., Cheng, J.: Tracking vital signs during sleep leveraging off-the-shelf WiFi. In: 16th ACM International Symposium on Mobile Ad Hoc Networking and Computing, pp. 267–276. ACM (2015)
15. Lv, J., Yang, W., Gong, L., Man, D., Du, X.: Robust WLAN-based indoor fine-grained intrusion detection. In: 2016 IEEE Global Communications Conference (GLOBECOM), pp 1–6 (2016)
16. Miikkulainen, R., Liang, J., Meyerson, E., Rawal, A., Fink, D., Francon, O., Raju, B., Shahrzad, H., Navruzyan, A., Duffy, N., et al.: Evolving deep neural networks. arXiv preprint arXiv:170300548 (2017)
17. Ohara, K., Maekawa, T., Matsushita, Y.: Detecting state changes of indoor everyday objects using Wi-Fi channel state information. Proc. ACM Interact Mob Wearable Ubiquitous Technol. **1**(3), 88:1–88:28 (2017). https://doi.org/10.1145/3131898

18. Oppermann, F.J., Boano, C.A., Römer, K.: A decade of wireless sensing applications: survey and taxonomy. In: The Art of Wireless Sensor Networks, vol 1: Fundamentals, pp. 11–50. Springer, New York, NY, USA (2014)
19. Pu, Q., Gupta, S., Gollakota, S., Patel, S.: Whole-home gesture recognition using wireless signals. In: 19th Annual International Conference on Mobile Computing & Networking, pp. 27–38. ACM (2013)
20. Qian, K., Wu, C., Yang, Z., Liu, Y., Zhou, Z.: PADS: passive detection of moving targets with dynamic speed using PHY layer information. In: 2014 20th IEEE International Conference on Parallel and Distributed Systems (ICPADS), pp 1–8 (2014)
21. Raja, M., Sigg, S.: Applicability of RF-based methods for emotion recognition: a survey. In: 2016 IEEE International Conference on Pervasive Computing and Communication Workshops (PerCom Workshops), pp. 1–6. IEEE (2016)
22. Schölkopf, B., Bartlett, P.L., Smola, A.J., Williamson, R.C.: Shrinking the tube: a new support vector regression algorithm. In: Advances in Neural Information Processing Systems, pp 330–336 (1999)
23. Seifeldin, M., Saeed, A., Kosba, A.E., El-Keyi, A., Youssef, M.: Nuzzer: a large-scale device-free passive localization system for wireless environments. IEEE Trans. Mob. Comput. **12**(7), 1321–1334 (2013)
24. Sobron, I., Del Ser, J., Eizmendi, I., Velez, M.: `EHUCOUNT` dataset (2017). www.ehu.eus/tsr_radio/index.php/research-areas/data-analytics-in-wireless-networks. Accessed 30th Nov 2017
25. Sobron, I., Del Ser, J., Eizmendi, I., Velez, M.: Device-free people counting in IoT environments: new insights, results and open challenges. IEEE Internet Things J. ,1 (2018). https://doi.org/10.1109/JIOT.2018.2806990, early Access
26. Such, F.P., Madhavan, V., Conti, E., Lehman, J., Stanley, K.O., Clune, J.: Deep neuroevolution: genetic algorithms are a competitive alternative for training deep neural networks for reinforcement learning. arXiv preprint arXiv:171206567 (2017)
27. Wang, X., Gao, L., Mao, S., Pandey, S.: Deepfi: deep learning for indoor fingerprinting using channel state information. In: IEEE Wireless Communications and Networking Conference (WCNC), pp 1666–1671 (2015). https://doi.org/10.1109/WCNC.2015.7127718
28. Wang, X., Gao, L., Mao, S.: CSI phase fingerprinting for indoor localization with a deep learning approach. IEEE Internet Things J. **3**(6), 1113–1123 (2016). https://doi.org/10.1109/JIOT.2016.2558659
29. Wang, Y., Wu, K., Ni, L.M.: Wifall: device-free fall detection by wireless networks. IEEE Trans. Mob. Comput. **16**(2), 581–594 (2017)
30. Wu, C., Yang, Z., Zhou, Z., Liu, X., Liu, Y., Cao, J.: Non-invasive detection of moving and stationary human with WiFi. IEEE J. Sel. Areas Commun. **33**(11), 2329–2342 (2015)
31. Zeng, Y., Pathak, P.H., Mohapatra, P.: Analyzing shopper's behavior through WiFi signals. In: Proceedings of the 2nd Workshop on Workshop on Physical Analytics, WPA 2015, New York, NY, USA, pp. 13–18. ACM (2015)
32. Zhang, C., Li, H., Wang, X., Yang, X.: Cross-scene crowd counting via deep convolutional neural networks. In: Proceedings of the IEEE Conference on Computer Vision and Pattern Recognition, pp. 833–841 (2015)
33. Zhang, Y., Zhou, D., Chen, S., Gao, S., Ma, Y.: Single-image crowd counting via multi-column convolutional neural network. In: Proceedings of the IEEE Conference on Computer Vision and Pattern Recognition, pp. 589–597 (2016)

Evolutionary Algorithms for Design of Virtual Private Networks

Igor Kotenko[1,2(✉)] and Igor Saenko[1,2]

[1] St. Petersburg Institute for Informatics and Automation of the Russian Academy of Sciences, 14-th Liniya, 39, St. Petersburg 199178, Russia
{ivkote,ibsaen}@comsec.spb.ru

[2] ITMO University, 49, Kronverkskiy prospekt, St. Petersburg, Russia

Abstract. Virtual Private Networks (VPNs) is a most known technology to create the protected communication links via the Internet. The paper offers a new approach to solve the problem of VPN network design based on evolutionary algorithms (genetic and differential evolution). The joint accounting of network bandwidth, reliability and cost, which indices are calculated on the basis of the offered queuing theory models, is the feature of the considered problem. The experimental assessment of the suggested decisions shows that the evolutionary algorithms can improve the VPN network efficiency up to 40% in comparison with the standard variants of its creation. The comparative assessment of the suggested evolutionary algorithms shows higher convergence of the differential evolution algorithm.

Keywords: Genetic algorithm · Differential evolution · Security
Virtual Private Network · Reliability

1 Introduction

The problem of creation of protected computer networks based on open communication channels in the Internet acquires great value now. VPN technology has high-priority among the most known security means supporting the protected information exchange via the Internet. This technology realizes idea of the secured "tunnel" created on open communication channels. On the initial and terminal ends of the "tunnel" the transferred messages are encrypted and decrypted. Messages are transferred in a ciphered form in the "tunnel" [1, 2]. The process of creation of the protected "tunnel" via open Internet channels is as follows. Crypto-routers are VPN nodes. By means of open Internet channels, the crypto-routers are connected with each other, forming structure of some topology. We will call the open Internet channels connecting crypto-routers and transferring IP packets between them as *IP channels*. The ciphered IP packets are transmitted on IP channels along the preset route through others crypto-routers in the direction of final crypto-router that is on the end of the "tunnel". The route can be simple, consisting of one IP channel (simple VPN channel), or composite, including other crypto-routers and IP channels connecting them (composite VPN channel). All VPN channels form the protected network which is laid out over an open network of IP channels of the Internet.

J. Del Ser et al. (Eds.): IDC 2018, SCI 798, pp. 287–297, 2018.
https://doi.org/10.1007/978-3-319-99626-4_25

The problem of searching the efficient VPN topology is important, and it attracts attention of many researchers for a long time. Such VPN properties as its bandwidth, reliability and cost are considered as the main optimization criteria. However, the approaches used to solve this problem earlier, for example, well-known "hose model" [3–5], are oriented on the accounting of the restrictions imposed on the most admissible value of bandwidth of IP channels between subscribers. Such problem statement becomes irrelevant now. The problem solved in the paper, on the contrary, has no restrictions on the maximum bandwidth of individual IP channels. Total network bandwidth appears as the main criterion. Indices of reliability and cost are additional criteria that are considered. Optimization task formulation is non-linear. By computational complexity, this task belongs to the class of NP-complete tasks. To solve the above optimization task the paper proposes to apply two kinds of evolutionary algorithms – the genetic algorithm (GA) and the differential evolution algorithm (DEA). Besides, the paper shows possibility to solve the task of the VPN design. The paper also outlines the comparative assessment of GA and DEA and proves higher convergence of DEA. These two results make *the main theoretical contribution.*

The paper has the following structure. Section 2 overviews related work. Section 3 presents the mathematical foundations for calculation of target indicators. Evolutionary algorithms for solving the problem are described in Sect. 4. The discussion of experimental results is presented in Sect. 5 and, at last, the conclusions and directions for further research are outlined in Sect. 6.

2 Related Work

The works on the VPN design are appeared in the late nineties. One of the first paper on this subject is [2]. It offered the problem definition called "hose model for VPNs". In this statement it was required to define the optimum network topology if the matrix of connectivity of subscribers (VPN terminals) and the maximum channel bandwidth are set. The cost index served as an optimization criterion. Some approaches were developed to solve the task defined by the hose model. [6] offered the approach based on partition of the set of VPN terminals into a set of groups. In this case the terminals of each group are not connected with each other. [7] suggested to apply an approach based on the partial-integer programming. The optimization criterion was the minimum of the cost to create the VPN structure. Possibility to use the reserve way in case of refusal of one of the main terminals was considered in [8]. The minimum of the general bandwidth was suggested as the optimization criterion. The VPN with ring topology was considered in [9]. All tasks were consolidated to the tasks of linear programming. A set of heuristics was offered to solve these tasks. [10] suggested an approach based on the analysis of a polyhedron which edges are network arcs. These tasks were consolidated to the linear form that allowed to offer efficient algorithms for their decision. On the basis of application of Hu's 2-commodity flow theorem, [11] received a new lower bound value of the optimal solution of the problem. To solve the task, authors offered a new randomized approximation algorithm. For VPN design on optical networks, [12] offered the approach based on linear integer programming. The compromise between the capacity efficiency and computation time was selected as the task criterion. [13] researched the problem of VPN design on mobile networks. As the task solution criterion the overall network performance was selected. The problem of designing VPN using cloud space was considered in [14] where solutions of the company "Juniper Networks" are described. The network security, performance and reliability were selected as the main properties defining the task solution criteria. The carried-out analysis of related works allows us to draw the following conclusions. The considered problem definitions are generally oriented not on end users, but on network service providers. As the task solutions were the routes which pass on the network controlled by provider. In the paper, on the contrary, the problem definition is oriented on the end user. The routes passing through the provider's responsibility zone determined by the Internet boundaries are not considered. It is required to find a network topology fitting to the given needs on the bandwidth, reliability and cost.

3 Mathematical Foundations

Basic types of topology of interconnected VPN channels are presented in Fig. 1. They are fully connected (Fig. 1a), radial (Fig. 1b), and ring (Fig. 1c) types.

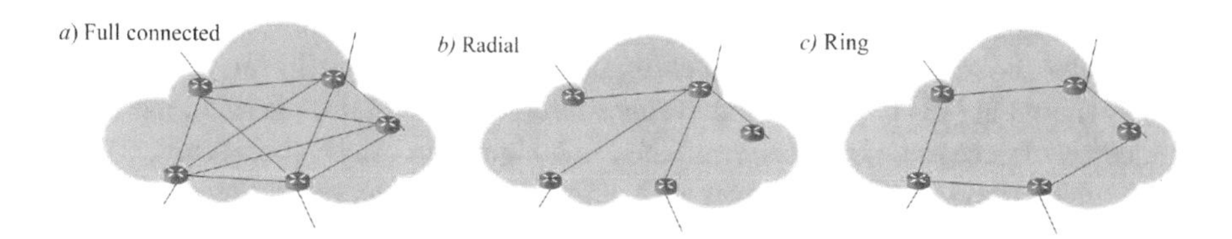

Fig. 1. Basic types of VPN topology.

Each type of network topology has advantages and disadvantages that affect the values of key performance indicators – bandwidth, reliability and cost (Table 1). In practice, a mixed topology should be implemented. In this case, the compromise combination of the values that meet the requirements shall be achieved.

Table 1. Compatible assessment for basic types of VPN topology

Indicators	Types of topology		
	Fully connected	Radial	Ring
Bandwidth	High	Middle	Middle
Reliability	High	Low	Middle
Cost	High	Low	Middle

The *initial data* for the VPN design is as follows: (1) the matrix of data flows, (2) the performance of crypto-routers, (3) the bandwidth of IP channels, (4) intensity of information flows between possible couples of subscribers. The *variables* of the problem are the elements of the matrix $\mathbf{X} = \{x_{ij}\}$ such that $x_{ij} = 1$, if between the local networks i and j there is a simple VPN channel, and $x_{ij} = 0$ otherwise. The *objective function* of the problem is a function that considers need for indexes of the VPN quality to reach the following best values: (1) *Bandwidth* must be maximal, (2) *Reliability* must be maximal, (3) *Cost* must be minimal.

We offer the normalized functions which values belong to an interval from 0 to 1 to assess the VPN quality indexes. We will designate such normalized functions as follows: F^{cap} is a function to assess the bandwidth (capacity), F^{rel} is a one to assess the reliability, F^{cost} is a one to estimate the cost. We will consider that for all three normalized functions the next condition is fair: the higher the value of the function, the better the value of the appropriate VPN property. The best option of the VPN creation is reached when all three normalized functions are equal to '1'.

Then *objective function* can be represented as a weighed additive convolution

$$\begin{aligned} &F \to \max, \\ &F = \alpha F^{cap} + \beta F^{rel} + \gamma F^{cost}, \end{aligned} \tag{1}$$

where α, β, and γ (α, β, γ >0) are the weight coefficients defining the comparative significance of F^{cap} and F^{rel} relatively to F^{cost}.

The objective function determined by expression (1) is non-linear. Besides, its variables are integers '0' or '1'. It allows us to claim that the task is NP-complete and requires for its solution to use the heuristic methods, as the methods of traditional mathematical programming are not effective for this task.

To solve the task we offer to use evolutionary algorithms, in particular, GA and DEA. GA, as shown in [15–17], allows us to solve complex optimization tasks in the field of generation of access control schemes. The DEA algorithm can be considered as a kind of GA which provides higher convergence of the algorithm because during evolutionary development the mutation operator automatically adapts to possible lowering of convergence [18]. Thus, the statement of the problem is in finding the maximum of the function (1) which is a generalized index of VPN quality and represents a convolution of individual indicators for the main network properties, i.e. bandwidth, reliability and cost. Operation of the evolutionary algorithms consists in an iterative evolutionary change of the originally created set of possible solutions. It is carried out to select the better solutions. Criterion of such selection is so-called *fitness function* which corresponds to expression (1). F^{cap}, F^{rel}, and F^{cost} are composite components of the fitness function.

We propose to estimate the bandwidth of VPN networks depending on values of the elements of the matrix **X**. Let us consider that the bandwidth of the VPN network is determined by the sum of the bandwidths of all its VPN channels [19]. To estimate the VPN bandwidth a model based on queuing theory is suggested [20]. Then, we select the following expression to calculate the utilization of the composite VPN channel between nodes i and j containing n simple channels:

$$\rho_{ij} = 1 - \frac{1}{1 + n\lambda_{ij}\left(2\nu + \mu - 3\lambda_{ij}\right)/\left(\left(\mu - \lambda_{ij}\right)\left(\nu - \lambda_{ij}\right)\right)}, \tag{2}$$

where ρ_{ij} is the load at the input for the composite VPN channel between nodes i and j, μ is a service intensity of crypto-rioters, and ν is a service intensity of IP channels. Thus, if we know the number of transits for each composite VPN channel and the load on each composite VPN channel, then we can find the bandwidth of the entire VPN network. It is defined as follows:

$$\mathrm{M}_{\mathrm{VPN}} = \sum_{l=1}^{n} \sum_{j=1}^{n} \frac{\lambda_{ij}}{\rho_{ij}}, \tag{3}$$

where ρ_{ij} is calculated according to (2).

To build the complex performance indicator we require a dimensionless indicator. As such indicator, we suggest a VPN utilization coefficient:

$$F^{\mathrm{cap}} = \sum_{i=1}^{n} \sum_{j}^{n} \lambda_{ij} / \mathrm{M}_{\mathrm{VPN}}. \tag{4}$$

Thus, the value of the index F^{cap} is always in the range [0; 1] and can be used to build a comprehensive performance indicator of the VPN network along with other indicators that have a probabilistic nature.

The reliability index should be dependent on the following factors: the VPN network topology specified by the matrix **X**, the probability $p_{\mathrm{fail},i}$ of failure of the crypto-router CR #*i* and the time $T_{\mathrm{recovery},i}$ of its recovery. If the VPN channel is fully loaded during time [0; *T*], the reliability index acts as the readiness index:

$$F^{\mathrm{rel}} = 1 - T_{\mathrm{recovery}} / T. \tag{5}$$

We assume that the load measure is the value $P^{-}_{VPN-ij,k}$, which lies in the range [0; 1]. In this case, the following expression will be used as a fitness indicator, characterizing the stability of the VPN channel between the *i*-th and *j*-th nodes of the network, in case there was a failure of the *k*-th node:

$$F^{\mathrm{rel}}_{ij,k} = 1 - \mathrm{P}^{-}_{\mathrm{VPN}-ij,k} \cdot T_{\mathrm{recovery},k} / T. \tag{6}$$

The physical meaning of the indicator (6) is the availability of the corresponding VPN channel. The following expression is proposed as the integral indicator of the reliability of the VPN network:

$$F^{\mathrm{rel}} = \left(\sum_{k=1}^{n} p_{\mathrm{fail},k} \sum_{i=1}^{n} \sum_{j=1}^{n} F^{\mathrm{rel}}_{ij,k} \right) / \sum_{k=1}^{n} p_{\mathrm{fail},k}. \tag{7}$$

The expression (7) is the additive convolution of indicators (6), characterizing the reliability of individual channels. The values $p_{\mathrm{fail},i}$ are used as weighting coefficients that are the probabilities of failure of individual crypto-routers.

The cost of the network should be dependent on the following factors: (1) *Cost of exploitation* of the network, (2) *Cost of renting* the IP network channels. The *Cost of exploitation* of the network is determined by the need to periodically change the key data on pairs of crypto routers, located at the ends of the simple VPN channel. Let us introduce the next parameters:

(1) $\mathrm{T} = \left\| \tau_{ij} \right\|$ is a matrix of frequencies of key data changes during [0; T],
(2) C_{key} is a cost of change of key data. Then the total cost of exploitation of the entire network is as follows:

$$F^{\mathrm{expl}} = C_{\mathrm{key}} \sum_{i=1}^{n} \sum_{j=1}^{n} \tau_{ij} x_{ij}, \tag{8}$$

where x_{ij} is an element of the matrix **X**. The expression

$$F^{\mathrm{rent}} = \sum_{i=1}^{n} \sum_{j=1}^{n} C_{\mathrm{IP},ij} x_{ij} \tag{9}$$

is suggested as an indicator of *Cost of renting* of the IP channels, where $C_{\mathrm{IP},ij}$ is the cost to rent one IP channel between nodes i and j during the period of time [0; T]. Both indicators, expressed by (8) and (9), can be combined into one indicator:

$$F^{\mathrm{cost}} = 1 - \left(\sum_{i=1}^{n} \sum_{j=1}^{n} \left(C_{\mathrm{key}} \cdot \tau_{ij} + C_{\mathrm{IP},ij} \right) \cdot x_{ij} \right) / \sum_{i=1}^{n} \sum_{j=1}^{n} \left(C_{\mathrm{key}} \cdot \tau_{ij} + C_{\mathrm{IP},ij} \right). \tag{10}$$

Thus, the model discussed above allows us to calculate the bandwidth, reliability and cost of the VPN as functions of the matrix **X**. It permits us to suggest the algorithms for solving the problems of VPN design presented in the next section.

4 Evolutionary Algorithms

Evolutionary algorithms (EAs), which we suggest to use for the VPN design task, are a kind of bio-inspired optimization methods [18]. Among EAs, to solve the tasks of discrete non-linear integer programming the task (1) belongs, the genetic algorithms (GA) and differential evolution algorithms (DEA) are best suited.

We will consider GA essence based on generalization of the modern views on its operation [21–24]. GA assumes execution of three main operations: selection, crossing and mutation. Crossing operation forms new solutions by exchanging parts of chromosomes, selected randomly from the population. Mutation randomly changes the genes selected from a population chromosome. Selection removes from the population the chromosomes with the worst values of fitness function. Operation of GA terminated when the population reached a steady state. To create the chromosome it is offered to use the elements of the matrix **X** lying above the main diagonal. As the matrix **X** is symmetric, it is enough to consider the elements located above the principal diagonal. Thus, the following type of the chromosome is offered:

$$[\mathbf{X}]_{\text{xp}} = \left[x_{12}, \ldots, x_{1N}; x_{23}, \ldots, x_{2N}; \ldots x_{N-1,N}\right], \tag{11}$$

where $x_{ij}(i, j = 1, \ldots, N)$ are the elements of the matrix $\mathbf{X}$.

To create the fitness function we use the expression (1). As a result we offer the following kind of the fitness function:

$$\textit{Fitness Function}\,(\mathbf{X}) = \alpha\, F^{\text{cap}}(\mathbf{X}) + \beta\, F^{\text{rel}}(\mathbf{X}) + \gamma\, F^{\text{cost}}(\mathbf{X}). \tag{12}$$

By means of weight coefficients, the direction of search of the task solution is determined by the set preference between bandwidth, reliability and cost of the VPN. If, for example, between the network properties the preference $F^{\text{cap}} \succ F^{\text{rel}} \succ F^{\text{cost}}$ is defined (i.e. high bandwidth is more preferable than reliability, and reliability is more preferable than cost) then $\alpha = 0.16$, $\beta = 0.30$, and $\gamma = 0.54$.

DEA, offered in [25], is similar to GA, as for search of the optimum solution it also uses the generations of individuals. In addition, the structures of chromosomes at both algorithms are identical. The main distinction between GA and DEA is that in GA the mutation is a result of little changes in chromosomes, while in DEA the mutation is a result of arithmetical combinations of solutions. Therefore, the mutation in DEA has no predetermined probability density function. DEA is characterized by simplicity of implementation, fast convergence, reliability and accuracy.

In its original form DEA can be described as follows. A set of the vectors called the population is initially generated. A new population of vectors is generated as follows. For each vector of the previous population (initial vector) three vectors belonging to the same population are selected in a random way. Over them a set of operations is executed; the result of these operations is a so-called mutant vector. Between mutant and initial vectors the crossing is executed. The result of crossing is a probe vector. If the probe vector is better, than initial one, the initial vector is replaced with the probe one. Otherwise the initial vector remains in population, and the probe vector is destroyed. In the offered DEA the mutation attempt is carried out over each vector of the current population. With probability P_{mut} each coordinate of this vector is changed on opposite value, creating a mutant vector. Further the basic DEA scheme works. With probability P_{cross} the coordinates of the mutant vector are replaced with coordinates of the initial vector, and basic selection is executed further.

5 Experimental Results

Experiments pursued two aims. The first one consisted in confirmation of possibility to solve the task (1) by means of evolutionary algorithms. The second aim consisted in carrying out a comparative assessment of the GA and DEA offered for the task solution. As the scenario for carrying out experiments a critical information infrastructure containing 10 local networks, that were necessary to communicate with each other via the Internet using VPN technology, was selected. For calculations the software written in the Delphi language was developed.

The solution of the optimization task (1) was carried out for three variants of initial data with different intensities of input loading and corresponding preference of target indices. Intensities had following values: 10, 100, or 1000. Variant 1 had 17 links between nodes, variant 2 and variant 3 – 29. Table 2 depicts values of preferences between components of the objective function (all components changed their values from 0.1 to 0.5) and the values of weight coefficients of the fitness function corresponding to them for different variants of the VPN.

Table 2. Preferences between the objective function components.

Variants	Preferences			Weigh coefficients		
	First	Second	Third	α	β	γ
Variant 1	F^{cost}	F^{rel}	F^{cap}	0.54	0.30	0.16
Variant 2	F^{cap}	F^{cost}	F^{rel}	0.16	0.54	0.30
Variant 3	F^{rel}	F^{cap}	F^{cost}	0.30	0.16	0.54

Results of the optimization task solution for different variants of initial data are depicted in Fig. 2. The dotted lines in Fig. 2 show new channels that were not available in previous variants.

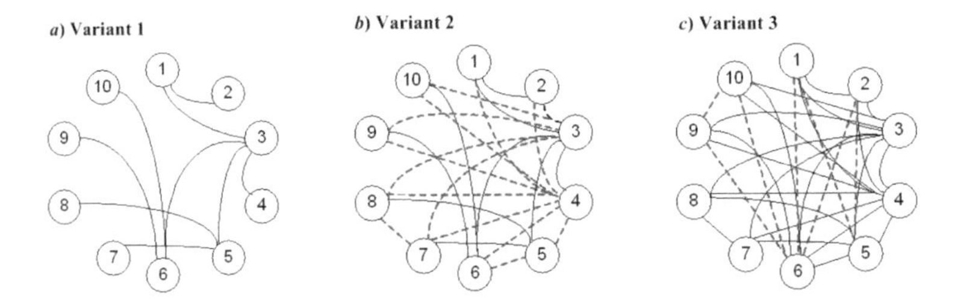

Fig. 2. Solutions for different variants.

As it can be seen from Fig. 2a, the matrix of minimum ranks, corresponding to the solution for Variant 1, contains the maximum rank equal to 3 (for example, for subscribers 2 and 10). The network consists of 9 simple VPN channels. For the second variant (Fig. 2b), the maximum rank in the matrix of minimum ranks is equal to 2. The network contains 25 direct channels. For the third variant (Fig. 2c), the matrix **X** defines several routes for all information directions.

The efficiency was calculated according (1) for three boundary variants of the VPN topology (Table 3). The VPN topologies obtained by EA have efficiency greater than for typical structures in all variants. The gain is from 10 to 40%.

The comparative assessment of the evolutionary algorithms GA and DEA was carried out for all three variants of initial data. For all algorithms identical parameters were used: the number of individuals in the population is equal 200, P^{cross} is equal 0.1, P^{mut} is equal 0.01. For each algorithm and each variant four experiments were carried out. As a result of these experiments the average number of executed generations was calculated.

Table 3. Comparative assessment of variants.

Variant	VPN topology			
	Fully connected	Radial	Ring	EA
1	0.64	0.45	0.36	0.76
2	0.62	0.68	0.58	0.80
3	0.54	0.68	0.74	0.88

The results of comparative assessment of algorithms GA and DEA are given in Table 4. For all variants the convergence of DEA is higher, than the convergence of GA. The gain is from 24.7% (variant 1) to 35.6% (variant 3).

Table 4. Comparative assessment of algorithms.

Variants	Average number of generations		Profit, %
	DEA	GA	
1	55.50	73.75	24.7
2	78.25	110.50	29.1
3	112.75	175.25	35.6

In general, the results of the experiments confirm the conclusion that the convergence of DEA is higher, than the convergence of GA, with identical accuracy of the solution of the task.

6 Conclusion

The paper suggests the new approach to solve the problem of the VPN design in which fragments of the distributed computer infrastructure are connected by the protected channels. The approach is based on GA and DEA. Variables are represented by the elements of the matrix of mutual connectivity of the VPN nodes. Fitness functions allow us to estimate the properties of bandwidth, reliability and cost for each possible variant of the VPN design and find the solution of the problem. This is the feature of the approach allowing us to expand the scope of the task solutions. Experiments showed that the optimal VPN structures, obtained on the basis of the proposed evolutionary algorithms, have gain from 10 to 40% compared to standard variants. The comparative assessment of GA and DEA demonstrated that the convergence of DEA was higher than one of GA. The gain reached 35%. Further research is related to consider heterogeneous crypto-routers and IP channels in VPN design.

Acknowledgment. This work was supported by grants of RFBR (projects No. 16-29-09482, 18-07-01369 and 18-07-01488), by the budget (the project No. AAAA-A16-116033110102-5), and by Government of Russian Federation (Grant 08-08).

References

1. Rehman, M.H.: Design and implementation of mobility for virtual private network users. Glob. J. Comput. Sci. Technol. Netw. Web Secur. **13**(9), 34–39 (2013)
2. Yurcik, W., Doss, D.: A planning framework for implementing virtual private networks. IT Prof. **3**, 41–44 (2001)
3. Duffield, N.G., Goyal, P., Greenberg, A.: A flexible model for resource management in virtual private networks. SIGCOMM Comput. Commun. Rev. **29**(4), 95–108 (1999)
4. Duffield, N.G., Goyal, P., Greenberg, A., Mishra, P., Ramakrishnan, K.K., Van der Merwe, J.E.: Resource management with hoses: point-to-cloud services for virtual private networks. IEEE/ACM Trans. Netw. **10**(5), 679–692 (2002)
5. Raghunath, S., Ramakrishnan, K.K.: Resource management for virtual private networks. Commun. Mag. **45**(4), 38–44 (2007)
6. Eisenbrand, F., Grandoni, F., Oriolo, G., Skutella, M.: New approaches for virtual private network design. In: Automata, Languages and Programming. Lecture Notes in Computer Science, vol. 3580, pp. 1151–1162. Springer (2005)
7. Srikitja, A., Tipper, D.: QoS-based Virtual Private Network Design for an MPLS network. http://www.pitt.edu/~dtipper/Apaper2002_1.pdf. Accessed 10 Apr 2018
8. Italiano, G.F., Rastogi, R., Yener, B.: Restoration Algorithms for Virtual Private Networks in the Hose Model. http://citeseerx.ist.psu.edu/viewdoc/summary?doi=10.1.1.16.7686. Accessed 10 Apr 2018
9. Hurkens, C.A.J., Keijsper, J.C.M., Stougie, L.: Virtual private network design: a proof of the tree routing conjecture on ring networks. SIAM J. Discrete Math. **21**(2), 482–503 (2004)
10. Altin, A., Amaldi, E., Belotti, P., Pinar, M.C.: Virtual private network design under traffic unce. Electron. Notes Discret. Math. **17**, 19–22 (2004)
11. Eisenbrand, F., Happ, E.: Provisioning a virtual private network under the presence of non-communicating groups. In: Algorithms and Complexity. Lecture Notes in Computer Science, vol. 3998, pp. 105–114. Springer (2006)
12. Haque, A., Ho, P.-H.: A study on the design of survivable optical virtual private networks (O-VPN). IEEE Trans. Reliab. **55**(3), 516–524 (2006)
13. Natarajan, M.C., Muthiah, R., Nachiappan, A.: Performance investigation of virtual private networks with different bandwidth allocations. IJCSI Int. J. Comput. Sci. **7**(1), 58–63 (2010)
14. Secure Cloud Connectivity for Virtual Private Networks (white paper). Next-Generation Virtualized Managed Services for the Enterprise with Secure-on-Network Links to the Cloud, Juniper Networks, Inc. (2014)
15. Saenko, I., Kotenko, I.: A genetic approach for virtual computer network design. In: Intelligent Distributed Computing VIII. Studies in Computational Intelligence, vol. 570, pp. 95–105. Springer (2015)
16. Saenko, I., Kotenko, I.: Genetic algorithms for role mining problem. In: Proceeding of the 19th International Euromicro Conference on Parallel, Distributed and Network-based Processing, Ayia Napa, Cyprus, 9–11 February, pp. 646–650 (2011)
17. Saenko, I., Kotenko, I.: Design of virtual local area network scheme based on genetic optimization and visual analysis. J. Wirel. Mob. Netw. Ubiquitous Comput. Dependable Appl. (JoWUA) **5**(4), 86–102 (2014)
18. Binitha, S., Sathya, S.S.: A survey of bio inspired optimization algorithms. Int. J. Soft Comput. Eng. (IJSCE) **2**(2), 137–151 (2012)
19. Kotenko, I., Saenko, I.: The genetic approach for design of virtual private networks. In: Proceedings of the 2015 IEEE 18th International Conference on Computational Science and Engineering (CSE-2015), Porto, 21–23 October, pp. 168–175 (2015)

20. Kleinrock, L.: Queueing Systems. Wiley, New York (1975)
21. Koza, J.R.: Genetic Programming: On the Programming of Computers by Means of Natural Selection. The MIT Press, Cambridge (1992)
22. Goldberg, D.E.: Genetic Algorithms in Search, Optimization, and Machine Learning, 1st edn. Addison-Wesley, Boston (1989)
23. Mitchell, M.: An Introduction to Genetic Algorithms. MIT Press, Massachusetts (1998)
24. Saenko, I., Kotenko, I.: Genetic optimization of access control schemes in virtual local area networks. In: Computer Network Security. Lecture Notes in Computer Science, vol. 6258, pp. 209–216. Springer (2010)
25. Storn, R., Price, K.: Differential evolution – a simple and efficient heuristic for global optimization over continuous spaces. J. Global Optim. **11**, 341–359 (1997)

Forming Groups in the Cloud of Things Using Trust Measures

Giancarlo Fortino[1], Lidia Fotia[2], Fabrizio Messina[3], Domenico Rosaci[2], and Giuseppe M. L. Sarné[4(✉)]

[1] DIMES, University of Calabria, Rende, CS, Italy
giancarlo.fortino@unical.it
[2] DIIES, University "Mediterranea", Reggio Calabria, Italy
{lidia.fotia,domenico.rosaci}@unirc.it
[3] DMI, University of Catania, Catania, Italy
messina@dmi.unict.it
[4] DICEAM, University "Mediterranea", Reggio Calabria, Italy
sarne@unirc.it

Abstract. The need of managing complex and interactive activities is becoming a key challenge in the "Internet of Things" (IoT) and leads to request large hardware and power resources. A possibility of facing such a problem is represented by the possibility of virtualizing physical IoT environments over the so called Cloud-of-Things (CoT), where each device is associated with one or more software agents working in the Cloud on its behalf. In this open and heterogeneous context, IoT devices obtain significant advantages by the social cooperation of software agents, and the selection of the most trustworthy partners for cooperating becomes a crucial issue, making necessary to use a suitable trust model. The cooperation activity can be further improved by clustering agents in different groups on the basis of trust measures, allowing each agent will to interact with the agents belonging to its own group. To this purpose, we designed an algorithm to form agent groups on the basis of information about reliability and reputation collected by the agents. In order to validate both the efficiency and effectiveness of our approach, we performed some experiments in a simulated scenario, which showed significant advantages introduces by the use of the trust measures.

1 Introduction

A key issue for the current "Internet of Things" (IoT) devices is given by the increasing complexity requiring more and more performing hardware and power resources. To face such an important issue, the "Information and Communications Technology" (ICT) world is developing both new technologies and standards [1,2]. At the same time, Cloud Computing (CC) has emerged as a mainstream in processing and storing data to form knowledge ubiquitously accessible in distributed environments and in an interoperable fashion [3]. To guarantee

J. Del Ser et al. (Eds.): IDC 2018, SCI 798, pp. 298–308, 2018.
https://doi.org/10.1007/978-3-319-99626-4_26

the scalability of computational and storing requirements coming by an overwhelming number of IoT devices, IoT and CC converged to realize the so called Cloud-of-Things [4] (CoT).

Moreover, by associating IoT devices with software agents, working on their behalf over the Cloud [5], it is possible to realize the independence of the agents from the physical hardware and power capabilities of IoT devices and also exploiting their social attitudes. In this scenario, when an agent has not suitable information to choose a reliable partner then, similarly to real communities, it can ask information to other agents it considers as trustworthy. To this purpose, the intuition underlying our proposal is that of supporting this process by encouraging agents to form groups of reliable recommenders.

A common viewpoint considers that groups should be formed based on both structural and semantic similarities [6] but, given the potential heterogeneity of IoT devices, different criteria need. We observe that in forming groups within a community a high level of mutual trustworthiness among the group members is crucial for cooperation [7], similarly such groups can significantly improve the IoT activities.

Now, we introduce a CoT environment ($\mathbf{C}$) where heterogeneous devices (D), obtain/release services and/or extract/exchange knowledge assisted over the cloud by personal software agents (S). The agent network is represented by a graph $G = \langle N, L \rangle$, where N is the set of the nodes, each one associated with an agent $a \in A$, and L is the set of oriented links between two agents.

An agent may belong to one or more groups. Each group is managed by an administrator that can join with or to remove from the group those agents resulted ineffective. Within the environment $\mathbf{C}$, the agents can obtain some data services (s) made available by other agents only for pay. To achieve their goals, they may benefit from their past experiences but, if this was not enough, it could require the opinions of other agents. In other words, when the agent a_i has not a suitable direct past experience about a producer agent a_j, it can ask a recommendation $rec \in [0, 1] \subseteq \mathbb{R}$ to another agent a_r; rec is free if a_r belongs to the same group of a_i, otherwise a_j pays a fee to a_r after the recommendation was provided. At this point, a_i assigns a feedback $f_{i,r} \in [0, 1] \subseteq \mathbb{R}$ to each agent recommended by a_r that it used: $f_{i,r} = 1$ (resp. $f_{i,r} = 0$) means that it perceived the maximum (resp. minimum) level of satisfaction. The difference between the feedback $f_{i,r}$ and the corresponding recommendation $rec_{i,r}$, in absolute value, provides a measure of the helpfulness of this recommendation; the average of the measures provided by a_r represents its helpfulness. Obviously, groups are interested in accepting those agents having a high reliability and helpfulness; at the same time agents are interested to be affiliated with groups formed by agents with a high reliability and helpfulness. To evaluate the helpfulness of an agent we consider the effectiveness of its recommendations, while that of a group is the average of the helpfulness of its members.

In this scenario, we introduce a trust measure to model a distributed group formation process aimed at maximizing both the advantage of an agent to join with a group and that of a group in accepting a new member. To obtain the reputation of the agent in C, we do not consider only the direct experience (i.e., the reliability) of an agent and also the opinions coming from other agents. Obviously, we observe that a_i can interact with a limited number of other agents so the most part of the population in A are not aware of its trustworthiness. For this reason, in place of the global reputation we adopt a *local reputation* [8] approach based only on the opinions coming from the friends of an agent. This approach allows to eliminate heavy computational tasks and communication overloads to collect opinions and evaluate the trustworthiness of their sources and, at the same time, to increase the system reactivity.

As a second contribution, we also deal with the modality to combine trust values to take a decision about accepting or refusing an affiliation request to a group, we propose to form groups in a simple way by adopting a voting mechanism, where each vote is represented by a trust value obtained by a suitable combination of reliability and local reputation.

The rest of the paper is organized as follows. Section 2 gives an overview on the related literature. Section 3 describes the adopted local trust model and voting mechanism, while Sect. 4 presents the algorithm to form groups. The experimental results are presented in Sect. 5 and in Sect. 6 some conclusions are drawn.

2 Related Work

More in general, in real and virtual communities, several approaches based on a trust criterion deal with the problem of suggesting (*i*) to a group if accepting a candidate for the affiliation or (*ii*) to a member of a community the groups which are the most suitable for joining with. These problems are more known, respectively, as *group recommendation* and *group affiliation* problems. In [9] a trust-based agreement procedure is adopted in a P2P system to form groups on the basis of the agent trustworthiness. This because the expectations to be engaged in satisfactory interactions is higher among the members of trust-formed groups and a group receives more benefits when trusted relationships occur among its members.

To compute trust in social communities, the most important factors to consider are the informative sources [10], the aggregation rules [11] and the modalities for inferring trust [12], but also if trust is computed in a local or a global way and by means of a centralized or a distributed approach. Some studies found that the accuracy of local trust is greater when a personal viewpoint is used [13], while the computational cost tightly depends by the horizon depth [14]. The predominance of local trust is particularly true in large communities where each actor usually interacts with a narrowest share of the community so that the most part of it is unknown and unreferenced like their opinions.

In social contexts, trust processes occurring in a community are usually represented by a graph, named *trust network*, where nodes represent the members and oriented edges represent trust relationships. Different techniques use the topological properties of the trust networks as, for example, in [15] a variant of the Breadth First Search is adopted to gather the reputation scores and, by using a voting, to compute an updated reputation rate for each user, while in [16] updated trust scores are propagated only by using fixed length paths. Local trust approaches are adopted by the TidalTrust [17], the MoleTrust [18] and the TrustWalker [19] algorithms.

Another characteristic of our proposal is to adopt a voting to reach a decision within a group. Voting mechanisms [20], in human and virtual communities, optimize the social utility [21] and avoid conflicts [22] by giving equal dignity to different interests and opinions. However, in huge communities a global voting could be difficult, or unfeasible, to realize and in these cases a local voting might represent the best alternative [23]. Unfortunately, any "ideal" global or local voting procedure exists because all them can be affected by manipulation, like strategic vote [24]. This aspect is very critical for software agent communities, indeed agents can efficiently and effectively examine manifold manipulation opportunities [25], but this problem is assumed as orthogonal with respect to the focus of our proposal.

3 The Trust Model

In this section, we describe the trust model which supports the group formation procedure discussed in the next section. In our scenario, we represent the information about the agent trust through a graph on which a directed edge between two nodes (i.e., agents) represents the level of trust that an agent has in another agent. In this context, the ego-network of an agent $a_i \in A$ is represented by the sub-graph $E_i \subseteq G$, formed by those nodes (i.e., agents) connected to a_i by oriented paths included in a fixed horizon. Now, we introduce a measure called local trust which takes into account reliability, local reputation and helpfulness measures ranging in the real interval $[0, 1]$, where $0/1$ represents their minimum/maximum values. For a pair of generic nodes $i, j \in G$, the local trust $\tau_{i,j}$ that i has about j is defined as a combination of the reliability $\eta_{i,j}$ and the *local reputation* $\mu_{i,j}$.

The reliability η_{ij} represents the confidence that the agent a_i has about the capability, at a certain time, of the agent a_j of providing good suggestions. η_{ij} is computed by averaging all the q feedbacks $f_{i,j} \in [0,1] \in \mathbb{R}$ that a_i assigned to a_j at the end of the q interactions carried out with it. It is an asymmetric measure (i.e., $\eta_{i,j} \neq \eta_{j,i}$). More formally:

$$\eta_{i,j} = \frac{1}{q} \cdot \sum_{k=1}^{q} f_{i,j}^{\,k} \tag{1}$$

The *local reputation* $\mu_{i,j}$ measures how much, on average, the agents belonging to the ego-network of a_i (i.e., E_i), estimate the capability of a_j of having good interactions. At a certain time, the agent a_i accepted m recommendations provided to it by a_r about other agents and the average helpfulness of a_r perceived by a_i will be computed as $\chi_{i,r} = \frac{1}{m} \cdot \sum_{s=1}^{m} |f_{i,r}^{s} - rec_r^s|$, with respect to each of the all m accepted suggestions coming by a_r (i.e., rec_r). Moreover, we introduce a weight denoted as $\omega_{i,r} = 2^{-(\,\widehat{l}_{(i,r)}-1)}$ which considers the distance between a_i and the recommender agent a_r belonging to the ego-network E_i, in order to give less importance to those recommender agents which are more "far" from a_i. $rec_{r,j} \in [0,1]$ is the recommendation provided by an agent a_r about another agent a_j and $\chi_{i,r} \in [0,1]$ is the (average) helpfulness of an agent a_r (in providing reliable suggestions) as it is perceived by the agent a_i on the basis of its experience (i.e., by means of the feedbacks it released after having interacted with the agent recommended by a_r). Finally, $\widehat{l}_{(i,r)}$ is the shortest path between a_i and the generic recommender agent a_r. By assuming that a_i is able to exploit a number p of recommenders in its ego-network to receive recommendations about a_j, then $\mu_{i,j}$ is calculated as:

$$\mu_{i,j} = \frac{1}{p} \cdot \sum_{r=1}^{p} \left(\chi_{i,r} \cdot \omega_{i,r} \cdot rec_{r,j} \right) \tag{2}$$

At the end, we compute the trust measure that an agent a_i has about an agent a_j by combining reliability and local reputation as:

$$\tau_{i,j} = \delta_i \cdot \eta_{i,j} + (1 - \delta_i) \cdot \gamma_{i,j} \cdot \mu_{i,j} \tag{3}$$

where δ and γ are two parameters ranging in $[0,1] \in \mathbb{R}$. The parameter δ weights reliability and local reputation for giving more or less relevance to one or other. The parameter γ is computed as $\gamma_{i,j} = p/\|E_i(x)\|$ and takes into account the dependability of $\mu_{i,j}$ on the number of p nodes belonging to E_i that contributed to compute $\mu_{i,j}$ (this because if the number of these nodes is too small then a_i will not receive a sufficient amount of information about a_j from its ego-network and the local reputation measure loses of relevance). Note that in presence of a newcomer agent then suitable "cold start" values of reliability, reputation and helpfulness are adopted.

Moreover, the "trustworthiness" of a group g, as perceived by a_i, can be determined by simply averaging all the trust measures computed by a_i for all the users belonging to g. Similarly, the "trustworthiness" of an agent a_i, as perceived by a group g, can be determined by simply averaging all the trust measures about a_i computed by all the users belonging to g.

Now, we introduce a simple voting mechanism [26], based on the local trust measures. When a new member y must be added to the group g, all the agents belonging to g will express a vote $v \in 0, 1$ about the acceptation of y in g (e.g., 0/1 means "not accept"/"accept"). The vote will depend from the recommendations coming from its ego-network (i.e., local trust measure) and a suitable threshold $\Gamma_g \in [0,1]$. More in detail, the voting process referred to a group g for a potential new member y is represented by the voting criterion v (e.g., see formula 4), as the output of a function $V(g, v, y)$. For instance, a reasonable strategy may be that of accepting a requester into a group only if the majority of the members of g vote for its acceptance.

$$v_{i,y} = \begin{cases} 0 & \mathit{if}\ \tau_{i,y} < \Gamma_g \\ 1 & \mathit{if}\ \tau_{i,y} \geq \Gamma_g \end{cases} \tag{4}$$

4 The Distributed Grouping Algorithm

In this section, we describe the proposed distributed algorithm designed to maximize the benefits of the single agent to join with a group and, at the same time, the benefits of a whole group when it decides to accept a new member.

The first part of the algorithm is designed to be executed by the single agent a_i. In particular, a_i improves its own "configuration" of groups to join with a group in terms of overall mutual trust with its own peers (the pseudocode is shown in Fig. 1). H_i is the set of the groups to which the agent a_i is affiliated, W is a parameter representing the maximum number of groups that an agent can join with and M represents the maximum number of groups the generic agent is capable to analyze. Also, we suppose that a_i stores the local trust measure $\tau_{i,g}$ of each group $g \in H_i$ contacted in the past and the time $\hat{t}_g$ elapsed from the last execution of the procedure for that group. Moreover, ϵ_i is a time thresholds fixed by the agent a_i, and $\sigma_i \in [0, 1]$ is a threshold on the level of trust between the agent a_i and the generic group $g \in H_i$. First of all, the algorithm calculates the values of local trust $\tau_{i,g}$ whenever the existing values are older than the threshold ϵ_i (Lines 1–3). Then, a set of candidate groups S_c, with $|S_c| < W$, is built on the basis of the local trust of the groups, and sorted in decreasing order, based on the trust values $\tau_{i,g}$, while Y is a set of groups randomly chosen and, finally, the set Z is given by union of Y and H. The set S_c might contain some groups already belonging to the existing set H_i, while some others might be new groups that were selected at random and put into the set Y. Then, based on the groups in the set S_c which are not in the set H_i, the agent a_i will be able to improve the quality of its choices by joining with those groups. The two loops in Lines 6–18 represents the kernel of the procedure, after which $H_i = S_c$. The second of the algorithm is performed by the agent *administrator* of a group for evaluating a request to join with the group (the pseudocode is listed in Fig. 2).

Let K be the set of the agents affiliated to the group g, where $||K|| \leq R$, being R the maximum number of agents allowed to be affiliated with the group g, while the set X is the union of the set X with the agent a_i candidate to be affiliated with the group g. The agent administrator a_g of a group g stores the values of the local trust measures computed by the members of its group for the agent a_i which has sent the request to join with, and the timestamp $\tilde{t}_i$ of its retrieval. Also, ψ is a time threshold fixed by the administrator a_g. When the agent a_i sends a join request to a_g, the administrator a_g asks to the members of its group the updated local trust values about the agent a_i (Lines 1–5). At this case, two different situation may occur. If the size of X has not reached the maximum R (Line 6), all the agents in g are invited to provide vote (i.e. a personal preference). All the vote are combined by means of the function $V(\cdot)$.

Input:
$E_i, H_i, W, \epsilon_i, \sigma_i$;
$Y = \{g \in G\}$ a set of groups randomly selected : $|Y| = M \leq W, H_i \bigcap Y = \{0\}, \quad Z = (H_i \bigcup Y)$

```
1:  for g ∈ Z : î > ε_i do
2:      Compute τ_{i,g} by exploiting the agents belonging to E_i.
3:  end for
4:  m ← 0
5:  Let be S_c = {g ∈ Z : τ_{i,g} ≥ σ_i}, with |S_c| ≤ W
6:  for all g ∈ S_c : g ∉ H_i do
7:      send a join request to the agent administrator of g
8:      if g accepts the request then
9:          m ← m + 1
10:     end if
11: end for
12: for all g ∈ H_i : g ∉ S_c do
13:     Sends a leave message to g
14:     m ← m − 1
15:     if (m==0) then break
16:     end if
17: end for
```

Fig. 1. The algorithm executed by the agent a_i.

Input: $K, R, a_i, \psi, X = K \bigcup \{a_i\}$;

```
1:  for all k ∈ K do
2:      if t̃_i ≥ ψ then
3:          ask to k for the updated local trust values about a_i
4:      end if
5:  end for
6:  if |X| < R then
7:      if V(g, v, a_i) == 1 then
8:          Send an accept message to a_i
9:      else
10:         Send a reject message to a_i
11:     end if
12:     return
13: else
14:     for all k ∈ X do
15:         compute τ_{k,w}
16:     end for
17:     Let X' = {k_1, k_2, ..., k_{K+1}} with k_i ∈ X ⋃{a_i} and τ_{g,a_i} ≥ τ_{g,a_k} if i < k
18:     if X[K + 1] == a_i then
19:         Send a reject message to a_i
20:     else
21:         Send a leave message to the node X[K + 1]
22:         Send an accept message to a_i
23:     end if
24: end if
```

Fig. 2. The algorithm executed by an agent administrator of the group g.

Then, agent a_i is admitted or not in the group on the basis of the result of the function $V(\cdot)$ (Lines 7–12). Otherwise, if the size of the set X has reached the maximum R (Line 13), a simple binary preference, e.g. that given by the function $V(\cdot)$ will not be enough. Indeed, since the size of the set X has reached the maximum allowed, if the agent a_i is admitted into the group, another agent has to be rejected from the group. As a consequence, the agents must be ranked to be comparable with its peers. In this case, a natural measure to rank the agents is the trust of the group vs the agent itself (Lines 14–23).

5 Experiments

In this section we report and discuss a few preliminary experimental results aimed at verifying the effectiveness of the approach described in the previous section, i.e. whether the approach is able to produce a group formation which satisfies the requirements discussed in the introductory section.

To this end, we generated a network of 1000 different agents and 1000 random trust relationships among agents, assuming that they live in a virtualized environment provided by the Cloud. In particular, the ratio between trusted and untrusted agents was set as 0.5, and the values of trust were simply generated on the basis of the normal distribution. The initial groups of the simulation were set by randomly placing the agents into them, and the maximum number of agents per group was set as 20, while the number of groups was set as 50. Finally, we set the parameter M – the maximum number of new groups the single agent is able to analyze – to span in the range $[5 \div 10]$ for different simulations. At each step of the simulation, a certain number of interactions among a subset of the agents was simulated, due to each interaction related feedbacks are released. The algorithm is also simulated by triggering the agent-side of the algorithm on 100 different agents randomly selected, as well as the corresponding administrator-side algorithm. Values of feedbacks were generated by means of a normal distribution with proper parameters: for unreliable agents we set $mean = 0.2$ and $stdDev = 0.1$ and for reliable agents we set $mean = 0.9$ and $stdDev = 0.1$, and the ratio between reliable and unreliable agents was set as 0.5.

In order to measure the performance of the algorithm, we define the *Average Mutual Trust* among the components of a group g as $AMT_g = 1/(2|g|)\sum_{i=1 i\neq j}^{|g|}(\tau_{i,j} + \tau_{j,i})$. Moreover, we defined the *Mean Average Mutual Trust*, for a certain configuration at a certain time-step, as $MAMT(Gr) = 1/(|Gr|)\sum_{i=1}^{|Gr|} AMT_{g_i}$.

The performance of the algorithm was tested one hundred of times always obtaining similar results, their average is shown in Fig. 3-a and b. Collected data span across one hundred steps of simulation, and were analyzed in order to obtain the median value of MAMT measured after each single step of the simulation for the different values of $M = [5 \div 10]$. Figure 3-b reports the first 30 steps of the simulation, and Fig. 3-a reports the results obtained until 100 steps of execution of the simulations. From Fig. 3-a it can be observed a slow convergence of the

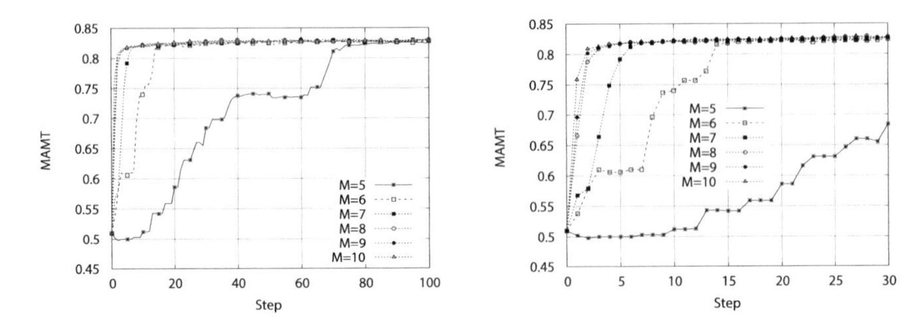

Fig. 3. (a) (Left) MAMT, results until 100 steps; (b) (Right) MAMT, results until 30 steps

MAMT values –towards its asymptotic value– when $M = 5$. Once $M = 6$ and, even better $M = 7$, the convergence is quite fast. More details can be caught in Fig. 3-b (steps 0–30 of the simulation) which shows a very slow increment of the MAMT for $M = 5$ during the first 30 steps, and a radical change of behavior for $M = 6$.

In the light of the experimental results shown in Fig. 3, it can be stated that the execution of the distributed algorithm for group formation – detailed in Sect. 4 – leads to a configuration of groups with a high level of (average) mutual trust among its members, as it was measured in terms of MAMT index.

6 Conclusions

In this paper, we introduce a CoT scenario where each device can be associated with one or more software agents working in its behalf over the cloud. Also, it is possible to exploit the social attitude of software agents to interact and cooperate in an easy way as well as to form complex agent social structures, as groups. To support the agents activities of their CoT devices, we propose to encourage agents to form groups of reliable recommenders by promoting a framework presenting both competition and cooperation traits. To this aim, we designed the distributed algorithm to form agent groups having a high mutual trust among their members. To verify efficiency and effectiveness of this algorithm, we carried out some experimentations, in a simulated agent CoT scenario, which confirmed the potential advantages deriving by the adoption of our proposal in improving individual and group satisfaction in terms of mutual trust.

References

1. Sheng, Z., Yang, S., Yu, Y., Vasilakos, A., Mccann, J., Leung, K.: A survey on the IETF protocol suite for the internet of things: standards, challenges, and opportunities. IEEE Wirel. Commun. **20**(6), 91–98 (2013)
2. Fortino, G., Trunfio, P.: Internet of Things Based on Smart Objects: Technology, Middleware and Applications. Springer (2014)

3. Wei, Y., Blake, M.B.: Service-oriented computing and cloud computing: challenges and opportunities. IEEE Internet Comput. **14**(6), 72–75 (2010)
4. Aazam, M., Khan, I., Alsaffar, A. A., Huh, E.-N.: Cloud of Things: integrating internet of things and cloud computing and the issues involved. In: 2014 11th International Bhurban Conference on Applied Sciences and Technology (IBCAST), pp. 414–419. IEEE (2014)
5. Fortino, G., Gravina, R., Russo, W., Savaglio, C.: Modeling and simulating Internet-of-Things systems: a hybrid agent-oriented approach. Comput. Sci. Eng. **19**(5), 68–76 (2017)
6. Doodson, J., Gavin, J., Joiner, R.: Getting acquainted with groups and individuals: information seeking, social uncertainty and social network sites. In: ICWSM (2013)
7. Fotia, L., Messina, F., Rosaci, D., Sarné, G.M.L.: Using local trust for forming cohesive social structures in virtual communities. Comput. J. **60**(11), 1717–1727 (2017)
8. De Meo, P., Messina, F., Rosaci, D., Sarné, G. M. L.: Recommending users in social networks by integrating local and global reputation. In: Proceedings of the 7th International Conference on Internet and Distributed Information Systems. LNCS, vol. 8729, pp. 437–446. Springer (2014)
9. Aikebaier, A., Enokido, T., Takizawa, M.: Trustworthy group making algorithm in distributed systems. Hum. Centric Comput. Inf. Sci. **1**(1), 6 (2011). https://doi.org/10.1186/2192-1962-1-6
10. Huynh, T., Jennings, N., Shadbolt, N.: An integrated trust and reputation model for open multi-agent systems. Auton. Agents Multi Agent Syst. **13**(2), 119–154 (2006)
11. Dellarocas, C.: Designing Reputation Systems for the Social Web. Boston University Questrom School of Business Research Paper Series. Paper No. 2010-18. SSRN: https://ssrn.com/abstract=1624697 (2010)
12. Kim, Y., Song, H.: Strategies for predicting local trust based on trust propagation in social networks. Knowl. Based Syst. **24**(8), 1360–1371 (2011)
13. Massa, P., Avesani, P.: Trust metrics on controversial users: balancing between tyranny of the majority. Int. J. Semant. Web Inf. Syst. (IJSWIS) **3**(1), 39–64 (2007)
14. Ziegler, C., Lausen, G.: Spreading activation models for trust propagation. In: 2004 IEEE International Conference on one-Technology, e-Commerce and e-Service, EEE 2004, pp. 83–97. IEEE (2004)
15. Golbeck, J., Hendler, J.: Inferring binary trust relationships in web-based social networks. ACM Trans. Internet Technol. **6**(4), 497–529 (2006)
16. Guha, R., Kumar, R., Raghavan, P., Tomkins, A.: Propagation of trust and distrust. In: Proceedings of the 13th International Conference on World Wide Web, pp. 403–412. ACM (2004)
17. Golbeck, J.: Computing and applying trust in web-based social networks. Ph.D. Thesis, University of Maryland, Department of Computer Science (2005)
18. Massa, P., Avesani, P.: Trust-aware recommender systems. In: Proceedings of the 2007 ACM Conference on Recommender systems, pp. 17–24. ACM (2007)
19. Jamali, M., Ester, M.: Trustwalker: a random walk model for combining trust-based and item-based recommendation. In: Proceedings of the 15th ACM SIGKDD International Conference on Knowledge Discovery and Data Mining, pp. 397–406. ACM (2009)
20. Council, N.R., et al.: Public Participation in Environmental Assessment and Decision Making. National Academies Press, Washington, DC (2008)

21. Xia, L.: Computational voting theory: game-theoretic and combinatorial aspects. Ph.D. thesis, Duke University (2011)
22. Beierle, T.C., Cayford, J.: Democracy in Practice: Public Participation in Environmental Decisions. Resources for the Future, Washington, DC (2002)
23. Chevaleyre, Y., Endriss, U., Lang, J., Maudet, N.: A short introduction to computational social choice, In: Theory and Practice of Computer Science, SOFSEM 2007, pp. 51–69 (2007)
24. Conitzer, V., Sandholm, T.: Universal voting protocol tweaks to make manipulation hard, arXiv preprint cs/0307018
25. Gibbard, A.: Manipulation of voting schemes: a general result. Econometrica J. Econ. Soc. **41**, 587–601 (1973)
26. Lai, L.S., Turban, E.: Groups formation and operations in the web 2.0 environment and social networks. Group Decis. Negot. **17**(5), 387–402 (2008)

Terrorism and War: Twitter Cascade Analysis

V. Carchiolo, A. Longheu(✉), M. Malgeri, G. Mangioni, and M. Previti

Dip. Ingegneria Elettrica, Elettronica e Informatica,
Università degli Studi di Catania, Catania, Italy
alessandro.longheu@dieei.unict.it

Abstract. Misinformation spreading over online social networks is becoming more and more critical due to the huge amount of information sources whose reliability is hard to establish; moreover, several humans psychology factors as echo chambers and biased searches, plus the intensive use of bot, makes the scenario difficult to cope with. Unprecedented opportunities of gathering data to enhance knowledge though raised, even if the threat of assuming a fake as real or viceversa has been hugely increased, so urgent questions are how to ascertain the truth, and how to somehow limit the flooding process of fakes. In this work, we investigate on the diffusion of true, false and mixed news through the Twitter network using a free large dataset of fact-checked rumor cascades, that were also categorized into specific topics (here, we focus on Terrorism and War). Our goal is to assess how news spread depending on their veracity and we also try to provide an analytic formulation of spreading process via a differential equation that approximates this phenomenon by properly setting the retweet rate.

1 Introduction

The ease of information lifecycle, from creating a news (true or false), to its spreading in online social networks, sharing among even millions of users without any limitation (neither control), lead to a sort of information *consumerism*, where the value of each single news is almost void and the whole amount of such information seems an ambient noise in a crowded room where everybody's speaking.

The emergency of misinformation is not a new question [1] but it got worse and worse mainly due to the huge amount of (possibly unreliable) information sources and to the raising of so-called *echo chambers* [2], where isolated groups of people with similar ideas tends to strengthen their opinions and avoid any debate with others, therefore increasing polarization effects. Another relevant factor is that most information does not come from being personal witness, rather we largely leverage on *trust* sources we get news from owns [3]; moreover, persons are generally considered biased searchers, i.e. we are likely to gather information that confirm our ideas [4].

J. Del Ser et al. (Eds.): IDC 2018, SCI 798, pp. 309–318, 2018.
https://doi.org/10.1007/978-3-319-99626-4_27

In addition to people behaviour and psychology, another challenge comes from *bots* (a.k.a. *cyborg users*), whose goal is to exploit humans social biases to strengthen fakes propagation [5,6].

This dark vision is greatly compensated by the unprecedented opportunity of gathering data to enhance knowledge, but the threat of assuming a fake as real or viceversa has been hugely increased, therefore urgent questions are (1) how to ascertain the truth, and then (2) how to somehow limit the flooding process of fakes or, conversely, how to facilitate the endorsement of true information. The former issue is not so simple to address, indeed often an information source (e.g., a left-wing party) aims to tag its opponents (e.g., a right-wing party) as unreliable in always providing true information, hence a verifiable and impartial criterion for labelling a news as true or false is needed. One possibility is to leverage the concept of *trust* (introduced above), whose definitions are consolidated, well-known, yet multifaceted [7,8] Even if several trustworthiness assessment mechanisms were introduced [9,10], the most adopted in online social networks is the distributed, recommendation-based one [11,12]. Its reliability however could be debated since many algorithm exist that allow to increase trust in such networks [13–15].

An alternative approach for truth assessment can be the exploitation of independent fact-checking organizations as factcheck.org, snopes.com, politifact.com, truthorfiction.com, hoax-slayer.net and many others. Both *distributed* (trust-based) and *centralized* (independent organizations) ideas offer advantages and drawbacks, as it can be easily conceived, e.g. in distributed approaches the risk of empowering or white-washing are well known [16,17], but it is also a risk to rely on some institution and its supposed 1independence'; some help comes from engaging both many and different ways of thinking, i.e. following a *bipartisan* approach.

The latter question, i.e. how to block fakes spreading or to foster the diffusion of true information, is still open [18] since in literature there is no a definitive answer about the spreading of misinformation, whether and how fakes and true news diffuse differently, and which factors affect this phenomenon, as human behaviour, technologies, network scale, type, structure and others. For instance, some researchers [19,20] claim that even if a news has been recognized as a fake, publicizing this fact may enforce that misinformation rather than definitively remove it from people's mind.

A mainstream consideration is that to *search and destroy* fake news proliferation, acting on two fronts is needed, i.e. both technological (as bots detection and stemming) as well as social (e.g. encouraging the culture of truth via collaboration between authoritative scientists and journalists/reporters).

The context of the work described in this paper is the analysis of news diffusion, aiming to stem fakes; in particular, we focus on the news spreading process in the Twitter social newtork [21]. We exploit the data set of rumor cascades in the range [2006, 2017] available at [18,22,23]. A cascade is a sequence of a tweet and related retweets; several cascades of different lengths for the same news actually exist. Cascades in the dataset where categorized into specific topics (here, we focus on *Terrorism and War*), and were also classified into *false, true, mixed*. The classification was carried out by six independent fact-checking organizations that agreed over 95% about the veracity of that news. We aim to assess how fakes spread with respect to true and mixed news; we also try to provide an analytic formulation of such spreading process via a differential equation that approximates this phenomenon by properly setting the retweet rate.

The paper is organized as follows. In Sect. 2 we introduce the dataset in detail, whereas in Sect. 3 we illustrate the analysis on data we carried out; subsequent considerations lead to the model proposed in Sect. 4. Our concluding remarks and future works are finally shown in Sect. 5.

2 Dataset Description: War and Terrorism

This work is based on part of the dataset used in [18]. The authors investigated the diffusion of true, false and mixed (i.e., partially true, partially false) news through the Twitter network using a dataset of fact-checked rumor cascades. Their data included 126,000 complete rumor cascades spread by 3 million people more than 4.5 million times.

To assess news veracity or falsity, authors accessed to the full Twitter historical archives to collect all English language tweets that contained at least a link to one of six independent rumor debunking websites (snopes.com, politifact.com, factcheck.org, truthorfiction.com, hoax-slayer.net and urbanlegends.about.com) from September 2006 to December 2016, then they reduced the number of tweets by taking only the ones that were replies to other tweets. For each reply tweet, they extracted the original tweet that they were replying to and then extracted all the retweets of the original tweet. To reconstruct the retweet graph, they used the time-inferred diffusion method [24] that compares the temporal order of the tweets with the follower graph in order to determine the correct path of each tweet over the network.

The fact-checking organizations behind the abovementioned websites, in addition to checking the news, tag them with a topic, so it is possible to divide them by subject areas. These tags have been reported for each tweet, so, thanks to a pre-filtering activity, we divide the huge initial dataset into 7 smaller sized datasets each one representing a topic (Table 1).

We started our analysis from *Terrorism and War* subset. We divided the tweets for truthfulness in true, false and mixed and then we rebuilt the cascade graphs, counting the tweets for each one (Table 2). Observing that many cascades were composed of a single tweet, we filtered the dataset in order to keep only the cascades with at least a retweet and finally we counted the rumors. In this paper, in fact, by "rumor" we mean all the tweets related to a specific subject and by "cascade" we mean the original tweet and all its retweets, so a rumor can be made up of one or more cascades.

Observing Table 2, we see the subset is imbalanced, because the goal of fact-checking organizations is to verify if controversial news contents are true or false, hence these organizations do not take into consideration surely true news contents and they do not appear in our subset. In some subset we see the number of

Table 1. Number of tweets for each category

Politics	2094775
Urban legends	801755
Business	151714
Science and technology	563291
Terrorism and war	**205588**
Entertainment	206926
Natural disasters	10687

Table 2. Terrorism and War subset characteristics

N. of...	True	Mixed	False
Tweets	12451	9590	183885
Cascades	3325	1057	8311
Filtered cascades	519	780	4542
Rumors	35	13	77

true news is much smaller, hence we select a medium-size subset, where all the typologies of veracity are present. However, in the next section, we analyze different kind of veracity separately before to perform a comparison among average curves, hence the subset unbalancing is not relevant for our kind of analysis.

3 Dataset Analysis

In this section we describe two types of analysis for each kind of truthfulness of the news: first we analyze the tweets per hour of each cascade, starting from the moment in which the original tweet was created (Figs. 1, 2 and 3) than all the retweets that starting from the original tweet (Figs. 4, 5 and 6). The execution of these two types of analysis was made to try to identify possible differences between true and false news in order to allow users to distinguish them and avoid false ones.

From Figs. 1 and 4, we observe that for true rumors, after the first two hours from the original tweet, the number of retweets for cascade decreases by about 80%. Only 3 of the 519 cascade analyzed exceed 200 retweets in the first hour and exceed 1000 total retweets, but exhaust their virality within 10 h of entry on the network. After the first day only 2 cascades have peaks above 50 retweets, while all the others are kept under 10 retweets per hour, in fact the total retweet curves become almost constant.

From Figs. 2 and 5, we observe that for the mixed rumors, perhaps due to the controversial nature of the content and the lack of reliability, there is not a great initial diffusion, in fact they all remain under 100 retweets during the first hour from the original tweet and have continuous oscillations of the same

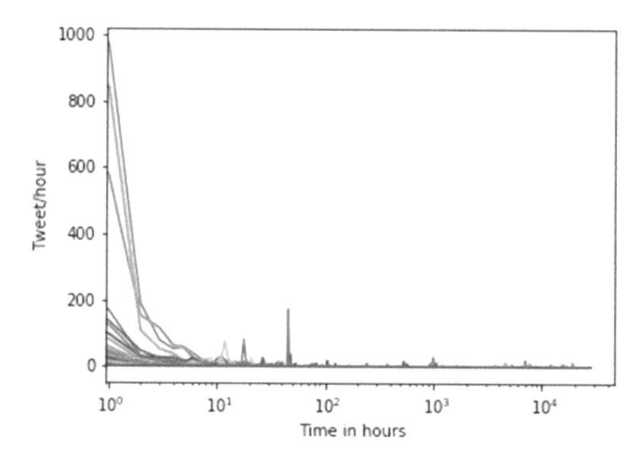

Fig. 1. Tweets per hour for true cascades

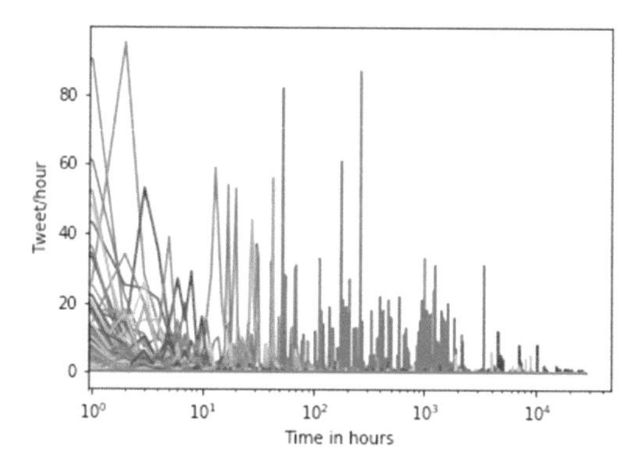

Fig. 2. Tweets per hour for mixed cascades

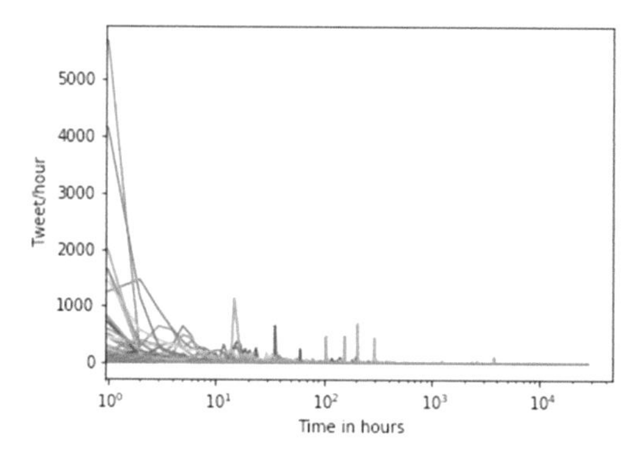

Fig. 3. Tweets per hour for false cascades

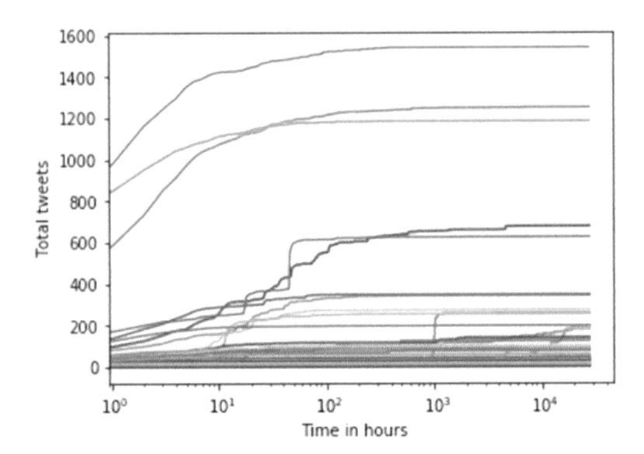

Fig. 4. Total tweets for true cascades

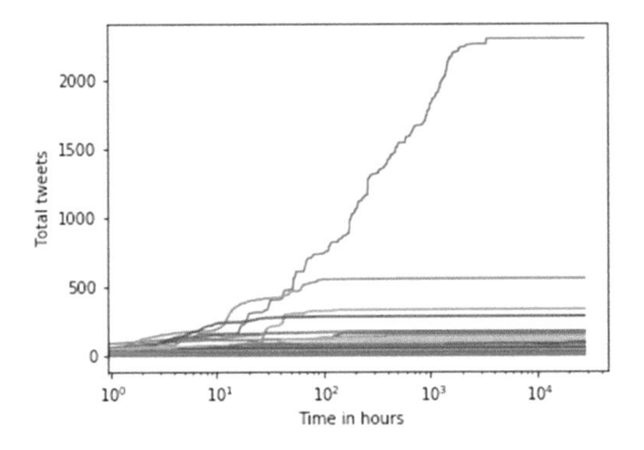

Fig. 5. Total tweets for mixed cascades

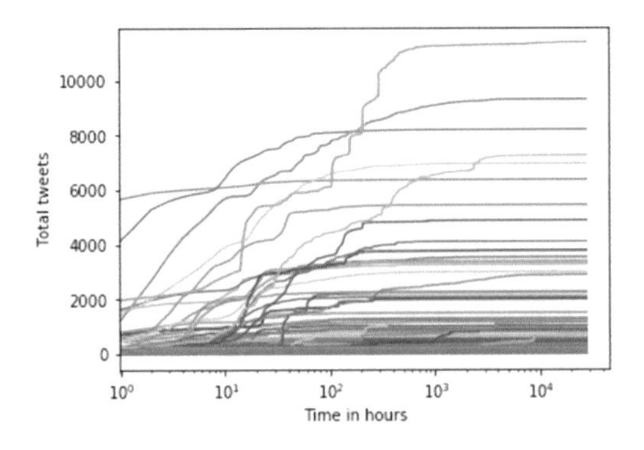

Fig. 6. Total tweets for false cascades

amount compared to the beginning of the observation even in the following two days. Except in one case, they never exceed 500 total retweets.

From Figs. 3 and 6, we observe that for the false rumors the number of tweets per cascade decreases by about 60% after the first two hours from the original tweet. Only 7 of the 4542 cascades exceed 1000 retweets in the first hour since the original tweet and only one keeps these sharing values for the next hour. These curves reach from 6000 to 11500 retweets and for all the other cascades one observes a greater spread than the real and mixed news, because unlike the two previous cases the news does not exhaust their virality during the first day. In fact, half of the cascade have peaks above 100 retweets per hour even after the first 24 h. For these reasons, it is clear that the false news propagate faster and stay longer on the network, reaching a much higher number of users.

From Figs. 1 and 4, we observe that for true rumors, after the first two hours from the original tweet, the number of retweets for cascade decreases by about 80%. Only 3 of the 519 cascade analyzed exceed 200 retweets in the first hour and exceed 1000 total retweets, but exhaust their virality within 10 h of entry on the network. After the first day only 2 cascades have peaks above 50 retweets, while all the others are kept under 10 retweets per hour, in fact the total retweet curves become almost constant.

From Figs. 2 and 5, we observe that for the mixed rumors, perhaps due to the controversial nature of the content and the lack of reliability, there is not a great initial diffusion, in fact they all remain under 100 retweets in the course of the first hour from the original tweet and have continuous oscillations of the same amount compared to the beginning of the observation even in the following two days. Except in one case they never exceed 500 total retweets.

From Figs. 3 and 6, we observe that for the false rumors, after the first two hours from the original tweet, the number of tweets per cascade decreases by about 60%. Only 7 of the 4542 cascades exceed 1000 retweets in the first hour since the original tweet and only one keeps these sharing values for the next hour, these curves reach from 6000 to 11500 retweets and for all the other cascades one observes a greater spread than the real and mixed news, because unlike the two previous cases the news does not exhaust their virality during the first day, in fact half of the cascade have peaks above 100 retweets per hour even after the first 24 h. For these reasons, it is clear that the false news propagate faster and stay longer on the network, reaching a much higher number of users.

4 Modeling the Retweet Dynamics

Observing the number of total tweets per cascade from Fig. 4, 5 and 6 we see that, despite the number of retweets is distinctly different (the total retweet curves of the false news have a higher slope and reach values much higher than other two graphs), all the curves are monotonic increasing with some steps in correspondence with the peaks present in the tweet per hour graphs. So it is possible to identify a parametrized function that approximates the value of each curve.

To get a more regular trend, we calculated the average mean curve of the total tweets by taking the number of tweets per hour of each cascade and calculating the average point by point (see Fig. 7).

In order to model the evolution of retweets over time, we generated a model starting from the observation of our data subset characteristics. In the following, we indicate with $R(t)$ the number of retweets at instant t. The number of retweets in a generic instant $t + \Delta t$ will be given by the sum between the ones at instant t and the ones that will be posted in the following instants with a certain retweet rate λ, but not exceeding the number of total retweet of the curve N, hence:

$$R(t + \Delta t) = R(t) + \lambda[N - R(t)]\Delta t \tag{1}$$

In the limit of a small δt, we have the following first-order ordinary differential equation:

$$R'(t) + \lambda R(t) = \lambda N \tag{2}$$

whose solution is:

$$R(t) = [R(t_0) - N]e^{-\lambda t} + N \tag{3}$$

where $R(t_0)$ indicates the number of tweets at the initial instant.

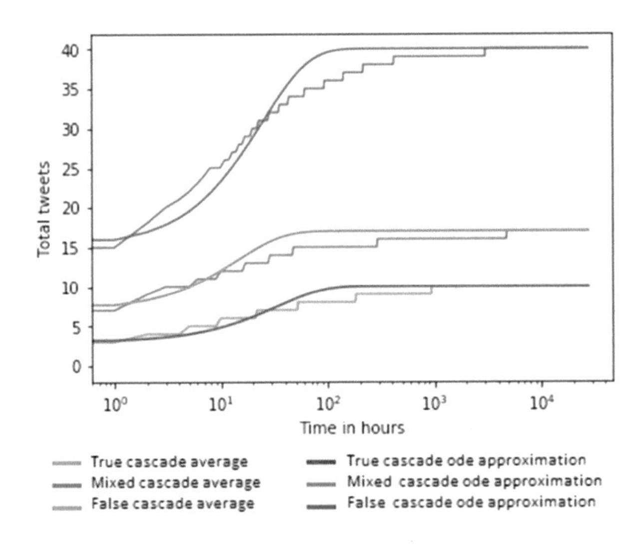

Fig. 7. Cascade averages and cascade ode approximations

By plotting this solution we obtained a set of curves with different slopes based on the λ value chosen. To find the ones that best fit the original curves, we calculated the mean square errors (mse) between the curves with different λ values and the real data obtained from the dataset and returns the one with the lowest mse. The result is shown in the Fig. 7.

The curves that best fit the real values ones are:

- for true retweet cascades $\lambda = 0.02$ with $mse = 0.1927$
- for mixed retweet cascades $\lambda = 0.01$ with $mse = 0.032$
- for false retweet cascades $\lambda = 0.03$ with $mse = 0.1931$

5 Conclusions

In this paper we considered news spreading process in the context of false, true and mixed news. We exploited a Twitter dataset with more than 200 thousand tweets, all related to the topic *Terrorism and War*, organized in 12 thousand complete rumor cascades. Results shown that false news spread faster and stay longer on the network, reaching a much higher number of users. We also provide a first simple model to analytically describe the spreading process. Future works include a deeper investigation of why fakes spread more than true news on networks, how to limit such diffusion in order to preserve correct information, and how analytic formulation can help to reach this goal.

References

1. McKernon, E.: Fake news and the public (1925)
2. Garimella, K., Morales, G.D.F., Gionis, A., Mathioudakis, M.: Quantifying controversy on social media. Trans. Soc. Comput. **1**(1), 3:1–3:27 (2018)
3. Metzger, M.J., Flanagin, A.J., Medders, R.B.: Social and heuristic approaches to credibility evaluation online. J. Commun. **60**(3) 413–439 (2010)
4. Schulz-Hardt, S., Frey, D., Lüthgens, C., Moscovici, S.: Biased information search in group decision making. J. Pers. Soc. Psychol. **78**(4), 655–669 (2000)
5. Ferrara, E., Varol, O., Davis, C., Menczer, F., Flammini, A.: The rise of social bots. Commun. ACM **59**(7), 96–104 (2016)
6. Varol, O., Ferrara, E., Davis, C.A., Menczer, F., Flammini, A.: Online human-bot interactions: detection, estimation, and characterization. CoRR abs/1703.03107 (2017)
7. Sztompka, P.: Trust: A Sociological Theory. Cambridge University Press, Cambridge (1999)
8. Artz, D., Gil, Y.: A survey of trust in computer science and the semantic web. Web Semant. Sci. Serv. Agents World Wide Web **5**(2), 58–71 (2007)
9. McKnight, D.H., Chervany, N.L.: The meanings of trust. Technical report, Minneapolis, USA (1996)
10. Gupta, M., Zhao, P., Han, J.: Evaluating Event Credibility on Twitter, pp. 153–164 (2012)
11. Massa, P., Avesani, P.: Trust-aware recommender systems. In: Proceedings of the 2007 ACM Conference on Recommender Systems, RecSys 2007, pp. 17–24. ACM, New York (2007)
12. Carchiolo, V., Longheu, A., Malgeri, M., Mangioni, G.: Trust assessment: a personalized, distributed, and secure approach. Concurr. Comput. Pract. Exp. **24**(6), 605–617 (2012)
13. Carchiolo, V., Longheu, A., Malgeri, M., Mangioni, G.: Users' attachment in trust networks: reputation vs. effort. IJBIC **5**(4), 199–209 (2013)

14. Carchiolo, V., Longheu, A., Malgeri, M., Mangioni, G.: The cost of trust in the dynamics of best attachment. Comput. Inform. **34**(1), 167–184 (2015)
15. Buzzanca, M., Carchiolo, V., Longheu, A., Malgeri, M., Mangioni, G.: Direct trust assignment using social reputation and aging. J. Ambient. Intell. Hum. Comput. **8**(2), 167–175 (2017)
16. Bedi, P., Vashisth, P.: Empowering recommender systems using trust and argumentation. Inf. Sci. **279**, 569–586 (2014)
17. Jiang, S., Zhang, J., Ong, Y.S.: An evolutionary model for constructing robust trust networks. In: Proceedings of the 2013 International Conference on Autonomous Agents and Multi-agent Systems, AAMAS 2013, Richland, SC, pp. 813–820. International Foundation for Autonomous Agents and Multiagent Systems (2013)
18. Vosoughi, S., Roy, D., Aral, S.: The spread of true and false news online. Science **359**(6380), 1146–1151 (2018)
19. Schwarz, N., Sanna, L.J., Skurnik, I., Yoon, C.: Metacognitive experiences and the intricacies of setting people straight: implications for debiasing and public information campaigns. In: Advances in Experimental Social Psychology, vol. 39, pp. 127–161. Academic Press (2007)
20. Ecker, U.K., Hogan, J.L., Lewandowsky, S.: Reminders and repetition of misinformation: Helping or hindering its retraction? J. Appl. Res. Mem. Cogn. **6**(2), 185–192 (2017)
21. Twitter. https://twitter.com/
22. Vosoughi, S., Roy, D., Aral, S.: Twitter data set. https://docs.google.com/forms/d/e/1faipqlsdvl9q8w3mg6myi4l8fi5x45smnrzgooedbroebonni5ibfkw/viewform
23. Vosoughi, S., Roy, D., Aral, S.: Twitter data set - supplementary material. http://science.sciencemag.org/content/suppl/2018/03/07/359.6380.1146.dc1
24. Goel, S., Watts, D.J., Goldstein, D.G.: The structure of online diffusion networks. In: Proceedings of the 13th ACM Conference on Electronic Commerce, pp. 623–638. ACM (2012)

A Genetic Algorithm with Local Search Based on Label Propagation for Detecting Dynamic Communities

A. Panizo(✉), G. Bello-Orgaz, and D. Camacho

Computer Science Department, Universidad Autónoma de Madrid, Madrid, Spain
{angel.panizo,gema.bello,david.camacho}@uam.es

Abstract. The interest in community detection problems on networks that evolves over time have experienced an increasing attention over the last years. Genetic Algorithms, and other bio-inspired methods, have been successfully applied to tackle the community finding problem in static networks. However, few research works have been done related to the improvement of these algorithms for temporal domains. This paper is focused on the design, implementation, and empirical analysis of a new Genetic Algorithm pair with a local search operator based on Label Propagation to identify communities on dynamic networks.

1 Introduction

The analysis of complex networks is a relevant research topic because many real world problems can be modeled as a complex network. Identifying communities within these networks have been applied to disciplines such as marketing [2,5], public healthcare [4], cybercrime [14,15,19] or ego-based social network analysis [10,16]. There are several definitions of a community, but no consensus has been reached between the scientific community about what is the correct one. One commonly accepted definition for a community is: a set of nodes that have stronger interactions between them than among the rest of nodes of the network. Two types of communities can be defined. The first ones are composed of nodes that can belong to several communities at the same time, and are called *overlapping communities*. The second ones are composed of nodes that can only belong simultaneously to a single community, and are called *non-overlapping communities*.

Dynamic networks (networks that change over time) has received a great deal of attention lately, mainly due to the temporal nature of real-world networks such as social networks or mobile communications. A dynamic network is usually modeled using a sequence of snapshots, where each snapshot represents the network at a given point in time. The communities found on these networks are called *dynamic communities*. To identify these type of communities not only we need to partition each snapshot but also we need to track the partitions between snapshots. Alhajj [1] defines a *static community* as a set of nodes and edges among them, that do not change in time. This author also introduces two

J. Del Ser et al. (Eds.): IDC 2018, SCI 798, pp. 319–328, 2018.
https://doi.org/10.1007/978-3-319-99626-4_28

possible definitions of a *dynamic community*. The first one as a sequence of *static communities*, one for each snapshot in the dynamic network. And a second one, as an initial *static community* and a series of modifications of it over time.

Dynamic community finding methods can be split in two groups. A first group that contains the methods that separate the community detection step from the temporal analysis, i.e. first, the communities are extracted and then the structural differences over time are analyzed; and a second one that follows the *evolutionary clustering* framework [8], to simultaneously optimize the two conflicting criteria community detection and temporal analysis of the communities at each snapshot. These methods assume that abrupt changes in the communities in a short time period do not happen, thus they look for *temporal smoothness* in the dynamic communities. For clarity purposes, we will call the methods in the first group *non temporal smooth* methods, and the second ones *temporal smooth* methods.

Genetic Algorithms (GAs) are a stochastic meta-heuristic inspired in the biological principles of Darwinian theory of evolution and Mendel's genetics. GAs are an effective method to solve combinatorial optimization problems, and have been used before to solve both the static [2,3,6] and the dynamic [9] community identification problem. GAs for static community detection have been paired with *local search strategies* to improve performance. The *local search strategies* are applied mostly during the *population initialization* [20] or during the *mutation process* [17]. In this paper, we introduce a GA with a *local search strategy* based on *Label Propagation* [24] that detects non-overlapping dynamic communities following a *non temporal smooth* approach. This work is focused on the community detection step, leaving the subsequent temporal analysis out of the scope. In *Label Propagation* each node starts in its own community and following an iterative process each node updates its community to match the community of the majority of its neighbors . The main idea behind our *local search strategy* is that consecutive snapshots in a dynamic network have a similar topology, so the results obtained for one snapshot can be reused by the next one. The proposed method run the GAs for each snapshot on a dynamic network to identify its community. The initial population of each GA is constructed by applying the update method of *Label Propagation* over the result population of the previous snapshot. Finally, an experimental phase using a real world dataset has been carried out in order to test the impact of the *local search strategy* when pair with a common GAs.

The main contribution of this paper is the introduction and testing of a new local search operator capable of extracting information from previously found solutions in order to boost the performance of a GAs that detects dynamic communities in dynamic networks modeled as a sequence of snapshots.

The rest of the article is structured as follows: Sect. 2 reviews other related methods for community finding. Section 3 describes the dynamic community detection problem and the proposed algorithm. Section 4 presents the procedure followed to test the algorithm and discusses the experimental results. Finally, the last section draws some conclusions and future research lines of the work.

2 Related Work

Two problems have to be solved for detecting dynamic communities: grouping nodes for each snapshot, and tracking these groups over time. Some works are only focused on how to track groups over time, and assume that the groups have already been discovered [7,11]. GED [7] defines seven independent events (Continuing, Shrinking, Growing, Splitting, Merging, Dissolving and Forming) that could change a group state, and a new measure called *inclusion* is used to decide which events occurs. Other work following this approach has been proposed by Greene et al. [11], that also defines seven independent events, very similar to the previous ones defined by GED. But in this work, the *Jaccard coefficient for binary sets* is used to decide which event occurs.

On the other hand, some research works are focused on solving both problems, detection and tracking of the groups or communities. Two of the most well-known methods related to this approach are *GraphScope* [25] and *FacetNet* [18]. The first one is a *non temporal smooth* method based on the Minimum Description Length (MDL) principle, and it is able to group nodes and track them in a streaming way. It reorganizes snapshots into segments, given a new incoming snapshot, it is combined with the current segment if there is a storage benefit, otherwise, the current segment is closed, and a new one is started with the new snapshot. A change of segment means that an event has occurred on the network. *FacetNet* is a *temporal smooth* method that uses probabilistic learning in order to identify dynamic communities. *FacetNet* follows Chakrabarti et al. [8] equation where a parameter α governs the importance between the *snapshot cost* and the *temporal cost*. To solve this equation they presents a low time complexity iterative algorithm which is guaranteed to converge to an optimal solution.

In particular, regarding GAs approaches to detect dynamic communities, Folino and Pizzuti present the DYMONGA [9] algorithm. This is a *temporal smooth* method based on a multi-objective approach, that uses the *modularity* to measure the quality of the communities, and the Normalized Mutual Information (NMI) to measure the change between two consecutive snapshots. DYMONGA is only focused on grouping the nodes in each snapshot. Kim et al. [12] also propose a multi-objective GA that uses immigrant schemes to track the best grouping of the network over time. In contrast to DYMONGA, the method proposed by Kim et al. returns the partition that best suits the last snapshot, taking into account all the previous snapshots, instead of returning the partitions for each snapshot.

Finally, there are several works that use GAs with local search in order to improve the performance of the method for detecting static communities. For example, Li et al. [17] introduce a new mutation operator, and apply a local search operator based on the small-world phenomena. The mutation operator creates n copies of the individual to be mutated. For each copy, several gens are randomly selected. For each selected gen, all the neighbors related to that gen move to the same community of the gen. The individual to be mutated is substituted by the copy with higher fitness. Wang et al. [26] proposed a modify crossover operator more suitable for community detection, and an heuristic mutation operator based on local modularity. When the mutation operator changes a gene, the node related to that gene is moved to the same community of the neighbor that has more edges in the local community. Finally, Pizzuti [23] applies local search after the GA has finished to boost community quality. The local search performs a greedy search over the best individual found by the GAs and tries to move each node of the network to the community of the neighbor that gets higher modularity after receiving it.

All the local search methods described in the paragraph above works on static environments, environments that never change. Their objective is taking an already valid solution and improving it if possible. On the contrary, our local search method works on dynamic environments, environments that change in time. In a dynamic environment is possible that a valid solution became invalid due to some changes in the environment, some nodes or edges have been removed or added to the graph. The objective of our local search operator is not improving a valid solution, but transforming an invalid solution, that have stopped being valid due to change, into a valid one maintaining the highest possible quality.

3 Algorithm Description

3.1 Problem Formulation

Let $\aleph = \{G^0..G^n\}$ be a dynamic network with n snapshots sorted in chronological order. Each snapshot G^t is modeled as a graph $G^t(V^t, E^t)$ where V^t is a set of vertices and E^t is a set of links, called edges, that connects two elements of V^t at time t. Our goal is to group the vertices in each G^t in such a way that vertices in the same group have more links among them than with the rest of vertices of G^t, and each vertex in G^t only belongs to one single group.

3.2 Genetic Algorithm with Local Search

The proposed algorithm is shown in Algorithm 1. It uses an elitist GA for detecting communities in each snapshot (G^t) of the dynamic network $\aleph$ (lines 1–13). The method evolves a random population for the first snapshot (G^0). The population is generated following an heuristic that will be described later. However, for the rest of snapshots ($G^t|t > 0$), instead of evolving a random population, it evolves the population generated by applying the local search operator to the population returned by the previous GA run. For each run of the GA, the best individual (C_{last}) of the last population generated is selected as final result (line 13), which will be used to apply the local search operator to the next run.

Each individual represents a possible grouping of the current snapshot (G^t) using the locus-based adjacency representation [22]. For the first snapshot (G^0), a random population, generated according to this encoding, is used as initial population (line 7). The random population is generated following the next heuristic: each random individual of the population is generated by filling each position in

Algorithm 1. GA with local search for detecting dynamic communities

```
1: function LocalSearchGA(ℵ)
2:     C_dynamic ← ∅
3:     P_last ← null
4:     C_last ← null
5:     for all G^t ∈ ℵ do
6:         if t = 0 then
7:             P_initial ← makeRandomPopulation(G^t, POPSIZE)
8:         else
9:             P_initial ← ∅
10:            for all individual ∈ P_last do
11:                P_initial ← P_initial ∪ labelLocalSearch(individual, G^t, C_last)
12:        P_last ← elitistGA(P_initial, G^t, ELITISM, CRXRATE, MUTRATE, MAXGEN, NCONV)
13:        C_last ← getBestSolution(P_last)
14:        C_dynamic ← C_dynamic ∪ C_last
15:    return C_dynamic
16: function labelLocalSearch(individual, G^t, C_last)
17:    for all i ∈ [1, size(individual)] do
18:        neighbors ← getNeighbors(i, G^t)
19:        if individual[i] ∉ neighbors then
20:            community ← getCommunityOfMost(neighbors, C_last)
21:            validNeighbors ← getNodesInCommunity(neighbors, community)
22:            individual[i] ← selectRandom(validNeighbors)
23:    return individual
```

the chromosome with a random neighbor, in G^0, of the node related with that position. On the other hand, for the rest of snapshots ($\{G^t | t > 0\}$), the individuals of the initial population are generated by applying the *labelLocalSearch* function to the individuals returned by the last run of the GA (lines 9–11). This function (lines 16–23) changes those genes inside the chromosome with values that refers to a node that no longer is a neighbor of that gene. To select a new value for it, first we find all the neighbors of the node represented by the gene in the actual snapshot (G^t) (line 18). Then, using the best solution (C_{last}) for the past snapshot (G^{t-1}), the algorithm selects the community of C_{last} which contains the higher number of the neighbors obtained previously (line 20). Finally, the value of the gene is swap with a random value among the neighbors that also belong to the mentioned community (line 22).

Once the initial population (random or not) has been created, it evolves using an elitist GA (line 12) with the next functions: *Two Points Crossover*, *Tournament Selection*, *Modularity* [21] as *Fitness function*. Finally, a specific mutation operator has been implemented, that randomly selects a series of genes, following the normal distribution, to change them with a random neighbor of the corresponding node. After testing with several combinations of the most common functions used in the literature, we have selected the combination described above.

4 Experimentation

4.1 Dataset Description

In this section we study the effectiveness of our method using the *Enron* [13] dataset. This dataset consists of emails sent by the workers of the Enron corporation between 1999 and 2003. In this dataset each node is an Enron worker and an edge connecting two nodes means that an email has been exchanged between those two workers. This communication by email has associated the date when it happened. So, a dynamic network can be generated, splinting the Enron network in several snapshots using these dates. In this case, each snapshot has all the emails sent over fifteen days, the weekends have been ignored because very little activity occurred on them. The snapshots are fifteen days long because shorter snapshots makes the network too dynamic to be analyzed. A total of 8 snapshot have been created, from the 1st of January of 2000 through the 30th of April of the same year. The number of nodes and edges of each snapshot in the network can be seen in the Fig. 1.

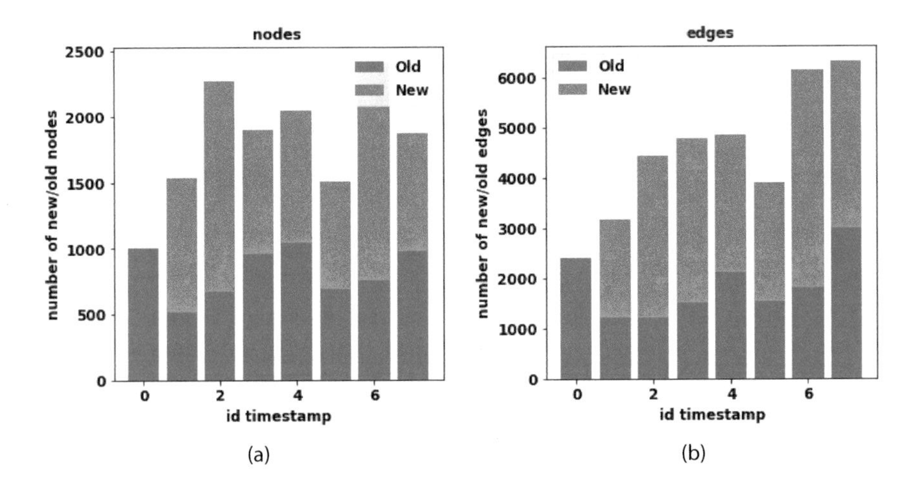

Fig. 1. Number of nodes and edges for each snapshot in the dynamic network. The blue part of the bars indicates the number of elements that existed on the previous snapshot, an the orange part represents the number of elements that are new in that particular snapshot.

Table 1. Genetic parameters of both GAs (with and without local search).

Parameter	Description	Value
POPSIZE	Population size	300
ELITISM	Number of individuals to select for the next generation	10%
MAXGEN	Maximum number of generations	500
NCONV	Number of generations with same best fitness (± 0.01)	10
CRXRATE	Crossover probability	1.0
MUTRATE	Mutation probability	0.1

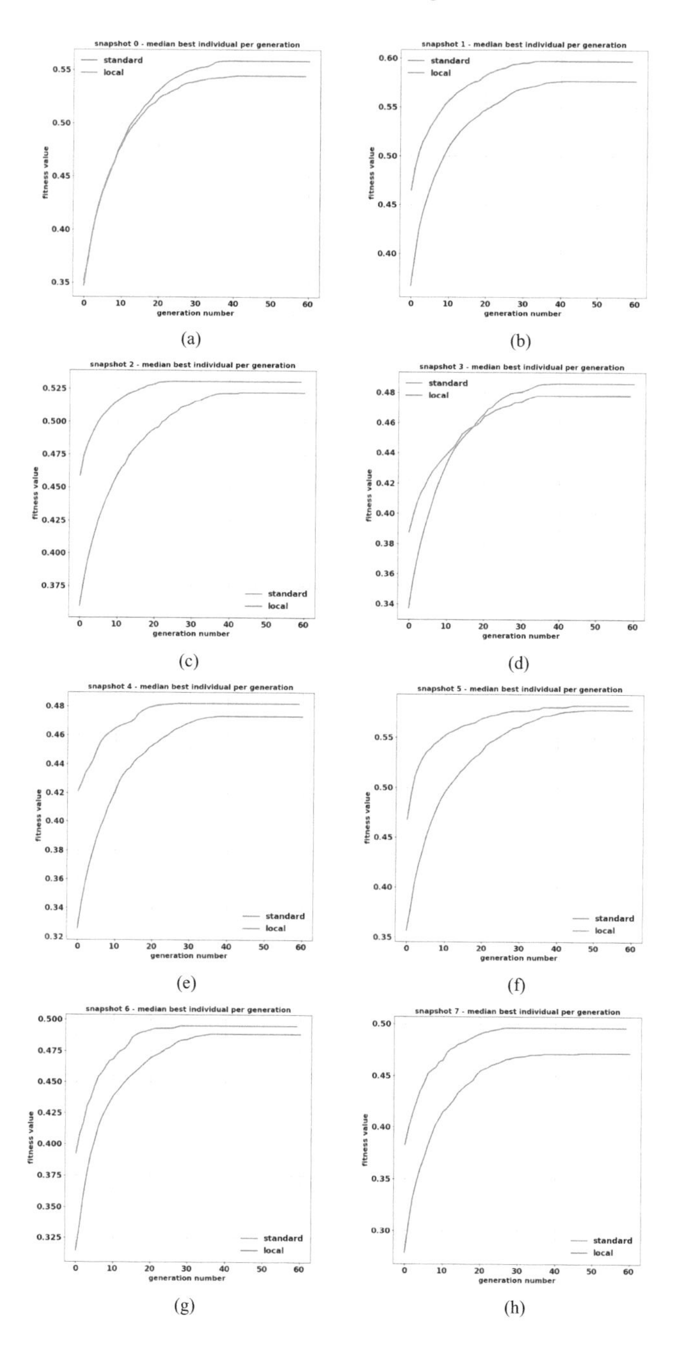

Fig. 2. Each graph corresponds to one snapshot of the dynamic network. All the graph show the fitness function value of the best individual of the population over the generations.

4.2 Experimental Results

In order to validate the effectiveness of our method we have compared it against an elitist GA without local search (evolves a new random population for every snapshot). For this comparison, the fitness of the best individual of every generation achieved by both methods is used. This allows us to check which method reaches a higher fitness value, and also how many generations are required to achieve it. Both GAs, the one with local search and the one without it, uses the same set-up. The parameter of the set-up can bee seen on Table 1. The different parameters of the algorithms have been tuned up experimentally. These settings have been used during all the experimentation phase.

Each experiment have been performed 30 times, and the median of all executions can be seen in the Fig. 2. Analyzing the results we can conclude that the elitist GA with local search achieves better results in almost every snapshot, excluding the *snapshot 3* (Fig. 2d). The *snapshot 0* (Fig. 2a) is a special case due to there is no previous solutions to perform the local search, and then both elitist GAs generate a random initial population. Therefore, in this particular case both GAs achieve similar results. For the rest of snapshots, beside the *snapshot 3*, the fitness value of the best individual is better during all the generations of the GAs. In the *snapshot 3* the population of the GA with local search starts being better but after the 15th generation the GA without local search surpasses it and achieves a better final result. In our opinion the main drawback of the local search method lies here: although the population of the different snapshots starts with better fitness values, the convergence speed of the algorithm tends to decrease. The local search method creates populations with lower diversity and reduces the exploration capacities of the GA. This makes the GA more likely to get trapped in a local optima.

5 Conclusions

In this paper we present a new evolutionary approach with a local search operator to detect communities in dynamic networks. The method executes an elitist GA to detect the communities in each snapshot. However, in order to improve the quality of the solutions, a local search operator is applied. This new operator is based on the *Label Propagation* method, and it uses the information of the last population generated by the GA at the previous snapshot, to generate the initial population of the current snapshot. This way, the knowledge generated in past populations can be reused to improve future ones. The experiments performed show that the method improves the global quality of the solutions at the expense of exploration capabilities in general terms. This could make the method more prone to getting trapped in a local optima, although in most of the tested cases this fact does not occur. Future research will focus on enhancing the exploration capabilities of the method. For this purpose, we will try to generate more diverse initial populations by using different local search operators over the same population. Besides, comparing our method with other well-know meta-heuristics found in the literature to determine if it is promising or not.

Acknowledgements. This work has been co-funded by the following research projects: EphemeCH (TIN2014-56494-C4-4-P) and DeepBio (TIN2017-85727-C4-3-P) projects (Spanish Ministry of Economy and Competitivity, under the European Regional Development Fund FEDER).

References

1. Alhajj, R., Rokne, J.: Encyclopedia of Social Network Analysis and Mining. Springer, New York (2014)
2. Bello, G., Menéndez, H., Okazaki, S., Camacho, D.: Extracting collective trends from Twitter using social-based data mining. In: Bădică, C., Nguyen, N.T., Brezovan, M. (eds.) Computational Collective Intelligence. Technologies and Applications, pp. 622–630. Springer, Heidelberg (2013)
3. Bello-Orgaz, G., Camacho, D.: Evolutionary clustering algorithm for community detection using graph-based information. In: 2014 IEEE Congress on Evolutionary Computation (CEC), pp. 930–937 (2014). https://doi.org/10.1109/CEC.2014.6900555
4. Bello-Orgaz, G., Hernandez-Castro, J., Camacho, D.: Detecting discussion communities on vaccination in Twitter. Futur. Gener. Comput. Syst. **66**, 125–136 (2016)
5. Bello-Orgaz, G., Menéndez, H., Okazaki, S., Camacho, D.: Combining social-based data mining techniques to extract collective trends from twitter. Malays. J. Comput. Sci. **27**(2), 95–111 (2014)
6. Bello-Orgaz, G., Menéndez, H.D., Camacho, D.: Adaptive k-means algorithm for overlapped graph clustering. Int. J. Neural Syst. **22**(05), 1250018 (2012)
7. Bródka, P., Saganowski, S., Kazienko, P.: Ged: the method for group evolution discovery in social networks. Soc. Netw. Anal. Min. **3**(1), 1–14 (2013)
8. Chakrabarti, D., Kumar, R., Tomkins, A.: Evolutionary clustering. In: Proceedings of the 12th ACM SIGKDD International Conference on Knowledge Discovery and Data Mining, pp. 554–560. ACM (2006)
9. Folino, F., Pizzuti, C.: An evolutionary multiobjective approach for community discovery in dynamic networks. IEEE Trans. Knowl. Data Eng. **26**(8), 1838–1852 (2014)
10. Gonzalez-Pardo, A., Jung, J.J., Camacho, D.: Aco-based clustering for ego network analysis. Futur. Gener. Comput. Syst. **66**, 160–170 (2017)
11. Greene, D., Doyle, D., Cunningham, P.: Tracking the evolution of communities in dynamic social networks. In: 2010 International Conference on Advances in Social Networks Analysis and Mining (ASONAM), pp. 176–183. IEEE (2010)
12. Kim, K., McKay, R.I., Moon, B.R.: Multiobjective evolutionary algorithms for dynamic social network clustering. In: Proceedings of the 12th Annual Conference on Genetic and Evolutionary Computation, pp. 1179–1186. ACM (2010)
13. Klimt, B., Yang, Y.: Introducing the enron corpus. In: CEAS (2004)
14. Lara-Cabrera, R., Gonzalez-Pardo, A., Barhamgi, M., Camacho, D.: Extracting radicalisation behavioural patterns from social network data. In: 2017 28th International Workshop on Database and Expert Systems Applications (DEXA), pp. 6–10 (2017). https://doi.org/10.1109/DEXA.2017.18
15. Lara-Cabrera, R., Gonzalez-Pardo, A., Camacho, D.: Statistical analysis of risk assessment factors and metrics to evaluate radicalisation in Twitter. Futur. Gener. Comput. Syst. (2017)

16. Lara-Cabrera, R., Pardo, A.G., Benouaret, K., Faci, N., Benslimane, D., Camacho, D.: Measuring the radicalisation risk in social networks. IEEE Access **5**, 10892–10900 (2017). https://doi.org/10.1109/ACCESS.2017.2706018
17. Li, S., Chen, Y., Du, H., Feldman, M.W.: A genetic algorithm with local search strategy for improved detection of community structure. Complexity **15**(4), 53–60 (2010)
18. Lin, Y.R., Chi, Y., Zhu, S., Sundaram, H., Tseng, B.L.: Facetnet: a framework for analyzing communities and their evolutions in dynamic networks. In: Proceedings of the 17th International Conference on World Wide Web, pp. 685–694. ACM (2008)
19. Malm, A., Bichler, G.: Networks of collaborating criminals: assessing the structural vulnerability of drug markets. J. Res. Crime Delinq. **48**(2), 271–297 (2011)
20. Mathias, S.B., Rosset, V., Nascimento, M.C.: Community detection by consensus genetic-based algorithm for directed networks. Procedia Comput. Sci. **96**, 90–99 (2016)
21. Newman, M.E., Girvan, M.: Finding and evaluating community structure in networks. Phys. Rev. E **69**(2), 026113 (2004)
22. Park, Y., Song, M.: A genetic algorithm for clustering problems. In: Proceedings of the Third Annual Conference on Genetic Programming, pp. 568–575 (1998)
23. Pizzuti, C.: Boosting the detection of modular community structure with genetic algorithms and local search. In: Proceedings of the 27th Annual ACM Symposium on Applied Computing, pp. 226–231. ACM (2012)
24. Raghavan, U.N., Albert, R., Kumara, S.: Near linear time algorithm to detect community structures in large-scale networks. Phys. Rev. E **76**(3), 036106 (2007)
25. Sun, J., Faloutsos, C., Papadimitriou, S., Yu, P.S.: Graphscope: parameter-free mining of large time-evolving graphs. In: Proceedings of the 13th ACM SIGKDD International Conference on Knowledge Discovery and Data Mining, pp. 687–696. ACM (2007)
26. Wang, S., Zou, H., Sun, Q., Zhu, X., Yang, F.: Community detection via improved genetic algorithm in complex network. Inf. Technol. J. **11**(3), 384–387 (2012)

On the Design and Tuning of Machine Learning Models for Language Toxicity Classification in Online Platforms

Maciej Rybinski[1(✉)], William Miller[2], Javier Del Ser[3,4,5], Miren Nekane Bilbao[5], and José F. Aldana-Montes[1]

[1] University of Málaga, 29071 Málaga, Spain
{maciek.rybinski,jfam}@lcc.uma.es
[2] Anami Precision, San Sebastián, Spain
william.miller@anamiprecision.com
[3] TECNALIA, Bizkaia, Spain
javier.delser@tecnalia.com
[4] Basque Center for Applied Mathematics (BCAM), Bizkaia, Spain
[5] University of the Basque Country (UPV/EHU), 48013 Bilbao, Spain
nekane.bilbao@ehu.eus

Abstract. One of the most concerning drawbacks derived from the lack of supervision in online platforms is their exploitation by misbehaving users to deliver offending (toxic) messages while remaining unknown themselves. Given the huge volumes of data handled by these platforms, the detection of toxicity in exchanged comments and messages has naturally called for the adoption of machine learning models to automate this task. In the last few years Deep Learning models and related techniques have played a major role in this regard due to their superior modeling capabilities, which have made them stand out as the prevailing choice in the related literature. By addressing a toxicity classification problem over a real dataset, this work aims at throwing light on two aspects of this noted dominance of Deep Learning models: (1) an empirical assessment of their predictive gains with respect to traditional Shallow Learning models; and (2) the impact of using different text embedding methods and data augmentation techniques in this classification task. Our findings reveal that in our case study the application of non-optimized Shallow and Deep Learning models attains very competitive accuracy scores, thus leaving a narrow improvement margin for the fine-grained refinement of the models or the addition of data augmentation techniques.

Keywords: Natural language processing · Deep learning
Online platforms · Multilabel classification

1 Introduction

There is no doubt that Internet has unleashed new ways for people to freely express their ideas and stimulate debate around topics of common interest, such

J. Del Ser et al. (Eds.): IDC 2018, SCI 798, pp. 329–343, 2018.
https://doi.org/10.1007/978-3-319-99626-4_29

as politics, economics and equality. Indeed, the genuine upsurge of technological means for this purpose – from discussion forums to blogs, social networks and alike – has lain at the core of the rise of well-known social movements that have found in these tools an efficient manner to reach ideological consensus, coordinate actions and gain momentum in the worldwide panorama. Examples of this noted role of Internet in the public sphere abound nowadays, such as the use of Twitter messages as an information source of politicians' statements, or the growing use of social platforms to advertise crowd-sourcing campaigns for social-aware initiatives.

However, the unregulated principle by which such platforms operate has lately come into hot controversy due to their lack of sensitiveness with respect to the intentions of the messages exchanged between their users. This absence of control and supervision has led to a myriad of misuses of these platforms, from intolerable threats among users to terrorist propaganda or gender-based hate speech. Unfortunately, the number of users intensively using these platforms makes it highly unfeasible to manually check through all the exchanged messages in search of inappropriate content. This challenge demands intelligent technologies and tools to automate the detection of offenses and misuses of these platforms, among which machine learning has emerged as a reportedly successful alternative in this regard. Computational models learned from the data posted in these platforms have been shown to attain notable performance when detecting radicalization [4], criminal recidivism [3], sex offenders [2] or impersonation [12,13].

Within the portfolio of machine learning models and techniques, it is widely agreed that Deep Learning has yielded a groundbreaking advance in predictive modeling, particularly for text processing and classification [5,7,11]. As a matter of fact, most contributions dealing with text analysis opt for different flavors of Recurrent Neural Networks (RNN) and Convolutional Neural Networks (CNN) due to their renowned capacity not only to detect patterns relating the documents at hand to a number of categories, but also to infer the most relevant features that discriminate between such categories. Beyond new neural network architectures and learning algorithms, this unprecedented capacity of Deep Learning models to learn from raw data has also been propelled by further neural processing layers and techniques with different objectives, such as overfitting prevention (Dropout), the integration of non-numerical data (text embeddings) or the addition of synthetic data to balance the class distribution of the entire dataset (data augmentation).

Unfortunately, the construction of Deep Learning models poses by itself a complex challenge due to the design parameters that must be set prior to training [6]. As a result of the last years of experimentation and practice, the community has reached a consensus on general guidelines to construct Deep Learning models [8,14]. However, in practice, the need for further tailoring the model structure to the specificities of the use case at hand remains. Nevertheless, in certain application domains the potential of Deep Learning to learn from raw data has overshadowed this model design complexity. Document (text) classification is arguably one of the applications where RNN models have become the typical *off-the-shelf* choice.

When dealing with text classification, the complexity inherent in the model design is due to manifold reasons. To begin with, models handle text, thereby requiring the derivation and implementation of methods to translate text into numerical feature vectors (vectorization). Such methods should preserve the label separability in the document representation, which should take into account the particularities of the classification problem at hand. When it comes to detecting offending (*toxic*) comments in online platforms, text indicators related to certain classes of toxicity do not necessarily occur distributed all over the document, but can also be localized in a certain part. The overall approach should somehow be designed to take into account the distribution of potential text indicators over the document (comment) to be classified. Furthermore, avant-garde techniques in data augmentation have been employed to enhance the detection of certain minority classes in multilabel/multiclass classification settings. In this particular case, empirical evidence should be sought for every problem to discern whether models endowed with data augmentation become specialized in the detection of the minority class for which the added data were created.

In this context, our manuscript elaborates on the identification of toxicity in comments posted in an online platform, a task that can be modeled as a multi-label classification problem and solved by diverse supervised learning methods. Specifically, the work presented in what follows provides informed insights on two main research questions, which are summarized next:

Q1. We provide empirical evidence that the effort normally (i.e. with *traditional* models) devoted to feature engineering in Shallow Learning is simply redirected to optimizing network structure/parameters and data augmentation in a Deep Learning approach. We show that in the multilabel classification problem at hand, competitive results can be obtained with a Logistic Regression (LR) model, and that without any further performance-driven fine tuning, surpassing these results with a proven *deep* model is not a trivial task. As we will later show, the high scores attained by LR leave a relatively narrow performance gain margin that a Deep Learning model could eventually achieve upon a costly design or tuning of its constituent elements. This is an important conclusion of utmost practical relevance in terms of design effort.

Q2. We evaluate how to achieve this performance gain by virtue of different combinations of embedding techniques and data augmentation methods. In particular, we inspect two different aspects: (1) the comparative performance of a simple oversampling approach aimed to avoid the loss of information when truncating comments in the text sample representation phase, while effectively serving as a balancing method for minority classes with sparse predictors along the text; and (2) the gains per class yielded by data augmentation and embedding techniques, towards grasping the intuition behind the use of *blends* (i.e. weighted model ensembles) in winning approaches of competitions and challenges related to text classification[1].

The final goal of our study is to underscore that in some practical cases, the application of Shallow and Deep Learning models without a fine-grained design optimization can report already solid accuracy results, leaving a relatively small margin for accuracy improvement. In the experimental evaluation presented here we showcase the process of building and fine-tuning of both shallow and deep classifiers. In the specific problem tackled in this manuscript the contribution of embeddings and data augmentation techniques is analyzed and found to be *local* rather than *global*; in other words, their application did not imply an improved model across all labels, but rather a specialized model for a certain label subset.

The rest of the paper is structured as follows: Sect. 2 describes the dataset, embeddings, classifiers and data augmentation techniques utilized in this study. Next, results are presented and discussed in Sect. 3. Finally, concluding remarks and future research lines are given in Sect. 4.

2 Materials and Methods

In this section we present a detailed overview of our experimental settings. As the multi-label classification task is closely matched to the dataset used in the experiments, we start by introducing a brief description of the data and the task itself. We then discuss the text representation methods used in our experiments (text-based features, embeddings) and the implication of using embedding-based text representations for the deep models. Then, we present the classification models used in the experiments (shallow and deep architectures). Finally, we present the data augmentation methods we evaluate in the context of deep learning models, i.e. the *double translation* and *sliding window oversampling*.

2.1 Data and Task

In all our experiments we use the dataset from the Toxic Comment Classification Challenge[2], which is the most complete real-world-sized text toxicity dataset to date. The original dataset has 159571 training instances and 153166 test instances, with each text annotated for toxicity by human evaluators. The task consists of deciding whether the input text (a Wikipedia discussion comment) belongs to any of the six toxicity classes: 'toxic', 'severe toxic', 'obscene', 'threat', 'insult', 'identity hate'. As the test target labels used in the competition have not been made public, we opt to use a static 9:1 split of the original training set stratified over the least imbalanced category ('toxic'). Therefore, in this work by *test* and *training* datasets we refer to this split (on original training data) of 143613 training samples and 15958 validation samples. Table 1 shows statistics of the data used in our experimental setup.

[1] See e.g. http://mlwave.com/kaggle-ensembling-guide/.

[2] http://www.kaggle.com/c/jigsaw-toxic-comment-classification-challenge.

Table 1. Overview of the dataset used in our evaluation. Comment lengths referred to in the table are lengths in words.

		Training	Test
Number of samples (%)	Total	143613	15958
	Toxic (TOXIC)	13764 (9.58%)	1530 (9.58%)
	Severe toxic (SEVERE)	1427 (0.99%)	168 (1.05%)
	Obscene (OSBCENE)	7606 (5.3%)	843 (5.28%)
	Threat (THREAT)	422 (0.29%)	56 (0.35%)
	Insult (INSULT)	7085 (4.93%)	792 (4.96%)
	Identity hate (IDENTITY)	1269 (0.88%)	136 (0.85)%
Comment length statistics	Average	70.29	70.4
	Standard deviation	103.65	106.96
	Maximum	1357	1404

Statistics and numbers in Table 1 show that in our split, despite strictly stratified only w.r.t. (with respect to) the 'toxic' class, test data roughly (rank correlation of 1) maintains the class distribution of the training data.

2.2 Text Representations

In our experiments we have used distinct text representation methods to match the models included in the study. 'Traditional' shallow learning models (i.e. logistic regression variants) are paired with 'traditional' text representations derived from the weighted bag-of-words/n-grams scheme. Deep Neural Network models (DNNs) operate on texts modeled as sequences of words, where each word is mapped to a real valued vector of a certain dimension (e.g. 200), its embedding. As a result, for DNN processing texts are represented as matrices whose i-th row is an embedding vector of an i-th word.

As far as the embeddings are concerned, we have adopted two widely accepted variants (and a combined approach): (A) we create stochastic embedding vectors at training time through the training of the embedding layer of our DNNs; (B) we use pre-trained word embeddings created through unsupervised learning of word representations derived from the co-occurrences in a large external corpus. The latter method is expected to maintain semantic affinity between words in a sense that semantically related words are expected to generate similar vector representations. For (B) we have used vectors pre-trained on Common Crawl corpus with GloVe algorithm [9]. A more complete description of the text representations involved in our experiments is given below.

2.2.1 Bag-of-Words (BoW) Representations

For the shallow learning models input texts (comments) are represented as vectors, i.e. each text is represented as one vector over the feature-space. The feature-space combines word and character based features. Formal definition of BoW model with word-based features is as follows. Given a collection of training samples $D = \{d_1, d_2, ...d_n\}$, let $V = \{w_1, w_2, ...w_m\}$ denote the set of distinct words present in the collection D. We can now define a term vector v to represent the document d: $v = (tf_d(w_1), tf_d(w_2), ...tf_d(w_m))$, where $tf_d(w)$ denotes the *frequency* of word (term) w in document d, i.e. the number of times w appears in d. A set of term vectors, one for each document, constitutes a vector space model (VSM). There are many flavors of vector space models. In this study we use one of the most popular, with term frequencies normalized by inverse document frequencies (IDF), in which: $v = (tf_d(w_1) \times idf(w_1), tf_d(w_2) \times idf(w_1), ...tf_d(w_m) \times idf(w_1))$, where $idf(w) = \log(n/|\{d : w \in d\}|)$. It is worth noting that at classification time the input documents are only represented w.r.t. terms *seen* in the training corpus.

Our character-based text representation also follows the above definition, with the only change being replacing words with n-gram ranges as tokens/terms. For example, 'sample text' has two tokens in the word-based model: {'sample', 'text'}. For a 1–2 n-gram range the same text is tokenized as 's', 'a', 'm', 'p', 'l', 'e', '', 't', 'e','x', 't', 'sa', 'am', 'mp', 'pl', 'le', 'e', 't', 'te', 'ex', 'xt'. To encode texts we use both of the above models, concatenating their features for each document (i.e. by horizontally stacking the matrices representing each of the vector space models).

Our implementation of this representation scheme involves two additional parameters: maximum number of features and n-gram range for the character-based tokenization. We use 15000 and 20000 most frequent[3] features for words and char n-grams respectively. The n-gram range we use for character-based feature is 1–5. We also apply sub-linear term frequency scaling, so $tf_d(w)$ in our formula is actually replaced with $1 + \log(tf_d(w))$.

2.2.2 Word Embedding

Word embedding is a collective name for a set of techniques, which map words from a vocabulary to a dense real-valued vector of a certain number of dimensions. The idea is conceptually closely related to that of the vector space model. VSMs allow creating numerical representations of documents (as vectors) to cater for document-document comparison and numerical processing. Word embeddings, deeply rooted in the principles of distributional linguistics (the famous '*you shall know a word by the company it keeps*', J.R. Firth), aim to achieve the same for words, while avoiding the vector sparsity, i.e. similar words should generate similar vectors.

Word embeddings can be included in a DNN through an embedding layer, which transforms input word-based one-hot representations (more on this below) into the dense vector representations. The inclusion of embeddings is related with two design decisions. We can define the embedding matrix M of dimensions $m \times k$, where m is the size of our vocabulary and k is the dimensionality of the embeddings. In a basic scheme, M is initialized stochastically **and** then modified

[3] Frequent in terms of total term frequency in the entire collection of training documents.

during DNN training (optimized through backpropagation), in which case we can talk about a *trainable* embedding layer. A common alternative design pattern involves the use of pre-trained embeddings obtained through an algorithm which has been previously validated w.r.t. its ability to maintain the affinity between words in the embedding process. In the second case it is commonly assumed that the M no longer needs optimization at training time. These two approaches are often referred to as *trainable* vs *pre-trained* embeddings (a denomination we will use further in this study), although technically trainability can be combined with the use of pre-trained vector dictionary (the latter will be referred to as *hot-started trainable embeddings*).

While trainable embeddings are not likely to maintain word affinity in general, they are optimized to do so when it matters to the output of the classifier. The use of pre-trained embeddings, on the other hand, eliminates a bulk ($m \times k$; consider an example vocabulary of 50K and a fairly standard embedding vector size 200) of trainable parameters from the classification model. In our experimental evaluation we consider both approaches, both with $k = 300$. For pre-trained embeddings we use GloVe [9] Common Crawl vector dictionary.

2.2.3 Text Representations in Deep Learning

As already stated in the introduction, DNNs have the capability of working directly with texts with little preprocessing. DNN models contemplated in this study train on encoded word sequences (and classify encoded word sequences), rather than on BoWs. We can refer to our previous definition of a vocabulary, being an ordered list of words present in the training data: $V = \{w_1, w_2, ...w_m\}$. Document d, can be now defined as a sequence of length l of words from V, which can be encoded as a $l \times m$ matrix of binary elements ($\{0, 1\}$) with at most one non-zero value per row. Rows of the matrix correspond to words in d (i.e. first row is the first word, etc.) and columns correspond to the positions in the entire vocabulary. So, for example, a non-zero element at position (i, j) would mean that w_j is the i-th word of document d. This text representation scheme provides more information than BoW-based approaches (i.e. preserves the word order), but it is computationally inefficient. This is where the embeddings play a key role. The embedding layer substitutes the one-hot m-dimensional sparse word vectors with their dense k-dimensional embeddings, transforming the input matrix to its $l \times k$ embedded representation.

For the sake of computational efficiency, input documents are often also trimmed w.r.t. to their length in words with a maximum length parameter (e.g. $l_{max} = 200$), which decreases the representation size along the other axis. This common approach is based on an assumption that it is enough to 'read' a certain portion of a document to categorize it correctly. In our parameters we use the value of $l_{max} = 200$, we also restrict the size of our vocabulary to 100000 (the same is done with BoW representations).

2.3 Shallow Learning Techniques

As anticipated in the introduction, question Q1 addressed in this work will be explored by resorting to a simple LR shallow learning model, whose simplicity will buttress even further one of the main conclusions of this work. The learning procedure of this model essentially reduces to the fitting of a parametric sigmoid function, which models the class probability given some linear combination of the input features. Furthermore, given the imbalance rate of some of the classes in the dataset we will utilize a well-known balancing technique for increasing the sensitivity of the LR classifier to minority labels. Among the portfolio of alternatives available for this purpose we opt for SMOTE (Synthetic Minority Over-sampling Technique, [1]), an oversampling technique that generates new synthetic instances in the neighborhood of already available data examples for the minority class at hand.

2.4 Deep Learning Techniques

From a wide selection of approaches evaluated and discussed during the competition we have chosen two, which appeared to have been the most successful single models: bidirectional GRU with convolution + pooling and CapsNet.

The bidirectional network used in our experiments[4] is conceptually similar to the one proposed in [15]. The model consists of an embedding layer, followed by a bidirectional GRU of 128 units. This layer is then analyzed with a one-dimensional convolutional layer of 64 filters. The output of the convolutional layer is passed to two pooling layers (average and max pooling), results of which are concatenated and passed on to the fully connected output layer directly. An overview of the architecture is depicted in Fig. 1.

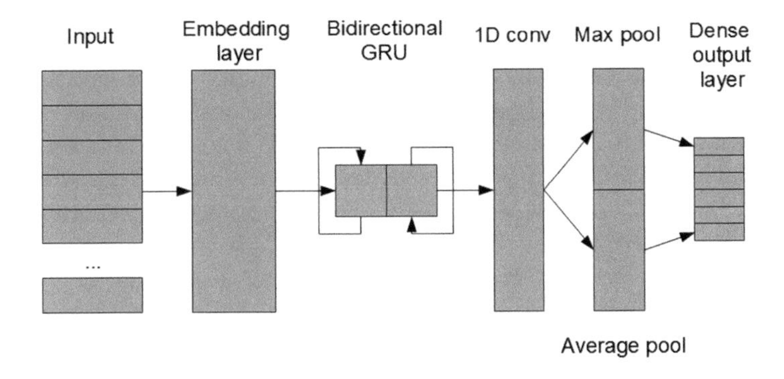

Fig. 1. Overview of the bidirectional GRU model.

On the other hand, CapsNet is a DNN architecture introduced recently in [10]. CapsNet is conceptually related to CNN, but it postulates using groups of

[4] https://www.kaggle.com/eashish/bidirectional-gru-with-convolution.

specialized neurons (capsules) to handle specific tasks. CapsNet uses a dynamic routing mechanism to chose the most appropriate capsule to handle a given problem. The intelligent routing is expected to provide an improvement over CNN, which route information through simple pooling operations. In our experiments we use a CapsNet variation similar to our GRU-based architecture[5]. CapsNet differs from the model described above in having the Capsule layer in place of convolutional layer with pooling. We use 10 capsules, with the capsule dimension of 16.

2.5 Data Augmentation

As stated previously, one of our main goals was to isolate and assess the benefits of textual data augmentation for the multi-label classification problem. We have evaluated two approaches and their combination.

First of the approaches, *double translation*, has been already used in the context of the competition[6]. It is based on augmenting the training dataset with comments automatically translated to another language and back to English. Training labels of the original comment are kept for the transformed one. In our experiments we have used three *intermediate* languages (translation is from English, to *intermediate*, to English): German, French, Spanish. We evaluate two ways of using the technique: as a general augmentation scheme (by applying the transformations to all training samples) and as a mean of balancing through augmentation (selective augmentation, we apply the transformations only to comments, which display some kind of toxicity).

Second method of data augmentation evaluated here is our simple oversampling scheme based on the idea of fully utilizing the 'toxic' training samples. As already mentioned, for practical reasons, it is common to truncate the comments at a certain number of (commonly) first words. Our idea is based on an observation that some labels are so rare, that using only first 200 words seems like discarding already scarce resources – instead of using the rest of the comment as well.

The method we propose is very simple and is based on sliding a window of size l_{max} through the comments longer than l_{max} to generate additional training samples. Sliding of the window is done with a step parameter, which determines when the next window starts w.r.t. to the start of the previous one. Particularly, windows overlap if the step is shorter than the window length. Our experiments compare overlapping and non-overlapping approaches.

3 Results and Discussion

Tables 2 and 3 present the summary of our experimental evaluation. We report mean and weighted ROC AUC scores for each model, apart from the 'per-class'

[5] https://www.kaggle.com/chongjiujjin/capsule-net-with-gru.
[6] https://github.com/PavelOstyakov/toxic/tree/master/tools.

results. To provide a full perspective on the per-class scores, for each toxicity class we report 3 typically used quality measures to assess the classification quality: ROC AUC, Hamming loss and F1 scores.

One of the most straightforward conclusions that can be drawn from our results is that a simple logistic regression baseline without 'per class' feature engineering provides very good overall performance (considering mean and weighted mean ROC-AUC scores; see result 47), being only slightly worse than some deep approaches (e.g. un-tuned CapsNet with randomly initialized embeddings, see result 33). One of the most important take-aways is that this single model already provides a decent approximation of what the single-model final performance might be, which is especially valuable given the fact that it is much faster to train than DNNs and needs little tuning to work.

Regarding the 'promise' of avoiding the need for feature engineering by 'going deep', we think it is worth pointing out that, at least in the context of text classification, the feature engineering effort is merely redirected to optimizing other parameters, network structure, augmentation strategy, embeddings, etc. Given that DNNs typically take much longer to train and have more degrees of freedom, this optimization becomes less straightforward. That being said, we have obtained the best results with deep networks (both architectures) with the mean ROC-AUC score (the same measure was used in the actual competition) surpassing 0.99 (in the case of CapsNet; result 7).

Moreover, the scores presented in Tables 2 and 3 seem to explain the surge in popularity of *blends*, *stacks* and other *ensembles* (e.g. a weighted average, for each class, of outputs of different classifiers or a different classifier for each class), at least in the context of classification competitions. It can be seen that different experiments (model, augmentation, embeddings combinations) varied significantly, so the best 'per-class' results are scattered between the rows of these tables. In this context blending seems to be a reasonable strategy, as it could allow us to get the best of top-performing configurations/classifiers. For the same reason, in the context of preparing a blend of classifiers, it is worthwhile to pursue significant improvements w.r.t. some classes at the expense of the others – the sum of improvements matters, rather than the overall performance of any given specific configuration.

The latter observation also sheds more light on the popularity of data augmentation techniques in the top contributions within the competition. While neither of the methods we evaluated improves the overall quality of any given model (in terms of weighted AUC), as any type of augmentation would cause a performance dip w.r.t. the most represented positive label (general 'toxic'; compare results 1, 9, 17, 25, 33, 39 against corresponding respective augmented configurations). Nonetheless, both techniques clearly are beneficial for the predictive capability on the labels that were leaning more towards the underrepresented side in the training dataset.

As far as the *double translation* augmentation is concerned, we have compared augmentation translating all comments vs translating only comments, which display some kind of toxicity. Augmentation of the entire dataset would tend

Table 2. Scores (AUC, Hamming Loss, F1) achieved by every model over each toxicity class in the dataset (1/2).

INDEX	MODEL	Avg. AUC	Weighted AUC	TOXIC	SEVERE	OSBCENE	THREAT	INSULT	IDENTITY
1	CapsNet - GloVe init (trainable) - Augmentation: none - Balancing: none	0.987375	0.986195	0.981316 0.035531 0.805755	0.992470 0.009588 0.331878	0.993097 0.017734 0.829415	0.983776 0.003447 0.266667	0.987057 0.025880 0.736102	0.986532 0.006392 0.560345
2	CapsNet - GloVe init (trainable) - Augmentation: double translation (toxic) - Balancing: none	0.988022	0.977464	0.957308 0.232548 0.443378	0.996416 0.004825 0.763077	0.991891 0.038163 0.724060	0.997687 0.001253 0.824561	0.993156 0.020805 0.816979	0.991674 0.004575 0.763754
3	CapsNet - GloVe init (trainable) - Augmentation: double translation (all) - Balancing: none	0.981484	0.983812	0.978753 0.036283 0.802187	0.991378 0.010465 0.379182	0.991965 0.018423 0.829070	0.961882 0.003509 0.000000	0.985630 0.026632 0.738139	0.979295 0.007206 0.410256
4	CapsNet - GloVe init (trainable) - Augmentation: sliding window (step=200) - Balancing: none	0.981963	0.981073	0.974094 0.038664 0.786431	0.990132 0.010402 0.481250	0.990188 0.018047 0.823313	0.975630 0.003133 0.285714	0.983870 0.025316 0.730667	0.977863 0.006893 0.455446
5	CapsNet - GloVe init (trainable) - Augmentation: sliding window (step=100) - Balancing: none	0.980924	0.980105	0.972954 0.037411 0.792059	0.990830 0.010277 0.405797	0.988960 0.018298 0.817272	0.973554 0.003321 0.311688	0.983418 0.025692 0.725569	0.975826 0.006329 0.471204
6	CapsNet - GloVe init (trainable) - Augmentation: double translation (toxic) + sliding window (step=200) - Balancing: none	0.985326	0.973740	0.948569 0.339203 0.350414	0.995314 0.007332 0.674095	0.992057 0.022747 0.811232	0.987197 0.002193 0.631579	0.993660 0.016794 0.836983	0.995161 0.004950 0.714801
7	CapsNet - GloVe init (trainable) - Augmentation: double translation (toxic) + sliding window (step=100) - Balancing: none	0.990135	0.980016	0.963348 0.156348 0.535122	0.998227 0.004324 0.796460	0.992255 0.032022 0.757014	0.999238 0.001441 0.818898	0.991112 0.025379 0.785601	0.996632 0.006580 0.700855
8	CapsNet - GloVe init (trainable) - Augmentation: double translation (all) + sliding window (step=200) - Balancing: none	0.983432	0.983913	0.979140 0.037536 0.801063	0.990963 0.010465 0.433898	0.991459 0.019175 0.822093	0.975885 0.003321 0.184615	0.985198 0.026006 0.738170	0.977949 0.007645 0.371134
9	CapsNet - GloVe init (non-trainable) - Augmentation: none - Balancing: none	0.987416	0.987666	0.984194 0.033463 0.829936	0.992110 0.009776 0.370968	0.993520 0.017546 0.837587	0.981109 0.002945 0.356164	0.988032 0.025442 0.756886	0.985528 0.006078 0.580087
10	CapsNet - GloVe init (non-trainable) - Augmentation: double translation (toxic) - Balancing: none	0.966319	0.942205	0.891341 0.473681 0.273243	0.992422 0.009086 0.607046	0.981785 0.021619 0.811166	0.977822 0.003133 0.456522	0.979532 0.027886 0.740525	0.975012 0.007332 0.502128
11	CapsNet - GloVe init (non-trainable) - Augmentation: double translation (all) - Balancing: none	0.982438	0.980167	0.973992 0.037035 0.795005	0.990105 0.009650 0.472603	0.988262 0.020805 0.803084	0.978348 0.003133 0.358974	0.981034 0.027572 0.725343	0.982888 0.007394 0.458716
12	CapsNet - GloVe init (non-trainable) - Augmentation: sliding window (step=200) - Balancing: none	0.980729	0.978816	0.971517 0.038852 0.781690	0.989821 0.010214 0.350598	0.988729 0.019865 0.808690	0.980980 0.002883 0.361111	0.980992 0.027322 0.720513	0.972337 0.006830 0.441026
13	CapsNet - GloVe init (non-trainable) - Augmentation: sliding window (step=100) - Balancing: none	0.981124	0.978624	0.970603 0.038225 0.788194	0.989916 0.009776 0.376000	0.989121 0.019865 0.802983	0.980745 0.002820 0.430380	0.980956 0.027196 0.706360	0.975403 0.006705 0.427807
14	CapsNet - GloVe init (non-trainable) - Augmentation: double translation (toxic) + sliding window (step=200) - Balancing: none	0.959927	0.937232	0.889368 0.500439 0.259733	0.992428 0.009149 0.578035	0.976841 0.020617 0.813598	0.960584 0.002820 0.366197	0.968178 0.029014 0.720242	0.972159 0.006893 0.438776
15	CapsNet - GloVe init (non-trainable) - Augmentation: double translation (toxic) + sliding window (step=100) - Balancing: none	0.963917	0.945325	0.903563 0.488282 0.265322	0.991818 0.008961 0.511945	0.979515 0.021118 0.804864	0.964302 0.002945 0.373333	0.974141 0.027760 0.731026	0.970165 0.006517 0.469388
16	CapsNet - GloVe init (non-trainable) - Augmentation: double translation (all) + sliding window (step=200) - Balancing: none	0.979778	0.976608	0.970227 0.039291 0.776948	0.988518 0.009776 0.458333	0.981744 0.022309 0.787336	0.976943 0.002695 0.410959	0.980149 0.028700 0.712673	0.981085 0.006830 0.493023
17	Bi-GRU - GloVe init (non-trainable) - Augmentation: none - Balancing: none	0.988949	0.987576	0.983399 0.032836 0.824632	0.991541 0.009838 0.384314	0.993811 0.016543 0.843602	0.987444 0.002757 0.541667	0.987813 0.025316 0.744627	0.989685 0.006204 0.512315
18	Bi-GRU - GloVe init (non-trainable) - Augmentation: double translation (toxic) - Balancing: none	0.964248	0.934249	0.874822 0.438589 0.284868	0.991964 0.009149 0.425197	0.979461 0.019238 0.816278	0.988196 0.002695 0.505747	0.978253 0.024690 0.741809	0.972793 0.006517 0.490196
19	Bi-GRU - GloVe init (non-trainable) - Augmentation: double translation (all) - Balancing: none	0.973328	0.97457	0.966293 0.038915 0.777658	0.989509 0.010465 0.365019	0.985179 0.020491 0.800731	0.953853 0.003321 0.361446	0.979017 0.028324 0.695418	0.966116 0.007770 0.373737
20	Bi-GRU - GloVe init (non-trainable) - Augmentation: sliding window (step=200) - Balancing: none	0.977318	0.9749767	0.966621 0.040231 0.772502	0.990315 0.009650 0.421053	0.985482 0.020115 0.809496	0.972418 0.002820 0.400000	0.977433 0.026382 0.721377	0.971639 0.006830 0.429319
21	Bi-GRU - GloVe init (non-trainable) - Augmentation: sliding window (step=100) - Balancing: none	0.973679	0.971520	0.963065 0.041233 0.758974	0.987318 0.010528 0.236364	0.982150 0.019927 0.804187	0.967118 0.002757 0.450000	0.974036 0.027134 0.717915	0.968386 0.006454 0.529680
22	Bi-GRU - GloVe init (non-trainable) - Augmentation: double translation (toxic) + sliding window (step=200) - Balancing: none	0.966473	0.934999	0.878400 0.416218 0.288865	0.994097 0.008585 0.529210	0.974509 0.021682 0.802059	0.986557 0.002131 0.613636	0.976945 0.023625 0.752787	0.988329 0.006392 0.548673
23	Bi-GRU - GloVe init (non-trainable) - Augmentation: double translation (toxic) + sliding window (step=100) - Balancing: none	0.970249	0.938829	0.882915 0.364143 0.323437	0.992894 0.008522 0.492537	0.975640 0.037035 0.714631	0.996115 0.002131 0.638298	0.983241 0.029264 0.736307	0.990687 0.005264 0.691176
24	Bi-GRU - GloVe init (non-trainable) - Augmentation: double translation (all) + sliding window (step=200) - Balancing: none	0.974095	0.973615	0.967236 0.039792 0.781110	0.985103 0.010152 0.429577	0.979138 0.022873 0.772869	0.963370 0.002883 0.378378	0.978813 0.028512 0.705882	0.970914 0.007018 0.445545
25	Bi-GRU - GloVe init (trainable) - Augmentation: none - Balancing: none	0.989180	0.987549	0.983525 0.033087 0.820774	0.992550 0.009713 0.279070	0.992998 0.016919 0.838323	0.989311 0.002883 0.410256	0.988171 0.024376 0.746910	0.988524 0.006454 0.572614

Table 3. Scores (AUC, Hamming Loss, F1) achieved by every model over each toxicity class in the dataset (2/2).

INDEX	MODEL	Avg. AUC	Weighted AUC	TOXIC	SEVERE	OSBCENE	THREAT	INSULT	IDENTITY
26	Bi-GRU - GloVe init (trainable) - Augmentation: double translation (toxic) - Balancing: none	0.986229	0.969349	0.936746 0.289823 0.388632	0.996688 0.004324 0.783699	0.993826 0.021995 0.817852	0.999769 0.001065 0.857143	0.993631 0.022998 0.804058	0.996713 0.003384 0.818792
27	Bi-GRU - GloVe init (trainable) - Augmentation: double translation (all) - Balancing: none	0.983077	0.982592	0.975787 0.038852 0.798308	0.990559 0.010465 0.418118	0.991235 0.019927 0.821749	0.972811 0.003447 0.035088	0.985553 0.026006 0.743669	0.982519 0.006454 0.507177
28	Bi-GRU - GloVe init (trainable) - Augmentation: sliding window (step=200) - Balancing: none	0.983281	0.981413	0.974470 0.036408 0.796782	0.988654 0.010528 0.348837	0.990117 0.017546 0.822560	0.985966 0.002883 0.439024	0.984663 0.024878 0.730116	0.975819 0.006642 0.490385
29	Bi-GRU - GloVe init (trainable) - Augmentation: sliding window (step=100) - Balancing: none	0.982856	0.980436	0.973488 0.038225 0.791382	0.989845 0.010026 0.148936	0.988411 0.018925 0.821302	0.985228 0.002695 0.410959	0.983721 0.026695 0.700000	0.976441 0.006893 0.427083
30	Bi-GRU - GloVe init (trainable) - Augmentation: double translation (toxic) + sliding window (step=200) - Balancing: none	0.983766	0.967697	0.939676 0.262063 0.411483	0.997735 0.004825 0.742475	0.987088 0.035656 0.730715	0.996845 0.002381 0.720588	0.988445 0.024815 0.781457	0.992804 0.004011 0.761194
31	Bi-GRU - GloVe init (trainable) - Augmentation: double translation (toxic) + sliding window (step=100) - Balancing: none	0.983101	0.965716	0.933639 0.266575 0.404202	0.996457 0.005890 0.711656	0.989432 0.028512 0.769853	0.998734 0.001629 0.771930	0.989224 0.025504 0.778202	0.991117 0.005389 0.718954
32	Bi-GRU - GloVe init (trainable) - Augmentation: double translation (all) + sliding window (step=200) - Balancing: none	0.983225	0.982099	0.975834 0.039165 0.788923	0.990422 0.010778 0.306452	0.990049 0.019426 0.808642	0.976550 0.003259 0.277778	0.984360 0.026131 0.719570	0.982139 0.006078 0.561086
33	CapsNet - Random init (trainable) - Augmentation: none - Balancing: none	0.976221	0.982580	0.979274 0.035593 0.805613	0.990275 0.010590 0.034286	0.990587 0.016857 0.839976	0.948764 0.003509 0.000000	0.984387 0.028074 0.714286	0.964042 0.008522 0.000000
34	CapsNet - Random init (trainable) - Augmentation: double translation (toxic) - Balancing: none	0.983944	0.970988	0.945519 0.304925 0.377590	0.995332 0.006329 0.687307	0.988818 0.045682 0.685369	0.990173 0.002131 0.645833	0.990931 0.022559 0.796610	0.992891 0.005765 0.714286
35	CapsNet - Random init (trainable) - Augmentation: double translation (all) - Balancing: none	0.978821	0.980386	0.973463 0.039040 0.791150	0.991056 0.009964 0.492013	0.990407 0.019175 0.817640	0.964168 0.003572 0.033898	0.983748 0.029202 0.703185	0.970082 0.008397 0.094595
36	CapsNet - Random init (trainable) - Augmentation: sliding window (step=200) - Balancing: none	0.977775	0.981984	0.976916 0.036784 0.800136	0.990533 0.010340 0.326531	0.991181 0.016669 0.834988	0.956560 0.003509 0.000000	0.984563 0.027008 0.715512	0.966895 0.008460 0.028777
37	CapsNet - Random init (trainable) - Augmentation: double translation (toxic) + sliding window (step=200) - Balancing: none	0.990244	0.982335	0.967481 0.128776 0.581552	0.995513 0.005953 0.712991	0.993684 0.022309 0.815353	0.997440 0.001880 0.722222	0.993032 0.020178 0.818489	0.994316 0.004324 0.762887
38	CapsNet - Random init (trainable) - Augmentation: double translation (all) + sliding window (step=200) - Balancing: none	0.977535	0.981265	0.975799 0.039855 0.790652	0.987112 0.010277 0.449664	0.990410 0.019865 0.816657	0.958716 0.003509 0.000000	0.984630 0.028262 0.720743	0.968540 0.008585 0.014388
39	Bi-GRU - Random init (trainable) - Augmentation: none - Balancing: none	0.980633	0.982951	0.977910 0.037097 0.799729	0.990881 0.010653 0.397163	0.991703 0.017609 0.829594	0.970133 0.003509 0.000000	0.985164 0.026758 0.722907	0.968007 0.008585 0.000000
40	Bi-GRU - Random init (trainable) - Augmentation: double translation (toxic) - Balancing: none	0.983200	0.965052	0.933082 0.285938 0.391843	0.998018 0.004011 0.782313	0.985627 0.064294 0.608098	0.999674 0.001316 0.823529	0.990851 0.028512 0.766786	0.991946 0.003008 0.827338
41	Bi-GRU - Random init (trainable) - Augmentation: double translation (all) - Balancing: none	0.984474	0.982499	0.975406 0.038539 0.795341	0.991038 0.010966 0.354244	0.991207 0.018611 0.820976	0.980084 0.003321 0.131148	0.985019 0.027071 0.716163	0.984088 0.007332 0.558491
42	Bi-GRU - Random init (trainable) - Augmentation: sliding window (step=200) - Balancing: none	0.983630	0.982031	0.974957 0.038100 0.786517	0.990915 0.010152 0.417266	0.990128 0.019363 0.804554	0.980261 0.003321 0.253521	0.985696 0.025692 0.723720	0.979822 0.005953 0.512821
43	Bi-GRU - Random init (trainable) - Augmentation: sliding window (step=100) - Balancing: none	0.982362	0.980701	0.973648 0.039291 0.788675	0.990479 0.010465 0.379182	0.989548 0.018925 0.807152	0.979600 0.003008 0.250000	0.983478 0.026131 0.730446	0.977420 0.006392 0.484848
44	Bi-GRU - Random init (trainable) - Augmentation: double translation (toxic) + sliding window (step=200) - Balancing: none	0.983651	0.966896	0.936130 0.223838 0.444997	0.996288 0.003948 0.817391	0.990685 0.044366 0.692975	0.999286 0.001441 0.813008	0.988311 0.036471 0.718569	0.991207 0.003635 0.797203
45	Bi-GRU - Random init (trainable) - Augmentation: double translation (toxic) + sliding window (step=100) - Balancing: none	0.983813	0.966568	0.934838 0.256423 0.417758	0.998043 0.003760 0.821429	0.988438 0.036972 0.724556	0.999334 0.001755 0.777778	0.991438 0.021995 0.807249	0.990786 0.003321 0.823920
46	Bi-GRU - Random init (trainable) - Augmentation: double translation (all) + sliding window (step=200) - Balancing: none	0.982441	0.981685	0.974793 0.035844 0.794096	0.989275 0.010841 0.352060	0.990368 0.018737 0.819771	0.972878 0.003008 0.272727	0.984587 0.026758 0.726457	0.982748 0.006642 0.495238
47	Logistic regression - Word and char n-gram features - Augmentation: none - Balancing: none	0.987026	0.984535	0.979695 0.055270 0.754864	0.990113 0.019426 0.486755	0.992265 0.024251 0.798542	0.992409 0.006141 0.461538	0.984078 0.039228 0.690099	0.983597 0.018486 0.431599
48	Logistic regression - Word and char n-gram features - Augmentation: none - Balancing: SMOTE	0.986116	0.983097	0.977337 0.041421 0.789289	0.989895 0.015353 0.531549	0.991559 0.019489 0.821980	0.992346 0.004449 0.523489	0.983261 0.0303297 0.729003	0.982299 0.015603 0.452747

to be less disruptive to the performance of the models on the 'toxic' class, but it did not contribute to any top scores within any of the labels, so it is safe to say that the classification on underrepresented labels benefits more from the selective approach (compare e.g. results 26 and 27).

While the augmentation technique we have proposed (sliding window oversampling) does not seem to contribute much in isolation, it does perform well when combined with the double translation. The combination of the methods contributed to some best 'scores per class' in our evaluation (e.g. result 7 of class *severe toxic*). While there is no conclusive evidence on whether the oversampling should or should not allow overlaps between the samples, we would say that the sliding window step should be treated as a hyper-parameter.

Both deep models seem to benefit from the hybrid approach towards word embedding, i.e. they work best with trainable embedding layer 'hot-started' with GloVe pre-trained embeddings, although the best result for the *toxic* class were achieved by a CapsNet model operating on non-trainable GloVe embeddings (see result 9).

As far as finding the best performing model, all the configurations involving GRU-based architecture with trainable hot-started embeddings and selective translation seem to provide very promising results w.r.t. the 5 minority labels, but the configuration without additional sliding window augmentation seems to work best (result 26). The overall (aggregated by mean or weighted mean) performance of the model is affected by its confusingly awful performance on the general *toxic* label, but if we were to compose a classifier ensemble we would probably start there.

4 Conclusions and Future Research Lines

Within this work we have elaborated on a comparative analysis between different machine learning models to undertake a language toxicity classification problem. This is a problem of utmost practical relevance for on-line platforms where the unregulated exchange of messages among their users opens up a risk for written offenses against certain collectives and individuals. The insights drawn in this study have gone beyond those typically drawn from a benchmark to assess different design choices. Indeed, we have empirically proven that in our case study a traditional logistic regression model is capable of achieving very competitive performance. More complex DNN approaches relying on advanced data augmentation techniques, neural processing layers and trainable word representations were found to only provide improvement in certain minority classes. This performance gain is decisive in score-based competition-style evaluations and other challenges alike, but in real deployment scenarios such performance differences can be out-weighted by the intensive design efforts needed to properly tune DNNs and their configurations.

In the near future we plan to invest efforts in several research directions. To begin with, new methods for data augmentation will be explored, with emphasis on recurrent Generative Adversarial Networks (GAN). In this regard we plan to derive alternative GAN models to selectively augment data in a class-by-class fashion. Likewise, the exploitation of formal relationships between labels could lead to avoiding the performance drop on the majority label when performing selective augmentation. Finally, we will inspect whether the lessons learned in this work can be extrapolated to other emerging text categorization problems (e.g. fake news detection).

Acknowledgements. The work of Maciej Rybinski has been partially funded by grant TIN2017-86049-R (Ministerio de Economa, Industria y Competitividad, Spain). Javier Del Ser also thanks the Basque Government for its funding support through the EMAITEK program.

References

1. Chawla, N.V., Bowyer, K.W., Hall, L.O., Kegelmeyer, W.P.: Smote: synthetic minority over-sampling technique. J. Artif. Intell. Res. **16**, 321–357 (2002)
2. Escalante, H.J., Villatoro-Tello, E., Garza, S.E., López-Monroy, A.P., Montes-y Gómez, M., Villaseñor-Pineda, L.: Early detection of deception and aggressiveness using profile-based representations. Expert. Syst. Appl. **89**, 99–111 (2017)
3. Hashim, E.N., Nohuddin, P.N.: Data mining techniques for recidivism prediction: a survey paper. Adv. Sci. Lett. **24**(3), 1616–1618 (2018)
4. Lara-Cabrera, R., Gonzalez-Pardo, A., Camacho, D.: Statistical analysis of risk assessment factors and metrics to evaluate radicalisation in twitter. Futur. Gener. Comput. Syst. (2017). https://www.sciencedirect.com/science/article/pii/S0167739X17308348
5. LeCun, Y., Bengio, Y., Hinton, G.: Deep learning. Nature **521**(7553), 436 (2015)
6. Miikkulainen, R., Liang, J., Meyerson, E., Rawal, A., Fink, D., Francon, O., Raju, B., Shahrzad, H., Navruzyan, A., Duffy, N., et al.: Evolving deep neural networks (2017). arXiv preprint arXiv:170300548
7. Nam, J., Kim, J., Mencía, E.L., Gurevych, I., Fürnkranz, J.: Large-scale multi-label text classification revisiting neural networks. In: Joint European Conference on Machine Learning and Knowledge Discovery in Databases, pp. 437–452. Springer (2014)
8. Pascanu, R., Gulcehre, C., Cho, K., Bengio, Y.: How to construct deep recurrent neural networks (2013). arXiv preprint arXiv:13126026
9. Pennington, J., Socher, R., Manning, C.: Glove: global vectors for word representation. In: Proceedings of the 2014 Conference on Empirical Methods in Natural Language Processing (EMNLP), pp. 1532–1543 (2014)
10. Sabour, S., Frosst, N., Hinton, G.E.: Dynamic routing between capsules. In: Advances in Neural Information Processing Systems, pp. 3859–3869 (2017)
11. Severyn, A., Moschitti, A.: Twitter sentiment analysis with deep convolutional neural networks. In: Proceedings of the 38th International ACM SIGIR Conference on Research and Development in Information Retrieval, pp. 959–962. ACM (2015)
12. Villar-Rodriguez, E., Del Ser, J., Bilbao, M.N., Salcedo-Sanz, S.: A feature selection method for author identification in interactive communications based on supervised learning and language typicality. Eng. Appl. Artif. Intell. **56**, 175–184 (2016)
13. Villar-Rodríguez, E., Del Ser, J., Torre-Bastida, A.I., Bilbao, M.N., Salcedo-Sanz, S.: A novel machine learning approach to the detection of identity theft in social networks based on emulated attack instances and support vector machines. Concurr. Comput. Pract. Exp. **28**(4), 1385–1395 (2016)

14. Zhang, Y., Wallace, B.: A sensitivity analysis of (and practitioners' guide to) convolutional neural networks for sentence classification (2015). arXiv preprint arXiv:151003820
15. Zhou, P., Qi, Z., Zheng, S., Xu, J., Bao, H., Xu, B.: Text classification improved by integrating bidirectional LSTM with two-dimensional max pooling (2016). arXiv preprint arXiv:161106639

Ensuring Availability of Wireless Mesh Networks for Crisis Management

Vasily Desnitsky[1,2], Igor Kotenko[1,2](✉), and Nikolay Rudavin[2]

[1] St. Petersburg Institute for Informatics and Automation of the Russian Academy of Sciences, 14-th Liniya, 39, St. Petersburg 199178, Russia
{desnitsky,ivkote}@comsec.spb.ru

[2] ITMO University, 49, Kronverkskiy prospekt, St. Petersburg, Russia
nikolay-rudavin@yandex.ru

Abstract. The paper is aimed at solving a problem of providing availability in wireless mesh networks for crisis management. It offers the concept of a system providing the availability by scanning the network and using drones to deliver emergency wireless communication modules to restore the network connectivity. The authors also offer a model of a crisis management network to assess availability of network nodes. A software prototype implementing a fragment of the crisis management network and a mechanism for ensuring its availability has been developed. The prototype is used to analyze security of the network and to perform experiments.

1 Introduction

Many wireless mesh networks (WMNs) are used today in crisis management for organization and coordinated interaction of various emergency services, such as fire protection, radiation protection, medical services, etc. They are becoming increasingly widespread. In case of an emergency, it is required to deploy a communication infrastructure to ensure the reliable and secure data transmission between the network nodes and operational headquarters. A spontaneous nature of the critical mission makes self-organization imposed frequently on the network. It is needed due to possible poorly predictable physical movements of network nodes, inability to pre-plan data transmission routes, and the varying signal level of network nodes. At the same time, data transmission is based on the use of mesh principles and dynamic generation of routes [1,2]. All these factors negatively affect the security and reliability, including violation of the availability of nodes as a result of situational rebuilds of the network topology. Availability violation can appear both in network connectivity failures and in the reduced bandwidth of communication channels that is insufficient for the mission objectives. Such availability violations can be either unintentional or performed purposefully by external or internal intruders deliberately disabling network nodes.

Note that availability violation in WMNs for crisis management can lead to critical and even catastrophic consequences, resulting in significant financial,

J. Del Ser et al. (Eds.): IDC 2018, SCI 798, pp. 344–353, 2018.
https://doi.org/10.1007/978-3-319-99626-4_30

technological, social, and humanitarian damage. Therefore, it stresses the importance of solving the problem of providing availability for this class of networks. The purpose of this paper is to construct a modeling apparatus and a technical prototype to ensure the availability of WMNs in emergency situations.

The main contribution of this work includes the proposed approach to ensuring the availability of the WMN nodes in emergency situations by (i) automatic scanning of the network nodes and construction of a network graph model; (ii) assessments of the dynamically changing graph model of a WMN and incoming data from its nodes, including detection of availability violation and potential violations of availability (network bottlenecks); (iii) supporting decision-making algorithms for reacting to the arising or potential availability violations through an automated movement of unmanned aerial vehicles (drones) to put extra nodes in areas of real and potential disruptions of the network connectivity.

Besides, as a proof-of-the-concept we developed a prototype of a system ensuring the availability of nodes of a wireless communication self-organizing crisis management network by using Arduino platform microcontrollers, Digi XBee s2 wireless modules, GPS sensors, Parrot AR.Drone 2.0 and other elements. On the basis of the developed prototype we performed a security analysis of the proposed approach. The novelty of the work includes, first, a specific research problem addressed, second, metrics for estimating the graph model of the network to reveal already arisen or potential connectivity disruptions and, third, results of the security analysis of the developed solution and experimental results. In particular besides the proof-of-the-concept the analysis allowed us to uncover possible attacks the proposed system may be subject to, which should be defeated by promising security mechanisms.

The rest of the paper is organized as follows. Section 2 gives a survey of related works in the field. Section 3 describes the scenario of using a wireless communication network for crisis management. Section 4 discloses the proposed conception of a system ensuring availability. Section 5 offers a graph based model for a crisis management network. Section 6 evaluates of availability of network nodes. Section 7 exposes experimental results and discussion. Section 8 concludes the paper.

2 Related Work

In a range of existing papers the modeling of processes and systems using the advantages of WMNs and drones solves rather specific tasks primarily connected to particular business processes and metrics of their effectiveness. Laarouchi et al. [3] provide an overview of the current state of the art and some use cases to identify major factors and challenges for reliability of drones.

Moribe et al. [4] offer an approach to a combined use of WMNs and drones in agriculture to improve the accuracy of collecting source data and ensuring the completeness of the gathered data. They developed a prototype for multiple temperature measurement of agro-crops (on the ground and in the air) and synchronization of these values by means of a suggested communication protocol.

They analyze ways to reduce energy consumption of a drone to increase the duration of the mission.

Kagawa et al. [5] propose a reliable multi-hop wireless system for controlling drones under conditions of their movement beyond-line-of-sight. They substantiate requirements for the reliability of data transmission between network nodes by calculating time delays depending on the speed of the data transfer. The use of drones is caused by the need to reduce presence of natural obstacles or artificial constructions preventing communication reliability along the wireless signal path.

Osanaiye et al. [6] address the problem of WMN availability, which violation can be associated with deliberate actions of intruders. They classify of DoS attacks and ways of counteracting such attacks in form of taxonomy. The analyzed ways of counteracting the attacks are evaluated on the basis of qualitative metrics of attack detection rate, efficiency and overhead costs. Zhang et al. [7] prove the criticality of Denial-of-Service (DoS) attacks in solving problems of ensuring the WMN availability. Thein et al. [8] propose a hierarchically organized cluster structure of wireless networks. Having been constructed on the basis of structural redundancy and stochastic processes it increases the network availability and survivability.

3 Scenario of Using WMN for Crisis Management

The object of the study is a self-organizing WMN, operatively delivered to the place of an emergency and deployed on the ground. The network is designed for reliable and secure messaging between its participants, namely employees of response services involved in the emergency liquidation. These services collaborate under a spontaneous nature of the mission goals and tasks, assuming poorly predictable changes in the topology, location of nodes and network composition. The network consists of the following main elements: (1) mobile headquarters, which include the coordinating node of the wireless network; (2) terminal nodes and router ones receiving and transmitting data from microcontrollers, UI units, sensors, actuators. A function of packet routing is imposed upon the router nodes, presuming uninterrupted operation of these nodes during the mission [1].

WMNs, whose nodes are integrated with moving objects in a social and technical environment, are subject to a problem of insufficient protection from threats of a loss of availability of their nodes. The loss of availability can occur due to the limited scope of the wireless signal, and as a result of deliberate illegal actions by personnel or external violators. Violation of availability can lead to a decrease in the bandwidth between the network nodes, the loss of data packets with forced re-transmission to the recipient, the loss of mission-critical business data of the system, loss of network manageability, including the arising address conflicts, incorrect assignment of roles to network nodes and other negative consequences [6]. Thus, the construction of services designed to ensure and improve availability and connectivity of network nodes will reduce risks of information security and reliability as well as decrease the amount of damage due to availability violation.

4 Conception of a System for Ensuring Availability

The proposed conception of a system for ensuring availability of WMNs for crisis management comprises the following elements: (i) means for automatically scanning the nodes and forming a graph model of the network; (ii) algorithms for analysis of the dynamically changing graph model of the network and incoming data from its nodes to solve the problem of monitoring of the network connectivity, detecting violations of availability of its nodes and identifying potential breaches of availability (bottlenecks); (iii) decision-making support for the automated spatial movement of unmanned aerial vehicles (drones) equipped with additional router nodes to places of real and potential violations of network availability in order to eliminate these violations. It is possible to suspend the drone for a short time gap in the area of the signal loss, and to unload a wireless module at a given point to restore disturbances in the network structure, increase the capacity of the channel between specific nodes and reduce probability of their connectivity disruption in the future.

Figure 1 shows the key elements used in the proposed approach. The center element of the figure represents a coordinating PC with a wireless XBee module as a node-coordinator. The operator observes the correctness of the WCN functioning, but all decision-making procedures are performed basically without direct involvement of the operator. The range of the signal is shown in a bright green circle. In the left part of the figure, a router node/terminal device is shown in a pale green circle, including an autonomous power source, a GPS receiver and a control module (for example, based on the Arduino UNO microcircuit). The node can be located at a fixed point (a tower) as well as on a mobile object (a car or a person). The right part shows an emergency communication node, namely a router-node attached to a drone (e.g. Parrot AR Drone 2.0). The range of its signal is indicated by a blue circle.

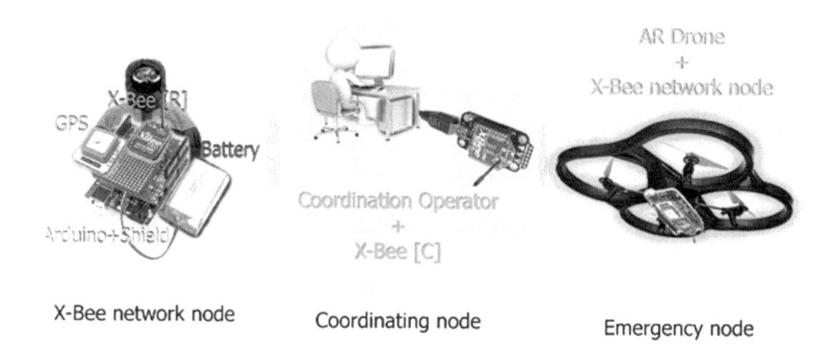

Fig. 1. Elements of a WMN.

Figure 2 shows a network scenario under a critical situation. It demonstrates a defeated router node on the right side (gray circle), that causes the loss of some other terminal device (red circle). On the left side a terminal device attached to

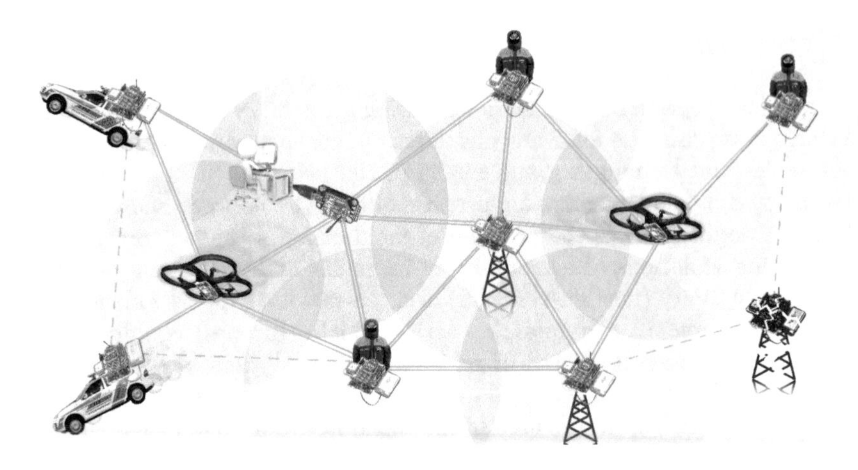

Fig. 2. Scenario of a network under a critical situation.

a moving car (orange circle) faces a risk of leaving the network coverage area, i.e. a potential availability violation. Two drones performing the routing task (blue circles) are targeted to solve both problems.

5 A Model of a Crisis Management System

To collect data on the time-varying structure of a WMN and data from its sensors we have built the following graph model of the network. Scanning the network allows getting data on available nodes, their addresses, SW/HW parameters, connections between the nodes, power of the node transmitters, signal quality, and GPS data of the nodes [9]. The network model represents an undirected graph, which vertices depict the nodes, the edges are bidirectional communication links. Tables 1 and 2 show the basic characteristics of the vertices and edges used the graph model.

The scanning procedure for the wireless XBee network is performed by using API packages and AT commands.

6 Assessment of Availability for Network Nodes

The LQI (Link Quality Indicator) values refer to the key characteristics used in assessing the availability of network nodes. LQI is determined by means of the used communication protocol, measured from 0 to 255, and denotes the reliability value of the established network connection as a rate of successfully delivered packets to the opposite side of the connection [10]. Table 3 shows some of the proposed metrics for assessing the network connections.

The procedure for assessing the availability of network nodes presents an algorithm for traversing the network graph and calculating metrics specified in

Table 1. Characteristics of vertices of the graph-based network model

Wireless network node
Network node name
Unique 64 bit address
16 bit network address used in the routing process
GPS location data (three coordinates)
Values of business data from analogues or digital sensors of the node
GUI elements of the node
Wire and wireless interfaces (of other types)
Power Level of the transmitter signal (PL0 PL4)
Binary state of availability/unavailability of the node from the coordinator
The last time the network node was turned on/off

Table 2. Characteristics of edges of the graph-based network model

Connection between nodes of wireless network
Node 1 signal source/receiver
Node 2 signal source/receiver
Binary state of the connection activity/inactivity
Current LQI level [10] of the signal from node 1 to node 2
Current LQI level of the signal from node 2 to node 1
History of LQI readings for the previous period of time t

Table 3 for identifying missing nodes and disturbed/unreliable connections. Decision on insufficient availability of network nodes is made based on the analysis of a graph of 5 states and transitions between them (Fig. 3). A state represents a network connection between two network nodes. Since it is possible from each state to move to each possible state, the graph is a fully connected. The notations are as follows: ‘(255)’ means ideal signal (0 the lowest, 255 the maximum); ‘$(\uparrow< 255)$’ means the signal is not ideal, but there is a tendency to recover; ‘$(?< 255)$’ means the signal is not ideal, but the trend is not definite; ‘$(\downarrow< 255)$’ means the signal is not ideal and its deterioration is observed; ‘(X)’ means the signal is gone. The last two states are considered as critical.

In the analyzes of the availability of some subsequent node, if all routes from the coordinator to it pass through at least one critical connection, then the decision component warns about the need to increase the availability of this node. If multiple routes are available, the one is chosen that optimizes some metric, e.g. one can choose the shortest route according to the number of hops, the route with the minimum sum of physical distances along all its connections, the route with the maximum value of the minimum LQI of all its connections. As a result, after the route is chosen, the GPS based coordinates of the target

Table 3.

	Metric	Description
C_{LQI}	Current LQI value for a connection	LQI value generated by the last API packet, AT command or a business message
M_{LQI}	LQI value, averaged over a time window	The metric allows tracing trends in the LQI changes (deterioration or improvement of the signal on average)
Q_{LQI}	Number of connections starting from this node	This topological characteristic of the node can be used to determine the total throughput of all routes to a given node
B_{LQI}	Presence of unreliable connections at this node (bottlenecks)	It is used to identify the number of unreliable links (e.g. there are lots of connections, but they are all unreliable)

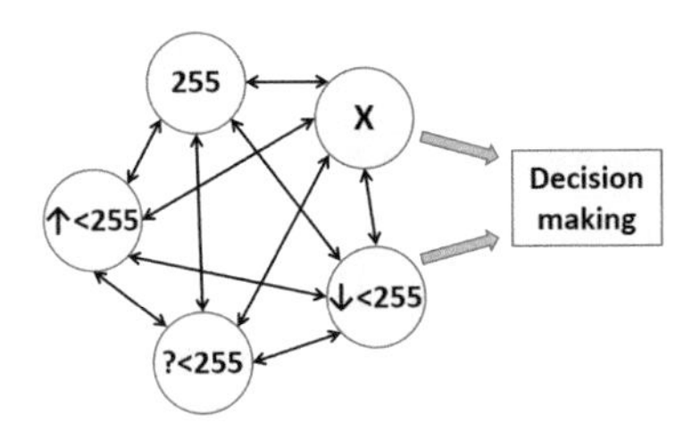

Fig. 3. Graph of states of network connections.

point for the drone are calculated as the midpoint between the coordinates of the given node and the nearest node of the selected route.

7 Experiments and Discussion

We developed a prototype of a WMN and a mechanism for ensuring availability of the network nodes on the basis of the proposed approach. A fragment of the user interface of the developed software for ensuring availability is shown in Fig. 4. This tool allows listing the available network nodes, connections between them and their characteristics as well as displaying the missing nodes/connections and determining the areas of network connectivity disruption. To assess the developed technical solution we analyzed information security threats. We analyzed attacks by the following classification features: attacked elements, source of the attack, means used in the attack, intent, pursued purpose and consequences of the attack.

As a result, we identified four most important types of attacks that can be performed by an external intruder to violate the availability of network nodes.

Attack 1 - spoofing by a false node penetrated into the XBee network. The node is exploited to mount an attack of capturing communication resources of

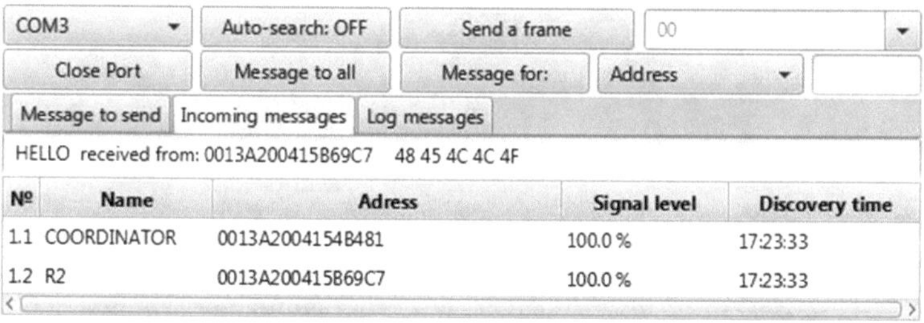

Fig. 4. GUI fragment of a tool supporting decision making.

the network by sending packets to other nodes, making it difficult to transfer packets in normal mode (i.e. DoS attack). Having got data on PAN ID network identifier, the intruder disables a node with a previously known MAC address and substitutes it by itself, cloning the victim's address. The attacker performs a replay attack, sending multiple requests in the form of AT command to the network, forcing other devices to generate packets and send replies. Iin case of sending an ND command (NetworkDiscovery), a broadcast-sending of a packet is performed, forcing all nodes of the network to respond by passing information on themselves. Taking into account that not all the devices are connected directly, there may occur critical delays in the transmission of packets. We conducted an experiment with the network loaded by messages from the attacking node to verify transmission of normal network packets. The coordinating device made a legitimate scan of the network state with an interval of 10 s (the search time was about 4 s). Further, the attacking module sent AT commands to the network with a frequency of about 10 instructions per second. Due to a confusion of the packets, the coordinator was stopped receiving a part of messages of the scanning algorithm correctly and in a good time.

Attack 2 - depletion of energy resources of an autonomous working terminal device of the network [11,12]. Having connected locally to the open elements of the node interface the intruder can initiate unnecessary operations within its hardware and unjustifiably increase the antenna power of the XBee module by internal AT commands. Having physical access to the device and the data channel between the Arduino Uno R3 microcontroller (on an 16u2 chip) and the XBee node, the intruder sends 'PowerLevel 4' AT command to the module, increasing the antenna power. Raising the power consumption from 33 W to 165 mW, the intruder transfers the XBee module from the sleep mode even by sending packet with the invalid addressing. Passing data on an input Arduino pin, the intruder forces it to quit the sleep mode, raising the power dissipation from 40 to 215 mW. Thus the attacker can effectively influence the node even without entering the XBee network, and acting locally, repeatedly increasing the power consumption. The scheme of the experiment on modeling the depletion of energy resources is shown in Fig. 5.

The Arduino microcontroller was put into the sleep mode with an awakening interval of 5 s (the full operation mode acts till the data is transmitted). The

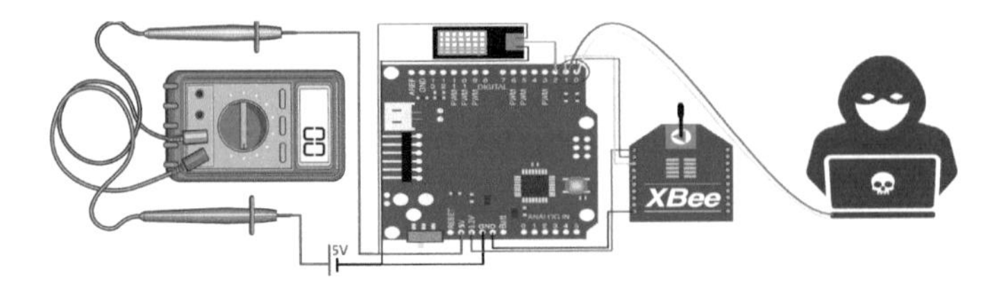

Fig. 5. A scheme of an experiment on modeling an energy exhaustion attack.

XBee module was working under the same sleep period. As a result its current consumption has been reduced to 0.03–0.04 A. The attacking AT-command 'SP', which returned the XBee module to the active mode, allowed it to increase the current consumption to 0.07 A. Therefore under continuous sending data to Arduino, which prevented it from falling asleep, the total current consumption of the node increased to 0.09A, more than 2 times reducing the lifetime of the device.

Attack 3 - software based failure of a node. Having an access to the coordinating computer, which has a direct connection to the coordinating XBee, it is possible to disconnect any of the network nodes. The intruder can obtain information on network nodes and their addresses as well as remotely manipulate them by sending API-packages containing AT commands that change parameters of the nodes. Thus, the intruder sends an AT-command 'PL 0x00' for setting the minimum signal strength (PowerLevel). As a result the communication with the node disappears completely, and it is possible to restore it either by connecting to the lost module locally or by sending an emergency node for the restoration of the connection and the subsequent transmission of a power return command. Experimentally obtained data confirmed that when the power of the node's antenna was reduced to a minimum, communication with the router disappeared, which was traced in the routing tables on the coordinating computer.

Attack 4 - generation of interference to an emergency drone (jamming attack). Since the drone periodically uses the XBee network to receive commands for managing the autopilot, under high interferences the control commands may be lost. Moreover, the power consumption of the control module is raised due to the forced increase in the power of the antenna. Furthermore, the control module receives data on the location via the GPS sensor set up on the drone, so if it is impossible to get the correct coordinates, the drone becomes useless and it can't return to its starting point. When an intruder defeats the navigation messages, the loss of the satellite signal causes the drone to stop following the route.

8 Conclusion

The paper proposed an approach for solving the problem of a targeted and unintentional loss of WMN availability. We suggested the metrics for assessing availability of network nodes and the method of their calculation. A prototype of a crisis management scenario has been implemented, and experiments have been performed. In our future research we are planning to go deeper inside mechanisms ensuring availability in the crisis management networks through interaction of drones as well as to analyze means for increasing security of WMSs against other types of attacks.

Acknowledgements. This work was partially supported by grants of RFBR (projects No. 16-29-09482, 18-07-01488), by the budget (the project No. AAAA-A16-116033110102-5), by Government of Russian Federation (Grant 08-08), and Grant of President of Russia No. MK-5848.2018.9.

References

1. Portmann, M., Pirzada, A.A.: Wireless mesh networks for public safety and crisis management applications. IEEE Eng. Manag. Rev. **39**(4), 114–122 (2011)
2. Dalmasso, I., Galletti, I., Giuliano, R., Mazzenga, F.: WiMAX networks for emergency management based on UAVs. In: IEEE ESTEL Conference, Rome, pp. 1–6 (2012)
3. Laarouchi, E., Cancila D., Chaouchi, H.: Safety and degraded mode in civilian applica-tions of unmanned aerial systems. In: IEEE/AIAA DASC Conference, St. Petersburg, pp. 1–7 (2017)
4. Moribe, T., Okada, H., Kobayashl, K., Katayama, M.: Combination of a wireless sensor network and drone using infrared thermometers for smart agriculture. In: IEEE Annual Consumer Communications & Networking Conference, Las Vegas, NV, USA, pp. 1–2 (2018)
5. Kagawa, T., Ono, F., Shan, L., Takizawa, K., Miura, R., Li, H-B., Kojima, F., Kato, S.: A study on latency-guaranteed multi-hop wireless communication system for control of robots and drones. In: WPMC Symposium, Bali, pp. 417–421 (2017)
6. Osanaiye, O.A., Alfa, A.S., Hancke, G.P.: Denial of service defence for resource availability in wireless sensor networks. IEEE Access **6**, 6975–7004 (2018)
7. Zhang, M., Liu, Y., Wang, J., Hu, Y.: A new approach to security analysis of wireless sensor networks for smart home systems. In: INCoS Conference, Ostrava, pp. 318–323 (2016)
8. Thein, T., Chi, S.D., Park, J.S.: Increasing availability and survivability of cluster head in WSN. In: International Conference on Grid and Pervasive Computing, Kunming, pp. 281–285 (2008)
9. Padalkar, S., Korlekar, A., Pacharaney, U.: Data gathering in wireless sensor network for energy efficiency with and without compressive sensing at sensor node. In: International Conference on Communication and Signal Processing, Melmaruvathur, pp. 1356–1359 (2016)
10. Diallo, C., Marot, M., Becker, M.: Using LQI to improve clusterhead locations in dense zigbee based wireless sensor networks. In: IEEE WiMob Conference, pp. 137–143 (2010)
11. Desnitsky, V., Kotenko, I.: Modeling and analysis of IoT energy resource exhaustion attacks. In: International Conference on Intelligent Distributed Computing (IDC 2017), Studies in Computational Intelligence, vol. 737, pp. 263–270. Springer, Heidelberg (2018)
12. Sharma, M.K., Joshi, B.K.: Detection & prevention of vampire attack in wireless sensor networks. In: ICICIC Conference, Indore, pp. 1–5 (2017)

Medicine and Biology

Automatic Fitting of Feature Points for Border Detection of Skin Lesions in Medical Images with Bat Algorithm

Akemi Gálvez[1,2], Iztok Fister[3], Iztok Fister Jr.[3], Eneko Osaba[4], Javier Del Ser[4,5,6], and Andrés Iglesias[1,2](✉)

[1] Toho University, 2-2-1 Miyama, Funabashi 274-8510, Japan
[2] Universidad de Cantabria, Avda. de los Castros, s/n, 39005 Santander, Spain
iglesias@unican.es
[3] University of Maribor, Smetanova, Maribor, Slovenia
[4] TECNALIA, Derio, Spain
[5] University of the Basque Country (UPV/EHU), Bilbao, Spain
[6] Basque Center for Applied Mathematics (BCAM), Bilbao, Spain

Abstract. This paper addresses the problem of automatic fitting of feature points for border detection of skin lesions. This problem is an important task in segmentation of dermoscopy images for semi-automatic early diagnosis of melanoma and other skin lesions. Given a set of feature points selected by a dermatologist, we apply a powerful nature-inspired metaheuristic optimization method called bat algorithm to obtain the free-form parametric Bézier curve that fits the points better in the least-squares sense. Our experimental results on two examples of skin lesions show that the method performs quite well and might be applied to automatic fitting of feature points for border detection in medical images.

Keywords: Computational intelligence · Medical images
Skin lesion · Border detection
Nature-inspired metaheuristic techniques · Bat algorithm

1 Introduction

1.1 Motivation

Early detection and efficient treatment of skin lesions have become a major concern in current health systems worldwide. While many skin lesions are simply disturbing because of aesthetical reasons (e.g., moles, nevus), others can be very dangerous (e.g., melanomas). According to the *World Cancer Report 2014*, melanoma is the most frequent and dangerous type of skin cancer [39]. The number of reported cases in 2015 was about 3.1 million people worldwide resulting in approximately 60,000 deaths. Furthermore, the number of reported cases and casualties due to malignant melanoma are increasing every year and this negative tendency is expected to continue in coming years.

J. Del Ser et al. (Eds.): IDC 2018, SCI 798, pp. 357–368, 2018.
https://doi.org/10.1007/978-3-319-99626-4_31

Visual inspection by a specialist is the most common diagnostic procedure. However, it is difficult to distinguish the melanoma from other skin lesions (such as moles or dysplastic nevus). Other diagnosis procedures include the ABCDE method, the Menzies scale, the 7-point checklist, and different types of biopsy [9,30]. However, they rely heavily on human intervention, leading to diagnostic results that can vary significantly even among experienced dermatologists.

Image-based methods are gaining traction in the field in recent years. The most classical one is dermoscopy, an epiluminescence microscopy diagnostic technique that uses a device called dermatoscope to distinguish between benign and malignant (cancerous) skin lesions, especially melanoma. Typically, this device uses a liquid medium or polarized light to cancel out skin surface reflections. During examination, the images are digitally captured in an process called digital epiluminescence dermoscopy. Dermoscopy is more precise than naked eye examination in about 20% for sensitivity (detection of melanomas) and about 10% for specificity (percentage of non-melanomas correctly diagnosed as benign). It also reduces screening errors as it enhances discrimination between real melanoma and other skin lesions [3]. However, it is also prone to errors due to the difficulty and subjectivity of the visual interpretation of images, and it is strongly dependent on the proficiency of the specialist.

Some computer-aided techniques have been developed for automatic analysis of dermoscopy images. An important step in the process is image segmentation, where the lesion is roughly separated from the background skin for better identification and classification of the skin lesion. Popular approaches for this problem include thresholding methods [5,16], edge-based methods [1], clustering methods [33,47], level set methods [29] and active contours [28], among others.

An important task in segmentation is the border detection of the skin lesion from the image. This is a valuable source of information for accurate diagnosis, as several clinical features can be computed directly from the detected border. Until recently, the border detection was handled manually by dermatologists by clicking with the mouse on different parts of the image on a computer screen to obtain an initial collection of feature points joined with linear segments. However, the resulting polyline is not well suited for this process, as the border of skin lesions rarely happens to be piecewise linear, but smooth. Given the input data, parametric approximation schemes are clearly better suited for this task. This is actually the main motivation of the present work.

1.2 Aims and Structure of the Paper

This paper addresses the problem of automatic fitting of feature points for border detection of skin lesions from dermoscopy images. Given a set of feature points selected by a specialist, the method applies a powerful nature-inspired metaheuristic called bat algorithm to obtain an accurate approximation of such feature points by using free-form parametric curves (in particular, polynomial Bézier curves). This problem can be mathematically expressed as a least-squares minimization problem. However, traditional mathematical techniques fail to solve the general case, as it leads to a difficult multimodal, multivariate, continuous nonlinear optimization problem. Our method solves this minimization problem without any previous knowledge about the problem beyond the data points. Two illustrative examples of skin lesions are used to discuss the performance of the method. Our results show that the developed method is a promising one to solve the tackled problem and it can be applied for semi-automatic segmentation of medical images.

The structure of this paper is as follows: Sect. 2 describes some previous work on parametric curve approximation. Then, Sect. 3 describes the bat algorithm. Our proposed method is described in detail in Sect. 4, and the experimental results are briefly discussed in Sect. 5. The paper closes with the main conclusions and some ideas for future work in the field.

2 Previous Work

Mathematically, the problem of data approximation with parametric curves can be formulated as an optimization problem, mostly solved through numerical procedures [6,22]. Other methods use error bounds [31], curvature-based squared distance minimization [38], or dominant points [32]. These methods require some particular constraints (such as high differentiability, noiseless data) which are not commonly met in real-world applications. Later on, it was shown that artificial intelligence techniques (mostly based on neural networks) can improve such results, either alone [17,18,26], combined with partial differential equations [4], generalized to functional networks [19], combining functional networks and genetic algorithms [15], or using support vector machines [21].

Other approaches are based on computational intelligence, mainly nature-inspired metaheuristic techniques [7,25,40]. Metaheuristic approaches applied to this problem include genetic algorithms [45,46], particle swarm optimization [10], artificial immune systems [14,37], firefly algorithm [11,12] and memetic approaches [13]. However, these methods are designed for explicit curves and are not applicable to the parametric case. In this paper we aim at filling this gap by applying a method called bat algorithm, described in next section. Given a set of feature points selected by a dermatologist, the method is applied to obtain the free-form parametric Bézier curve that fits the points better in the least-squares sense, thus performing semi-automatic segmentation of dermoscopy images.

3 The Bat Algorithm

The *bat algorithm* is a bio-inspired swarm intelligence algorithm originally proposed by Xin-She Yang in 2010 to solve optimization problems [41–43]. The algorithm is based on the echolocation behavior of microbats, which use a type of sonar called *echolocation*, with varying pulse rates of emission and loudness, to detect prey, avoid obstacles, and locate their roosting crevices in the dark. The idealization of the echolocation of microbats is as follows:

1. Bats use echolocation to sense distance and distinguish between food, prey and background barriers.
2. Each virtual bat flies randomly with a velocity $\mathbf{v}_i$ at position (solution) $\mathbf{x}_i$ with a fixed frequency f_{min}, varying wavelength λ and loudness A_0 to search for prey. As it searches and finds its prey, it changes wavelength (or frequency) of their emitted pulses and adjust the rate of pulse emission r, depending on the proximity of the target.
3. It is assumed that the loudness will vary from an (initially large and positive) value A_0 to a minimum constant value A_{min}.

Require: (Initial Parameters)
  Population size: $\mathscr{P}$; Maximum number of iterations: $\mathscr{G}_{max}$; Loudness: $\mathscr{A}$
  Pulse rate: r ; Maximum frequency: f_{max} ; Dimension of the problem: d
  Objective function: $\phi(\mathbf{x})$, with $\mathbf{x} = (x_1, \ldots, x_d)^T$; Random number: $\theta \in U(0,1)$

1: $g \leftarrow 0$
2: Initialize the bat population $\mathbf{x}_i$ and $\mathbf{v}_i$, $(i = 1, \ldots, n)$
3: Define pulse frequency f_i at $\mathbf{x}_i$
4: Initialize pulse rates r_i and loudness $\mathscr{A}_i$
5: **while** $g < \mathscr{G}_{max}$ **do**
6:   **for** $i = 1$ **to** $\mathscr{P}$ **do**
7:     Generate new solutions by using eqns. (1)-(3)
8:     **if** $\theta > r_i$ **then**
9:       $\mathbf{s}^{best} \leftarrow \mathbf{s}^g$ //select the best current solution
10:       $\mathbf{ls}^{best} \leftarrow \mathbf{ls}^g$ //generate a local solution around $\mathbf{s}^{best}$
11:     **end if**
12:     Generate a new solution by local random walk
13:     **if** $\theta < \mathscr{A}_i$ *and* $\phi(\mathbf{x_i}) < \phi(\mathbf{x}^*)$ **then**
14:       Accept new solutions, increase r_i and decrease $\mathscr{A}_i$
15:     **end if**
16:   **end for**
17:   $g \leftarrow g + 1$
18: **end while**
19: Rank the bats and find current best $\mathbf{x}^*$
20: **return** $\mathbf{x}^*$

Algorithm 1. Bat algorithm pseudocode

Some additional assumptions are advisable for further efficiency. For instance, we assume that the frequency f evolves on a bounded interval $[f_{min}, f_{max}]$. This means that the wavelength λ is also bounded, because f and λ are related to each other by the fact that the product $\lambda.f$ is constant. For practical reasons, it is also convenient that the largest wavelength is chosen such that it is comparable to the size of the domain of interest (the search space for optimization problems). For simplicity, we can assume that $f_{min} = 0$, so $f \in [0, f_{max}]$. The rate of pulse can simply be in the range $r \in [0, 1]$, where 0 means no pulses at all, and 1 means the maximum rate of pulse emission. With these idealized rules indicated above, the basic pseudo-code of the bat algorithm is shown in Algorithm 1. Basically, the algorithm considers an initial population of $\mathscr{P}$ individuals (bats). Each bat, representing a potential solution of the optimization problem, has a location $\mathbf{x}_i$ and velocity $\mathbf{v}_i$. The algorithm initializes these variables with random values within the search space. Then, the pulse frequency, pulse rate, and loudness are computed for each individual bat. Then, the swarm evolves in a discrete way over iterations, like time instances until the maximum number of iterations, $\mathscr{G}_{max}$, is reached. For each generation g and each bat, new frequency, location and velocity are computed according to the following evolution equations:

$$f_i^g = f_{min}^g + \beta(f_{max}^g - f_{min}^g) \quad (1)$$

$$\mathbf{v}_i^g = \mathbf{v}_i^{g-1} + [\mathbf{x}_i^{g-1} - \mathbf{x}^*]\, f_i^g \quad (2)$$

$$\mathbf{x}_i^g = \mathbf{x}_i^{g-1} + \mathbf{v}_i^g \quad (3)$$

where $\beta \in [0,1]$ follows the random uniform distribution, and $\mathbf{x}^*$ represents the current global best location (solution), which is obtained through evaluation of the objective function at all bats and ranking of their fitness values. The superscript $(.)^g$ is used to denote the current generation g. The best current solution and a local solution around it are probabilistically selected according to some given criteria. Then, search is intensified by a local random walk. For this local search, once a solution is selected among the current best solutions, it is perturbed locally through a random walk of the form: $\mathbf{x}_{new} = \mathbf{x}_{old} + \varepsilon \mathscr{A}^g$, where ε is a uniform random number on $[-1,1]$ and $\mathscr{A}^g = <\mathscr{A}_i^g>$, is the average loudness of all the bats at generation g. If the new solution achieved is better than the previous best one, it is probabilistically accepted depending on the value of the loudness. In that case, the algorithm increases the pulse rate and decreases the loudness. This process is repeated for the given number of iterations. In general, the loudness decreases once a new best solution is found, while the rate of pulse emission decreases. For simplicity, the following values are commonly used: $\mathscr{A}_0 = 1$ and $\mathscr{A}_{min} = 0$, assuming that this latter value means that a bat has found the prey and temporarily stop emitting any sound. The evolution rules for loudness and pulse rate are as: $\mathscr{A}_i^{g+1} = \alpha \mathscr{A}_i^g$ and $r_i^{g+1} = r_i^0[1 - exp(-\gamma g)]$ where α and γ are constants. Note that for any $0 < \alpha < 1$ and any $\gamma > 0$ we have: $\mathscr{A}_i^g \to 0$, $r_i^g \to r_i^0$ as $g \to \infty$. Generally, each bat should have different values for loudness and pulse emission rate, which can be achieved by randomization. To this aim, we can take an initial loudness $\mathscr{A}_i^0 \in (0,2)$ while the initial emission rate r_i^0 can be any value in the interval $[0,1]$. Loudness and emission rates will be updated only if the new solutions are improved, an indication that the bats are moving towards the optimal solution.

Bat algorithm is a very promising method that has already been successfully applied to several problems, such as multilevel image thresholding [2], economic dispatch [27], B-spline curve reconstruction [20], optimal design of structures in civil engineering [24], robotics [34–36], fuel arrangement optimization [23], planning of sport training sessions [8], and many others. The interested reader is also referred to the general paper in [44] for a comprehensive review of the bat algorithm, its variants and other interesting applications.

4 The Proposed Method

For data fitting we consider a *free-form parametric Bézier curve* $\mathbf{C}(\tau)$ *of degree* η, defined as:

$$\mathbf{C}(\tau) = \sum_{j=0}^{\eta} \mathbf{P}_j \phi_j^{\eta}(\tau) \tag{4}$$

where $\mathbf{P}_j$ are vector coefficients called *control points*, $\phi_j^{\eta}(\tau)$ are the *Bernstein polynomials of index* j *and degree* η, given by:

$$\phi_j^{\eta}(\tau) = \binom{\eta}{j} \tau^j \ (1-\tau)^{\eta-j}$$

and τ is the *curve parameter*, defined on a finite interval $[0, 1]$. Note that in this paper vectors are denoted in bold. By convention, $0! = 1$.

Suppose that we are provided with a list of data points $\{\mathbf{Q}_\mu\}_{\mu=\mathbf{1},\ldots,\chi}$ in $\mathbb{R}^2$ selected by a trained dermatologist from dermoscopy images. This list $\{\mathbf{Q}_\mu\}$ is always sorted and then joined to define a border between a skin lesion and the skin background, thus enclosing a region that contains the full skin lesion under analysis. Our goal is to compute the curve $\mathbf{C}(\tau)$ approximating the data points better in the least-squares sense. This approximation scheme is particularly adequate to describe the border through a smooth mathematical equation instead of a (possibly large) list of data points connected by straight lines. To do so, we have to minimize the least-squares error, Υ, defined as the sum of squares of the residuals:

$$\Upsilon = \sum_{\mu=1}^{\chi} \left(\mathbf{Q}_\mu - \sum_{j=0}^{\eta} \mathbf{P}_j \phi_j^{\eta}(\tau_\mu) \right)^2 \tag{5}$$

where we need a parameter value τ_μ to be associated with each data point $\mathbf{Q}_\mu$, $\mu = 1, \ldots, \chi$. Considering the column vectors $\mathbf{C}_j = (\phi_j^{\eta}(\tau_1), \ldots, \phi_j^{\eta}(\tau_\chi))^T$, $j = 0, \ldots, \eta$, where $(.)^T$ means transposition, and $\bar{\mathbf{Q}} = (\mathbf{Q}_1, \ldots, \mathbf{Q}_\chi)$, Eq. (5) becomes the following system of equations (called the *normal equation*):

$$\begin{pmatrix} \mathbf{C}_0^T.\mathbf{C}_0 & \ldots & \mathbf{C}_\eta^T.\mathbf{C}_0 \\ \vdots & \vdots & \vdots \\ \mathbf{C}_0^T.\mathbf{C}_\eta & \ldots & \mathbf{C}_\eta^T.\mathbf{C}_\eta \end{pmatrix} \begin{pmatrix} \mathbf{P}_0 \\ \vdots \\ \mathbf{P}_\eta \end{pmatrix} = \begin{pmatrix} \bar{\mathbf{Q}}.\mathbf{C}_0 \\ \vdots \\ \bar{\mathbf{Q}}.\mathbf{C}_\eta \end{pmatrix} \tag{6}$$

which can be compacted as:

$$\mathscr{M}\,\mathscr{D} = \mathscr{R} \tag{7}$$

with $\mathscr{M} = \left[\sum_{j=1}^{\chi} \phi_l^{\eta}(\tau_j) \phi_i^{\eta}(\tau_j) \right]$, $\mathscr{R} = \left[\sum_{j=1}^{\chi} \mathbf{Q}_j \phi_l^{\eta}(\tau_j) \right]$ for $i, l = 0, \ldots, \eta$. If values are assigned to the τ_i, Eq. (7) is a classical linear least-squares minimization that can readily be solved by standard numerical techniques. But if the τ_i are unknowns, the problem becomes more difficult. Since the blending functions

$\phi_j^\eta(\tau)$ are nonlinear in τ, it is a nonlinear continuous optimization problem. It is also a multimodal problem, since there might be arguably more than one data parameterization vector leading to the optimal solution. As a conclusion, solving the parameterization problem leads to a difficult multimodal, multivariate, continuous, nonlinear optimization problem.

We solve this general problem by applying the bat algorithm described above to determine suitable parameter values for the least-squares minimization of functional Υ according to (5). To this aim, each bat (potential solution) is a parametric vector $\mathscr{T}^k = (\tau_1^k, \tau_2^k, \ldots, \tau_\chi^k) \in [0,1]^\chi$, where the $\{\tau_\mu^k\}_{\mu=1,\ldots,\chi}$ are strictly increasing parameters. These parametric vectors are initialized with random values and then sorted. Application of our method yields new bats representing the potential solutions of this optimization problem. The process is performed iteratively for a given number of iterations $\mathscr{G}_{max}$, until the convergence of the minimization of the error is eventually achieved. The bat with the best global value for our fitness function is taken as the final solution of our problem.

A critical issue for bat algorithm is the parameter tuning, which is well-known to be problem-dependent. Our choice has been fully empirical. We set the population size to 100 bats, as larger values increase the computational time without any significant improvement of the fitness function. The initial and minimum loudness, f_{max}, and parameter α are set to 0.5, 0, 2, and 0.6, respectively. We also set the initial pulse rate and parameter γ to 0.5 and 0.4, respectively. However, our results do not change significantly when varying these values slightly. All executions are performed for a total of 10,000 iterations.

All the computational work in this paper has been performed on a personal PC with a 2.6 GHz. Intel Core i7 processor and 8 GB. of RAM. The source code has been implemented by the authors in the programming language of the popular numerical program *Matlab*, version 2015b.

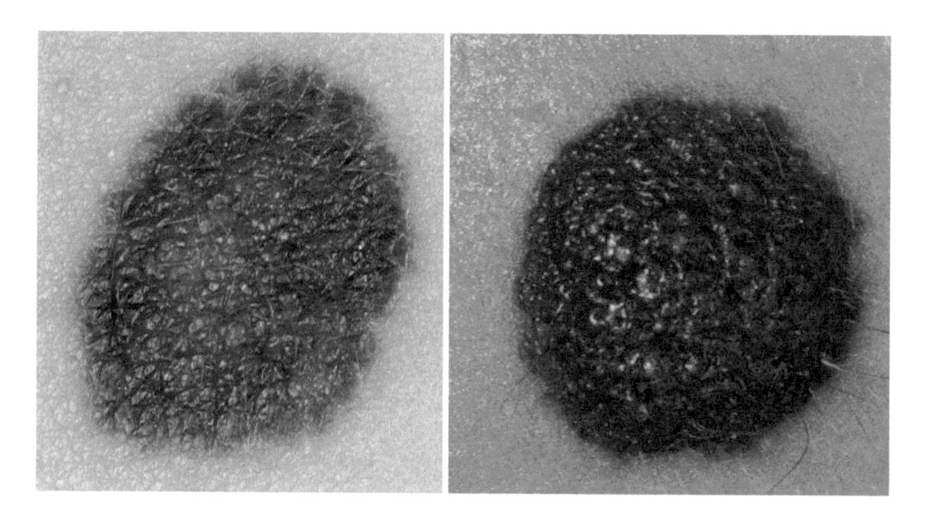

Fig. 1. Dermoscopy images of skin lesions used in this paper.

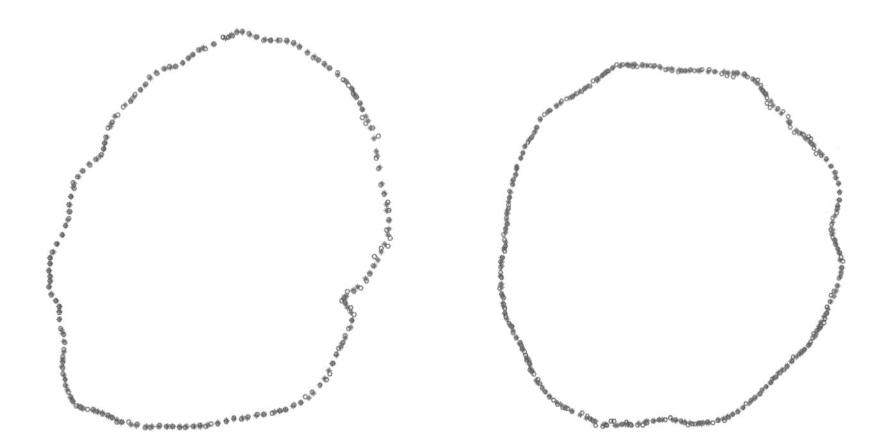

Fig. 2. Automatic fitting of the feature points for border detection for the examples in Fig. 1: feature points (red empty circles), best fitting Bézier curve (solid blue line) and reconstructed feature points (blue stars).

5 Experimental Results

Our method has been applied to several examples of skin lesion images obtained from a digital image archive of the University Medical Center of Groningen, The Netherlands, which is currently open access for research purposes. In this paper we analyze only two of them because of limitations of space. They have been carefully chosen so that they have a good visual appearance in terms of brightness and contrast, thus making it easier to appreciate our graphical results. They are displayed in Fig. 1 and correspond to two dermoscopy images from which a collection of 162 and 220 feature points respectively have been marked by a specialist. These feature points are displayed as red empty circles in Fig. 2 for both cases. We applied our method to these examples by using Bézier curves of different degrees, and then selecting those minimizing the least-squares functional in Eq. (5). The best fitting Bézier curves obtained by our method, corresponding to $n = 55$ and $n = 40$ respectively, are displayed as a blue solid line in Fig. 2 left and right, respectively. The figure also displays the feature points reconstructed by our method as blue stars. As the reader can see, the method obtains a very good fitting of the data points. This observation is confirmed by our numerical results, where we obtain a value of $\Upsilon = 1.72959$ for the first example and $\Upsilon = 1.80875$ for the second one. We also computed the RMSE (root-mean square error), given by: $RMSE = \sqrt{\Upsilon/\chi}$ and obtained a value of 1.0321×10^{-1} and 9.0672×10^{-2}, respectively. We also noticed that the approximation is not optimal yet, as expected from an approximation method. In particular, the original data seems to be slightly more oscillating than the reconstructed curve in both cases. We remark, however, that a perfect matching between the original and the reconstructed features points is not actually required for clinical practice; the accuracy of our results (even if not totally optimal) is generally considered adequate for early diagnosis purposes. Figure 3 shows the convergence diagram of our method (i.e. the fitness function value Υ vs. the number of iterations) for both examples.

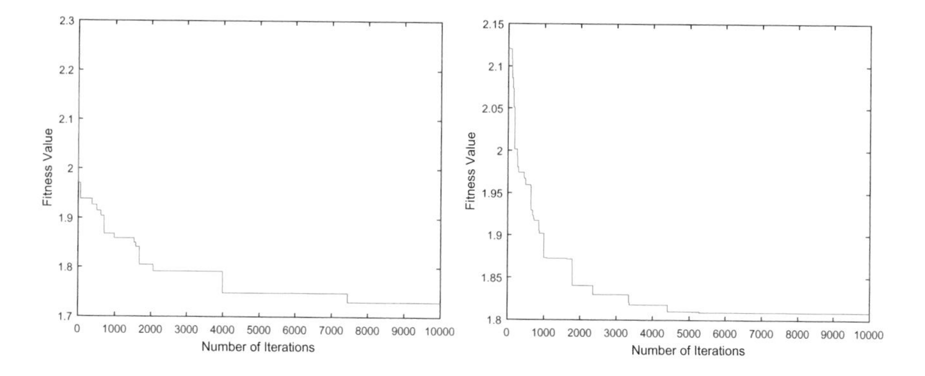

Fig. 3. Convergence diagrams for the examples in Fig. 2.

6 Conclusions and Future Work

In this paper we address the problem of automatic fitting of feature points for border detection of skin lesions. Given a set of feature points selected by a dermatologist from a dermoscopy image, we apply the bat algorithm to obtain the free-form parametric Bézier curve that fits such points better in the least-squares sense. Our experimental results on two examples of skin lesions show that the method performs quite well and might be applied to automatic fitting of feature points for border detection in medical images. This segmentation procedure is very useful for early diagnosis of melanoma and other skin diseases. Regarding the computation times, the two examples in this paper can be reconstructed in just 2–3 min, depending on the individual execution. Although this CPU time is larger than with linear interpolation, the quality of the fitting is much better with our method. Note that this comparison is based on the same initial set of feature points. This is a very important remark, since the selection of feature points is more critical than any other factor for a good fitting to the ground truth. This task is currently carried out manually, making the whole process strongly dependent on human expertise.

About our plans for future work in the field, the present method can be improved in several ways. We plan to extend this approach by using piecewise polynomial schemes for better accuracy. We also plan to include some recent methods based on fusion thresholding and related techniques to obtain the initial feature points without (or with minimal) human intervention to the aim to perform image segmentation in a more automatic way. We also plan to apply this technique to other interesting problems in other fields of medical imaging, not only dermoscopy images.

Acknowledgments. This research work has been kindly supported by the project PDE-GIR of the European Union's Horizon 2020 research and innovation programme, Marie Sklodowska-Curie grant agreement No 778035, the Spanish Ministry of Economy and Competitiveness (Computer Science National Program), grant #TIN2017-89275-R of the Agencia Estatal de Investigación and European Regional Development Funds (AEI/FEDER-UE), the project #JU12 of SODERCAN and European Regional Development Funds (SODERCAN/FEDER-UE), the Slovenian Research Agency (Research Core Funding No. P2-0057), and the project EMAITEK of the Basque Government.

References

1. Abbas, A.A., Guo, X., Tan, W.H., Jalab, H.A.: Combined spline and B-spline for an improved automatic skin lesion segmentation in dermoscopic images using optimal color channel. J. Med. Syst. **38**, 80–80 (2014)
2. Alihodzic, A., Tuba, M.: Improved bat algorithm applied to multilevel image thresholding. Sci. World J. **2014**, 16 pages (2014). article ID 176718
3. Argenziano, G., Soyer, H.P., De Giorgi, V.: Dermoscopy: A Tutorial. EDRA Medical Publishing & New Media, Milan (2002)
4. Barhak, J., Fischer, A.: Parameterization and reconstruction from 3D scattered points based on neural network and PDE techniques. IEEE Trans. Vis. Comput. Graph. **7**(1), 1–16 (2001)
5. Celebi, M.E., H. Iyatomi, H., Schaefer, G., Stoecker, W.V.: Lesion border detection in dermoscopy images. Comput. Med. Imaging Graph. **33**(2), 148–153 (2009)
6. Dierckx, P.: Curve and Surface Fitting with Splines. Oxford University Press, Oxford (1993)
7. Engelbrecht, A.P.: Fundamentals of Computational Swarm Intelligence. Wiley, Chichester (2005)
8. Fister, I., Rauter, S., Yang, X.-S., Ljubic, K., Fister Jr., I.: Planning the sports training sessions with the bat algorithm. Neurocomputing **149**, Part B, 993–1002 (2015)
9. Friedman, R.J., Rigel, D.S., Kopf, A.W.: Early detection of malignant melanoma: the role of physician examination and self-examination of the skin. Cancer J. Clin. **35**(3), 130–151 (1985)
10. Gálvez, A., Iglesias, A.: Efficient particle swarm optimization approach for data fitting with free knot B-splines. Comput. Aided Des. **43**(12), 1683–1692 (2011)
11. Gálvez, A., Iglesias, A.: Firefly algorithm for explicit B-Spline curve fitting to data points. Math. Probl. Eng., Article ID 528215, 12 pages (2013)
12. Gálvez A., Iglesias A.: From nonlinear optimization to convex optimization through firefly algorithm and indirect approach with applications to CAD/CAM. Sci. World J. Article ID 283919, 10 pages (2013)
13. Gálvez, A., Iglesias, A.: New memetic self-adaptive firefly algorithm for continuous optimization. Int. J. Bio Inspired Comput. **8**(5), 300–317 (2016)
14. Gálvez, A., Iglesias, A., Avila, A., Otero, C., Arias, R., Manchado, C.: Elitist clonal selection algorithm for optimal choice of free knots in B-spline data fitting. Appl. Soft Comput. **26**, 90–106 (2015)
15. Gálvez, A., Iglesias, A., Cobo, A., Puig-Pey, J., Espinola, J.: Bézier curve and surface fitting of 3D point clouds through genetic algorithms, functional networks and least-squares approximation. Lectures Notes in Computer Science, vol. 4706, pp. 680–693 (2007)

16. Garnavi, R., Aldeen, M., Celebi, M.E., Varigos, G., Finch, S.: Border detection in dermoscopy images using hybrid thresholding on optimized color channels. Comput. Med. Imaging Graph. **35**(2), 105–115 (2011)
17. Gu, P., Yan, X.: Neural network approach to the reconstruction of free-form surfaces for reverse engineering. Comput. Aided Des. **27**(1), 59–64 (1995)
18. Hoffmann, M.: Numerical control of Kohonen neural network for scattered data approximation. Numer. Algorithms **39**, 175–186 (2005)
19. Iglesias, A., Echevarría, G., Gálvez, A.: Functional networks for B-spline surface reconstruction. Futur. Gener. Comput. Syst. **20**(8), 1337–1353 (2004)
20. Iglesias, A., Gálvez, A., Collantes, M.: Multilayer embedded bat algorithm for B-spline curve reconstruction. Integr. Comput. Aided Eng. **24**(4), 385–399 (2017)
21. Jing, L., Sun, L.: Fitting B-spline curves by least squares support vector machines. In: Proceedings of the 2nd International Conference on Neural Networks & Brain, Beijing (China), pp. 905–909. IEEE Press (2005)
22. Jupp, D.L.B.: Approximation to data by splines with free knots. SIAM J. Numer. Anal. **15**, 328–343 (1978)
23. Kashi, S., Minuchehr, A., Poursalehi, N., Zolfaghari, A.: Bat algorithm for the fuel arrangement optimization of reactor core. Ann. Nucl. Energy **64**, 144–151 (2014)
24. Kaveh, A., Zakian, P.: Enhanced bat algorithm for optimal design of skeletal structures. Asian J. Civ. Eng. **15**(2), 179–212 (2014)
25. Kennedy, J., Eberhart, R.C., Shi, Y.: Swarm Intelligence. Morgan Kaufmann Publishers, San Francisco (2001)
26. Knopf, G.K., Kofman, J.: Adaptive reconstruction of free-form surfaces using Bernstein basis function networks. Eng. Appl. Artif. Intell. **14**(5), 577–588 (2001)
27. Latif, A., Palensky, P.: Economic dispatch using modified bat algorithm. Algorithms **7**(3), 328–338 (2014)
28. Ma, Z., Tavares, J.M.: A novel approach to segment skin lesions in dermoscopic images based on a deformable model. IEEE J. Biomed. Health Inform. **20**, 615–623 (2016)
29. Machado, D.A., Giraldi, G., Novotny, A.A.: Multi-object segmentation approach based on topological derivative and level set method. Integr. Comput. Aided Eng. **18**, 301–311 (2011)
30. Nachbar, F., Stolz, W., Merkle, T., Cognetta, A.B., Vogt, T., Landthaler, M., Bilek, P., Braun-Falco, O., Plewig, G.: The ABCD rule of dermatoscopy. High prospective value in the diagnosis of doubtful melanocytic skin lesions. J. Am. Acad. Dermatol. **30**(4), 551–559 (1994)
31. Park, H.: An error-bounded approximate method for representing planar curves in B-splines. Comput. Aided Geom. Des. **21**, 479–497 (2004)
32. Park, H., Lee, J.H.: B-spline curve fitting based on adaptive curve refinement using dominant points. Comput. Aided Des. **39**, 439–451 (2007)
33. Schmid, P.: Segmentation of digitized dermatoscopic images by two-dimensional color clustering. IEEE Trans. Med. Imaging **18**(2), 164–171 (1999)
34. Suárez, P., Iglesias, A.: Bat algorithm for coordinated exploration in swarm robotics. Adv. Intell. Syst. Comput. **514**, 134–144 (2017)
35. Suárez, P., Gálvez, A., Iglesias, A.: Autonomous coordinated navigation of virtual swarm bots in dynamic indoor environments by bat algorithm. In: International Conference in Swarm Intelligence, ICSI 2017. Lecture Notes in Computer Science, vol. 10386, pp. 176–184 (2017)
36. Suárez, P., Iglesias, A., Gálvez, A.: Make robots be bats: specializing robotic swarms to the bat algorithm. Swarm Evol. Comput. (2018, in press). https://www.sciencedirect.com/science/article/abs/pii/S2210650217306338

37. Ulker, E., Arslan, A.: Automatic knot adjustment using an artificial immune system for B-spline curve approximation. Inf. Sci. **179**, 1483–1494 (2009)
38. Wang, W.P., Pottmann, H., Liu, Y.: Fitting B-spline curves to point clouds by curvature-based squared distance minimization. ACM Trans. Graph. **25**(2), 214–238 (2006)
39. World Cancer Report 2014. World Health Organization. Chapter 5.14 (2014)
40. Yang, X.-S.: Nature-Inspired Metaheuristic Algorithms, 2nd edn. Luniver Press, Frome (2010)
41. Yang, X.S.: A new metaheuristic bat-inspired algorithm. Stud. Comput. Intell. **284**, 65–74 (2010)
42. Yang, X.S.: Bat algorithm for multiobjective optimization. Int. J. Bio Inspired Comput. **3**(5), 267–274 (2011)
43. Yang, X.S., Gandomi, A.H.: Bat algorithm: a novel approach for global engineering optimization. Eng. Comput. **29**(5), 464–483 (2012)
44. Yang, X.S.: Bat algorithm: literature review and applications. Int. J. Bio Inspired Comput. **5**(3), 141–149 (2013)
45. Yoshimoto, F., Moriyama, M., Harada, T.: Automatic knot adjustment by a genetic algorithm for data fitting with a spline. In: Proceedings of Shape Modeling International 1999, pp. 162–169. IEEE Computer Society Press (1999)
46. Yoshimoto, F., Harada, T., Yoshimoto, Y.: Data fitting with a spline using a real-coded algorithm. Comput. Aided Des. **35**, 751–760 (2003)
47. Zhou, H., Schaefer, G., Sadka, A., Celebi, M.E.: Anisotropic mean shift based fuzzy c-means segmentation of dermoscopy images. IEEE J. Sel. Top. Signal Process. **3**(1), 26–34 (2009)

Multi-objective Metaheuristics for a Flexible Ligand-Macromolecule Docking Problem in Computational Biology

Esteban López Camacho[1(✉)], María Jesús García-Godoy[1], Javier Del Ser[2], Antonio J. Nebro[1], and José F. Aldana-Montes[1]

[1] Departamento de Lenguajes y Ciencias de la Computación, University of Málaga, Málaga 29071, Spain
{esteban,mjgarciag,antonio,jfam}@lcc.uma.es
[2] TECNALIA, Univ. of the Basque Country (UPV/EHU) and Basque Center for Applied Mathematics (BCAM), Bizkaia, Spain
javier.delser@tecnalia.com

Abstract. The problem of molecular docking focuses on minimizing the binding energy of a complex composed by a ligand and a receptor. In this paper, we propose a new approach based on the joint optimization of three conflicting objectives: E_{inter} that relates to the ligand-receptor affinity, the E_{intra} characterizing the ligand deformity and the RMSD score (Root Mean Square Deviation), which measures the difference of atomic distances between the co-crystallized ligand and the computed ligand. In order to deal with this multi-objective problem, three different metaheuristic solvers (SMPSO, MOEA/D and MPSO/D) are used to evolve a numerical representation of the ligand's conformation. An experimental benchmark is designed to shed light on the comparative performance of these multi-objective heuristics, comprising a set of HIV-proteases/inhibitors complexes where flexibility was applied. The obtained results are promising, and pave the way towards embracing the proposed algorithms for practical multi-criteria in the docking problem.

Keywords: Molecular docking · Multi-objective optimization Flexibility modeling · SMPSO · MOEA/D · MPSO/D

1 Introduction

The so-called molecular docking research field studies how a small molecule (ligand) behaves in the binding site of a macromolecule (receptor). Since the number of ligand-protein structures extracted from X-crystallography and nuclear magnetic resonance has increased, molecular docking tools have been increasingly used in drug discovery, having a great importance in the pharmaceutical industry [11].

The quality of the results obtained in molecular docking design problems depends on the energy function that evaluates the computed ligands' conformations, as well as the search method to minimize the energy associated to

J. Del Ser et al. (Eds.): IDC 2018, SCI 798, pp. 369–379, 2018.
https://doi.org/10.1007/978-3-319-99626-4_32

the ligand-protein complex. Over the last two decades, more than 60 molecular docking tools have been developed according to the related literature, many of them providing new energy functions and search methods [15]. One of the most popular molecular docking tools amongst the scientific community is AutoDock [12]. This docking program hinges on an empirical free-energy force field that allows the application of flexibility to the ligand and the macromolecule and provides two search methods: a GA (Genetic algorithm) and a LGA (Lamarckian Genetic algorithm). Some recent studies have yielded new single- and multi-objective metaheuristic solvers relying on the AutoDock energy function. However, there are only a few studies that apply multi-objective approaches to solve the molecular docking problem by minimizing two or more objectives, such as Janson *et al.* that implemented ClustMPSO, an algorithm based on the Particle Swam Optimization algorithm using AutoDock 3.05 [6]. This work was indeed the first attempt to tackle the molecular docking problem as a multi-objective problem where the intermolecular E_{inter} and intramolecular E_{intra} energies were optimized. Thereafter, other promising studies [5] have elaborated on the application of recently contributed multi-objective metaheuristics have been applied to minimize two objectives – e.g. E_{inter} and E_{intra} – with AutoDock. In [10], the same authors provided an approach to minimize the E_{inter} and the Root Mean Square Deviation (RMSD), the latter quantifying the distance (in Å) of the atomic coordinates from the co-crystallized and computed ligands.

In this paper we propose a new approach based on optimizing three conflicting objectives: (1) E_{inter}, which is computed by the difference between bound and unbound states of the ligand and the receptor, (2) E_{intra}, which corresponds to the difference between the bound and unbound stated of the ligand-receptor complex; and (3) the RMSD score. Three multi-objective algorithms are used to efficiently tackle this many-objective problem, namely, speed modulation multi-objective particle swarm optimization (SMPSO) [13], multi-objective evolutionary algorithm based on decomposition (MOEA/D) [19] and a new multi-objective particle swarm optimization algorithm based on decomposition (MPSO/D) [20], whose application to molecular docking problems has been shown to be very promising [4]. In order to assess the performance of these algorithms, we have selected a set of instances that involve HIV-proteases, which are therapeutic targets in research studies of the HIV virus and inhibitors with a wide range of sizes (small, medium, large). To carry out the experiments, we have used jMetalcpp [8] that integrates the AutoDock 4.2 evaluation function and the multi-objective algorithms. As the energy function of this AutoDock version allows for flexibility, we have included up to 16 torsion degrees to the ligands and aminoacids' side chains, thus making docking simulations more realistic.

At this point it is important to highlight that in the literature related to single- and multi-objective studies applied to the molecular docking, instances used in the benchmark were computed as *rigid*, making the problem to solve less complex by decreasing the space of search. Therefore, the work presented in what follows covers this lack of research to solve the molecular docking problem by minimizing three conflicting objectives with three different metaheuristic solvers using, in addition, *flexible* instances. The obtained results provide a different perspective to interpret molecular docking results by the practitioners, and can improve the way to find interesting candidate drugs for therapeutic targets in the pharmacological industry.

The remainder of this paper is structured as follows: Sect. 2 surveys briefly the literature related to the application of multi-objective algorithms to solve the molecular docking problem, and Sect. 3 formulates the problem itself. Section 5 describes the applied methodology to perform the experiments and analyzes the obtained results. Finally, Sect. 6 ends the paper and outlines new research lines rooted on our findings.

2 Related Work

In the last decades, studies related to improved search methods for the molecular docking problem have focused on the minimization of one single objective: the final free binding energy ΔG. Most of this background uses the AutoDock energy function provided by [12] using single-objective metaheuristics. For example, Atilgan *et al.* proposed a new algorithm program called AutoDockX that applies a GA, which was tested over a set of ligand-protein complexes [18]. In [9] the authors analyzed a set of standard mono-objective metaheuristics such as PSO, Differential Evolution (DE), gGA and ssGA. Results therein revealed that DE obtained the best overall results in terms of energy and RMSD scores. In other studies such as [2,16] the authors implemented two GAs with local search inspired in the LGA algorithm of AutoDock. In [18] a new GA was proposed based on the analysis of the backbone dihedral angles of conformations of amino acids (AAs) and dipeptides on the receptor. Recently, in [7] the authors implemented DockThor, a single-objective algorithm to optimize the binding energy using jMetal [8].

In the last few years, some studies have focused on applying multi-objective approaches to solve the molecular docking problem. Grosdidier *et al.* used a search method called EADock that minimizes the solvation free energy of the complex and the number of energetic evaluations to the convergence of the problem [14]. Likewise, in [1] the authors proposed a multi-objective approach based on the combination of an energy term with the surface term. In 2015 several studies involved the application of flexibility in ligands and receptors by using a set of representative multi-objective strategies. For example, García-Godoy *et al.* [5] proposed a multi-objective strategy based on minimizing E_{inter} and E_{intra}, similarly to what was proposed in [6].

However, despite the upsurge of prior research, there is only a study based on formulating and solving the molecular docking problem as the joint minimization of three conflicting objectives. In [14] three evolutionary multi-objective optimization algorithms were evaluated to minimize three objectives: E_{intra}, the protein-compound couple's Van der Waal's/electrostatic energy of interactions and the shape of the macromolecule. To evaluate the solutions returned by the multi-objective algorithm, an energy function was applied. The multi-objective approach was applied to predict the ligand's conformation of three rigid ligand-protein complexes. According to this context, the lack of studies based on solving the molecular docking problem by optimizing more than two objectives in flexible instances has been the underlying motivation for the work presented in this paper.

3 Problem Statement

Any multi-objective problem consists of two elements: the decision and the objective spaces. The decision space involves all the possible solutions and the objective space that includes all the objectives values:

- **Decision space**: The objective of molecular docking problem is to minimize the ligand (L)-receptor (R) complex binding energy. The ligand-receptor interaction is evaluated by an energy function composed by three components. The first component refers to the ligand's translation, that includes three values (x, y, z) in the Cartesian space. The second element corresponds to the ligand's orientation modeled as four-variable quaternions including the angle slope (θ). The third component involves the degrees on freedom (torsion angles) from the receptor's and macromolecule's bonds. In this paper, we have added 10 torsion angles to the ligand and 6 to the macromolecule aminoacids' side chains. Each solution is encoded as a real-value vector of $7+n$ variables (in the selected instances, the total number of variables equals 23). In addition, a grid-based approach [12] has been applied to calculate the potential associated to each point of the grid (the 3-dimension space where L and R interact with each other).
- **Objective space**: as mentioned before, the three objectives correspond to E_{inter}, E_{intra} and RMSD. The first and two objectives are terms of the AutoDock energy function (see the score function formulation in [12]). Specifically, E_{inter} and E_{intra} are calculated as follows:

$$E_{inter} = Q^{R-L}_{bound} - Q^{R-L}_{unbound}, \tag{1}$$

$$E_{intra} = Q^{R-R}_{bound} - Q^{R-R}_{unbound} + Q^{L-L}_{bound} - Q^{L-L}_{unbound}, \tag{2}$$

  where Q_{bound} and $Q_{unbound}$ are the different bound and unbound status for the Ligand (L) and the receptor (R). Each pair (Q) includes evaluations of dispersion/repulsion (vdw), hydrogen bonds ($hbond$), electrostatics ($elec$) and desolvation (sol). On the other hand, the RMSD score is a measure of similarity between two superimposed atomic coordinates of the co-crystallized and the computed ligands. The lower this value is, the better the solution is. A ligand conformation with an RMSD score of 2Å or lower is considered as a good prediction by the molecular docking technique. However, it is worth noting that a solution with the lowest RMSD score is not always the best solution. There can also be solutions with higher RMSD scores and lower E_{inter} that correspond with ligand-macromolecule interactions, which can be interesting from a pharmacological point of view given that new regulatory ligand binding sites can be detected. This third objective in the formulated docking problem is given by:

$$RMSD = \sqrt{\frac{1}{N}\sum_{i}^{N} d_i^2}, \quad (3)$$

where the average is done over the N pairs of equivalent atoms from the co-crystallized and computed ligand, being d the distance between two atoms in the i-th pair.

4 Multi-objective Metaheuristics

In order to efficiently tackle the problem formulated above, three multi-objective algorithms have been considered: SMPSO and MOEA/D, which have shown good results when dealing with the molecular docking problem [5]; and MPSO/D, recently applied to the problem optimizing two objectives such as E_{inter} and the RMSD score, but using rigid ligand-protein complexes [4]. We next describe them briefly for the sake of completeness:

- **SMPSO** is a multi-objective particle swarm optimization (PSO) algorithm that incorporates a velocity restriction mechanism. The approach allows producing new effective particle positions in those cases where the velocity becomes too high. This solver also includes additional features such as the use of polynomial mutation as a turbulence factor, and an external archive to store non-dominated solutions found during the search. In [13] a comparative analysis of SMPSO with five representative multi-objective solvers was performed in terms of the quality of the resulting approximation sets and the convergence speed to the Pareto front. The obtained results showed that SMPSO returned the best overall results in such terms.
- **MOEA/D** is a multi-objective evolutionary algorithm based on decomposition [19]. This algorithm divides the multi-objective problem into M scalar optimization subproblems, which are solved simultaneously by evolving the population. At each generation, the best solution for each subproblem is obtained. The neighborhood relations among these subproblems are defined based on the distances between their aggregation coefficient vectors. The optimization of each subproblem is based on the information from the subproblems within its neighborhood.
- **MPSO/D** is a relatively new multi-objective PSO algorithm based on decomposition [3]. This algorithm decomposes the objective problem into a set of sub-regions on the basis of a set of direction vectors. Each subregion has a solution that maintains a level of diversity. MPSO/D also exploits the crowding distance to calculate the fitness of the solutions for the selection operator, as well as the neighboring particles (solutions) of a given particle to infer the global best historical position found by the algorithm.

5 Experiments and Results

In order to assess the performance of the considered solvers when facing the casted three-objective molecular docking problem, we have considered a set of structures used by Morris *et al.* in [12], where the AutoDock 4.2 energy function was tested. Table 1 shows the selected instances that correspond to a set of molecular docking problems where flexibility was applied to the ligands and the ARG-8 side chains of the two monomers A and B of the HIV-proteases. All these instances were previously preprocessed (preparation of the macromolecule and ligand applying flexibility to the bonds), and potentials were calculated for a three-dimensional square (x, y, z) grid of 120 Å per dimension and a grid spacing of 0.375 Å using the AutoGrid4 software.

After the macromolecule and ligand preparations and the grid calculations, we run 30 times the algorithms to solve the 11 flexible instances using jMetalcpp, which integrates the AutoDock energy function [8]. In Table 2, we have summarized the parameter settings for each algorithm. The swarm sizes for SMPSO, MPSO/D are set to 100 particles and the MOEAD/D population size is set to 100 individuals. The stopping condition is set to 1,000,000 evaluations. These parameters present default settings, which are used in the molecular docking problem [12].

Table 1. Accession codes, inhibitor-HIV protease crystal structure, and PDB resolution

PDB Code	Protein-ligand complexes	Resolution (Å)
1AJV	HIV-1 protease/Cyclic sulfamide inhibitor AHA006	2.00
1AJX	HIV-1 protease/Cyclic urea inhibitor AHA001	2.00
1BV9	HIV-1 protease/α-D-glucose	2.20
1D4K	HIV-1 protease/Macrocyclic peptidomimetic inhibitor 8	1.85
1G2K	HIV-1 protease/Cyclic sulfamide inhibitor AHA047	1.95
1HIV	HIV-1 protease/Dihydroethylene-containing inhibitor	2.00
1HPX	HIV-1 protease/KNI-272	2.00
1HTF	HIV-1 protease/Penicillin-derived asymmetric inhibitor GR126045	2.20
1HTG	HIV-1 protease/Penicillin-derived asymmetric inhibitor GR137615	2.00
1HVH	HIV-1 protease/Nonpeptide cyclic cyanoguanidines Q8261	1.80
2UPJ	HIV-1 protease/Nonpeptide inhibitor U100313	3.00

For these executions, values for the median and the interquartile range were calculated as statistical measures of central tendency and dispersion, respectively. Two quality indicators were used to evaluate the algorithms' performance:

the Hypervolume (I_{HV}) [21]) and Unary Additive Epsilon Indicator ($I_{\epsilon+}$) [22]). I_{HV} takes into account both convergence and diversity, whereas the $I_{\epsilon+}$ gives a measure of the convergence degree of the obtained Pareto front approximations. Given that the molecular docking problem is a real-world problem, the Pareto front to calculate these two measures to each instance was obtained from the set of non-dominated solutions in all executions of the selected algorithms.

Table 2. Parameter settings of SMPSO, MPSO/D and MOEA/D

Algorithm	Parameter	Value
All	*Swarm size/Population size*	100 Particles/Individuals
	Maximum number of evaluations	1,000,000
SMPSO [13]	*Archive Size*	100
	C_1, C_2	1.5
	ω	0.9
	Mutation	Polynomial mutation
	Mutation probability	1.66
	Mutation distribution index η_m	20
	Selection method	Rounds
MPSO/D [20]	C_1, C_2	$rand(1.5, 2.0)$
	ω	$rand(0.1, 0.5)$
MOEA/D [20]	μ	0.5
	C_r	1.0
	Mutation probability	$p_m = 1/(\#$ decision variables)

A first analysis is performed with the results obtained for I_{HV} and $I_{\epsilon+}$. Table 3 reports the median and interquartile range of the computed distributions of both quality measures. As shown in this table, MPSO/D obtains the best I_{HV} for almost all the instances, except for the instances 1HIV and 1HTG that are characterized by containing large inhibitors according to Morris *et al.* [12]. For the $I_{\epsilon+}$, 6 out of 11 instances obtained the best results for MPSO/D. SMPSO scores the best results for the instances 1HIV, 1HPX and 2UPJ and MOEA/D for the instances 1HTG and 1HVH.

We have also applied the Friedman's ranking and Holm's post-hoc multicompare tests to assess which algorithms are statistically worse than the control one (i.e. the one ranking the best) to evaluate the results by applying a p-value of 0.05. Table 4 evinces that MPSO/D is the best ranked algorithm according to Friedman's test for the two indicators, followed by SMPSO. Therefore, MPSO/D was used as control algorithm in the post-hoc Holm tests. In terms of I_{HV}, the adjusted p-values show that SMPSO and MPSO/D are statistically better than the MOEA/D. On the other hand, the adjusted p-values for $I_{\epsilon+}$ provide evidence that the obtained results for MPSO/D are not statistically significant.

Table 3. Median and interquartile range of the I_{HV} and $I_{\epsilon+}$.

	MOEA/D	SMPSO	MPSO/D
	Hypervolume (I_{HV})		
1AJV	$2.13e-01_{6.0e-01}$	$4.91e-01_{4.8e-02}$	$6.75e-01_{7.8e-02}$
1AJX	$9.71e-01_{2.8e-02}$	$9.82e-01_{3.4e-03}$	$9.90e-01_{4.3e-03}$
1D4K	$8.94e-01_{1.1e-01}$	$9.80e-01_{1.2e-02}$	$9.82e-01_{3.5e-02}$
1G2K	$4.41e-01_{5.9e-01}$	$6.22e-01_{5.8e-02}$	$8.13e-01_{2.4e-02}$
1HIV	$8.41e-01_{8.0e-02}$	$9.26e-01_{2.3e-02}$	$6.16e-01_{3.5e-01}$
1HPX	$6.22e-01_{2.8e-01}$	$7.88e-01_{4.7e-02}$	$8.15e-01_{4.4e-02}$
1HTF	$8.87e-01_{9.5e-02}$	$8.67e-01_{2.5e-02}$	$9.36e-01_{5.3e-02}$
1HTG	$1.00e+00_{5.1e-02}$	$9.87e-01_{6.5e-03}$	$9.98e-01_{2.2e-03}$
1HVH	$7.51e-01_{5.5e-01}$	$6.61e-01_{4.7e-02}$	$7.58e-01_{5.8e-02}$
1VB9	$1.87e-01_{9.7e-01}$	$7.72e-01_{8.4e-02}$	$8.69e-01_{8.5e-01}$
2UPJ	$9.80e-01_{8.2e-02}$	$9.88e-01_{5.8e-03}$	$9.82e-01_{7.9e-03}$
	Epsilon ($I_{\epsilon+}$)		
1AJV	$5.32e-01_{4.9e-01}$	$4.05e-01_{7.7e-02}$	$3.00e-01_{6.2e-02}$
1AJX	$2.93e-02_{2.8e-02}$	$1.84e-02_{3.4e-03}$	$1.01e-02_{4.3e-03}$
1D4K	$9.83e-02_{1.0e-01}$	$1.71e-02_{1.2e-02}$	$1.34e-02_{3.3e-02}$
1G2K	$3.75e-01_{4.0e-01}$	$2.74e-01_{4.1e-02}$	$1.52e-01_{2.5e-02}$
1HIV	$1.52e-01_{8.6e-02}$	$6.67e-02_{2.4e-02}$	$3.85e-01_{3.3e-01}$
1HPX	$2.31e-01_{1.5e-01}$	$1.44e-01_{5.0e-02}$	$1.65e-01_{2.7e-02}$
1HTF	$9.44e-02_{1.0e-01}$	$1.18e-01_{2.3e-02}$	$4.09e-02_{4.2e-02}$
1HTG	$3.22e-04_{5.1e-02}$	$1.33e-02_{6.5e-03}$	$2.07e-03_{2.2e-03}$
1HVH	$1.29e-01_{5.6e-01}$	$2.30e-01_{6.3e-02}$	$1.45e-01_{3.2e-02}$
1VB9	$7.78e-01_{1.3e+00}$	$2.07e-01_{8.4e-02}$	$1.10e-01_{7.7e-01}$
2UPJ	$1.95e-02_{8.2e-02}$	$1.17e-02_{5.8e-03}$	$1.78e-02_{7.9e-03}$

Table 4. Average Friedman's rankings with Holm's Adjusted p-values (0.05) of compared algorithms for the test set of 11 docking instances. Symbol * indicates the control algorithm and column at right contains the overall ranking of positions with regards to I_{HV} and $I_{\epsilon+}$.

Hypervolume (I_{HV})			Epsilon ($I_{\epsilon+}$)		
Algorithm	Fri_{Rank}	$Holm_{Ap}$	Algorithm	Fri_{Rank}	$Holm_{Ap}$
***MPSO/D**	**1.36**	-	***MPSO/D**	**1.54**	-
SMPSO	2.09	8.80e-02	SMPSO	2.06	2.86e-01
MOEA/D	2.54	1.11e-02	MOEA/D	2.39	6.60e-02

6 Conclusions and Future Work

This work has presented a new approach to solve a molecular docking problem based on the optimization of three conflicting objectives: E_{inter}, E_{intra} and RMSD. We have carried out the experiments using a set of HIV-proteases/inhibitors complexes where flexibility was applied and three multi-objective solvers (SMPSO, MOEA/D and MPSO/D) that have previously shown good results solving the molecular docking problem with two objectives. The analyses performed lead us to the following conclusions:

- MPSO/D obtains the best overall results in terms of I_{HV} (convergence and diversity). Therefore, for practitioners in studies *in silico* to test new candidate drugs for therapeutic targets, MPSO/D could be very useful when three objective are applied as it obtains the best results taking into account the convergence and the diversity. The diversity in the ligand's conformations by using this three-objective approach can be very interesting from a pharmacological point of view. For example, if the expert is interested in a set of ligands' conformations that present an RMSD lower than 2Å a low value of E_{intra} and a high value of E_{intra}, a possible solution for this could be a ligand's conformation, characterized by its energetic stability, whose interaction to the macromolecule is weak and close to the crystallographic ligand binding site.
- According to the Friedman's ranking and Holm's post-hoc multicompare tests, the results show that MPSO/D, SMPSO and MOEA/D render the same performance in terms of converge, but SMPSO and MOEA/D obtain better results according to their diversity. This means that the solutions returned by MPSO/D, SMPSO and MOEA/D cover the set of non-dominated solutions and SMPSO and MOEA/D return a more diverse set of results. According to these analyzed results, it would be necessary to perform a more extensive study that would cover a larger number of instances of the HIV-proteases with different complexity.
- The selected multi-objective solvers have attained a good performance by optimizing three objectives for flexible molecular docking instances. As future research lines it could be interesting to apply a selective flexibility in different residues than ARG-8, similar to the approach presented in [17]. Likewise, other energy scoring functions will be explored towards allowing more degrees of freedom to be applied.

Acknowledgments. This work has been partially supported by the proyect grants TIN2014-58304 y TIN2017-86049-R (Ministerio de Economía, Industria y Competividad) and P12-TIC-1519 (Plan Andaluz de Investigación, Desarrollo e Innovación). Javier Del Ser would also like to thank the Basque Government for its support through the EMAITEK Funding Program.

References

1. Boisson, J.C., Jourdan, L., Talbi, E.G., Horvath, D.: Parallel multi-objective algorithms for the molecular docking problem. In: 2008 IEEE Symposium on Computational Intelligence in Bioinformatics and Computational Biology, pp. 187–194 (2008). https://doi.org/10.1109/CIBCB.2008.4675777
2. Boxin, G., Changsheng, Z., Jiaxu, N.: Edga: a population evolution direction-guided genetic algorithm for proteinligand docking. J. Comput. Chem. **23**(7), 585–596 (2016). https://doi.org/10.1089/cmb.2015.0190
3. Dai, C., Wang, Y., Ye, M.: A new multi-objective particle swarm optimization algorithm based on decomposition. Inf. Sci. **325**(C), 541–557 (2015). https://doi.org/10.1016/j.ins.2015.07.018

4. Garca-Nieto, J., Lpez-Camacho, E., Garca-Godoy, M.J., Nebro, A.J., Aldana-Montes, J.F.: Multi-objective ligand-protein docking with particle swarm optimizers. Swarm Evol. Comput. (2018). https://doi.org/10.1016/j.swevo.2018.05.007
5. García-Godoy, M.J., López-Camacho, E., García Nieto, J., Nebro, A.J., Aldana-Montes, J.F.: Solving molecular docking problems with multi-objective metaheuristics. Molecules **20**(6), 10,154–10,183 (2015)
6. Janson, S., Merkle, D., Middendorf, M.: Molecular docking with multi-objective particle swarm optimization. Appl. Soft Comput. **8**(1), 666–675 (2008). https://doi.org/10.1016/j.asoc.2007.05.005
7. Leonhart, P.F., Spieler, E., Ligabue-Braun, R., Dorn, M.: A biased random key genetic algorithm for the protein-ligand docking problem. Soft Comput. (2018). https://doi.org/10.1007/s00500-018-3065-5
8. López-Camacho, E., García Godoy, M.J., Nebro, A.J., Aldana-Montes, J.F.: JMETALCPP: optimizing molecular docking problems with a C++ metaheuristic framework. Bioinformatics **30**(3), 437–438 (2014)
9. López-Camacho, E., García Godoy, M.J., García-Nieto, J., Nebro, A.J., Aldana-Montes, J.F.: Solving molecular flexible docking problems with metaheuristics: a comparative study. Appl. Soft Comput. **28**, 379–393 (2015). https://doi.org/10.1016/j.asoc.2014.10.049
10. López-Camacho, E., García-Godoy, M.J., García-Nieto, J., Nebro, A.J., Aldana-Montes, J.F.: A New Multi-objective Approach for Molecular Docking Based on RMSD and Binding Energy, pp. 65–77. Springer International Publishing, Cham (2016)
11. Meng, X.Y., Zhang, H.X., Mezei, M., Cui, M.: Molecular docking: a powerful approach for structure-based drug discovery. Curr. Comput. Aided Drug Des. **7**(2), 146–157 (2011)
12. Morris, G.M., Huey, R., Lindstrom, W., Sanner, M.F., Belew, R.K., Goodsell, D.S., Olson, A.J.: AutoDock4 and AutoDockTools4: automated docking with selective receptor flexibility. J. Comput. Chem. **30**(16), 2785–2791 (2009)
13. Nebro, A.J., Durillo, J.J., Garcia-Nieto, J., Coello Coello, C.A., Luna, F., Alba, E.: SMPSO: A new PSO-based metaheuristic for multi-objective optimization. In: IEEE Symposium on Computational Intelligence in Multi-Criteria Decision-Making, pp 66–73 (2009). https://doi.org/10.1109/MCDM.2009.4938830
14. Oduguwa, A., Tiwari, A., Fiorentino, S., Roy, R.: Multi-objective optimisation of the protein-ligand docking problem in drug discovery. In: Proceedings of the 8th Annual Conference on Genetic and Evolutionary Computation, GECCO 2006, pp. 1793–1800 (2006). https://doi.org/10.1145/1143997.1144287
15. Pagadala, N.S., Syed, K., Tuszynski, J.: Software for molecular docking: a review. Biophys. Rev. **9**(2), 91–102 (2017). https://doi.org/10.1007/s12551-016-0247-1
16. Peh, S.C.W., Hong, J.L.: Glsdock - drug design using guided local search. In: Gervasi, O., Murgante, B., Misra, S., Rocha, A.M.A., Torre, C.M., Taniar, D., Apduhan, B.O., Stankova, E., Wang, S. (eds.) Computational Science and Its Applications - ICCSA 2016, pp. 11–21. Springer International Publishing, Cham (2016)
17. Abreu, R.M., Froufe, H.J., Queiroz, M.J., Ferreira, I.C.: Selective flexibility of side-chain residues improves VEGFR-2 docking score using autodock vina. Chem. Biol. Drug. Des. **79**(4), 530–4 (2012)
18. Ru, X., Song, C., Lin, Z.: A genetic algorithm encoded with the structural information of amino acids and dipeptides for efficient conformational searches of oligopeptides. J. Comput. Chem. **37**(13), 1214–1222 (2016). https://doi.org/10.1002/jcc.24311

19. Zhang, Q., Li, H.: Moea/d: a multiobjective evolutionary algorithm based on decomposition. IEEE Trans. Evol. Comput. **11**(6), 712–731 (2007). https://doi.org/10.1109/TEVC.2007.892759
20. Zhao, Y., Liu, H.L.: Multi-objective particle swarm optimization algorithm based on population decomposition. In: Yin, H., Tang, K., Gao, Y., Klawonn, F., Lee, M., Weise, T., Li, B., Yao, X. (eds.) Intelligent Data Engineering and Automated Learning - IDEAL 2013, pp. 463–470. Springer, Heidelberg (2013)
21. Zitzler, E., Thiele, L.: Multiobjective evolutionary algorithms: a comparative case study and the strength pareto approach. IEEE Trans. Evol. Comput. **3**(4), 257–271 (1999)
22. Zitzler, E., Thiele, L., Laumanns, M., Fonseca, C.M., Da Fonseca, V.: Performance assessment ofmultiobjective optimizers: an analysis and review. IEEE Trans. Evol. Comput. **7**(2), 117–132 (2003)

Identifying the Polypharmacy Side-Effects in Daily Life Activities of Elders with Dementia

Viorica Chifu(✉), Cristina Pop, Tudor Cioara, Ionut Anghel, Dorin Moldovan, and Ioan Salomie

Computer Science Department, Technical University of Cluj-Napoca, 26-28 G. Baritiu Str., Cluj-Napoca, Romania
{Viorica.Chifu,Cristina.Pop,Tudor.Cioara,Ionut.Anghel, Dorin.Moldovan,Ioan.Salomie}@cs.utcluj.ro

Abstract. This paper addresses the problem of polypharmacy management in older patients with dementia. We propose a technique that combines semantic technologies with big data machine learning techniques to detect deviations in daily activities which may signal the side effects of a drug-drug interaction. A polypharmacy management knowledge base was developed and used to semantically define drug-drug interactions and to annotate with the help of doctors significant registered deviations from the elders' routines. The Random Forest Classifier is used to detect the days with significant deviations, while the k-means clustering algorithm is used to automate the deviations annotation process. The results are promising showing that such an approach can be successfully applied for assisting doctors in identifying the effects of polypharmacy in the case of patients with dementia.

1 Introduction and Related Work

According to the World Health Organization there are around 50 million people diagnosed with dementia in the world and this number increases as around 10 million new cases are reported each year [12]. Management of this syndrome requires both pharmacological and non-pharmacological interventions. The treatment of dementia is a huge challenge as over 21% of the elders suffering from dementia are exposed to polypharmacy. In the pharmaceutical domain, polypharmacy management is referred to as medication review, and includes both medication intake and monitoring of medication plan adherence as well as medication side effects detection. Most older adults with dementia take over 6 drugs/day such as anti-dementia drugs, drugs controlling underlying risk factors, or drugs dealing with symptoms of dementia. Medication review is utmost important because unfortunately, due to age-related drug metabolism and other problems related to frailty or due to transitions towards home healthcare, the elderly is at risk of experiencing drug-related problems. These problems are usually caused by drug use, choice and adverse reactions, interactions or contraindications. Strictly following the medication plan tends to be a huge challenge in case

J. Del Ser et al. (Eds.): IDC 2018, SCI 798, pp. 380–389, 2018.
https://doi.org/10.1007/978-3-319-99626-4_33

of dementia: (1) it can be difficult for medical professionals to properly assess the behavioral and psychological symptoms of the elderly patient, since changing is gradually and is likely to be multifactorial, and (2) furthermore due to cognitive decline, it tends to be challenging for the patient to take the appropriate medication at the right time. Also, medication side effects detection in elderly people with dementia is even more challenging because of the changes associated with cognitive, behavioral and health decline that need to be continuously addressed. Drug related problems may occur in various steps of the medication process (from prescription till evaluation) and they are dependent of the pharmaceutical care chain's actors (e.g. physician, nurse, pharmacist, patient). In our vision these problems can be minimized by increasing the awareness and coordination among elderly patient, caregiver, pharmacist and healthcare personal in dementia care, leveraging on novel ICT technologies to lower the barriers imposed by the dementia symptoms. This will offer physicians the means of proactively identifying adverse drug problems and to provide appropriate intervention and support for elders and their caregivers to plan ahead and adjust their care and treatment as symptoms progress.

To identify the polypharmacy side effects affecting the daily life activities of patients with dementia, two issues need to be addressed: first the significant behavioral deviations from the patient's baseline must be detected, and second the correlation of these deviations with side effects of drug-drug interactions must be performed. The research literature reports several approaches for detecting behavioral deviations from a person's baseline. For example, in [2] the authors propose a system which detects the activities performed by elder persons as well as their wellness state using association rules mining and frequent behavior patterns. The proposed system extracts associations between the duration of each sub activity and the location of the sub activity by using the FP-Growth algorithm, identifies the correlations between the sub activities and the time in 24 h by using the Generalized Sequential Pattern mining algorithm, and extracts patterns of vital parameters for a specific time of the day based on a set of association rules. The detection of the wellness status of the elderly people is based on the data collected from sensors, the information stored in a knowledge base and the association rules. In [3] the authors propose a Bayesian statistical approach for detecting the abnormal behavior of a person. For behavioral modeling and anomaly detection the authors propose three types of probabilistic features: the likelihood of the sensor activation - provides information about long-term health status of the monitored person, the likelihood of the sensor sequence - detects the confusing states or delirium, and the likelihood of the sensor event - provides information about the physical conditions (i.e. weaknesses, falls, and unresponsive statuses) of the monitored person. In [4], authors present a semantic-based approach for detecting the deviations from the normal behavior of people living in their homes. A context-aware hierarchical clustering algorithm is used to identify the activities that are specific for particular days of the week. Also, sequential pattern mining and itemset mining algorithms are used to identify temporal relationships between the normal activities. To identify the deviations from the normal behavior, the proposed approach uses a set of semantic rules which are continuously extended by the systems users in case new deviations are identified. In [5], authors propose a technique for detecting abnormal behaviors of people monitored in their homes by analyzing the manipulation of household objects, besides the interaction of the monitored person with its home environment. Machine learning techniques are used to recognize the specific activities performed with the objects, while symbolical reasoning is used to detect activity-dependent/independent anomalies as well as subject-dependent anomalies.

In this paper we present techniques for detecting the effects of drug-drug interactions upon the daily life activities of patients with dementia by using machine learning algorithms and semantic knowledge base. The information about the patient's daily activities are provided by IoT sensors and a Random Forest classifier is employed to detect significant deviations from the patients baseline which may signal side effects of drugs usage. A Polypharmacy Management Knowledge base has been created to capture the potential side-effects of drug-drug interactions and has been used to label historical data. Then, a K-Means algorithm is used to cluster similar days containing significant deviations from the baseline and the results of the clustering algorithm have been used to correlate future monitored days with potential drug-drug interactions. The novelty of our approach, in contrast to other state of the art approaches, is the fact that besides detecting the days containing significant deviations from an elder routine, we also identify the polypharmacy causes that lead to these deviations. The paper is structured as follows: Sect. 2 presents the technique for identifying significant deviations from the baseline and the possible drug-drug interactions causing the significant deviations; Section 3 presents preliminary experimental results; Section 4 concludes the paper.

2 The Technique for Assessing the Effects of Polypharmacy in the Daily Life Activities

Our goal is to analyze the heterogeneous and distributed streams of monitored data regarding patients with dementia, to establish the baseline daily life activities and to detect deviations that represent sudden/gradual changes in the patient's routines, which may signal progression of symptoms, well-being decline or medication side effects. To achieve this, we have designed a stack of technological resources (see Fig. 1) on top of which analytics are implemented. We have implemented a service that deals with monitoring and integration of various variables and distributed data sources describing the elders daily life activities, medication intake and adherence to the prescribed therapy. The service generates streams of data as AVRO Messages using data collected from sensors or Daily Life Activities data sets allowing for testing our techniques prior to actual deployment in real life contexts. The data flow management is achieved using a Confluent streaming platform [14] featuring: (*i*) Zookeeper [13] - centralized service for maintaining configuration information, distributed consistent states and synchronization, (*ii*) Kafka [13] - for building real-time data streaming pipelines, (*iii*) Kafka Connect [14] - for integration with the master data set kept in the Cassandra database [13] on which our technique will be executed through the Apache Spark machine learning engine. The Schema Registry [14] stores all the enforceable schemas, AVRO [13] is used as serialization framework, while the exchange of messages is done through Kafka REST proxies. Our technique uses the Polypharmacy Management Knowledge Base that we have defined and is implemented using two Spark Jobs: (*i*) Random Forest for detecting the deviations of the daily life activities from the patient's baseline and (*ii*) K-Means for automatically labeling the deviations with drug-drug interactions.

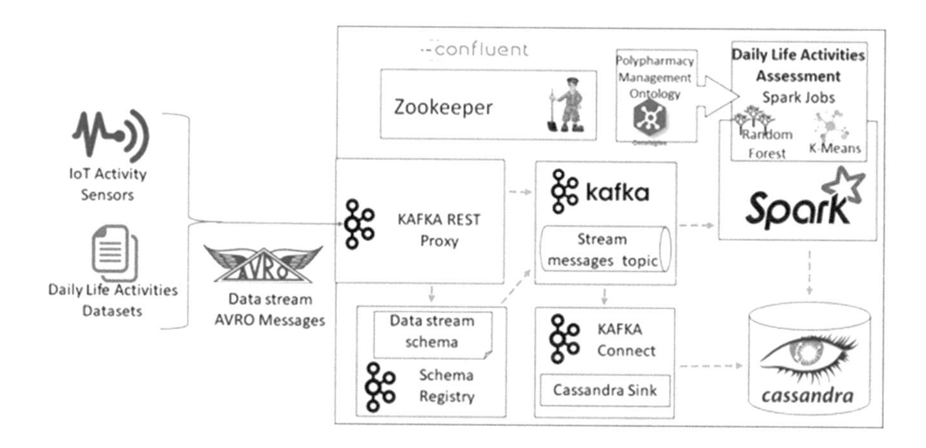

Fig. 1. Technological stack of resources used in the assessment process

2.1 Polypharmacy Management Knowledge Base

We developed a knowledge base which models the medication and drugs interaction for dementia treatment which was implemented as an ontology. In the implementation of the Polypharmacy Management Ontology (PMO) we have followed the methodology proposed in [6] consisting of 6 steps: (1) define the ontology domain and objectives; (2) identify a similar ontology to reuse the concepts; (3) identify a list of terms that will belong to the ontology; (4) define classes and class hierarchies by using a top-down approach, a bottom-up approach, or a mixed approach; (5) define the properties of the classes; (6) define the instances of the classes.

The domain modeled through the PMO is the management of polypharmacy in the case of dementia, and the objectives of the ontology are to provide knowledge about the drugs administered in the case of dementia, the interactions between these drugs, the adverse effects of these drugs, and the adverse effects caused by the interactions between these drugs. In the design of our ontology, we have reused and adapted the Drug-Drug Interactions Ontology

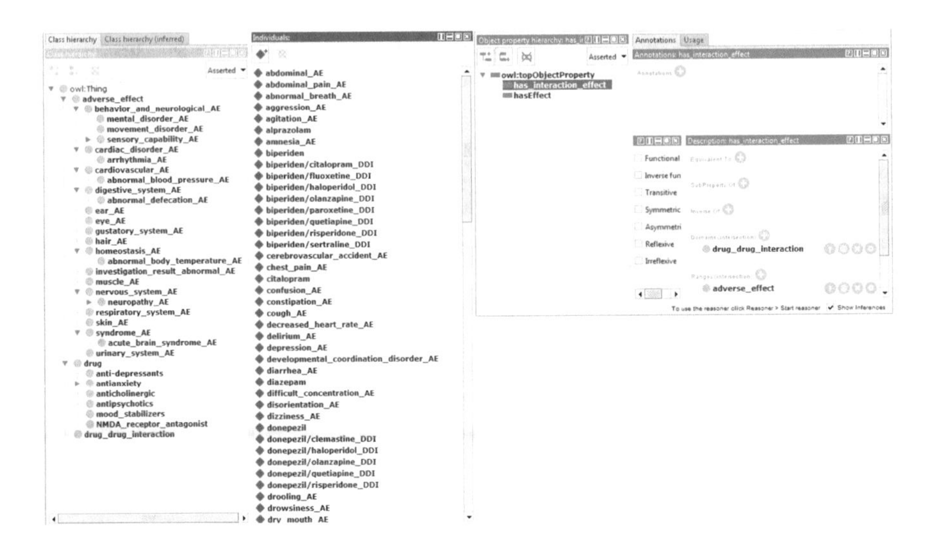

Fig. 2. Core ontology concepts and properties of the PMO

(DINTO) ontology [1,7] while the drugs side effects were taken from [8]. DINTO models the pharmacological effect of drugs, the pharmacodynamic actions of drugs, the mechanisms by which these actions are performed, the processes of absorption, transportation, distribution, metabolism and elimination of drugs, the recommended dose, and the interactions between drugs. Since in the case of people with dementia there are only few drugs that are administered to them, we decided to design a simplified version of this ontology, in which we include, besides the concepts that describe the drugs administered in the case of dementia and their individual adverse effects also the concepts describing the adverse effects of drug-drug interactions. The list of drugs related terms used when building the PMO was extracted from DINTO and from electronic medical documents [7–11]. The list contains only those terms that refer to concepts describing the drugs administered in dementia and the possible interactions between these drugs. For developing the PMO, we used a top-down development approach. We have started with the definitions of the most general concepts in the domain and then we have specialized these concepts into more specific concepts. In our ontology there are 3 categories of taxonomic trees, namely *Drug*, *Adverse_Effects*, *Drug_Drug_Interaction* (see Fig. 2). The *Drug* concept corresponds to the medication prescribed in the case of dementia. According to [7–11] we have classified the drugs prescribed in treating dementia in 7 main categories: *Anti – depressants*, *Antianxiety*, *Anticholinergic*, *Antipsychotics*, *Mood stabilizers*, and *NMDA receptor antagonists*. The *Adverse_Effect* tree contains a classification of the side effects of individual drugs for dementia. The *Drug_Drug_Interaction* tree models the potential interaction between two drugs that are simultaneously administered in dementia. Next, we have defined the properties of the concepts as class attributes. The *hasEffect* and *has_interaction_effect* object properties have been used to represent: the adverse effects that may occur in the case of administrating drugs for dementia, and the adverse effects that may occur because of the interaction between two dementia drugs that are simultaneously administrated. Finally, we defined instances for the classes belonging to the PMO taxonomic trees.

2.2 Detecting Significant Deviations and Assessing the Effects of Polypharmacy

To classify the activities throughout a day as normal routine or as exhibiting significant deviations from a baseline we used the implementation of the Random Forest algorithm suitable for big data stream classification provided by Apache Spark [13]. Random Forest is a classification algorithm which works by building a set of decision trees. Each decision tree is trained on a sub-set of the training data set. During the testing phase, each decision tree part of the Random Forest algorithm votes the class to which a test instance belongs, and the class with most of the votes is assigned to the associated test instance. In our case, we have used the Random Forest implementation from Spark MLlib for binary classification. For training the Random Forest classification algorithm, we have used a data set consisting of days annotated as having or not having a deviation from the normal baseline. Additionally, the days deviating from the baseline are annotated with semantic information about the possible drug-drug interaction and its adverse effects that might cause the deviation from the baseline. We defined a day containing monitored activities as:

$$day = (t_{start_1}, t_{end_1}, a_1), (t_{start_2}, t_{end_2}, a_2), ..., (t_{start_n}, t_{end_n}, a_n) \quad (1)$$

where t_{start_i} and t_{end_i} are the start and the end *timestamp* of the activity a_i.

The baseline for a patient with dementia is represented by the daily activities which the patient is normally carrying out. The baseline for a patient is defined by the doctor based on discussions with the patient, its family and caregivers. Also, based on discussions with doctors, we have considered five types of daily life activities as significant enough for allowing the detection of polypharmacy side effects: sleeping, feeding, toilet hygiene, functional mobility and community mobility. The number of features extracted from the data set that will be provided as input for the Random Forest Classifier is equal with the number of distinct types of activities considered. Consequently, each day will be represented as a vector of five features and the value of each feature represents the total time corresponding to an activity type performed by the patient in a day (see Table 1).

Figure 3 - (a) shows the activities performed by the patient in a monitored day together with their duration, while Fig. 3 - (b) shows the extracted vector of features corresponding to this day.

We consider that a day, $day_{monitored}$, has significant deviations from the patient's baseline, *baseline* if the total time corresponding to at least one activity type performed by the patient in the $day_{monitored}$ is higher or lower than a

Table 1. Vector of features representing a monitored day

Features					Label
Sleeping	Feeding	Toilet hygiene	Functional mobility	Community mobility	Normal/ Deviation
Total time in 24 h	Total time in 24 h	Total Time 24 h	Total time in 24 h	Total time in 24 h	0/1

Activities | Hour of the Day
0 1 2 3 4 5 6 7 8 9 10 11 12 13 14 15 16 17 18 19 20 21 22 23 24
Sleeping
Feeding
Functional Mobility
Toilet Hygiene
Community Mobility

a)

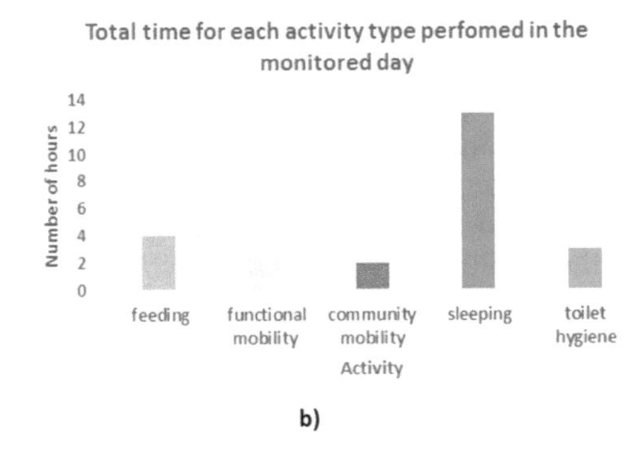

Fig. 3. Example of features extraction

pre-defined threshold compared to the same activity type in the *baseline*. The semantic annotation of a monitored day containing significant deviations from the baseline is defined as:

$$annotation = [((d_i - d_j), eff_1, eff_2, ..., eff_n), flag] \tag{2}$$

where (*i*) $(d_i - d_j)$ is an ontology concept from the PMO describing the interaction between two drugs administrated in the case of dementia; (*ii*) eff_i is an ontology concept describing an adverse effect that could appear as a result of the interaction between the two drugs, and (*iii*) $flag$ is a boolean value indicating that the day has a significant deviation (i.e. flag = true). A day without deviations from the baseline is annotated only with the flag set to false. The annotation is manually performed by a doctor who analyzes the deviations of a monitored day from the patient's baseline, and decides, in case of significant deviations, which drug-drug interaction (from the patient's medication plan) and side-effects cause the deviation.

To automatically detect, based on historical data collected from sensors, the drug-drug interaction causing the deviations of the current day, day_c, from the baseline we have defined a K-Means clustering method consisting of two steps. First, we have applied the K-Means clustering algorithm implementation available in Spark MLlib [13], on a set of initially annotated days containing deviations from the baseline. Each cluster will contain similar annotated days, and the label of the cluster is given by the annotation (i.e. drug-drug interaction and its adverse effects) of the cluster's centroid. In our case, the similarity between two days is defined as the Euclidean distance computed between the features vectors associated to the two days. Then, in the second step, day_c, classified by the Random Forest algorithm as having a deviation from the baseline, is assigned to an existing cluster for which the Euclidean distance between day_c and the cluster's centroid is minimum (see Fig. 4). day_c will be also annotated with the drug-drug interaction and its adverse effects corresponding to the assigned cluster's centroid.

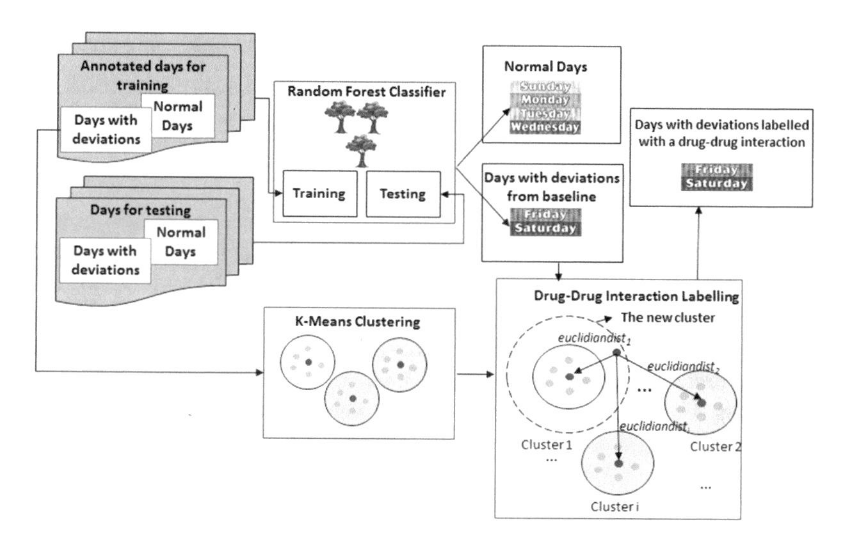

Fig. 4. Labeling the days containing deviations from the baseline with a drug-drug interaction

3 Lab Evaluation

At this stage of our work, we tested the algorithms individually on an in-house developed data set of activities. This data set adapts the one available at [15] to the activities considered in our approach and extends it so as to contain monitored data for a whole year. The information about the activities performed in each day from the data set has been represented using Formula 1 and has been manually annotated by the doctor (see Formula 2). 80% of the annotated data was used for training the Random Forest Classifier and 20% for testing the classifier (when testing, the annotation was omitted and was used only to evaluate the classification correctness). We performed a set of preliminary tests to identify the configuration of the classifier's parameters which lead to a learned model with high precision and recall. Each test consists of running the classifier for different values of its parameters: (*i*) $noTrees$, the number of trees, (*ii*) $maxBins$, the number of bins used when discretizing continuous features, and (*iii*) $maxDepth$, the maximum depth of a tree. Table 2 presents a fragment of the best experimental results obtained when varying the classifier's parameters.

Table 2. Top ten experimental results obtained for the Random Forest Classifier

Parameter	Values									
$NoTrees$	20	5	6	2	3	2	12	12	14	6
$MaxDepth$	5	7	5	4	5	6	8	9	3	3
$maxBins$	90	40	90	10	70	10	10	10	40	20
$F-measure$	0.9861	0.9723	0.9723	0.95853	0.9444	0.9306	0.9300	0.9300	0.9020	0.8876
$Precision$	0.9865	0.9739	0.9739	0.9620	0.9444	0.9311	0.9328	0.9328	0.9045	0.8921
$Recall$	0.9861	0.9722	0.9722	0.9583	0.9444	0.9305	0.9305	0.9305	0.9027	0.8888

From the annotated data set, only the days annotated as containing deviations from the patient's baseline were clustered using the K-Means Clustering Algorithm (see Sect. 2.2). In this case we analyzed how the number of clusters influences the Set Sum of Squared Error (WSSSE) measure (see Fig. 5). We have varied the number of clusters between 2 and 6, 6 corresponding to the maximum number of possible drug-drug interactions that could appear in the case of a patient taking 4 drugs. In our case, we have observed that the optimal value for the number of clusters is 4 because at this value the WSSSE graph forms an elbow.

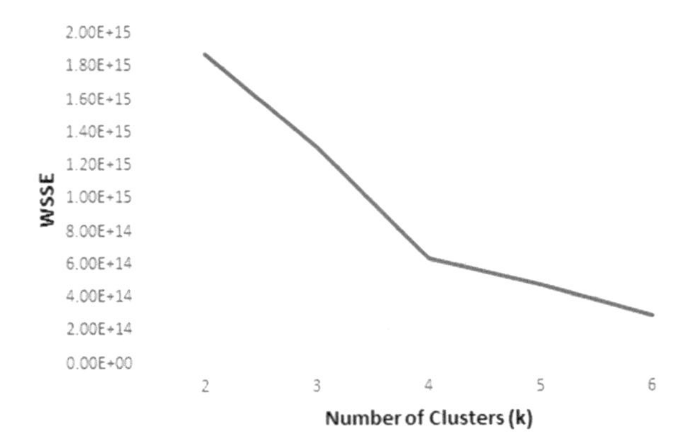

Fig. 5. WSSSE values when varying the number of clusters

4 Conclusions

In this paper we have proposed a two-step method for detecting the effects of polypharmacy in the case of patients with dementia applied on a set of data containing information about the daily activities performed by a patient. First, the Spark MLlib Random Forest Classifier is applied to classify days as being normal or as having significant deviations from a baseline behavior, while second, the Spark MLlib K-Means Clustering Algorithm is used to identify the drug-drug interactions causing the deviation of a day from the baseline. In the case of the Random Forest Classifier we have performed a series of experiments to identify the configuration of the classifier's specific parameters in order to obtain a model with a high precision and recall. In the case of the K-Means Clustering Algorithm we have used the Set Sum of Squared Error measure to identify the optimal value for the k parameter.

Acknowledgements. This work was supported by a grant of the Romanian National Authority for Scientific Research and Innovation, CCCDI UEFISCDI and of the AAL Programme with co-funding from the European Union's Horizon 2020 research and innovation programme project number AAL 44 / 2017 within PNCDI III.

References

1. Zazo, M.H., et al.: An ontology for drug-drug interactions. In: Proceedings of the 6th International Workshop on Semantic Web Applications and Tools for Life Sciences (2014)
2. Ujager, F.S.: Evaluation of wellness detection techniques using complex activities association for smart home ambient. Int. J. Adv. Comput. Sci. Appl. **7**(8), 243–253 (2016)
3. Ordonez, F.J.: Sensor-based bayesian detection of anomalous living patterns in a home setting. Pers. Ubiquitous Comput. J. **19**(2), 259–270 (2015)
4. Hoque, E., et al.: Holmes: a comprehensive anomaly detection system for daily in-home activities. In: Proceedings of the 2015 International Conference on Distributed Computing in Sensor Systems, pp. 40-51 (2015)
5. Civitarese, G., Bettini, C.: Monitoring objects manipulations to detect abnormal behaviors. In: Proceedings of the International Conference on Pervasive Computing and Communications Workshops (2017)
6. Noy, N.F., McGuinness, D.L.: Ontology development 101: a guide to creating your first ontology (2001). https://protegewiki.stanford.edu/wiki/Ontology101
7. https://bioportal.bioontology.org/ontologies/DINTO
8. https://www.drugs.com/
9. https://betterhealthwhileaging.net/medications-to-treat-difficult-alzheimers-behaviors/
10. https://www.dementia.org.au/national/about-dementia/how-is-dementia-treated/drug-treatments-and-dementia
11. https://www.alz.org/alzheimers_disease_standard_prescriptions.asp
12. http://www.who.int/mediacentre/factsheets/fs362/en/
13. http://apache.org/index.html#projects-list
14. https://www.confluent.io/
15. https://archive.ics.uci.edu/ml/datasets/Activities+of+Daily+Living+(ADLs)+Recognition+Using+Binary+Sensors

Other applications

A Study of the Predictive Earliness of Traffic Flow Characterization for Software Defined Networking

Hegoi Garitaonandia[1(✉)], Javier Del Ser[2], Juanjo Unzilla[1], and Eduardo Jacob[1]

[1] University of the Basque Country (UPV/EHU), 48013 Bilbao, Spain
{hegoi.garitaonandia,juanjo.unzilla,eduardo.jacob}@ehu.eus
[2] TECNALIA, Univ. of the Basque Country (UPV/EHU) and Basque Center for Applied Mathematics (BCAM), Bizkaia, Spain
javier.delser@tecnalia.com

Abstract. Software Defined Networking (SDN) is a new network paradigm that decouples the control from the data plane in order to provide a more structured approach to develop applications and services. In traditional networks the routing of flows is defined by masks and tends to be rather static. With SDN, the granularity of routing decisions can be downscaled to single TCP sessions, and can be performed dynamically within a single data stream. In this context We propose a novel approach – coined as micro flow aware routing – aimed at implementing routing of flows based on the properties of transport-level information, which is closely related to the type of application. Our proposed scheme relies on the early characterization of the flow based on statistical predictors, which are computed over a time window spanning the first exchanged packets over the session. We evaluate different window lengths over real traffic data to examine the Pareto trade-off between the earliness of flow characterization and its predictive accuracy. These results stimulate further research towards ensuring the practicality of the scheme.

Keywords: Software defined networking · Traffic engineering
Internet traffic classification · Machine learning

1 Introduction

Software Defined Networking (SDN) has emerged as an efficient network technology capable of supporting the dynamic nature of future network functions and intelligent applications through simplified hardware, software, and management [8]. The advent of this new networking paradigm has certainly revolutionized the Telecommunications field due to its cost effectiveness, flexibility and adaptability to the ever-growing IT needs of companies and end users. Remarkably, SDN has been widely identified as one of the major technological breakthroughs of the latest years in computer communications, with straightforward synergies

J. Del Ser et al. (Eds.): IDC 2018, SCI 798, pp. 393–403, 2018.
https://doi.org/10.1007/978-3-319-99626-4_34

and connections to other technological trends such as Cloud Computing [3] and Big Data [1].

In a typical SDN architecture the control plane is decoupled from the data plane. Therefore, network intelligence and state are logically centralized, so that the underlying network infrastructure is abstracted from the applications deployed on upper layers of the stack [2]. Among the main arguments for this approach is that it provides a more structured software environment for developing network-wide abstractions, while potentially simplifying the data plane [9]. Furthermore, the flexibility and scalability provided by programmatic network configurations allows for the development of unprecedented network functionalities and applications, such as software-defined mobile networking [5] or network-related security solutions [7]. Although SDN has been at the core of a vibrant research activity in the last few years, several recent studies have noted that several challenges remain insufficiently addressed to date in regards to performance, security and energy efficiency [6].

SDN technology allows taking decisions on many fields of the packet header, including those corresponding to the transport protocol; this work focuses on routing decisions for TCP sessions. In this regard, routing in traditional networks takes into account flows that are defined by IP masks. By properly configuring these masks in the routing tables of the different routers of the network, routes among subnets defined by such masks can be defined and established. However, this static routing procedure, very much related to the way routers classify traffic, does not consider the diversity of traffic flows that may hold within streams transported over a given TCP session. With SDN the granularity of the flow definition can be certainly extended to almost any header field combination, as opposed to the classical approach which deals with IP addresses and protocols. This finer granularity allows targeting a single TCP session, so that an ad-hoc dynamic routing decision can be made on a per session basis. We refer to this postulated paradigm as *micro flow aware routing*.

Micro flow aware routing involves two main ingredients. The first is the definition of flows or categories that group different TCP sessions. This characterization can be based on statistical properties of packets and data exchanged over the session at hand. We postulate that the discovery of groups of similar sessions in terms of these properties could be exploited for different purposes once routing decisions were defined for every such group, e.g. elastic resource provisioning or priority-based flow forwarding. Furthermore, this strategy to determine the flow type based on the transported traffic data has been acknowledged to overcome unpredictable and obscure port assignation [4]. The second ingredient of micro flow aware routing hinges on the prediction of this flow type based on statistics of the session computed over the initial inbound and outbound packets of the session. This prediction can be regarded as an *early* characterization of the session, which can be utilized to dynamically apply route policies suited for the estimated properties of the traffic conveyed over the session.

Intuitively, a trade-off should be expected between the earliness of the prediction and its quality [4]. The former is given by the time from the establishment of the session to the prediction itself, whereas the latter yields from the comparison between the predicted session type and that computed over the statistics of the entire session. This being said, the contribution of this manuscript is twofold:

1. We propose an initial design of a micro flow aware routing approach for SDN that embraces machine learning at its core. Specifically, traffic flow prediction is addressed by formulating a supervised learning problem over a labeled record of labeled session traces.
2. By using real traffic data captured from an University network, we provide an informed insight on the balance between predictive accuracy and earliness of the aforementioned approach achieved by different machine learning models. Albeit preliminary, our results are encouraging and foster further developments aimed at achieving a better balance – in the Pareto sense – between such goals.

The remainder of the paper is organized as follows. The proposed micro flow aware routing approach is described in Sect. 2, whereas details on the data collection campaign are given in Sect. 3. Results are presented and discussed in Sect. 4. We finish this work with some concluding remarks and future research outlined in Sect. 5.

2 Proposed Approach

The proposed approach is schematically described in Fig. 1, which is designed to deal with the issues presented in Sect. 1. The SDN controller and the SDN switch are part of a typical SDN infrastructure. In a SDN network the control and the data plane are not only decoupled but the switch can react, in fact, dynamically to the traffic it is actually receiving. This is called reactive mode operation: when an SDN switch receives a data-packet, it searches for the flow rules installed in the switch's TCAMs (Ternary Content-Addressable Memories) and acts according to the rule that matches the actual packet header. In case there is no matching rule, the switch will forward the packet to the Controller for inspection.

After inspection in the Controller, a new packet processing decision will be made. This will generate a new set of flow-rules that will be installed by the controller on the SDN Switch, to ensure that future packets that comply with the flow definition will be adequately treated (for example, routed). The packet that was received will also be forwarded to the switch to ensure that the whole session is forwarded according to the new flow-rules. In the example depicted previously, the SDN Switch is used to interconnect two networks, that is, to implement a router.

To improve the functionalities of traditional SDN Switches, we introduce a new functionality that allows it to act, not only upon the usual rules, but on rules based on the statistical properties of the flow as well. For this purpose, the system needs to establish the set of categories or flows that the SDN software switch is going to consider. This task is performed off-line and based on an

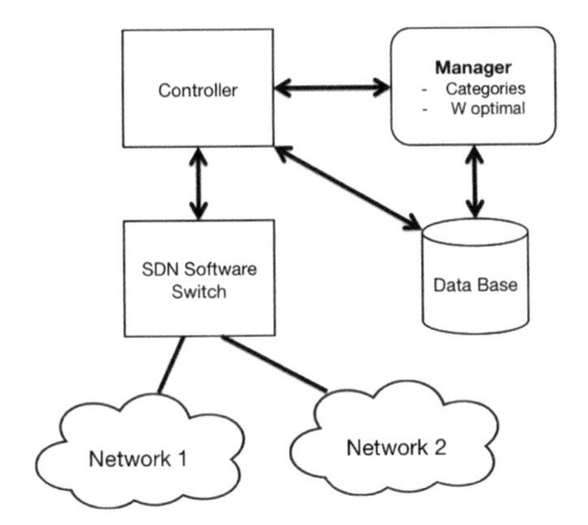

Fig. 1. Block diagram of the proposed design for a micro flow aware routing approach for SDN setups.

observation of network traces captured during a time frame (e.g. 1 h), which must be representative of the traffic expected to be processed through the SDN switch (correspondingly, a labor day). This task is performed by the Manager. The Manager captures networks traces from the SDN switch, via the Controller. Afterwards, a clustering algorithm is applied on the captured sessions using the statistical properties of transport-level information as their characterizing features. At this point it is important to remark that this first characterization is done by considering statistics computed over all packets of the session.

Second, our objective is that the SDN software switch determines as early as possible the flow-type for a session under study based on the statistical profile of its exchanged data. For this purpose, we use classification algorithms [4]. Different earliness windows W for the flow type prediction can be used in the Manager, which is used to select the subset of initial packets over which statistics are computed and fed to the predictive model. The selection of a proper value of W can be driven by different criteria, e.g. computing capacity of the SDN controller and switch, the need for more agile rerouting decisions (as in mobile networking environments) or a minimum accuracy threshold set for the predictive model. Anyhow, once the optimal prediction window is established, the Manager communicates with the Controller, and the Controller installs the rules in the SDN Software Switch, so that it can apply them on-line. The analyzed traces used for cluster analysis are kept in the Data Base, as well as the routing rules.

3 Data Collection Process

We now seek to assess the trade-off between earliness and predictive quality identified in Sect. 1. To this end, we capture all TCP traffic going through a 10 Gbps link of the University of the Basque Country. Specifically, traces of all sessions were recorded for one full hour during work time. We take all the cautions to ensure that data remain real throughout all subsequent processing steps. We reconstruct the TCP sessions by using a simple algorithm that indexes the packets by a 4-tuple formed by IP (client/server) and port (client/server).

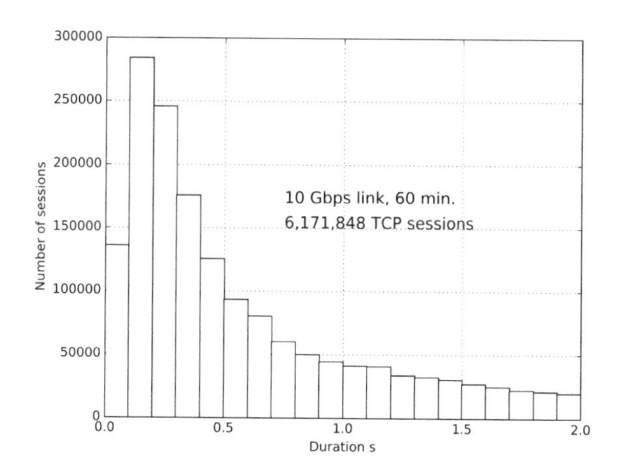

Fig. 2. Histogram of the duration of the recorded TCP sessions.

During the processing, data about the TCP sessions is kept in memory until the analyzed block is completely processed. The capture is performed by the `tcpdump` tool. We use two orthogonal methods to extract the variables, one based on libpcap, faster, and another one based on regular expressions on top of `tcpdump` visualization, for cross-checking results. The results obtained with both methods are compatible. As shown in Fig. 2, we found that most sessions last less than one second. This is the reason why experiments consider the corresponding W interval.

There are two reconstruction stages. In the first one, each session is represented by four vectors. Two vectors contain the arrival times of the packets for the client and for the server, and two vectors contain the lengths of the packets for the client and for the server. The second stage, which involves feature extraction, is described in the following section. After each stage the data is serialized so that it can be easily accessed in the remaining analysis.

4 Experimental Results and Discussion

Several computer experiments have been carried out to examine the predictability of different traffic classes over the aforementioned set of captured network traces. First we emphasize that in this preliminary work we restrict our experimentation to the predictive part of the proposed micro flow aware routing approach, thus deferring the prescriptive part of the architecture for future work.

Nevertheless, in order to grasp informed insights on the viability of micro flow aware routing it is of utmost importance to assess the predictive performance of several supervised learning model to characterize real traffic data. This is indeed the targeted goal of the results discussed in what follows. This being said, we retrieve the number of sessions captured in the aforementioned dataset, from which we extract features related to statistical properties of transport-level information. According to the related literature, we choose a similar set to [4], namely:

- Volume of the session (bytes).
- Duration of the session (milliseconds).
- Different statistics of the length of packets exchanged during the windowed start of the session (min, max, mean, standard deviation).
- Different Inter Arrival Time (IAT) between packets exchanged during the windowed start of the session (min, max, mean, standard deviation).

These features are calculated for both inbound and outbound data streams between server and client. We consider window sized equally spaced in 50 ms intervals from 50 ms (early predictive characterization) to 1 s (correspondingly, late predictive characterization). This accounts to 26 features (predictors) computed for a total of 6, 171, 848 sessions.

Once features have been extracted, we label the entire dataset by considering the Euclidean similarity between sessions. Such a similarity is computed over the previous set of features, computed over the entire duration of each session rather than on a windowed time frame. The dataset is previously standardized (Z-score) to make all features weight statistically the same in the computation of the similarity. The well-known KMeans++ clustering algorithm is utilized to find groups of similar sessions in the dataset, varying the sought number of clusters as a design parameter. Although heuristics to properly set this design parameter abound in the machine learning literature (e.g. X-Means or G-Means), in this preliminary work we opt for the Elbow method, which permits to visually inspect how the percentage of variance explained in the dataset evolves as a function of the number of clusters. As shown in Fig. 3, an apparent turning range occurs around 20 clusters; consequently, the cluster arrangement found for this particular number of clusters will be selected for labeling the sessions.

Once sessions within the dataset have been labeled with the cluster information (20 different categories or *flow types*), we proceed by testing the performance of a number of different classifiers. An emphasis is placed on the earliness by which labels can be reliably predicted by varying the window size W over which features in the above list are computed. Specifically, the following models have been utilized in our benchmark:

- A Logistic Regression (LR) model, used in a one-vs-rest (OvR) scheme for multiclass classification.
- A Linear Discriminant Analysis (LDA) model, which fits class conditional densities to the data and uses Bayes' rule to yield a linear decision boundary over the feature space.

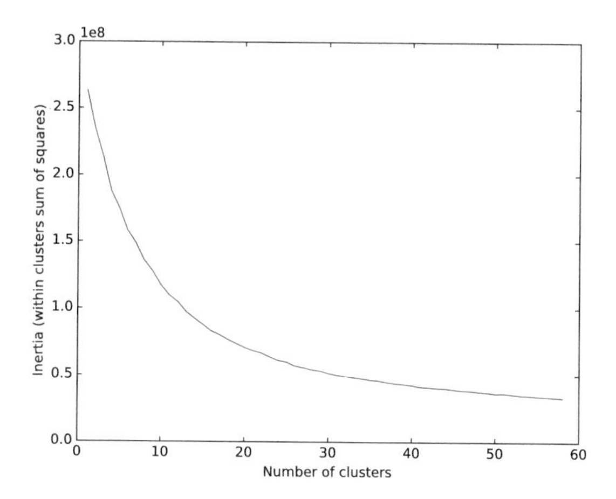

Fig. 3. Inertia (within-cluster sum of squares) of the cluster arrangement found by KMeans++ as a function of the number of clusters.

- A Gaussian Naive Bayes (GNB) model, which approximates the class conditional probability of every feature by assuming conditional independence between them.
- A kNN classifier (kNN), based on majority voting of the class labels of the k most similar neighbors to every test example, wherein similarity is given by the Euclidean distance measured over the standardized feature space.
- Two different boosting ensembles: AdaBoost (ADA) and Gradient Boosting Classifier (GBC), each relying on decision trees as their core learners, which learn from the input data in a chain fashion so as to progressively focus on those instances misclassified by previous classifiers.
- A Decision Tree classifier (DT), which recursively seeks the most discriminating split over each feature of the dataset. This model choice is particularly suitable for the micro flow aware routing approach described in Sect. 2 as its learned structure over the dataset directly produces a set of IF-THEN rules that could be deployed on the SDN Software Switch in a straightforward manner.
- A Random Forest (RF) classifier, which essentially reduces to a bagging ensemble of DT classifiers, learned over random *views* of the input data. As it relies on several DT learners, the learned knowledge can be described as a multi-set of rules, whose outputs (one per rule set) are voted to produce the finally predicted label.

The learning algorithms characterizing the above classifiers is driven by different parameters, whose values must be tuned appropriately to yield a trained model with good generalization properties. For this to occur and in order to fairly compare all the above algorithms, the entire dataset is shuffled and split into two stratified subsets of data instances: a training dataset comprising 4,937,478 sessions (80%), and a test dataset with the remaining 1,234,370 sessions (20%). Parameter tuning is then performed by 10-fold cross-validation over the training subset. For every model in the benchmark a fine-grained value grid of its parameters is efficiently explored by a Genetic Algorithm wrapper configured with crossover and mutation rate equal to 0.9 and 0.1, respectively. The search is driven by the maximization of the weighted F1 score averaged over the 10 folds of the cross-validation procedure. After parameter tuning, the configured model is trained over the entire training set and queried to predict the data instances in the test set. Finally, scores are computed by comparing predicted and true labels in the form of accuracy, Cohen's Kappa and the aforementioned weighted F1 score.

The results of this benchmark are summarized in Table 1 for different values of the earliness window W. A first inspection of the scores in this table reveals two effects that conform to intuition. On one hand, as the value of W increases,

Table 1. Prediction scores of the supervised learning models considered in the benchmark over the test set, for different values of the earliness window W.

		Score	Classification model							
			LR	LDA	GNB	kNN	ADA	GBC	DT	RF
Earliness window W (ms)	50	Accuracy	0.505	0.506	0.571	0.567	0.615	0.491	0.592	0.641
		Cohen's Kappa	0.259	0.256	0.231	0.352	0.346	0.416	0.383	0.456
		Weighted F1	0.501	0.506	0.458	0.571	0.567	0.616	0.593	0.642
	150	Accuracy	0.548	0.553	0.619	0.588	0.644	0.528	0.616	0.687
		Cohen's Kappa	0.316	0.320	0.278	0.426	0.376	0.462	0.417	0.527
		Weighted F1	0.542	0.548	0.492	0.620	0.586	0.644	0.614	0.688
	300	Accuracy	0.569	0.575	0.662	0.614	0.664	0.550	0.636	0.712
		Cohen's Kappa	0.347	0.353	0.310	0.490	0.417	0.493	0.448	0.565
		Weighted F1	0.563	0.572	0.522	0.662	0.613	0.664	0.634	0.712
	450	Accuracy	0.577	0.581	0.674	0.622	0.679	0.558	0.646	0.725
		Cohen's Kappa	0.357	0.361	0.322	0.508	0.429	0.515	0.464	0.585
		Weighted F1	0.572	0.578	0.532	0.675	0.620	0.679	0.647	0.726
	600	Accuracy	0.590	0.589	0.684	0.631	0.690	0.566	0.652	0.736
		Cohen's Kappa	0.376	0.373	0.334	0.524	0.442	0.531	0.472	0.602
		Weighted F1	0.585	0.587	0.544	0.685	0.629	0.689	0.652	0.737
	750	Accuracy	0.588	0.587	0.690	0.633	0.692	0.571	0.654	0.741
		Cohen's Kappa	0.374	0.370	0.341	0.532	0.445	0.535	0.476	0.608
		Weighted F1	0.584	0.586	0.550	0.690	0.631	0.691	0.653	0.741
	900	Accuracy	0.600	0.593	0.693	0.638	0.695	0.574	0.661	0.741
		Cohen's Kappa	0.392	0.379	0.345	0.537	0.453	0.539	0.488	0.609
		Weighted F1	0.595	0.591	0.554	0.693	0.637	0.694	0.662	0.742
	1000	Accuracy	0.598	0.594	0.695	0.636	0.697	0.575	0.658	0.745
		Cohen's Kappa	0.389	0.380	0.347	0.540	0.449	0.542	0.480	0.615
		Weighted F1	0.593	0.592	0.553	0.696	0.634	0.696	0.659	0.746

the features of the session computed in the time frame $[start, start + W]$ become more predictive of the class of session to which the started session belongs. Therefore scores, in general, improve as we wait for more information to feed the computation of features, an intuitive effect that can be visually observed in Fig. 4 for the Cohen's Kappa score. This can be noted in the monotonically increasing score values with W increasing for any model. In comparative terms, RF emerges as the outperforming classifier in the benchmark, with Kappa and weighted F1 values surpassing 0.6 and 0.7 for $W \geq 600$ ms.

Although these numbers can be regarded as an arguably good result in terms of data modeling, the high prediction latency needed to attain them jeopardizes the practical implementation of this predictive model in the context of micro flow aware routing. In light of the session duration histogram in Fig. 2 it should be clear that a large fraction of the sessions end up sooner than $W = 600$ ms. Consequently, the networking benefits postulated by the application of micro flow aware routing would not be applicable to these sessions if the predictive model were not queried before W. Reducing this threshold, however, would imply a degradation of the accuracy of the model, which would eventually outweigh the improved earliness of the overall approach. As explained in the closing section of this paper, further research will be conducted towards achieving a better balance between these two conflicting aspects, namely, earliness and model performance.

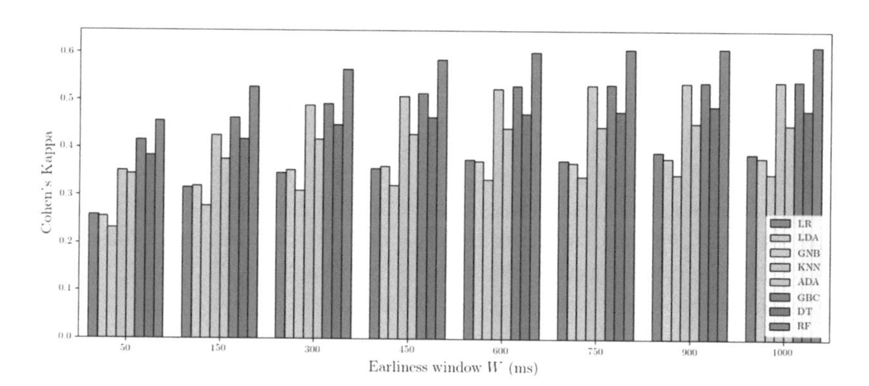

Fig. 4. Bar plot depicting Cohen's Kappa scores in Table 1 as a function of the window size W. It can be observed that as W increases, so does the quality of the predicted labels.

5 Concluding Remarks and Future Research Lines

This work has elaborated on a micro flow aware routing approach for SDN environments. Specifically, our envisaged setup performs an early predictive classification of Internet traffic in different categories or flows. This characterization is based on the statistical properties of transport level information, which is linked to the application type itself. Both packet processing and computing power capabilities are far beyond what is currently available in commercial switches, but they could be achieved by using Network Function Virtualization (NFV) technology and software switches like OpenVSwitch on commodity x86 servers, along with Data Plane Development Kit (DPDK) supported by high speed Network Interface Cards (NICs).

Two key elements lie at the core of our proposed procedure: first, a clustering process assigns traces to a set of categories or flows, which is performed off-line by the Manager over a history of captured data traces. This clustering process permits not only to infer which session types can be discriminated in sessions handled by the SDN Software Switch, but also to label the collected traces for subsequent predictive modeling. Indeed, this second core element of the proposed scheme aims at learning the pattern relating a set of early predictors computed over the initial packets exchanged over a session and its type (flow). The duration W of this initial time window has been shown to be a critical choice for the practicality of the overall scheme, as it drives the trade-off between the earliness of the predictive characterization of the session and its quality. All in all, once the value of W is determined the Manager can apply statistical routing rules suited to deal with the characteristics of the predicted cluster, which is done by deploying them on the SDN Software Switch via the Controller. A Random Forest classifier was found to perform best in the benchmark, followed by Decision Trees. An additional benefits of these learning algorithms is that IF-THEN rules can be mapped easily to SDN routing decisions.

For the case under study, a window size of 500–600 ms was determined to balance properly the trade-off between earliness and predictive performance. However, such a window length would discard more than 50% of the total sessions. Although the subset of eventually discarded sessions result to be those with less traffic volume (and hence, less critical in regards to network management), this issue remains to be studied in detail in the future. In particular we will delve on the derivation of new predictors that potentially attain a better Pareto balance between these design objectives.

Acknowledgements. This work was supported in part by the Basque Government through the EMAITEK program, and by the Spanish Ministry of Economy, Industry and Competitiveness through the State Secretariat for Research, Development and Innovation under the "Adaptive Management of 5G Services to Support Critical Events in Cities" (5G-City) project (TEC2016-76795-C6-5-R). Finally authors would like to thank you the University of the Basque Country IT services for allowing us to conduct this experiment on their premises with real production data.

References

1. Cui, L., Yu, F.R., Yan, Q.: When big data meets software-defined networking: SDN for big data and big data for SDN. IEEE Network **30**(1), 58–65 (2016)
2. Fundation ON: Software-defined networking: the new norm for networks. ONF White Paper **2**, 2–6 (2012)

3. Jain, R., Paul, S.: Network virtualization and software defined networking for cloud computing: a survey. IEEE Commun. Mag. **51**(11), 24–31 (2013)
4. Nguyen, T.T., Armitage, G.: Training on multiple sub-flows to optimise the use of machine learning classifiers in real-world ip networks. In: Proceedings 2006 31st IEEE Conference on Local Computer Networks, pp. 369–376. IEEE (2006)
5. Pentikousis, K., Wang, Y., Hu, W.: Mobileflow: toward software-defined mobile networks. IEEE Commun. Mag. **51**(7), 44–53 (2013)
6. Rawat, D.B., Reddy, S.R.: Software defined networking architecture, security and energy efficiency: a survey. IEEE Commun. Surv. Tutor. **19**(1), 325–346 (2017)
7. Scott-Hayward, S., Natarajan, S., Sezer, S.: A survey of security in software defined networks. IEEE Commun. Surv. Tutor. **18**(1), 623–654 (2016)
8. Sezer, S., Scott-Hayward, S., Chouhan, P.K., Fraser, B., Lake, D., Finnegan, J., Viljoen, N., Miller, M., Rao, N.: Are we ready for SDN? Implementation challenges for software-defined networks. IEEE Commun. Mag. **51**(7), 36–43 (2013)
9. Tootoonchian, A., Gorbunov, S., Ganjali, Y., Casado, M., Sherwood, R.: On controller performance in software-defined networks. Hot-ICE **12**, 1–6 (2012)

An e-Exam Platform Approach to Enhance University Academic Student's Learning Performance

Radu Albastroiu[1](✉), Anisia Iova[1](✉), Filipe Gonçalves[2](✉), Marian Cristian Mihaescu[1](✉), and Paulo Novais[2](✉)

[1] Department of Computer Science and Information Technologies, University of Craiova, Craiova, Romania
albastroiuradug@gmail.com, anisiai@yahoo.com, mihaescu@software.ucv.ro
[2] Algoritmi Research Centre/Department of Informatics, University of Minho, Braga, Portugal
fgoncalves@algoritmi.uminho.pt, pjon@di.uminho.pt

Abstract. Nowadays it is common for higher education institutions to use computer-based exams, partly or integrally, in their evaluation processes. The fact that exams are undertaken in a computer allows for new features to be acquired that may provide more reliable insights into the behaviour and state of the student during the exam. Current performance monitoring approaches are either intrusive or based on productivity measures and are thus often dreaded by workers. Moreover, these approaches do not take into account the importance and role of the numerous external factors that influence productivity. In this paper, we outline a non-intrusive and non-invasive performance monitoring approach developed, as a stress detection system. It is based on guidance from psychological stress studies, as well as from the nature of stress detection during high-end exams, through real-time analysis of mouse movements and decision-making behavioural patterns during the execution of high-end exams, in order to enhance university academic students' learning performance.

Keywords: Psychological stress classification
Human-computer interaction · Biometric analysis
Performance assessment · Machine Learning

1 Introduction

The term stress is commonly used to describe a set of physiological responses that emerge as a reaction to a challenging stimulus that alter an organism's environment [1]. Indeed, prolonged exposure to stress-inducing factors is a growing concern, especially in complex activities, which demands great responsibility and reliability. Lack of cumulative stress management can lead to states of emotional exhaustion (burnout) and other mental disorders (e.g. depression, chronic

J. Del Ser et al. (Eds.): IDC 2018, SCI 798, pp. 404–413, 2018.
https://doi.org/10.1007/978-3-319-99626-4_35

stress, chronic diseases) with potential consequences at the personal, professional, family, social and economic levels [2]. These cases are most commonly observed in stressful environments (e.g. educational/workplace environments), where the need for more efficient and discrete psychological monitoring systems are required to prevent student's/worker's emotional collapse (through the application of cooping techniques) and to improve their daily performance.

This work proposes the development of a non-intrusive and non-invasive performance monitoring environment with the means to assess the user's performance state in real-time. Specifically, the paper discusses how a group of medical students were monitored as a way of finding a predictive relationship between a student's performance and biometric behaviours (mouse dynamics and decision-making behaviours) in high-end exams. Students were monitored in terms of the efficiency of their interaction patterns with the computer, an approach that can be included in the so-called behavioural biometrics. The main aim of this line of research is to provide additional sources of contextual information about students during evaluation tasks that can allow the educational institution to design better and more individualised teaching strategies.

It's important to remind that this work presents part of the project that uses the mouse performance behaviour analysis and the decision making performance analysis of individuals under pressure in workplace/study environments (developed by the research team of the University of Minho in the past [3,4]).

2 Related Work

Studies suggest that academic students experience a high incidence of personal distress, with potentially adverse consequences on academic performance, competency, professionalism, and health [5]. It is critical for educators to understand the prevalence and causes of student distress, potential adverse personal and professional consequences, and institutional factors that can positively and negatively influence student health.

Existing stress monitoring approaches rely on the use of complicated or expensive hardware, or in the collection of biological variables, all of which alter the routine of the student in some way, which is not desirable [6].

A particularly promising method is mouse behaviour analysis. This method measures and assesses a user's mouse behaviour characteristics for use as a biometric. Compared with other biometrics such as the face, fingerprint and voice [7], mouse dynamics is less intrusive and requires no specialised hardware to capture biometric information. Also, a growing segment of the population is already engaged in this tool, which facilitates the process of gathering baseline data for comparison, requires no additional equipment, adjusts to individual differences, scales easily, and may be used to detect early signs of behavioural pattern changes.

Another method is based on the assessment of a user's decision-making patterns during high-risk assignments. Everything an individual does, consciously or unconsciously, is the result of some decision. To make a decision, an individual need to know the problem, the need and purpose of the decision, the criteria of the decision, and the alternative actions to take. Then there's the need to determine the best alternative [8]. Particularly decision-making analysis aims to evaluate the performance of an individual, based on the time between decisions, the correctness of the selected decisions, the objectiveness of the decisions taken, among other features.

In this paper, we make use of mouse dynamics and decision-making biometrics to propose a set of different features to characterise the interaction of medical students with the computer while taking exams. These methods will be essential to provide a faster and more effective response to students whose performance can be improved through the use of coping techniques that improve an individual's stress management and consequently their academic performance. The proposed system presents the following innovative key elements: (1) Multi-modal approach (parallel analysis of a set of biometric behaviours features as a way to enable a more comprehensive analysis); (2) Exclusive use of Soft Sensors (only hardware devices commonly seen on a computer/laptop - e.g. computer mouse device - are required, decreasing the system's operational cost); (3) Distributed and Scalable (the proposed system can monitor multiple students simultaneously, not significantly influencing its complexity).

3 Architecture

Taking into account the requirements outlined, it is essential the analysis of two types of information: an individual's mouse behaviour and decision-making behaviour. Based on an individual's behaviour, a set of biometric metrics are processed automatically. The proposed framework presents the following components (as shown in Fig. 1):

- *MedQuizz*: An e-assessment management system that enables the accomplishment of the formative and summative evaluation of the students throughout the learning process and the final evaluation of the knowledge obtained in each curricular unit, respectively. It is a useful tool to create item banks and logs of the individual's decision-making action[1]. This component is used to study the cognitive performance and the behaviour decision making patterns of an individual [4];
- *MouseDynamics*: A behavioural biometric framework that extracts and analyses the movement characteristics of the mouse input device when a computer user interacts with a graphical user interface for behavioural pattern analysis purposes. Includes the sheer acquisition and classification of the mouse input data that supports the human-based or autonomous decision-making mechanisms [3,9];

Assessment of stress is performed through the Perceived Stress Scale tool (PSS-13), where each student performs one every month. PSS-13 is a 13-item scale that assesses the perception of stressful experiences by asking the respondent to rate the frequency of his/her feelings and thoughts related to events

[1] MedQuizz website can be assessed through the website http://www.medquizz.com/.

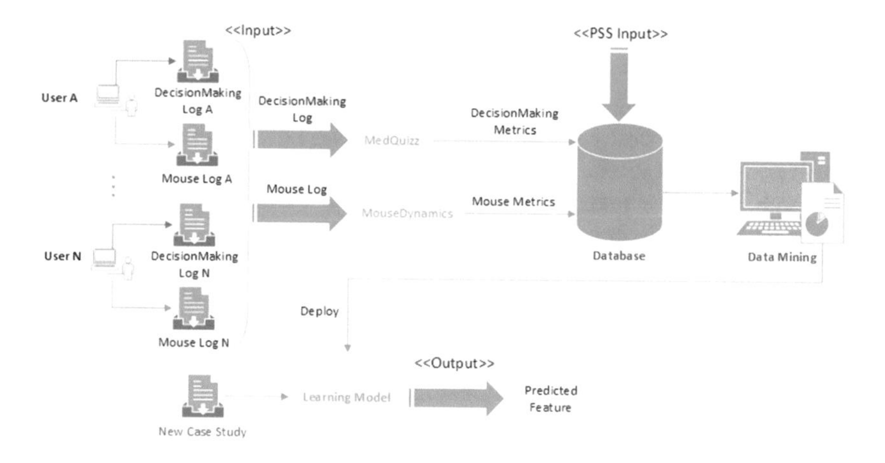

Fig. 1. Architecture framework.

and situations that occurred over the previous month [10]. PSS-13 scores are obtained by reversing the scores on the six positive items (e.g. $0 = 4$, $1 = 3$, $2 = 2$, etc.) and then summing across all 13 items, where items 4, 5, 6, 7, 9 and 10 are the positively stated items. The final PSS-13 score varies between [0–52]. Notably, high PSS scores have been correlated with higher biomarkers of stress [11]. Overall, the PSS is an easy-to-use questionnaire with established acceptable psychometric properties.

Since the study of behavioural features is mostly related to the individual's conduct habits, this procedure considers the individual's ID, actions of mouse and decisions made during the exams [12].

4 Biometric Features

Through the mentioned framework, the system allows to capture and process the mouse behaviour and decision-making biometric features of students, during high-end exams. As such, the calculated features can be grouped into two different sets: Mouse behaviour metrics (processed in the MouseDynamics module) and Decision-Making Behaviour Metrics (handled in the MedQuizz module). Since the framework computes a large set of behavioural biometrics, only the most relevant will be presented in this section:

1. *Mouse Behaviour Metrics*
   - *Absolute Sum of Degrees (ASD)*: This feature seeks to find how much the mouse turned, independently of the direction to which it turned (in degree units);
   - *Distance of the Mouse to the Straight Line (DMSL)*: This feature quantifies the sum of the successive distances of the mouse to the straight line defined by two consecutive clicks (in pixels);

- *Click Duration (CD)*: Measures the time span between mouse click usage. The longer the clicks, the less efficient the interaction is (in milliseconds);
- *Distance Between Clicks (DBC)*: Represents the total distance travelled by the mouse between two consecutive mouse clicks (in pixels);
- *Mouse Velocity*: The distance travelled by the mouse (in pixels) over the time (in milliseconds);

2. *Decision-Making Behaviour Metrics*
   - *Average/Median Time Between Decision (ATBD)*: This feature seeks to find the average/median time it takes each individual to take a decision. The decisions analysed vary from entering/leaving a question, inserting/changing/removing answers, marking/unmarking questions for review, among others (in milliseconds);
   - *Average Time Between Questions (ATBQ)*: This feature measures the average time the individual spent on all visualised questions (in milliseconds);
   - *Correct Decision-Making Ratio (CDMR)*: Measures the ratio between the number of decisions considered correct and the number of answers inserted, changed and removed (in percentage);

5 Study Design

Data was collected from the participation of a group of 52 3rd-year medical students during 3 different computer-based high-stake exams, based on their mouse behaviour performance and decision-making behaviour performance. Through the evaluation of exams, these students test their academic knowledge in a monthly period. In these exams, students are indicated to their seats, and at the designated time they log in the exam platform using their personal credentials and the exam begins. The participation in the data-collection process does not imply any change in the student's routine, and all monitored metrics are calculated through background processes, making the collection data process completely transparent from the student's point of view, just like a normal routine exam. These exams consist mostly of single-best-answer multiple choice questions, where the students only use the mouse as an interaction means. When the exams end, students are allowed to leave the room.

To complement the data recorded, students provide each month their perceived stress state, by means of a self-reported questionnaire (PSS-13). This tool offers a way to self-evaluate the degree to which they believe their life has been unpredictable, uncontrollable, and overloaded during the previous month. The assessed items are general in nature rather than focusing on specific events or experiences.

All methods used for data collection were considered since any external factor can in influence variations in the decision making and mouse performance behaviour of the individuals. As such, it was important to include non-intrusive and non-invasive measures as essential requirements during the execution of the exams.

6 Data Preparation and Analysis

In this experimental study, the input dataset consisting of the results for the mouse behaviour performance and decision-making were initially labelled through the PSS-13 score questionnaire and split into three categories (numeric categorisation process), where the problem was designed as a multi-class classification task, i.e. each training point belongs to one of N different classes (in this study we used 3 classes - Low Stress: 0–17; Moderate Stress: 18–35; High stress: 36–52). The goal was to construct a ML model which, given a new data point, would correctly predict the class to which the new point belongs. Due to this process, it was observed the presence of a higher number of moderate stress instances, in comparison to the low stress and high stress (specifically 76% of total cases were labelled as moderate stress cases). As a result, a number of pre-processing techniques were required to ensure the resolution of the dataset imbalance, enabling a more appropriate analysis of the study cases.

Oversampling was one of the techniques used to adjust the dataset class distribution. Its main goal was to adjust the bias found in the original dataset. In order to balance the dataset, this method randomly replicates observations from a minority class (i.e. the low and high samples) which will lead to an equal number of each type of stress level. Undersampling is the direct opposite of oversampling, working with the majority class by reducing its number of observations to make the dataset balanced. This technique was used to randomly choose observations from the moderate entries, which were eliminated until the dataset became balanced [13]. By cause of the big difference of entries between the majority class and minority class, a hybrid dataset was created by choosing a middle number of entries. This meant reducing the number of majority labels (moderate stress), while increasing the number of minority labels (low and high-stress), being helpful in both cases when under-sampling would reduce too much of the samples from the original dataset and the oversampling would have a bigger amount of repeated observations from the minority classes.

Another method applied was ROSE (Random Over-Sampling Examples), a smoothed bootstrap-based technique which has been recently proposed by Menardi and Torelli [14]. ROSE helps to relieve the seriousness of the effects of an imbalanced distribution of classes by aiding both the phases of model estimation and model assessment. Artificial balanced samples are generated according to a smoothed bootstrap approach and allow for aiding both the phases of estimation and accuracy evaluation of a binary classifier in the presence of a rare class.

The SMOTE (Synthetic Minority Oversampling Technique) method is an algorithm which creates Artificial data samples based on feature space similarities from minority entries. In other words, the minority class is oversampled by taking each minority class sample and introducing synthetic examples along the line segments joining any/all of the 'k' minority classes nearest neighbours (low and high stress study cases). Depending upon the amount of oversampling required, neighbours from the k nearest neighbours are randomly chosen [15]. Based on its potential, this technique was applied to shift the classifier learning bias towards the minority cases while keeping a similar distance between the samples' features.

Through the use of the mentioned methods, the group intended to compare the learning process and classification variation performance when testing different unbalance solving techniques for the same study case. Given the fact that most of the features are scale-variant, the next step was to normalise the datasets. This was achieved by applying two different methods: min-max scaling and Z-score normalisation.

Min-Max scaling (normalisation) refers to the process of converging all values of a feature to a fixed range. When the minimum and maximum values are estimated from the given set of matching scores, this method is not robust (i.e. the method is highly sensitive to outliers in the data used for estimation) [16]. This method retains the original distribution of scores except for a scaling factor and transforms all the scores into a common range between 0 and 1. Distance scores can be transformed into similarity scores by subtracting the min–max normalised score from 1. Moreover, another commonly used score normalisation technique is the Z-score normalisation which is calculated using the arithmetic mean and standard deviation of the given data. This scheme can be expected to perform well if prior knowledge about the average score and the score variations of the matcher is available, in which a feature is re-scaled [16].

Both approaches were performed on the previously obtained datasets, resulting in three separate datasets categories: the unscaled datasets, the normalised datasets and the standardised datasets (resulting in a total number of 18 pre-processed datasets). Their performance was observed when using classical ML techniques (i.e., NB, Decision Trees, Neural Networks, Random Forest and SVM) and averaging their results to reveal 3 overall accuracies, showing that the min-max scaling and Z-score normalisation have an increase of 4,33% and 4,44% respectively when compared to the unscaled data. To increase the model's performance, a series of variable combinations were obtained through various feature selection methods.

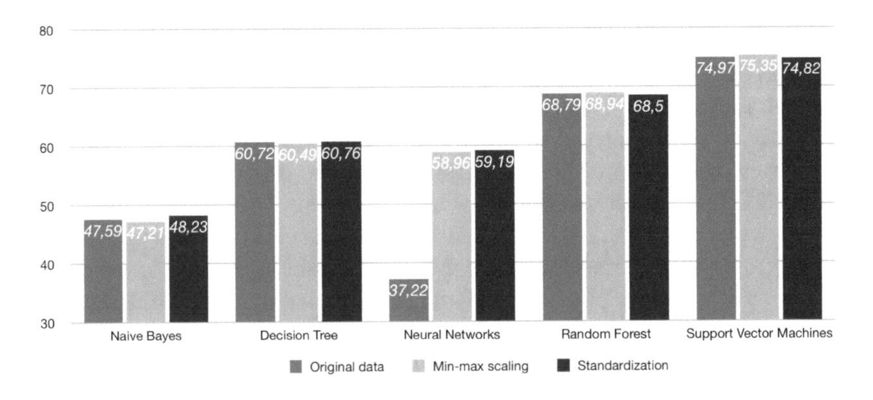

Fig. 2. Original vs. min-max scaling vs. standardised ML performance comparison.

Filter methods are defined as an induction pre-processing step that can remove irrelevant attributes. These methods select variables regardless of the model. Although they are robust to over-fitting, they tend to select variables separately, meaning they do not consider the relationships between certain variables. The Pearson's correlation coefficient is a measure of the correlation between two variables X and Y [15]. A number of feature combinations analysed have been obtained through this method. The wrapper method, on the other hand, is defined as a search through the space of feature subsets using the estimated accuracy from an induction algorithm as a measure of goodness of a particular feature subset, unlike filter methods. Wrapper methods use a learning machine to measure the quality of subsets of features without incorporating knowledge about the specific structure of the classification or regression function, and can, therefore, be combined with any learning machine methods [17]. In contrast to filter and wrapper approaches, in embedded methods the learning part and the feature selection part cannot be separated, i.e. feature selection and classification are performed simultaneously. Logistic Regression and Random Forest (RF) models were applied as the embedded methods to provide an evaluation of the features' importance to predict the user's perceived stress. To further explain this approach, the group fitted the Random Forest and Logistic Regression models on all datasets, which then provided an importance evaluation of each feature for each dataset, ranked according to their relevance to predict the stress class category for each specified models.

Based on the features selection of these techniques, it was settled that the most relevant features for this case study are: Correct Decision-Making Ratio (CDMR), Distance of the Mouse to the Straight Line (DMSL), Average/Median Time Between Decision (ATBD) and Distance Between Clicks (DBC).

7 Research Results

With the completion of the data preparation and feature selection processes, the group went on to the next phase: development and selection of the most appropriate ML techniques to predict a student's stress levels during high-end exams. The analysis of the data began by selecting a set of ML techniques that would better fit the case study. The selected techniques included: Decision Tree (DT), RF, Naive Bayes (NB) and Support Vector Machines (SVM). These models were trained on each dataset prepared in the previous steps. Moreover, in order to provide a more reliable result, two different methods were used to train/test the ML models: splitting the dataset into a training and testing subsets (70%–30% split); through the use of the 10-folds cross-validation.

Based on the ML models results, two aspects are highlighted: although the oversampled and hybrid datasets offer the higher ML classification performance accuracy (average accuracy of 76.51% on hybrid datasets, respectively 79,34% on oversampled datasets), this is due to the amount of replicated observations, leading into an overfitting model; due to the classes unbalance in the original dataset, most tested cases were categorised wrongly as moderate stress study cases (with an accuracy of 68,91% since most instances are labelled with the moderate stress). According to these observations, we decided that the most reliable datasets were the ones provided by the ROSE and SMOTE methods(average accuracy of 55,55% on ROSE datasets and 70,16% on SMOTE datasets). Also, a significant improvement of the ML model's classification performance was observed (5% on average) when the data was standardised or min-max scaling was involved (as shown in Fig. 2). In a general way, the group concluded that the SVM and RF techniques presented the best categorisation performance, with SVM presenting an average accuracy of 75,05% and RF an average accuracy of 68.74% on all datasets.

Taking into account these results, some fitting techniques were used to improve both SVM and RF models accuracy performance. The Random Forest model proved to offer the best results when using the standard number of 500 trees and number of leafs 6, while the SVM model showed significant improvement after the parameter tuning, averaging an increase of 5%. For tuning, exhaustive grid search was used, enabling the model to tune its hyperparameters to improve the categorisation of future case studies. The respective confusion matrices from the tuned SVM and RF models are shown in Fig. 3.

Actual \ Predicted	lowStress	moderateStress	highStress	Accuracy
lowStress	14	2	0	
moderateStress	0	17	0	84,61%
highStress	1	2	13	

(a) SVM Confusion Matrix.

Actual \ Predicted	lowStress	moderateStress	highStress	Accuracy
lowStress	15	1	0	
moderateStress	6	10	1	81,63%
highStress	1	0	15	

(b) Random Forest Confusion Matrix.

Fig. 3. Resulting confusion matrices from tuned SVM and RF models.

8 Conclusion and Future Work

In this paper, we present a technological approach for a non-intrusive and non-invasive analysis of performance in groups of students during high-end exams. It is common sense that stress significantly influences learning capacities, thus learning success. To cope with this factor, a Stress Monitoring Framework was presented, which proposes the use of usual equipment as sensors, without the user being aware of them, giving particular attention to the mouse and decision-making patterns of medical students during the execution of e-exams.

By monitoring human behaviour, our aim is to study the effect of stress on the performance of high demand tasks and to point out how each individual is affected by stress. This will allow the educational institution to act on each student, through personalised teaching and coping strategies, and thus improve the quality of the future professionals that are being trained.

Significant work is still needed however. Future work will be guided by the aim of using this knowledge to useful ends. Specifically, we envision the use of a similar system in stressful environments such as workplaces. In addition, an analysis based on statistical test results should be required to best understand the current variation of medical students stress levels. The main aim is to provide a faster and more effective response to users whose performance can be improved through the use of coping techniques which improve an individual's stress management and consequently their work performance.

Acknowledgements. This work is part-funded by ERDF–European Regional Development Fund and by National Funds through the FCT–Portuguese Foundation for Science and Technology within project NORTE-01-0247-FEDER-017832. The work of Filipe Gonçalves is supported by a FCT grant with the reference ICVS-BI-2016-005.

References

1. O'Sullivan, G.: Soc. Indic. Res. **101**(1), 155 (2011)
2. Colligan, T.W., Higgins, E.M.: J. Work. Behav. Health **21**(2), 89 (2006)
3. Carneiro, D., Novais, P., Pêgo, J.M., Sousa, N., Neves, J.: In: International Conference on Hybrid Artificial Intelligence Systems, pp. 345–356. Springer (2015)
4. Gonçalves, F., Carneiro, D., Novais, P., Pêgo, J.: In: International Symposium on Intelligent and Distributed Computing, pp. 137–147. Springer (2017)
5. Woloschuk, W., Harasym, P.H., Temple, W.: Med. Educ. **38**(5), 522 (2004)
6. Cohen, S., Kessler, R.C., Gordon, L.U.: Measuring Stress: A Guide for Health and Social Scientists. Oxford University Press on Demand, New York (1997)
7. Jain, A.K., Ross, A., Pankanti, S.: IEEE Trans. Inf. Forensics Secur. **1**(2), 125 (2006)
8. Saaty, T.L.: Int. J. Serv. Sci. **1**(1), 83 (2008)
9. Carneiro, D., Pimenta, A., Neves, J., Novais, P.: Soft Comput. **21**(17), 4917 (2017)
10. Pais Ribeiro, J., Marques, T.: Psicologia, Saúde & Doenças **10**(2), 237 (2009)
11. Pruessner, J.C., Hellhammer, D.H., Kirschbaum, C.: Psychosom. Med. **61**(2), 197 (1999)
12. Zhou, X., Dai, G., Huang, S., Sun, X., Hu, F., Hu, H., Ivanović, M.: Comput. Intell. Neurosci. **2015**, 12 (2015)
13. Yap, B.W., Rani, K.A., Rahman, H.A.A., Fong, S., Khairudin, Z., Abdullah, N.N.: In: Proceedings of the first International Conference on Advanced Data and Information Engineering (DaEng-2013), pp. 13–22. Springer (2014)
14. Menardi, G., Torelli, N.: Data Min. Knowl. Discov. **28**(1), 92 (2014)
15. Chawla, N.V., Bowyer, K.W., Hall, L.O., Kegelmeyer, W.P.: J. Artif. Intell. Res. **16**, 321 (2002)
16. Al Shalabi, L., Shaaban, Z., Kasasbeh, B.: J. Comput. Sci. **2**(9), 735 (2006)
17. Weston, J., Mukherjee, S., Chapelle, O., Pontil, M., Poggio, T., Vapnik, V.: In: Advances in neural Information Processing Systems, pp. 668–674 (2001)

Collective Profitability of DAG-Based Selling-Buying Intermediation Processes

Amelia Bădică[1], Costin Bădică[1(✉)], Mirjana Ivanović[2], and Doina Logofătu[3]

[1] University of Craiova, A.I.Cuza 13, 200530 Craiova, Romania
ameliabd@yahoo.com, cbadica@software.ucv.ro
[2] University of Novi Sad, Faculty of Sciences, Novi Sad, Serbia
mira@dmi.uns.ac.rs
[3] University of Applied Sciences, Frankfurt, Germany
logofatu@fb2.fra-uas.de

Abstract. We revisit our formal model of intermediation business processes and propose its generalization from trees to DAGs. With the new model, a company can use multiple sellers to better reach the market of potential buyers interested in purchasing its products. The sellers are engaged in transactions via a set of intermediaries that help connecting with end customers, rather than acting directly in the market. This process can be represented by a complex DAG-structured business transaction. In this work we present a formal model based on DAGs of such transactions and we generalize our results regarding collectively profitable intermediation transactions.

1 Introduction

Manufacturing and wholesale businesses are using complex business processes to manage their distribution activities, rather than directly providing their products to the customers. Such processes define distribution channels of groups of individuals or organizations responsible for directing the flow of products to the potential customers [8]. Intermediation networks are found in finance and logistics [10–12].

A distribution channel contains market intermediaries that link a seller to a customer or to another intermediary with the overall goal of linking the initial seller to the ultimate buyers. Some requirements are set on an intermediation process:

- R1. A seller can simultaneously use multiple different distribution channels, or equivalently multiple different distribution channels can serve the same seller. This approach enables sellers to expand their horizon by increasing the set of reachable customers.
- R2. For efficiency reasons, distribution channels can share market intermediaries or equivalently a market intermediary can serve multiple distribution channels.

J. Del Ser et al. (Eds.): IDC 2018, SCI 798, pp. 414–424, 2018.
https://doi.org/10.1007/978-3-319-99626-4_36

– R3. A business can use multiple seller agents to better reach the market of potential buyers that are interested in purchasing its products for both reasons of efficiency and market horizon expansion.

The focus of this paper in on defining a simple formal model of intermediation that is suitable for multi-agent business processes [1,7]. Using this model, sellers, buyers and market intermediaries are represented by self-interested agents engaged in an intermediation business process, seeking for collaborative advantage [9]. We are interested to define correctness criteria that guarantee the collective profitability as an incentive for all the business agents to engage in a collaborative intermediation process, thus ensuring systems's robustness and sustainability.

In our previous work we have proposed a model addressing requirement R1. The idea was to capture the structure of the intermediation process with the help of a rooted tree [3]. Each node of this tree represents a self-interested agent. The root of this tree represents a seller, each internal node of this tree represents a market intermediary, while each tree leaf represents a customer (buyer in our case). The agents interact using the interconnection links defined by the tree structure, with the collective goal of solving the intermediation problem, while each agent is interested to achieve its private goal of obtaining a profit.

In this paper we address all the requirements R1-R3 by generalizing the structure of the intermediation process to a directed acyclic graph (DAG hereafter). Using this approach, the business can use multiple sellers, thus addressing requirement R3, while market intermediaries can be shared by distinct distribution channels (requirement R2). Then, using linear algebra techniques we were able to establish necessary and (in some cases) sufficient conditions of the existence of profitable pricing strategies for all the participants at the intermediation business process.

2 A DAG-based Formal Model of Intermediation

Let us assume that a business is interested to bring its set of products $\mathcal{P} = \{1, 2, \ldots, k\}$ to the market of potential customers. The company can use a set of seller agents $\mathcal{S}$ such that each seller agent $S \in \mathcal{S}$ is responsible with distributing a nonempty subset of products $P_S \subseteq \mathcal{P}$. We assume that there are no overlapping duties of the seller agents of $\mathcal{S}$, so the family of sets $\{P_S\}_{S \in \mathcal{S}}$ defines a partition of $\mathcal{P}$.

Let us denote with $\mathcal{B}$ the set of potential customers (representing buyer agents) that are interested to purchase the products from $\mathcal{P}$ such that each buyer agent $B \in \mathcal{B}$ is interested to purchase a nonempty subset of products $P_B \subseteq \mathcal{P}$. We assume that there are no overlapping purchases of the buyer agents of $\mathcal{B}$, so the family of sets $\{P_B\}_{B \in \mathcal{B}}$ defines a partition of $\mathcal{P}$.

Let us assume that the company is using a set of intermediaries $\mathcal{I}$ that help connecting the seller agents with the end customers. Each intermediary $I \in \mathcal{I}$ has a dual role of buyer, as well as of seller. Agent I buys one or more products from one or more provider agents that can be either seller agents from the set $\mathcal{S}$ or other intermediaries from the set $\mathcal{I}$. Then agent I re-sells the acquired products to other agents that can be either customers from the set $\mathcal{B}$ or other intermediaries from the set $\mathcal{I}$.

The complex intermediation activity can be described with a business process that defines a complex structure of distribution channels. This process enables the business to distribute their products by engaging seller agents $\mathcal{S}$ through a complex intermediation structure connecting them to the potential customer agents $\mathcal{B}$. In what follows, let us call this structure an *intermediation DAG*.

Let us consider a directed graph $G = \langle V, A \rangle$ such that V is the set of vertices and A is the set of directed edges or arcs. We introduce the following functions:

$$in : V \to 2^V \text{ such that } in(v) = \{u | (u, v) \in A\} \text{ for all } v \in V$$
$$out : V \to 2^V \text{ such that } out(v) = \{u | (v, u) \in A\} \text{ for all } v \in V$$

Intuitively, an intermediation DAG can be defined with the help of a DAG structure $G = \langle V, A \rangle$. Vertices v with no incoming arcs, i.e. $in(v) = \emptyset$, represent the seller agents. Vertices v with no outgoing arcs, i.e. $out(v) = \emptyset$, represent the buyer agents. Vertices v with at least one incoming arc and with at least one outgoing arc, i.e. $in(v) \neq \emptyset$ and $out(v) \neq \emptyset$, represent intermediary agents. Note that an intermediary agent has both buyer and seller (actually re-seller) role.

We require the intermediation graph to be a DAG as it is both inefficient and counter-intuitive to allow an agent to sell a product and then re-purchase it latter in a distribution channel where the overall goal is to distribute the product to the end customers as efficiently as possible.

Definition 1 (Intermediation DAG). Let us consider three finite, nonempty and pairwise disjoint sets $\mathcal{S}$, $\mathcal{B}$, and $\mathcal{I}$ of seller, buyer and respectively intermediary agents, as well as a finite set $\mathcal{P}$ of products. An *intermediation DAG* is a triple $G = \langle V, A, p, g \rangle$ such that:

(i) $\langle V, A \rangle$ is a DAG with the set of vertices defined as $V = \mathcal{S} \cup \mathcal{B} \cup \mathcal{I}$ such that:
   1. $in(v) = \emptyset$ for all $v \in \mathcal{S}$ and $out(v) = \emptyset$ for all $v \in \mathcal{B}$
   2. $in(v) \neq \emptyset$ and $out(v) \neq \emptyset$ for all $v \in \mathcal{I}$.

(ii) $p : V \to 2^{\mathcal{P}} \setminus \{\emptyset\}$ and $g : A \to 2^{\mathcal{P}} \setminus \{\emptyset\}$ are two functions assigning to each agent (vertex) and to each transaction (arc), a nonempty set of products such that:
   1. $\{g((u, v))\}_{v \in out(u)}$ defines a nontrivial partition of set $p(u)$ for all $u \in V$
   2. $\{g((v, u))\}_{v \in in(u)}$ defines a nontrivial partition of set $p(u)$ for all $u \in V$
   3. $\cup_{S \in \mathcal{S}} p(S) = \cup_{B \in \mathcal{B}} p(B) = \mathcal{P}$.

For example, Fig. 1 presents an intermediation DAG involving 8 agents such that: $\mathcal{S} = \{1, 2\}$, $\mathcal{B} = \{5, 7, 8\}$, and $\mathcal{I} = \{3, 4, 6\}$.

The set $g((u, v))$ represents the products that are transacted in the interaction between agents u (with seller role) and v (with buyer role). For example $g((6, 7)) = \{1\}$ means that in the transaction between agents 6 (with seller role) and 7 (with buyer role), the exchanged product is 1.

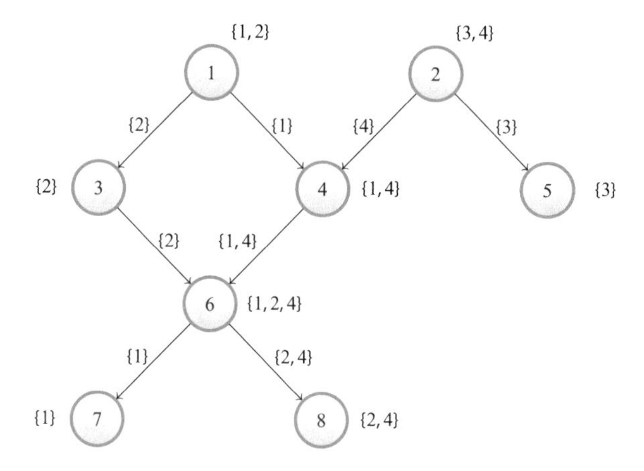

Fig. 1. DAG-based intermediation process.

The set $p(u)$ represents the products that are managed by agent u. If u is a seller then $p(u)$ represents the products sold by u. If u is a buyer then $p(u)$ represents the products bought by u. If u is an intermediary agent then $p(u)$ represents the products bought and re-sold by u.

Note that, using the notation introduced at the beginning of this section, if $S \in \mathcal{S}$ then $p(S) = P_S$ and if $B \in \mathcal{B}$ then $p(B) = P_B$. Moreover, observe in Fig. 1 that $p(1) = P_1 = \{1, 2\}$ and $p(7) = P_7 = \{1\}$.

Referring to Definition 1, condition i.1 requires that each seller agent is just involved in selling its assigned set of products to other agents. For example, seller agent 2 sells product 2 to agent 3 and product 1 to agent 4.

Similarly, condition i.2 requires that each buyer agent is just involved in purchasing its assigned set of products from other agents. For example buyer agent 7 purchases product 1 from agent 6.

Finally, condition i.3 requires that each intermediary agent is actually intermediating at least one transaction by purchasing and then re-selling at least one product. For example, intermediary agent 4 purchases product 1 from agent 1 and product 4 from agent 2, while also selling both products $\{1, 4\}$ in a single shot to agent 6.

Moreover, condition ii.1 requires that each agent which is not an end customer is involved in selling activities such that: it sells all its products to its outgoing agents and never sells the same product twice to different agents. For example agent 6 sells all its products $\{1, 2, 4\}$ to agents 7 and 8 such that it sells product $\{1\}$ to agent 7 and products $\{2, 4\}$ to agent 8.

Similarly, condition ii.2 requires that each agent which is not company seller is involved in buying activities such that: it buys all the products of its incoming agents and never buys the same product twice from different agents. For example agent 4 buys products $\{1, 4\}$ from agents 1 and 2 such that it buys product $\{1\}$ to agent 1 and product $\{2\}$ from agent 2.

Last but not least, condition ii.3 requires that all the products sold by the seller agents eventually reach the end customers. For example, for the intermediation DAG presented in Fig. 1 we have: $p(1) \cup p(2) = \{1,2\} \cup \{3,4\} = \{1,2,3,4\}$, while $p(5) \cup p(7) \cup p(8) = \{1\} \cup \{3\} \cup \{2,4\} = \{1,2,3,4\}$, so condition ii.6 clearly holds.

A directed graph is weakly connected if its underlying undirected graph is connected. Very often we focus on weakly connected intermediation DAGs. If the DAG is not weakly connected then it can be decomposed into connected components, with each component representing a separated weakly connected intermediation DAG.

Proposition 1 (Number of transactions). *The total number of transactions t of a weakly connected intermediation DAG satisfies the inequality: $t \geq |\mathcal{S}| + |\mathcal{B}| + |\mathcal{I}| - 1$.*

This result follows easily from the observations that t equals to the number of arcs and that the minimum number of arcs of a weakly connected directed graph with n vertices is $n - 1$. The equality is obtained when the DAG is a tree [3].

3 Profitability

An intermediation DAG defines rigorously the structure of a complex intermediation transaction. The creation of such structures is beyond the scope of the present paper. Nevertheless, we can speculate that participating seller, intermediation and buyer agents can use the computational techniques of middle-agents [2,5], service composition [7], and interaction protocols [13] combined with suitable collaborative e-business models based on customer relations and trust management [4], to incrementally define the business process that represents the intermediation DAG.

Following the model previously introduced in [3], we now assign economic information to an intermediation DAG.

Let $P \subseteq \mathcal{P}$ be a nonempty subset of products. We denote with σ_P^S the limit price of seller S for selling the subset P of products. This means that S will agree to sell the subset P of products only for a price p such that $p \geq \sigma_P^S$. Similarly, we denote with β_P^B the limit price of an end customer B for agreeing to pay and buy the whole subset P of products. This means that B will agree to purchase the subset P of products only for a price p such that $p \leq \beta_P^B$.

In what follows we augment an intermediation DAG with annotations about limit and transaction prices. Limit prices are assigned to seller and buyer agents, while transaction prices are assigned to transactions of the intermediation DAG.

Definition 2 (Annotated intermediation DAG). Let us consider an intermediation DAG defined by $G = \langle V, A, p \rangle$. The annotated version of G is $G^a = \langle V, A, p, \pi \rangle$ such that the additional parameter $\pi : \mathcal{S} \cup \mathcal{B} \cup A \rightarrow (0, +\infty)$ is the annotation function used for assigning economic information to the elements of G as follows:

(i) If $S \in \mathcal{S}$ is a seller agent then $\pi(S) = \sigma^S_{p(S)} > 0$ representing the limit price of seller S for selling the subset of products $p(S)$
(ii) If $B \in \mathcal{B}$ is a buyer agent then $\pi(B) = \beta^B_{p(B)} > 0$ representing the limit price of buyer B for purchasing the subset of products $p(B)$
(iii) If $T = (u, v) \in A$ denotes a transaction then $\pi(T) = \pi^{u,v}_{g((u,v))} > 0$ represents the transaction price for which agent u agrees to sell the products $g((u, v))$ to agent v.

For example, referring to Fig. 1, the arc linking node 4 to node 6 is annotated with transaction price $\pi^{4,6}_{\{1,4\}}$.

Following the observation from [3], a potential selling-buying transaction between an agent S in seller role with limit price σ and an agent B in buyer role with limit price β is profitable if and only if $\beta \geq \sigma$. If this condition holds then the transaction price can be set to any arbitrary value $\pi \in [\sigma, \beta]$.

Definition 3 (Collective profitability). An intermediation DAG is called *collectively profitable* if and only if it can be annotated with transaction prices such that each transaction participant is profitable, i.e. it gains by performing the transaction.

A participant $v \in V$ is profitable if and only if its utility $u(v) > 0$. The utility u_v can be computed as follows:

(i) If $v = S \in \mathcal{S}$ is a seller agent then its utility is defined as follows:

$$u(S) = -\sigma^S_{p(S)} + \sum_{v \in out(S)} \pi^{S,v}_{g((S,v))} \tag{1}$$

(ii) If $v = B \in \mathcal{B}$ is a buyer agent representing an end customer then its utility is:

$$u(B) = \beta^B_{p(B)} - \sum_{v \in in(B)} \pi^{v,B}_{g((v,B))} \tag{2}$$

(iii) If $v = I \in \mathcal{I}$ is an intermediary agent then its utility is defined as follows:

$$u(I) = \sum_{v \in out(I)} \pi^{I,v}_{g((I,v))} - \sum_{v \in in(I)} \pi^{v,I}_{g((v,I))} \tag{3}$$

Let us consider the following system of $|\mathcal{S}| + |\mathcal{B}| + |\mathcal{I}|$ inequations and t variables (t represents the number of transactions, see Proposition 1):

$$\begin{aligned} -\sigma^S_{p(S)} + \sum_{v \in out(S)} \pi^{S,v}_{g((S,v))} &\geq 0 \quad S \in \mathcal{S} \\ \beta^B_{p(B)} - \sum_{v \in in(B)} \pi^{v,B}_{g((v,B))} &\geq 0 \quad B \in \mathcal{B} \\ \sum_{v \in out(I)} \pi^{I,v}_{g((I,v))} - \sum_{v \in in(I)} \pi^{v,I}_{g((v,I))} &\geq 0 \quad I \in \mathcal{I} \end{aligned} \tag{4}$$

We can now formulate a necessary and sufficient condition for the collective profitability of an intermediation DAG, using the following lemma.

Lemma 1. *An intermediation DAG is collectively profitable if and only if there exists an annotation with transaction prices that satisfies the system (4) of inequations.*

The result stated by Lemma 1 follows directly from the Eqs. (1), (2), and (3) that define utilities $u(S)$, $u(B)$, and $u(I)$ of the participating agents.

Applying Lemma 1 we can obtain a simpler necessary condition of collective profitability of an intermediation DAG – see Proposition 2.

Proposition 2 (Necessary condition for collective profitability). *If an intermediation DAG is collectively profitable then:*

$$\sum_{B \in \mathcal{B}} \beta^B_{p(B)} \geq \sum_{S \in \mathcal{S}} \sigma^S_{p(S)} \tag{5}$$

Proposition 2 represents a partial generalization of our previous result that stated a similar necessary condition for the simpler case of an intermediation tree [3].

4 Discussion

In this section we present some reflections about the result of Proposition 2. Firstly, we show that generally, the condition stated by this proposition is necessary but not sufficient. For that, it is enough to provide a counter-example (see Sect. 4.1). Secondly, we revisit our result for intermediation trees formulated in [3]. There we have only presented a proof for a specific example, without giving a general proof. Here, we outline a general proof based on Farkas' lemma [6] (see Sect. 4.2).

4.1 Case Study

Inequations (6) are obtained from general inequations (4) particularized for the example intermediation DAG shown in Fig. 1, by introducing the positive constants $\alpha_i > 0$ for $i = 1, \ldots, 8$ representing node utilities.

$$\begin{aligned}
\pi_2^{1,3} + \pi_1^{1,4} &= \sigma_{1,2}^1 + \alpha_1 \\
\pi_3^{2,5} + \pi_4^{2,4} &= \sigma_{3,4}^2 + \alpha_2 \\
-\pi_2^{1,3} + \pi_2^{3,6} &= \alpha_3 \\
-\pi_1^{1,4} - \pi_4^{2,4} + \pi_{1,4}^{4,6} &= \alpha_4 \\
-\pi_3^{2,5} &= -\beta_3^5 + \alpha_5 \\
-\pi_2^{3,6} - \pi_{1,4}^{4,4} + \pi_1^{6,7} + \pi_{2,4}^{6,8} &= \alpha_6 \\
-\pi_1^{6,7} &= -\beta_1^7 + \alpha_7 \\
-\pi_3^{6,8} &= -\beta_{2,4}^8 + \alpha_8
\end{aligned} \tag{6}$$

Proposition 3. *If* $\beta_1^7 + \beta_{2,4}^8 < \sigma_{1,2}^1$ *then the example intermediation DAG shown in Fig. 1 is not collectively profitable.*

Proof Sketch. It is not difficult to see that:

$$\begin{aligned} \pi_4^{2,4} &= \alpha_2 + \alpha_5 - (\beta_3^5 - \sigma_{3,4}^2) \\ \alpha_4 &= [(\beta_1^7 + \beta_{2,4}^8 - \sigma_{1,2}^1) - (\alpha_1 + \alpha_3 + \alpha_6 + \alpha_7 + \alpha_8)] + [(\beta_3^5 - \sigma_{3,4}^2) - (\alpha_2 + \alpha_5)] \end{aligned} \tag{7}$$

Now, under the assumption of Proposition 3, the conditions $\alpha_i \geq 0$ and $\pi_4^{2,4} \geq 0$ cannot simultaneously hold, thus concluding the proof.

However, if the sufficient condition (8) following Proposition 2 for the DAG from Fig. 1, is augmented with the supplementary condition (9), then this intermediation DAG is collectively profitable. So under assumptions (8) and (9) it is possible to choose $\alpha_i > 0$, for all $i = 1, \ldots, 8$ such that system (6) has positive solutions.

$$\beta_1^7 + \beta_{2,4}^8 + \beta_3^5 \geq \sigma_{1,2}^1 + \sigma_{3,4}^2 \tag{8}$$

$$\beta_1^7 + \beta_{2,4}^8 \geq \sigma_{1,2}^1 \tag{9}$$

4.2 Intermediation Tree

Let us now assume that our intermediation DAG is a rooted tree. This means that:

(i) The set $\mathcal{S}$ is a singleton and its unique element S is the root of the tree.
(ii) For each buyer node $B \in \mathcal{B}$ there is a unique path in the graph from S to B.

We focus now on the reciprocal of Proposition 2 for intermediation trees.

Proposition 4 (Sufficient condition for collective profitability). *If an intermediation tree satisfies the following condition:*

$$\sum_{B \in \mathcal{B}} \beta_{p(B)}^B \geq \sigma_{p(S)}^S \tag{10}$$

then it is collectively profitable.

The proof of Proposition 4 uses a well-established result of linear algebra known as Farkas' lemma [6]. We firstly review this result before we outline our proof.

Theorem 1 (Farkas' lemma [6]). *Let* $A \in \mathbb{R}^m \times \mathbb{R}^n$ *and let* $b \in \mathbb{R}^m$ *be nonzero. Then exactly one of the following holds:*

(i) There is a strictly positive solution $u \in \mathbb{R}^n$ *to the system* $Ax = b$.
(ii) There is a vector $v \in \mathbb{R}^m$ *for which* $v^T A \leq 0$ *and* $v^T b > 0$.

Proof Sketch (of Proposition 4). We transform inequations (4) into Eqs. (11):

$$A\pi = b + \alpha \tag{11}$$

such that if n is the number of nodes and e is the number of arcs of the intermediation DAG then: $A \in \mathbb{R}^n \times \mathbb{R}^e$, $b, \alpha \in \mathbb{R}^n$, and $\pi \in \mathbb{R}^e$. Moreover, A is the incidence matrix of the DAG, b is defined as follows:

$$\begin{aligned} b_S &= \sigma^S_{p(S)} && S \in \mathcal{S} \\ b_B &= -\beta^B_{p(B)} && B \in \mathcal{B} \\ b_I &= 0 && I \in \mathcal{I} \end{aligned} \tag{12}$$

and the following condition holds:

$$\sum_{B \in \mathcal{B}} \beta^B_{p(B)} = \sigma^S_{p(S)} + \sum_{i=1}^{n} \alpha_i \tag{13}$$

Then, following Farkas' lemma, we must find some values $\alpha_i \geq 0$ for which there is no vector $v \in \mathbb{R}^n$ such that $v^T A \leq 0$ and $v^T(b + \alpha) > 0$.

It is not difficult to observe that we can define a partition $(\mathcal{X}_B)_{B \in \mathcal{B}}$ of the set of nodes $\mathcal{B} \cup \mathcal{I}$ such that:

(i) For all $B \in \mathcal{B}$ we have $B \in \mathcal{X}_B$.
(ii) For all $B \in \mathcal{B}$ and $X \in \mathcal{X}_B$ then X is a member of the path from root S to B.

Following (13) we obtain:

$$\sum_{B \in \mathcal{B}} (\beta^B_{p(B)} - \alpha_B) > \sum_{B \in \mathcal{B}} \sum_{I \in \mathcal{X}_B \setminus \{B\}} \alpha_I \tag{14}$$

Following (14), we can easily define $\alpha_i > 0$ for all $i = 1, \ldots, n$ such that for each $B \in \mathcal{B}$ the following holds:

$$\beta^B_{p(B)} - \alpha_B > \sum_{I \in \mathcal{X}_B \setminus \{B\}} \alpha_I \tag{15}$$

Let us now assume by contradiction that there exists a vector $v \in \mathbb{R}^n$ such that $v^T A \leq 0$ and $v^T(b + \alpha) > 0$.

From $v^T A \leq 0$ it follows that for all nodes $B \in \mathcal{B}$ and for each node $I \in \mathcal{I}$ on a path from S to I we have:

$$v_B \geq v_I \tag{16}$$

With a simple algebraic manipulation, substituting $\sum_{i=1}^{n} \alpha_i$ from Eq. (13) into inequality $v^T(b + \alpha) > 0$, we obtain:

$$\sum_{B \in \mathcal{B}} (v_B - v_S)(\beta^B_{p(B)} - \alpha_B) < \sum_{I \in \mathcal{I}} (v_I - v_S)\alpha_I \tag{17}$$

However, one can easily observe that inequality (17) contradicts (15) and (16), thus concluding the proof of Proposition 4.

5 Conclusion

In this work we have proposed an extension of our results on tree-based intermediation business processes to DAG-based structures. More exactly, we formally captured the new model of DAG-based intermediation, and we also provided a necessary condition for collective profitability of participants. Although it turned out that this condition is generally not sufficient, we provided a formal proof (missing from our previous publication) that in fact this condition is necessary and sufficient for the special case of intermediation trees. According to our counter-example, we were able to conclude that, in the general case, collective profitability depends both on the DAG structure, as well as on the agents' private parameters – limit prices.

References

1. Bădică, C., Budimac, Z., Burkhard, H.-D., Ivanović, M.: Software agents: languages, tools, platforms. Comput. Sci. Inf. Syst. **8**(2), 255–298 (2011). https://doi.org/10.2298/CSIS110214013B
2. Bădică, A., Bădică, C.: FSP and FLTL framework for specification and verification of middle-agents. Appl. Math. Comput. Sci. **21**(1), 9–25 (2011). https://doi.org/10.2478/v10006-011-0001-6
3. Bădică, A., Bădică, C., Ivanović, M., Buligiu, I.: Collective profitability and welfare in selling-buying intermediation processes. In: Computational Collective Intelligence, ICCCI 2016, part II. Lecture Notes in Computer Science, vol. 9876, pp. 14–24. Springer, Cham (2016). https://doi.org/10.1007/978-3-319-45246-32
4. Kravari, K., Kontopoulos, E., Bassiliades, N.: Trusted reasoning services for semantic web agents. Informatica (Slovenia) **34**(4), 429–440 (2010)
5. Dignum, V., Tranier, J., Dignum, F.: Simulation of intermediation using rich cognitive agents. Simul. Model. Pract. Theory **38**(10), 1526–1536 (2010). https://doi.org/10.1016/j.simpat.2010.05.011
6. (Gyula) Farkas, J.: Über die Theorie der Einfachen Ungleichungen. Journal für die Reine und Angewandte Mathematik **124**(124), 1–27 (1902). https://doi.org/10.1515/crll.1902.124.1
7. Coria, J.A.G., Castellanos-Garzón, J.A., Corchado, J.M.: Intelligent business processes composition based on multi-agent systems. Expert. Syst. Appl. **41**(4), 1189–1205 (2014). Part 1. https://doi.org/10.1016/j.eswa.2013.08.003
8. Jehle, G.A., Reny, P.J.: Advanced Microeconomic Theory, 3rd edn. Pearson Education Limited, Newmarket (2011)
9. Kanter, R.M.: Collaborative advantage: the art of alliances. Harv. Bus. Rev. **72**(4), 96–108 (1994)
10. Leon, F., Bădică, C.: An optimization web service for a freight brokering system. Serv. Sci. **9**(4), 324–337 (2017). Informs. https://doi.org/10.1287/serv.2017.0191

11. Nagurney, A., Ke, K.: Financial networks with intermediation: risk management with variable weights. Eur. J. Oper. Res. **172**(1), 40–63 (2006). https://doi.org/10.1016/j.ejor.2004.09.035
12. Nault, B.R., Dexter, A.S.: Agent-intermediated electronic markets in international freight transportation. Decis. Support. Syst. **41**(4), 787–802 (2006). https://doi.org/10.1016/j.dss.2004.10.008
13. Poslad, S.: Specifying protocols for multi-agent systems interaction. ACM Trans. On Auton. Adapt. Syst. **2**(4) (2007). Article 15. https://doi.org/10.1145/1293731.1293735

Knowledge-Based Metrics for Document Classification: Online Reviews Experiments

Mihaela Colhon[1(✉)] and Costin Bădică[2]

[1] Department of Computer Science, University of Craiova, Alexandru Ioan Cuza, 13, 200585 Craiova, Romania
mcolhon@inf.ucv.ro

[2] Department of Computer and Information Technology, University of Craiova, Alexandru Ioan Cuza, 13, 200585 Craiova, Romania
cbadica@software.ucv.ro

Abstract. In this paper we propose a new method that addresses the documents classification problem with respect to their topic. The presented method takes into consideration only textual measures. We exemplify the method by considering three sets of documents of gradually different topics: (i) the first two sets contain reviews that comment the published entity features characteristics representing electronic devices – laptops and mobile phones; (ii) the third set contains reviews about touristic locations. All the review texts are written in Romanian and were extracted by crawling popular Romanian sites. The paper presents and discusses the obtained evaluation scores after the application of textual measures.

1 Introduction

Automatic text classification has gained huge popularity with the advancement of information technology [15] as the constant growing of the online reviews volume asks for further development and refinement of automatic tools [8] that can deal with this kind of texts. We claim that in order to perform well on text categorisation problems, the computer needs to access extensive and deep knowledge. Over a decade ago, Lenat and Feigenbaum formulated in [19] the *knowledge principle*, which postulates that "If a program is to perform a complex task well, it must know a great deal about the world it operates in". Text categorisation is certainly a complex task [13] that can help to organise a large quantity of unordered text documents into a small number of meaningful and coherent groups, thereby providing a basis for intuitive and informative navigation and browsing mechanisms [17].

This type of work is becoming even more interesting and demanding with the development of the World Wide Web and the evolution of Web 2.0. For example, popular applications including: spam filtering, opinion detection [26] and opinion mining from online reviews products [1], text sentiment mining [21],

J. Del Ser et al. (Eds.): IDC 2018, SCI 798, pp. 425–435, 2018.
https://doi.org/10.1007/978-3-319-99626-4_37

allow the grouping of the results returned by search engines in order to help users to quickly identify and focus on relevant results. Another possible usage of this type of application is represented by the automation of the definition of conferences' schedule programs. This assumes learning of a suitable distance function between each pair of papers in order to allow their automatically grouping into programme sessions, depending on their addressed topics. Under this assumption, the grouping of research papers assumes that any two papers are regarded as similar if they share similar thematic topics.

The automatic text classification method is based on automatically analysing the text content in order to allocate it into a pre-determined catalogue or taxonomy. The methods of automatic text classification mainly use information retrieval techniques, initially developed for digital libraries. Traditional information retrieval is mainly interested about the retrieval of relevant documents by employing either keyword-based or statistic-based computational techniques. One central step in automatic text classification is to identify the major topics of the texts.

This paper presents a simple and yet effective knowledge-based approach for the task of text classification. The approach uses the log-likelihood measure in order to identify the most relevant topic terms of the texts based on which the classification is designed. The structure of the paper is as follows: in the next section, related words concerning the subject debated in the paper are given. Section 3 presents the methodology we have used to implement the task of texts classification, while Sect. 4 presents case studies in which the designed method is exemplified on particular problems addressing the classification of a set of three types of reviews together with the obtained evaluation scores. The Conclusions and Future Works Section summarizes the results presented in this paper, covering both the strengths and the weaknesses of our approach, and finally outlines some possible directions for continuing this research.

2 Related Works

Text classification is a supervised machine learning task which involves the automatic assignment of a text document to a class from a set of pre-defined classes. The classification can be either *hard* if a unique label is explicitly assigned to each of the test instances or *soft* if, instead of a label, a probability distribution is determined which enables the assignment of multiple labels to each text instance. There have been developed many classification algorithms, and some of them are already implemented in publicly available software packages such as BOW toolkit [20] or WEKA [25].

Text document classification requires the application of natural language processing (NLP) approaches that usually include the following pre-processing tasks: 1. removal of stop words; 2. removal of infrequent words; 3. stemming of the words using stemmer programs such as Porter Stemmer for English; 4. feature selection using statistical or semantic approaches, and 5. choice and application of a suitable machine learning technique such as Naïve Bayes, Decision Tree,

Neural Network, Support Vector Machines, Hybrid techniques, etc., for training a text classifier.

The Naïve Bayes classifier [7,11] is probably the mostly used and simplest classifier. It models the distribution of documents in each class by using a probability distribution which assumes that probabilities of different classes are mutually independent. This is clearly a false assumption, although the Naïve Bayes model performs surprisingly well in many practical scenarios.

Decision trees [12] are basically capturing hierarchical representations of subsets of instances (examples) such that each instance can be classified by starting at the root node, sequentially testing its attributes and following down to the branch corresponding to the value of each instance attribute. Support Vector Machines (initially introduced in [6]) were extensively used in text classification, pattern recognition and face detection applications. There have been done also a number of efforts to add external *a priori* knowledge to supervised machine learning techniques [10,22].

The knowledge-based approaches rely on syntactic and/or semantic parsing, knowledge bases such as scripts or machine readable dictionaries, without using any corpus statistics. These approaches can include: word annotation by their part of speeches, noun-phrase chunking using chunker programs, and dependency links identification using Dependency Parsers (such as Stanford Parser for English); identification of bigrams or trigrams instead of unigrams and keeping track of frequent bigrams (see [18]).

The research presented in this paper addresses the problem of short document classification. We apply our proposed classification method on a set of user reviews extracted from the Internet that address different topics. This work is thus a continuation of our ongoing research on automatic analysis of Romanian social media data. This work was started in 2014 with the proposal of a sentiment analysis method, that was applied to a set of touristic reviews [9]. Then, we developed a graph-based approach for the analysis of the user interactions and reviews (the reviewers) from Web site of a touristic community [2–4]. This work was followed by the proposal of a supervised classification method of the reviewers based on their portfolio data [8]. In this paper we propose an extended analysis based on a new classification method of active users in an online community, using new data crawled from very popular Romanian online stores.

3 The Proposed Classification Model

The methodology presented in this section aims to discover a finite number of terms, called keywords in what follows, with the intended scope of designing a topical classification model of a set of short documents representing real user reviews expressed in online communities of different domains – e-commerce and tourism.

In the literature, the term "keywords" means units that can carry or participate in discovering the meaning of a text or the text-type as they are statistically significant in texts when compared with a reference corpus. In this way, the keywords tend to indicate the aboutness of the texts as well as the texts style [16], thus making them appropriate for texts classification tasks.

A keywords list is mainly used in order to store some terms that can indicate the topic concept primarily described in a text, based on which the classification is performed. In order to detect these terms, a comparison of word frequencies between the *study texts* and a neutral document, called in what follows the *reference document*, must be performed. This approach follows the idea that a word that is significantly more frequent in the analyzed texts than in the reference corpus could be related to the topic of those texts. In our work we have used a concordancing program, the freeware AntConc[1], to extract keywords and then calculate the keyness values.

The *keyness strength* is a measure of the difference between the frequency of a word in analyzed texts versus the reference document. It is then a quality possessed by words, word clusters or phrases which is not language-dependent, but text-dependent [23]. The manual finding of words' keyness is a laborious work, but the development of automatic keyword analysis method successfully solved the issue. However, it has been claimed that the results obtained using automatic tools are less accurate than the results obtained by manual analysis [24]. Nevertheless, the appropriate choice of a reference corpus in what concerns its size and content can solve the inaccuracies of the automatic analysis. The larger the reference corpus is, the better the keywords detection will be (some researchers suggest more than five times larger then the texts under analysis).

By default, the AntConc software uses the log-likelihood measure to acquire the high-frequency terms from the text and the obtained keywords are ranked by default according to their keyness values. On the resulted data, a *Bonferroni* correction [5] is applied in order to reduce the cases in which the keywords are appearing by chance.

The fact that the meanings of the words in texts can be different from their meaning in isolation might influence keywords-based approaches. The weakness of the keywords-based classifications relies in the fact that the terms are words that must be explicitly mentioned in the training documents, and the only knowledge we have about them consists in their occurrence frequencies.

Because this approach cannot generalize, words in the test set that do not appear in the training set will be ignored. In order to exemplify this case, let us take the example given in [13]: "Suppose we have a collection of pharmaceutical documents and we are trying to learn the concept of antibiotics. If a particular training document" misses the word "antibiotic" and also its forms, then "the system will likely miss an important piece of evidence". A possible solution for this problem can consist in providing, along with the terms extracted from the training corpus, also their synonymous words.

But, in the case of short texts – as reviews are, this weakness is not present as we will demonstrate in what follows. That is due to the fact that reviews are short pieces of text that usually include the word-form of the described topic features.

---

[1] http://www.laurenceanthony.net/.

4 Experiments and Discussions

In this section we present the experimental results that were obtained by applying our classification method on three different types of online reviews.

The Data Set. We crawled several Romanian sites, very popular in the Romanian online community, in order to extract a significant number of online reviews with the goal of evaluating our proposed document classification method. We have focused on reviews as it was shown that they contain user opinions that are more trustable then any other information posted online[2]

One Web site – AmFostAcolo[3], which was used in our previous works, acts in the touristic field, while the rest of the crawled sites sell products from various categories including also electronic devices[4].

Our data set consists of three types of reviews. All these pieces of text represent unstructured natural language constructions in which the users express their impressions and opinions about various items of interest.

From AmFostAcolo site we extracted 5000 reviews, each review containing an user impression about a specific touristic location. These texts, named in what follows *touristic reviews* usually refer to a set of touristic aspects which are debated in their contents such as accommodation, kitchen or services [8].

From the online store sites we extracted 79485 reviews, representing user opinions about purchased mobile phones and 2471 reviews containing user opinions about purchased laptops.

All the extracted reviews are accompanied by a user rating which were normalized on an unique 1–5 scale where 1 means the worst value and 5 represents the best value.

The histogram of review length (determined as number of words) corresponding to the mobile phones data set is shown in Fig. 1(a). Most of the reviews have less than 100 words (70882 of the total of 79486 reviews).

Our analysis revealed that the collected mobile phone reviews are mostly (74%) positive – i.e. more than two-thirds of the reviews have the maximum rating (see Fig. 1(b)). Counting together the reviews with 4 and 5 rating points resulted in almost 87% of the total number of mobile phone reviews.

The histogram of the laptops reviews length (determined as number of words) is shown in Fig. 1(c). As it can be seen, this graphic is very similar with the histogram of the mobile phone reviews length: most of the reviews have less than 100 words (2095 of a total of 2471 reviews). Only the total number of reviews differs. Thus an observation immediately results: in both cases the review length is distributed according to the Power law rule, i.e. most reviews contain few

[2] "Research shows that 91% of people regularly or occasionally read online reviews, and 84% trust online reviews as much as a personal recommendation." says Craig Bloem, Founder and CEO at FreeLogoServices.com.

[3] http://amfostacolo.ro.

[4] Because the activity domain of these sites is in business we chose not to mention the sites' names and also to anonymize all the data that could have a potential commercial characteristic.

words (less than 100 words), while only a few number of reviews contain a larger amount of words.

As concerning the ratings, only 62% (the smallest percent of the three data sets) correspond to 5 rating points reviews (see Fig. 1(d)), while counting together the 5 and 4 rating points reviews represents 80% of the total number of laptop reviews.

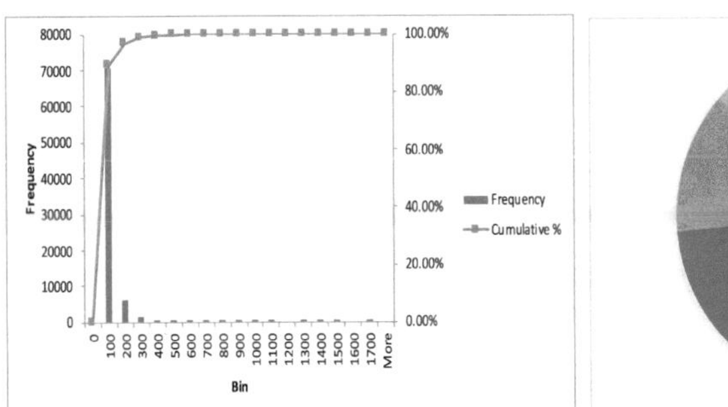

(a) The histogram of mobile phones reviews length

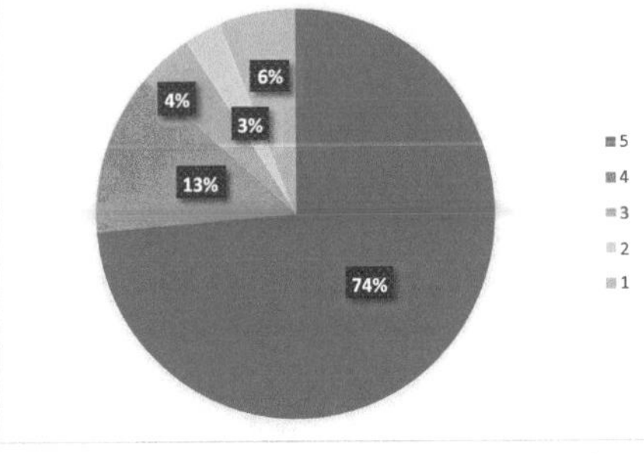

(b) The user rating of the mobile phones reviews

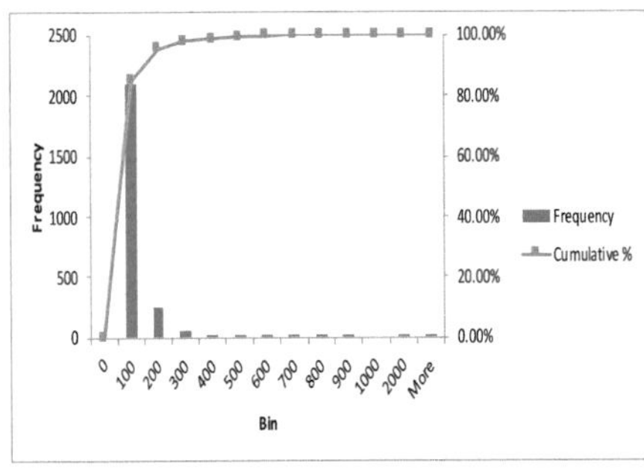

(c) The histogram of laptops reviews length

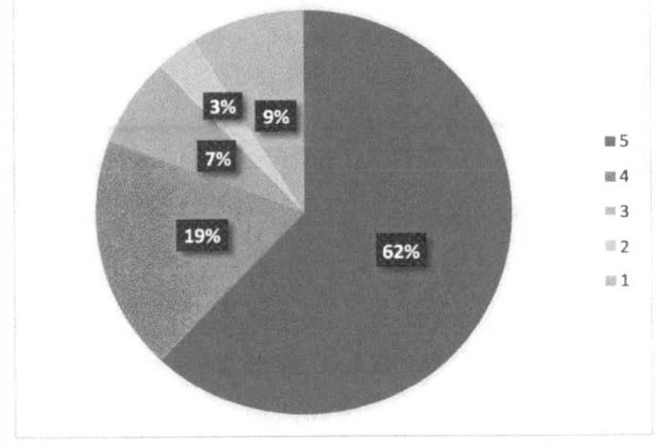

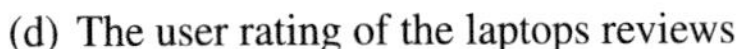

(d) The user rating of the laptops reviews

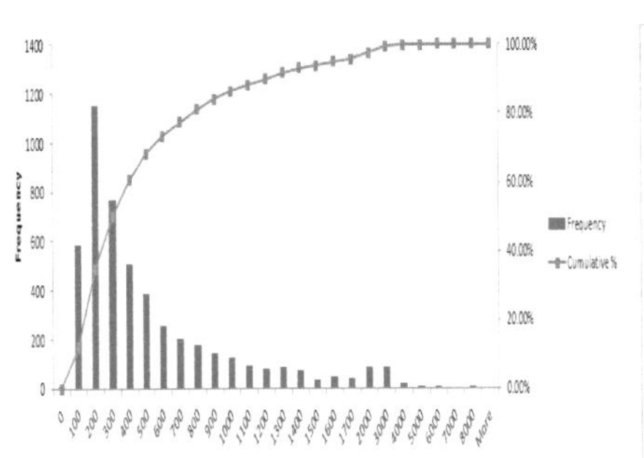

(e) The histogram of touristic reviews length

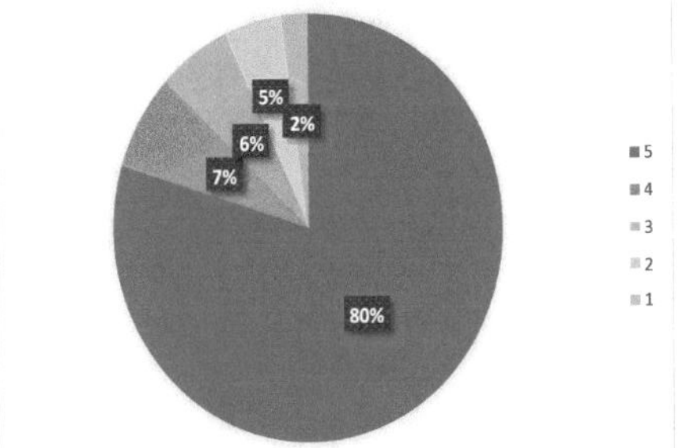

(f) The user rating of the touristic reviews

Fig. 1. Statistics for the extracted reviews.

The histogram of the review length (determined as number of words) of the touristic data set is shown in Fig. 1(e). As it can be seen, most of these reviews contain more than 100 words, opposite to the lengths of the mobile phones and laptops reviews which have, on average, less than 100 words. Most of the touristic data set reviews have 5 rating points (80%), and together, the 4 and 5 rating points reviews represent 87% of the total number of the touristic reviews.

Experimental Results. In order to identify the list of keywords for each review class, the review texts were analysed and compared against the most common words of the Romanian language. As a reference document for the performed study, we chosen a Romanian Geography manual [14].

From the list of keywords returned by AntConc program with its default configuration we selected the most frequent 10 entries. Tables 1, 2, and 3 present the evaluation scores corresponding to the keywords considered for each data set. The keywords are listed in the following tables using their base form and in a decreasing order upon their keyness value.

Table 1. Evaluation scores for the touristic reviews

Keywords	Actual class = yes		Actual class = no	
	True Positive	False Negative	False Positive	True Negative
camera (room$^{en.}$)	.54	.45	.13	.86
hotel	.33	.66	.00	1.00
pensiune (guest house$^{en.}$)	.32	.67	.00	1.00
masa (food$^{en.}$)	.39	.60	.00	.99
plaja (beach$^{en.}$)	.17	.82	.00	.99
a ajunge (to arrive$^{en.}$)	.35	.64	.01	.98
a recomanda (to recommend$^{en.}$)	.33	.66	.16	.83
drum (journey$^{en.}$)	.38	.61	.00	.99
cazare (accommodation$^{en.}$)	.28	.71	.00	1.00
ziua (day$^{en.}$)	.55	.44	.14	.85
All10TuristicKeywords	**.89**	**.10**	**.20**	**.79**

$^{en.}$ English translation

As one can see, the touristic reviews share 3 common keywords with the mobile phones reviews: *camera* which in Romanian means *room* but also *(video) camera*, *ziua* (in English *day*) and *a recomanda* (in English *to recommend*) (see Tables 1 and 3). In the evaluation process we pay attention to the review lengths by restricting the ambiguous cases (when at least one common keyword is found in the review text) to obey the *length rule*: at least 100 words for the touristic reviews and maximum 100 words for the mobile phones and laptops reviews.

Table 2. Evaluation scores for the laptop reviews

Keywords	Actual class = yes		Actual class = no	
	True Positive	False Negative	False Positive	True Negative
laptop	.34	.65	.00	.99
a recomanda (to recommend[en.])	.20	.79	.17	.82
ssd	.13	.86	.00	.99
windows	.17	.82	.00	.99
tastatura (keyboard[en.])	.10	.89	.00	.99
a plăcea (to like[en.])	.07	.92	.00	.99
hdd	.07	.93	.00	.99
driver	.06	.93	.00	.99
gaming	.06	.93	.00	.99
a instala (to install[en.])	.08	.91	.01	.98
All10LaptopsKeywords	**.62**	**.37**	**.19**	**.80**

[en.] English translation

Table 3. Evaluation scores for the reviews about mobile phones

Keywords	Actual class = yes		Actual class = no	
	True Positive	False Negative	False Positive	True Negative
telefon (phone[en.])	.46	.53	.07	.92
iphone	.09	.90	.00	1.00
camera (camera[en.])	.12	.87	.25	.74
samsung	.06	.93	.00	.99
android	.05	.94	.00	.99
ziua (day[en.])	.14	.85	.38	.61
poza (photo[en.])	.08	.91	.10	.89
apple	.04	.95	.00	.99
bateria (battery[en.])	.16	.83	.03	.96
a recomanda (to recommend[en.])	.16	.83	.29	.70
All10PhonesKeywords	**.57**	**.42**	**.21**	**.78**

[en.] English translation

The scores of Table 1 suggest that the touristic reviews are correctly predicted within their class and, by considering the length rule, they can be also successfully differentiated from the reviews of the other two topics.

The scores of Table 2 suggest that laptops reviews are not so accurately identified within their class (which means that the chosen keywords do not cover all the possible descriptions). Still, these poor results are balanced with very good ones when reviews on other topics are provided, which means that the considered keywords for the laptop reviews are highly specific for texts on this topic.

The scores of Table 3 are with five percents smaller than the scores of Table 2 for the case in which the classification is performed against similar topics texts. The good results obtained for the laptops keywords when other topic texts are considered are preserved also for the mobile phones keywords.

From the results shown in Tables 1, 2 and 3, it can be also observed that the "a recomanda" keyword (in English "to recommend") which has a significant keyness value is also a highly ambigous keyword as it appears in all the three types of reviews. Therefore, it is a keyword for the review texts compared with non-review texts, and perhaps it should not be taken into account when a classification among several types of reviews is performed.

5 Conclusions and Future Works

This paper presented a simple and yet effective keyword-based text classification enrich with quantitative data about the processed texts, that is the texts' length.

The results we obtain sustain the effectiveness of this approach in medium difficulty categorisation tasks where the topic of a document can be identified by several distinguishable keywords. For future works we propose to explore other knowledge that can be used in this study and to apply this knowledge-based categorisation method on other language reviews, more precisely on English reviews.

References

1. Balahur, P., Balahur, A.: What does the world think about you? Opinion mining and sentiment analysis in the social web. The Scientific Annals of "Alexandru Ioan Cuza" University of Iaşi Communication Science (2015). ISSN 2068-1143
2. Becheru, A., Bădică, C.: A deeper perspective of online tourism reviews analysis using natural language processing and complex networks techniques. In: Proceedings of the 12th International Conference Linguistic Resources and Tools for Processing the Romanian Language, ConsILR 2016, pp. 189–192 (2016)
3. Becheru, A., Buşe, F., Colhon, M., Bădică, C.: Tourist review analytics using complex networks. In: Proceedings of the 7th Balkan Conference on Informatics Conference, BCI 2015, pp. 25:1–25:8 (2015)
4. Becheru, A., Bădică, C., Antonie, M.: Towards social data analytics for smart tourism: a network science perspective. In: Trandabăţ, D., Gîfu, D. (eds.) Linguistic Linked Open Data, RUMOUR 2015. Communications in Computer and Information Science, vol. 588, pp. 35–48. Springer, Cham (2016)

5. Bonferroni, C.E.: Il calcolo delle assicurazioni su gruppi di teste. In: Studi in Onore del Professore Salvatore Ortu Carboni, Rome, Italy, pp. 13–60 (1935)
6. Boser, B.E., Guyon, I.M., Vapnik, V.N.: A training algorithm for optimal margin classifiers. In: Proceedings of the Fifth Annual Workshop on Computational Learning Theory - COLT 1992, p. 144 (1992). https://doi.org/10.1145/130385.130401
7. Chai, K.M.A., Chieu, H.L., Ng, H.T.: Bayesian online classifiers for text classification and filtering. In: Proceedings of the 25th Annual International ACM SIGIR Conference on Research and Development in Information Retrieval (SIGIR 2002), pp. 97–104. ACM, New York (2002)
8. Colhon, M., Bădică, C.: Users classification in an online community of Romanian tourists. In: Joint Proceedings of the 1st Workshop on Temporal Dynamics in Digital Libraries (TDDL 2017), the (Meta)-Data Quality Workshop (MDQual 2017) and the Workshop on Modeling Societal Future (Futurity 2017) co-located with 21st International Conference on Theory and Practice of Digital Libraries (TPLD 2017), Thessaloniki, Greece, paper 8 (2017)
9. Colhon, M., Bădică, C., Şendre, A.: Relating the opinion holder and the review accuracy in sentiment analysis of tourist reviews. In: Proceedings of 7th International Conference on Knowledge Science, Engineering and Management, KSEM 2014, pp. 246–257 (2014)
10. Do, C., Ng, A.: Transfer learning for text classification. In: Proceedings of Neural Information Processing Systems (NIPS) (2005)
11. Feng, G., Guo, J., Jing, B.-Y., Hao, L.: A Bayesian feature selection paradigm for text classification. Inf. Process. Manag. **48**(2), 283–302 (2012)
12. Friedl, M.A., Brodley, C.E.: Decision tree classification of land cover from remotely sensed data. Remote Sens. Environ. **61**(3), 399–409 (1997)
13. Evgeniy, G., Markovitch, S.: Harnessing the expertise of 70,000 human editors: knowledge-based feature generation for text categorization. J. Mach. Learn. Res. **8**, 2297–2345 (2007)
14. Gîfu, D., Sălăvăstru, A.: A study of geographic data annotation. In: Proceedings of the Summer School on Linguistic Linked Open Data, EUROLAN-2015, Sibiu, Romania (2015)
15. Goyal, R.D.: Knowledge based neural network for text classification. In: Proceedings of the IEEE International Conference on Granular Computing, pp. 542–547 (2007). https://doi.org/10.1109/GrC.2007.108
16. Groom, N.: Closed-class keywords and corpus-driven discourse analysis. In: Bondi, M., Scott, M. (eds.) Keyness in Texts, pp. 59–78 (2010)
17. Huang, A.: Similarity measures for text document clustering. In: Proceedings of the New Zealand Computer Science Research Student. Conference (NZCSRSC 2008), Christchurch, New Zealand (2008)
18. Jingbo, Z., Yao, T.: A knowledge-based approach to text classification. In: Proceedings of the first SIGHAN Workshop on Chinese Language Processing (SIGHAN 2002), vol. 18, pp. 1–5. Association for Computational Linguistics, Stroudsburg (2002). https://doi.org/10.3115/1118824.1118844
19. Lenat, D., Feigenbaum, E.: On the thresholds of knowledge. Artif. Intell. **47**, 185–250 (1990)
20. McCallum, A.K.: Bow: a toolkit for statistical language modeling, text retrieval, classification and clustering (1996)

21. Ng, V., Dasgupta, S., Arifin, S.M.N.: Examining the role of linguistic knowledge sources in the automatic identification and classification of reviews. In: Proceedings of the 21st International Conference on Computational Linguistics and 44th Annual Meeting of the Association for Computational Linguistics, pp. 611–618 (2006)
22. Raina, R., Ng, A., Koller, D.: Constructing informative priors using transfer learning. In: Proceedings of the 23rd International Conference on Machine Learning (ICML), Pittsburgh, PA (2006)
23. Scott, M.: Problems in investigating keyness, or clearing the undergrowth and marking out trails. In: Bondi, M., Scott, M. (eds.) Keyness in Texts, pp. 43–57. John Benjamins, Amsterdam (2010)
24. Scott, M., Tribble, C.: Textual Patterns: Key Words and Corpus Analysis In Language Education. John Benjamins, Philadelphia (2006)
25. Weka 3: Data Mining with Open Source Machine Learning Software in Java, Machine Learning Group at the University of Waikato (2017)
26. Yadollahi, A., Shahraki, A.G., Zaiane, O.R.: Current state of text sentiment analysis from opinion to emotion mining. ACM Comput. Surv. **50**(2), Article 25 (2017)

Experiments of Distributed Ledger Technologies Based on Global Clock Mechanisms

Yuki Yamada and Tatsuo Nakajima(✉)

Waseda University, 3-4-1 Okubo, Shinjuku, Tokyo, Japan
{yuki.yamada,tatsuo}@dcl.cs.waseda.ac.jp

Abstract. This paper reports on some experiments using different global clock mechanisms in distributed ledger technologies. Recently, using global clocks in distributed systems has become practical due to the progress of small atomic clock devices. However, current distributed systems such as typical distributed ledger technologies assume traditional loosely synchronized clocks. In this paper, we have implemented logical and physical global clock mechanisms in a distributed ledger system and investigated how different clock mechanisms influence the performance and scalability of distributed ledger technologies. When comparing these clocks, we found that the number of messages exchanged among the nodes is increased due to the number of the nodes required when using logical global clocks; thus, physical global clocks are more suitable than are logical global clocks for use in distributed ledger systems. We also found that the guarantee of transaction ordering based on the global time and the transaction throughput become a tradeoff in distributed ledger systems.

1 Introduction

Time management is a fundamental mechanism in distributed systems. However, despite the fact that the concept of time is a very essential factor for guaranteeing consistency in distributed systems, such as in consistent event ordering, it is actually difficult to acquire a time that is accurately synchronized in distributed systems [18]. Therefore, based on the premise that it is typically necessary to guarantee causal event ordering even if it cannot acquire the accurate synchronized time, a method to guarantee causal event ordering was invented by a method called a logical global clock [4,13]. A logical global clock is widely used now in various types of distributed systems, and fully synchronized physical global clocks have not been considered seriously.

However, the premise of global clock usage is changing in recent years. One consideration is time management based on physical global clocks using atomic clock devices. An atomic clock device is a timepiece that clocks accurate time by applying the properties of atoms. For example, according to [19], a precision of approximately 10^{-16} s is realized with clocks using an optical lattice.

J. Del Ser et al. (Eds.): IDC 2018, SCI 798, pp. 436–445, 2018.
https://doi.org/10.1007/978-3-319-99626-4_38

In addition, it has become possible to realize an atomic clock device termed a chip-scale atomic clock device [12] at low cost and small size in recent years. If a small atomic clock becomes popular and widely used, the computer system can acquire precisely synchronized time at a lower cost. Under such circumstances, it is worth reinvestigating the design of distributed systems to adopt completely synchronized physical clocks.

In recent years, distributed ledger technologies have attracted attention as an example of a novel distributed system. The distributed ledger technology is a peer-to-peer (P2P) distributed database technology generalizing the basic technology proposed in Bitcoin [14]. A distributed ledger is also called a blockchain due to its adopted data structure. A major feature of a distributed ledger is that it can constitute autonomous highly reliable networks called non-centralized systems without assuming the existence of specific operating groups and nodes. The order of transactions to change a distributed blockchain data structure is determined by the order of added blocks generated by the transaction. By misusing this property, it is possible to intentionally make double payments [5]. Certainly, a distributed ledger discards illegal transactions over time, but the frequent occurrence of delayed transactions causes a large serious delay in the entire network. This phenomenon implies that the system causes a scalability issue [17].

In particular, the scalability issue is related to a problem named the blockchain branching problem. Consider, for example, a case where two nodes that succeeded in mining on different blocks appear. In this case, respective nodes add the mined blocks to the blockchain data structure independently. At this time, there are two newest blocks in the blockchain data structure, and then, it can be said that the blockchain data structure is temporarily branched. In this case, the distributed ledger system needs to adopt one of the branches according to some rules. For example, in the case of Bitcoin, a branch in which more blocks are connected, that is, a longer branch, is adopted. In other words, even if a branch is created, it is designed to converge to the best branch over time. However, if a block in another branch is invalidated, the invalid block will not be fixed until the block is successfully mined again. Therefore, the final commitment of the block has to wait until the probability to generate branches becomes almost zero. Based on the above discussion, incorporating the concept of a global clock offers a possibility to improve the performance and scalability of distributed ledger technologies.

The purpose of this research is to investigate whether the performance and scalability improvement is possible by adopting logical and physical global clocks in distributed ledger technologies. In particular, the paper investigates the following two issues: the first issue is whether the scalability can be improved by applying global clocks, and the second issue is how the performance and scalability are influenced according to different global clock mechanisms. We conducted experiments to investigate the differences between logical and physical global clocks and discuss the merits and demerits of embracing atomic clock devices for adopting physical global clocks.

2 Related Work

2.1 Logical and Physical Global Clock Mechanisms

When a distributed service needs to take into account the correct causal relationship among events, a logical global clock is widely used. A logical global clock such as a Lamport clock and a vector clock [4,13] is a mechanism for capturing chronological and causal event relationships in distributed systems. Since distributed systems may have no accurately synchronized global clocks, logical global clocks allow global ordering on events among different nodes. For realizing a logical global clock, each process has two data structures: logical local time and logical global time. Logical local time is used by a process to mark its own events, and logical global time is the local information about global time. Logical local time is updated after each local event, and logical global time is updated when processing exchange messages.

As presented in Sect. 1, physical global clocks based on atomic clock devices such as chip-scale atomic clock devices [12] have become realistic in recent years. This makes the atomic clock device smaller and thus possible to install on any computer-based device. Additionally, recently, there have been several studies that realize atomic clock devices that can be employed on smartphones [1]. Additionally, examples of services of atomic clock devices include the TrueTime API of a distributed database called Spanner [7] and Time Sync Service in Amazon Web Services [11]. Our purpose is to investigate the differences when applying different global clocks to distributed ledger technologies. There are recent examples of applying global clocks in distributed systems, as shown above, but there has been no research to exploit global clocks in distributed ledger technologies.

2.2 Distributed Ledger Technologies and Its Scalability Problem

Scalability problems are seriously concerned in distributed ledger technologies [17]. If the problems cannot be solved, distributed ledger technologies cannot withstand the scale-up with the number of nodes and transactions. In Bitcoin, recent examples to solve scalability problems include the Bitcoin Cash project [3] that relaxes the block size limitation, the Lightning network [15] that processes transactions outside the blockchain (off-chain), and Sidechains [2] that process transactions by connecting a blockchain to other blockchains. Additionally, Bitcoin-NG [8] is trying to solve the problem by significantly changing the underlying Bitcoin protocol.

Addtionally, scalability problems occur not only at the transaction level but also at the block level. As a recent example, Algorand [10] does not generate a blockchain branch by creating a block only when a group of nodes approve it through a distributed algorithm. Additionally, IOTA [16] includes a mechanism that allows branches to exist to some extent by introducing a directed acyclic graph (DAG) data structure into the blockchain data structures.

3 Experiments and Discussions

3.1 Designing Experiments

As described before, we define the following two mechanisms to realize global clocks. One is a *logical global clock*, which implements a global clock using a conventional distributed algorithm such as a Lamport clock, and the second is a *physical global clock*, which assumes an environment where an accurate time can be acquired independently by using chip-scale atomic clock devices. The main objective of this research is to investigate whether these global clock mechanisms are effective for distributed ledger technologies because most of the current distributed ledger technologies simply use the loosely synchronized local time in each node. For example, in Bitcoin, it is possible to use the blockchain data structure generated at regular intervals by Proof of Work as a timestamp server [14]. However, this method is based on a native time concept in which the order of blocks is substantially determined by the local time of each node.

Based on the open source code explained in [9][1], our distributed ledger system has been enhanced by incorporating different global clock mechanisms. We created codes to implement two logical global clocks: the first is a vector clock, and the second is Causal Multicast (CBCAST) [4]. The implementation of a vector clock simply compares the causal relationships among transactions, but CBCAST extends the vector clock and provides a mechanism for delaying out-of-order transactions and recovering them in the correct causal order. With the two logical global clocks, we would like to investigate the overhead to guarantee the strict transaction ordering based on global time. In our distributed ledger system, a physical global clock using a chip-scale atomic clock device is realized by a simulation. The distributed ledger system is virtualized on a single machine and all nodes in the distributed ledger system share the same clock of the machine for simulating a global clock.

The experiments take into account the following three points. The first point is to investigate how the number of messages is changed depending on the number of nodes participating in the distributed ledger system. The result shows the number of messages that generates the overhead due to the global clock mechanisms. The second point is to investigate the throughput of actually completed transactions against the throughput of issued transactions. The result shows the degradation of the transaction throughput due to the overhead. These two experiments executed programs to generate traffic loads for 10 min on 50 Docker container instances. The third point is to investigate the guarantee of event ordering at the block level in the distributed ledger system. As described in Sect. 1, the blockchain branching problem significantly influences the scalability of a distributed ledger system. For investigating the point, we consider how global clocks affect the blockchain branching problem.

In this research, we conducted the experiments by launching the Docker container on an Amazon EC Services instance (c4.4xlarge) of Amazon Web Services and by executing scripts to conduct the experiments.

[1] An actual Python code can be found in https://github.com/dvf/blockchain.

3.2 Results of the Experiments

In this section, we present results to conduct the experiments described in the previous section. Figure 1 shows the result on the transaction throughput with respect to the number of nodes. In this figure, the horizontal axis represents the number of nodes and the vertical axis represents the transaction throughput, where *no-clock* is a version without a global clock, *vector* is a version that implements a vector clock, *cbcast* is a version that implements CBCAST, and *physical* is a version that implements a physical global clock.

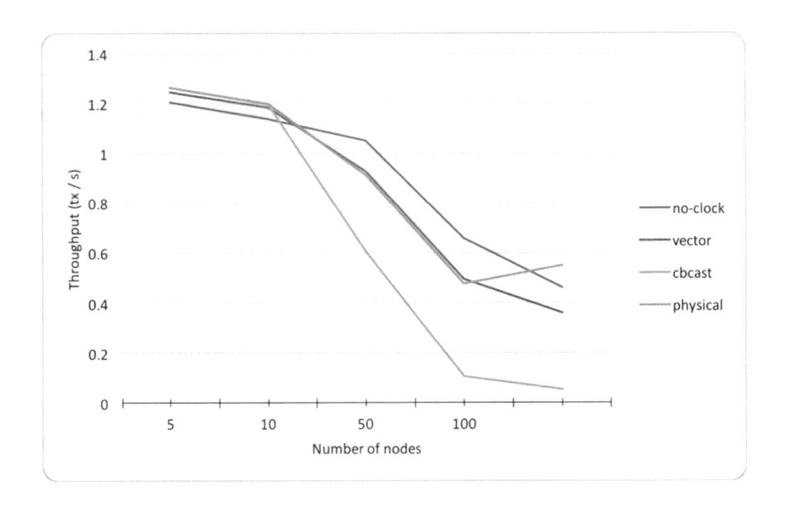

Fig. 1. Transaction throughput with respect to the number of nodes.

The results shown in Fig. 2 present the ratio of committed transactions to issued transactions. In this figure, the horizontal axis represents the number of issued transactions per second and the vertical axis represents the ratio of the number of completed transactions per second to the actual number of issued transactions per second. This result represents the number of blocks processed by the actual distributed ledger system in time. In other words, it shows the actual measured transaction throughput compared with the originally assumed transaction throughput.

As shown in Figs. 1 and 2, the version without global clocks provides a higher transaction throughput than the version to incorporate physical global clocks or logical global clocks. The clock management based on logical global clocks requires message exchanges for executing clock synchronization protocols, and the processing becomes an overhead to reduce the transaction throughput of the entire system. Additionally, since physical global clocks require access to system clocks in our implementation, it is considered that the processing has become an overhead.

When guaranteeing the transaction ordering based on the global time, the transaction throughput of the entire system is reduced. For example, when using

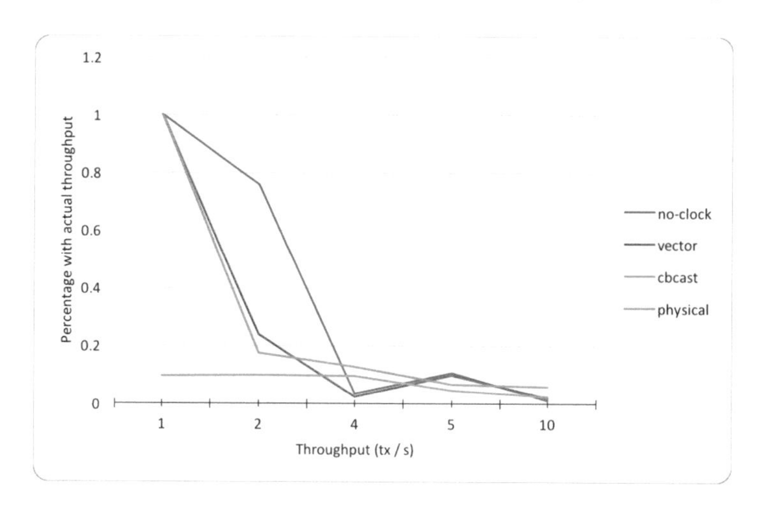

Fig. 2. Transaction throughout degradation with different global clocks.

physical and logical global clocks the transaction throughput is reduced approximately 0.05 tx/s. As a result, introducing the global clocks cannot be said to greatly contribute to scalability. However, guaranteeing the transaction ordering based on the global time itself decreases to generate blockchain branches in the distributed ledger system as we will explain later. We need to investigate the tradeoffs when the consistency of transaction ordering is necessary.

Next, we consider how to address the blockchain branching problem by using the global clock concept. The precise timestamps allocated to blocks make it possible to allocate a clear order to the blocks in each branch even in situations where a network is partitioned. However, a blockchain data structure is considered to be a collection of many transactions event histories. For example, since there is a causal relationship between transactions, the consistency cannot be maintained by a method simply adopting the block 101 as the branch A and the block 102 as the branch B alternately. In addition, if there is a case where a transaction is newly generated by using a reward generated in Proof-of-Work, the transaction becomes invalid and the succeeding transactions are also invalidated at the same time. Even though it is possible to judge the order of blocks and the order of transactions, it is very difficult to maintain the consistency of the entire blockchain data structure. Due to this reason, it is difficult to fundamentally solve the blockchain branching problem only by using the global clock realistically.

On the other hand, in the case of a small-scaled branch, it may be possible to improve the transaction throughput. In resolving the blockchain branching problem, the block generated earlier in time is always adopted. The fact that the generation of the block early means that the blocks are typically continued after this block, and it can be said that the number of blocks generated per unit of time is larger in the branches of the previously generated block.

Since the average generation time of the block is adjusted to be constant, the value of the number of blocks generated per unit time in the entire system may be improved by adopting the branch that generated a slightly earlier time. For example, consider the selection of a branch when there are two blocks 101 that are next to the block 100. Here, it is assumed that the generation time of the block 100 is 1000 in the Unix time and that the average generation time of the block is 10 min. The time of the two blocks 101 is 1100 and 1500 in the Unix time. At this time, the expected times at which generation of the next block 102 is completed are 1700 and 2100, respectively. In that case, it takes 700 and 1100 to generate two blocks respectively.

When the number of transactions that can be included in one block is 100, and the respective transactions throughput per unit time is $\frac{200}{700} = 0.286$ (tx/s) and $\frac{200}{1100} = 0.182$ (tx/s), then there is a difference of the throughput of approximately 0.1 (tx/s). The transaction throughput at the completion of block 200 is $\frac{100 \times 100}{(1100-1000+600 \times 100)} = 0.166$ (tx/s) and $\frac{100 \times 100}{(1500+600 \times 100)-1000} = 0.165$ (tx/s), and thus the difference is approximately 0.001 (tx/s). Although this is not a large difference, it can claim that it is possible to improve transaction throughput. To generalize, when the average value of the number of transactions included in a block is s, the average of the block generation time is m, and the block generation time of the block candidate to be adopted is t_1, t_2 ($t_1 < t_2$), respectively. The difference of the transaction throughput in the entire distributed ledger at the time when n blocks are generated after a branch occurs becomes the following.

$$\frac{sn}{t_1 + m(n-1)} - \frac{sn}{t_2 + m(n-1)}$$

We would like to apply to the discussion in the actual Bitcoin system. According to Blockchain.info [6], the equation variables above become $s \approx 2000$ and $m \approx 600$. At this time, when $t_1 = 0$ and $t_2 = 600$, the throughput becomes about 0.667 tx/s at six blocks ahead. The Bitcoin transaction throughput is said to be 7 tx/s at the maximum, and the throughput is 3.33 tx/s where $s = 2000$ and $m = 600$. Thus, the transaction throughput of Bitcoin can be improved by adopting global clocks.

By using the version of the distributed ledger system that uses physical global clocks, we would like to calculate the number of block generation per unit time in the branch that has completed the mining earlier. In this experiment, we prepared two versions of a distributed ledger system: the first one uses the previously obtained correct answer value to the initial value and instantly completes the first mining, and the other version performs usual mining. Similarly, in the second and subsequent times, mining is performed as usual for both versions. In this experiment, the comparison between two versions is performed after acquiring the average block generation time.

First, we obtain the average time taken for mining. In this experiment, a hash in which five consecutive 0s at the head of the hash value is regarded as a correct answer (probability of 0.5^{20}). The average mining completion time was 1.295 s. With this setting we measured the time to mine ten blocks with the two versions: one with and another without physical global clocks. The results are shown in Table 1.

Table 1. Time to be used for block mining.

	Time used for mining 10 blocks (s)	Number of blocks generated per second (block/s)
With an initial value	18.56	0.539
Without initial value	21.16	0.473

From the discussion, it can also be said that the method of selecting the branch earlier in the global clock order improves the number of block generation per unit time and contributes to the improvement of the scalability at the block level.

3.3 Discussions

In this section, we discuss the differences according to respective global clock mechanisms. It can be said that transaction throughput of the distributed ledger system itself decreases with processing the overhead by incorporating global clocks from the results shown in the previous section. However, what we found when conducting the experiments was that using logical global clocks increased the number of messages with the scaling up of a network. The vector clock and CBCAST need to hold the counter in each node. That is, the amount of data per message becomes larger in proportion to the number of nodes participating in the network. Thus, it is practically difficult to utilize logical global clocks due to the overhead. However, conversely, when operating a distributed ledger system in a small network or a private network, the influence of the problem caused by the amount of data is not so serious, so logical global clocks are considered to be usable in this case. The difference between the vector clock and CBCAST is to decide whether or not to receive a message by CBCAST and to move the message to the queue when it is rejected, and the processing waits until it becomes receivable again. As seen from the results in the previous section, CBCAST is generally much more inferior to the vector clock due to the transaction throughput. Considering the actual Bitcoin network, since the broadcast message from a large number of nodes arrives, the overhead due to the consistent transaction ordering becomes enormous.

Next, we would like to consider a case to use physical global clocks. From the results of the previous section, the version to use physical global clocks showed a lower transaction throughput than that of the no clock version. From this fact, it was found that the overhead of the time acquisition process is an important factor when implementing physical global clocks. Since the acquisition and comparison of time is frequently performed in the processing in which the distributed ledger system operates, the delay of the time acquisition processing leads to the delay in the entire system. Thus, it is important to offer a mechanism to access an atomic clock device with a small cost, for example, through a CPU register.

We next discuss the necessity of physical global clocks. The physical global clock can be treated as a trustworthy time-stamping mechanism in practice, but the consistency of event ordering cannot be guaranteed with timestamp data alone. However, by ordering transactions according to the approval of the transactions at the time of generating blocks, it is possible to speed up the verification process of illegal transactions. It is also possible to prevent double payments by preferentially prioritizing in the global clock order.

Next, we like to compare the results of logical global clocks and physical global clocks. As mentioned earlier, there is a problem caused by increasing the amount of data due to the scale up of a network when adopting logical global clocks. However, it can be said that this phenomenon does not occur when using physical global clocks. The reason is because transactions do not carry any information of other nodes similar to that in logical global clocks. The difference between the physical global clock and the implementation without a global clock is whether there is timestamp information in the transaction messages. The fact that the amount of data per transaction does not increase even if the number of nodes is scaled up means that it can be scalable, as it does not suffer the scale of the number of nodes greatly with respect to the clock information. However, as we can see from the results of the experiments, the overhead of processing simply by receiving information from a large number of nodes is proportional, and the result is that the distributed ledger system itself becomes scalable simply by introducing physical global clocks.

4 Conclusion and Future Work

In this paper, we investigated whether it is possible to solve scalability problems by adopting global clocks in distributed ledger technologies. Our results showed that applying the concept of global clocks to distributed ledger technologies does not sufficiently contribute to the improvement of the scalability because there is a tradeoff relationship between the guarantee of transaction ordering based on global time and the number of messages at the transaction level. However, at the block level, the results showed that it is possible to contribute to the improvement of the scalability by accelerating the overall block creation time by adopting global clocks in the branch selection process.

There are two issues to be addressed in the future. In this research, we assumed that the values representing global clocks are not altered. However, it is necessary to consider tamper resistance of the clock values. The second issue is that it is necessary to verify our results in existing real distributed ledger systems such as Ethereum and HyperLedger.

References

1. A New Architecture for Miniaturization of Atomic Clocks Exploiting a Piezoelectric-Thin-Film Vibration. https://www.nict.go.jp/en/press/2018/01/23-1.html. Accessed Mar 2018
2. Back, A., Corallo, M., Dashjr, L., Friedenbach, M., Maxwell, G., Miller, A., Poelstra, A., Timon, J., Wuille, P.: Enabling Blockchain Innovations with Pegged Sidechains (2014). https://blockstream.com/sidechains.pdf. Accessed Mar 2018
3. Bitcoin Cash. https://www.bitcoincash.org. Accessed Mar 2018
4. Birman, K., Joseph, T.: Exploiting virtual synchrony in distributed systems. In: Proceedings of the Eleventh ACM Symposium on Operating Systems Principles (SOSP 1987), pp. 123–138 (1987)
5. Bitcoin, Wiki: Irreversible Transactions (2017). https://en.bitcoin.it/wiki/Irreversible_Transactions. Accessed Mar 2018
6. BLOCKCHAIN LUXEMBOURG S.A. Blockchain Charts: The Most Trusted Source for Data on the Bitcoin Blockchain. https://blockchain.info/charts. Accessed Mar 2018
7. Corbett, J.C., Dean, J., Epstein, M., Fikes, A., Frost, C., Furman, J.J., Ghemawat, S., Gubarev, S., Heiser, C., Hochschild, P., Hsieh, W., Kanthak, S., Kogan, E., Li, H., Lloyd, A., Melnik, S., Mwaura, D., Nagle, D., Quinlan, S., Rao, R., Rolig, L., Saito, Y., Szymaniak, M., Taylor, C., Wang, R., Woodford, D.: Spanner: Googles globally distributed database. ACM Trans. Comput. Syst. **31**(3), Article 8 (2013)
8. Eyal, I., Gencer, A.E., Sirer, E.G., Van Renesse R.: Bitcoin-NG: a scalable blockchain protocol. In: Proceedings of the 13th USENIX Conference on Networked Systems Design and Implementation (NSDI 2016), pp. 45–59 (2016)
9. Daniel, F.: Learn Blockchains by Building One (2017). https://hackernoon.com/learn-blockchains-by-building-one-117428612f46. Accessed Mar 2018
10. Gilad, Y., Hemo, R., Micali, S., Vlachos, G., Zeldovich, N.: Algorand: scaling byzantine agreements for cryptocurrencies. In: Proceedings of the 26th Symposium on Operating Systems Principles (SOSP 2017), pp. 51–68 (2017)
11. Introducing the Amazon Time Sync Service. https://aws.amazon.com/about-aws/whats-new/2017/11/introducing-the-amazon-time-sync-service/. Accessed Mar 2018
12. Knappe, S., Schwindt, P., Shah, V., Hollberg, L., Kitching, J., Liew, L., Moreland, J.: A chip-scale atomic clock based on 87Rb with improved frequency stability. Opt. Express **13**(4), 1249–53 (2005)
13. Lamport, L.: Time, clocks, and the ordering of events in a distributed system. Commun. ACM **21**(7), 558–565 (1978)
14. Nakamoto, S.: Bitcoin: A Peer-to-Peer Electric Cash System (2008). https://blockchain.org/bitcoin.pdf. Accessed Mar 2018
15. Poon, J., Dryja, T.: The Bitcoin Lightning Network: Scalable Off-chain Instant Payments, Technical report (draft) (2015). Accessed Mar 2018
16. Serguei, P.: The Tangle (2017). https://iota.org/IOTA_Whitepaper.pdf. Accessed Mar 2018
17. Sompolinsky, Y., Zohar, A.: Accelerating Bitcoin's transaction processing. Fast money grows on trees, not chains. IACR Cryptol. ePrint Arch. **2013**, 881 (2013)
18. Tanenbaum, A.S., Steen, M.V.: Distributed Systems: Principles and Paradigms. Prentice-Hall, Upper Saddle River (2007)
19. Yamaguchi, A., Fujieda, M., Kumagai, M., Hachisu, H., Nagano, S., Li, Y., Ido, T., Takano, T., Takamoto, M., Katori, H.: Direct comparison of distant optical lattice clocks at the 10^{-16} uncertainty. Appl. Phys. Express **4**(8), 082203 (2011)

Author Index

J. Del Ser et al. (Eds.): IDC 2018, SCI 798, pp. 447–448, 2018.
https://doi.org/10.1007/978-3-319-99626-4

Printed in the United States
By Bookmasters